Monte Carlo Techniques
in Radiation Therapy

IMAGING IN MEDICAL DIAGNOSIS AND THERAPY

William R. Hendee, Series Editor

Forthcoming titles in the series

Monte Carlo Techniques in Radiation Therapy

Edited by
Joao Seco
Frank Verhaegen

CRC Press
Taylor & Francis Group
Boca Raton London New York

CRC Press is an imprint of the
Taylor & Francis Group, an **informa** business

A TAYLOR & FRANCIS BOOK

CRC Press
Taylor & Francis Group
6000 Broken Sound Parkway NW, Suite 300
Boca Raton, FL 33487-2742

© 2013 by Taylor & Francis Group, LLC
CRC Press is an imprint of Taylor & Francis Group, an Informa business

No claim to original U.S. Government works

Version Date: 20130102

International Standard Book Number: 978-1-4665-0792-0 (Hardback)

Library of Congress Cataloging-in-Publication Data

Monte Carlo techniques in radiation therapy / editors, Joao Seco, Frank Verhaegen.
    p. ; cm. -- (Imaging in medical diagnosis and therapy)
  Includes bibliographical references and index.
  Summary: "Preface Monte Carlo simulation techniques made a slow entry in the field of radiotherapy in the late seventies. Since then they have gained enormous popularity, judging by the number of papers published and PhD degrees obtained on the topic. Calculation power has always been an issue, so initially only simple problems could be addressed. They led to insights, though, that could not have been obtained by any other method. Recently, fast forwarding some thirty years, Monte Carlo-based treatment planning tools have now begun to be available from some commercial treatment planning vendors, and it can be anticipated that a complete transition to Monte Carlo-based dose calculation methods may take place over the next decade. The progress of image-guided radiotherapy further advances the need for Monte Carlo simulations, in order to better understand and compute radiation dose from imaging devices and make full use of the four-dimensional information now available. Exciting new developments in in-beam imaging in light ion beams are now also being investigated vigorously. Many new discoveries await the use of Monte Carlo technique in radiotherapy in the coming decades. The book addresses the application of the Monte Carlo particle transport simulation technique in radiation therapy, mostly focusing on external beam radiotherapy and brachytherapy. It includes a presentation of the mathematical and technical aspects of the technique in particle transport simulations. It gives practical guidance relevant for clinical use in radiation therapy, discussing modeling of medical linacs and other irradiation devices, issues speci?c to electron, photon, proton/particle beams, and brachytherapy, utilization in the optimization of treatment planning, radiation dosimetry"--Provided by publisher.
  ISBN 978-1-4665-0792-0 (hardback : alk. paper)
  I. Seco, Joao. II. Verhaegen, Frank. III. Series: Imaging in medical diagnosis and therapy.
  [DNLM: 1. Monte Carlo Method. 2. Radiotherapy Planning, Computer-Assisted. 3. Radiometry--methods. 4. Radiotherapy, Image-Guided. WN 250.5.R2]

  615.8'42--dc23   2012050709

Visit the Taylor & Francis Web site at
http://www.taylorandfrancis.com

and the CRC Press Web site at
http://www.crcpress.com

To all who made this book possible

Contents

Series Preface

Advances in the science and technology of medical imaging and radiation therapy are more profound and rapid than ever before, since their inception over a century ago. Further, the disciplines are increasingly cross-linked as imaging methods become more widely used to plan, guide, monitor, and assess treatments in radiation therapy. Today the technologies of medical imaging and radiation therapy are so complex and so computer driven that it is difficult for the persons (physicians and technologists) responsible for their clinical use to know exactly what is happening at the point of care, when a patient is being examined or treated. The persons best equipped to understand the technologies and their applications are medical physicists, and these individuals are assuming greater responsibilities in the clinical arena to ensure that what is intended for the patient is actually delivered in a safe and effective manner.

The growing responsibilities of medical physicists in the clinical arenas of medical imaging and radiation therapy are not without their challenges, however. Most medical physicists are knowledgeable in either radiation therapy or medical imaging, and expert in one or a small number of areas within their discipline. They sustain their expertise in these areas by reading scientific articles and attending scientific talks at meetings. In contrast, their responsibilities increasingly extend beyond their specific areas of expertise. To meet these responsibilities, medical physicists periodically must refresh their knowledge of advances in medical imaging or radiation therapy, and they must be prepared to function at the intersection of these two fields. How to accomplish these objectives is a challenge.

At the 2007 annual meeting of the American Association of Physicists in Medicine in Minneapolis, this challenge was the topic of conversation during a lunch hosted by Taylor & Francis and involving a group of senior medical physicists (Arthur L. Boyer, Joseph O. Deasy, C.-M. Charlie Ma, Todd A. Pawlicki, Ervin B. Podgorsak, Elke Reitzel, Anthony B. Wolbarst, and Ellen D. Yorke). The conclusion of this discussion was that a book series should be launched under the Taylor & Francis banner, with each volume in the series addressing a rapidly advancing area of medical imaging or radiation therapy of importance to medical physicists. The aim would be for each volume to provide medical physicists with the information needed to understand technologies driving a rapid advance and their applications to safe and effective delivery of patient care.

Each volume in the series is edited by one or more individuals with recognized expertise in the technological area encompassed by the book. The editors are responsible for selecting the authors of individual chapters and ensuring that the chapters are comprehensive and intelligible to someone without such expertise. The enthusiasm of volume editors and chapter authors has been gratifying and reinforces the conclusion of the Minneapolis luncheon that this series of books addresses a major need of medical physicists.

Imaging in Medical Diagnosis and Therapy would not have been possible without the encouragement and support of the series manager, Luna Han of Taylor & Francis. The editors and authors, and most of all I, are indebted to her steady guidance of the entire project.

William Hendee
Series Editor
Rochester, Minnesota

Preface

Monte Carlo simulation techniques made a slow entry in the field of radiotherapy in the late 1970s. Since then they have gained enormous popularity, judging by the number of papers published and PhDs obtained on the topic. Calculation power has always been an issue, so initially only simple problems could be addressed. They led to insights, though, that could not have been obtained by any other method. Recently, fast forwarding some 30 years, Monte Carlo–based treatment planning tools have now begun to be available from some commercial treatment planning vendors, and it can be anticipated that a complete transition to Monte Carlo–based dose calculation methods may take place over the next decade. The progress of image-guided radiotherapy further advances the need for Monte Carlo simulations, in order to better understand and compute radiation dose from imaging devices and make full use of the four-dimensional information now available. Exciting new developments in in-beam imaging in light ion beams are now also being vigorously investigated. Many new discoveries await the use of the Monte Carlo technique in radiotherapy in the coming decades.

The book addresses the application of the Monte Carlo particle transport simulation technique in radiation therapy, mostly focusing on external beam radiotherapy and brachytherapy. It includes a presentation of the mathematical and technical aspects of the technique in particle transport simulations. It gives practical guidance relevant for clinical use in radiation therapy, discussing modeling of medical linacs and other irradiation devices, issues specific to electron, photon, proton/particle beams, and brachytherapy, utilization in the optimization of treatment planning, radiation dosimetry, and quality assurance (QA).

We have assembled this book—a first of its kind—to be useful to clinical physicists, graduate students, and researchers who want to learn about the Monte Carlo method; we hope you will benefit from the collective knowledge presented herein. In addition, the editors wish to sincerely thank all the outstanding contributing authors, without whom this work would not have been possible. They would also like to thank the series editor, Dr. Bill Hendee, for the opportunity. We sincerely thank the editorial and production staff at Taylor & Francis for the smooth collaboration and for the pleasant interactions.

Editors

 **Joao Seco** is assistant professor of radiation oncology at Harvard Medical School and Massachusetts General Hospital in Boston, Massachusetts. He earned his PhD from the Institute of Cancer Research, University of London, UK in 2002. He held research positions at the Royal Marsden Hospital, UK and the Harvard Medical School, Boston, Massachusetts for several years. His group has published more than 50 research papers and he was the recipient of the Harvard Club of Australia Foundation Award in 2012 on the development of Monte Carlo and optimization techniques for use in radiation therapy of lung cancer. His interests range from proton imaging and therapy to photon beam modeling Monte Carlo, electronic portal imaging, and 4D Monte Carlo proton and photon dosimetry. He started working on Monte Carlo simulations while a master's student at the Laboratory of Instrumentation and Experimental Particle Physics (LIP), a Portuguese research institute part of the CERN (the European Organization for Nuclear Research) worldwide network for particle physics research.

 **Frank Verhaegen** is head of clinical physics research at the Maastro Clinic in Maastricht, the Netherlands. He holds a professorship at the University of Maastricht. Formerly, he was an associate professor at McGill University in Montréal, Canada. He earned his PhD from the University of Ghent in Belgium in 1996. He held research positions at the Royal Marsden Hospital and the National Physical Laboratory, UK for several years. Dr. Verhaegen is a Fellow of the Institute of Physics and Engineering in Medicine and the Institute of Physics. His group has published more than 120 research papers and was the recipient of the Sylvia Fedoruk Prize for best Canadian Medical Physics paper in 2007. His interests range broadly in imaging and dosimetry for photon and electron therapy, brachytherapy, particle therapy, and small animal radiotherapy. Dr. Verhaegen has been passionate about Monte Carlo simulations since the days of his master's thesis in the late 1980s.

Contributors

Alex F. Bielajew
Department of Nuclear Engineering
 and Radiological Sciences
University of Michigan
Ann Arbor, Michigan

Peter J. Biggs
Department of Radiation Oncology
Massachusetts General Hospital
Boston, Massachusetts

Hugo Bouchard
Centre hospitalier de l'Université
 de Montréal
Montréal, Québec, Canada

Joanna E. Cygler
Department of Medical Physics
The Ottawa Hospital Cancer Centre
Ottawa, Ontario, Canada

Nicholas Depauw
Department of Radiation Oncology
Massachusetts General Hospital
Boston, Massachusetts

George X. Ding
Department of Radiation Oncology
Vanderbilt University School of
 Medicine
Nashville, Tennessee

Bruce A. Faddegon
Radiation Oncology Department
University of California,
 San Francisco
San Francisco, California

Andrew Fielding
School of Physical and Chemical Sciences
Queensland University of Technology
Queensland, Australia

Matthias Fippel
Brainlab AG
Bavaria, Germany

Michael K. Fix
Division of Medical Radiation Physics
Inselspital
University of Bern
Bern, Switzerland

Maggy Fragoso
Department of Communication and
 Strategy
Alfa-Comunicações
Praia, Santiago Island, Cape Verde

Emily Heath
Department of Physics
Ryerson University
Toronto, Ontario, Canada

Sami Hissoiny
Département de Génie Informatique
 et Génie Logiciel
École Polytechnique de Montréal
Montréal, Québec, Canada

Xun Jia
Department of Radiation Medicine
 and Applied Sciences
University of California, San Diego
La Jolla, California

Steve B. Jiang
Department of Radiation Medicine
 and Applied Sciences
University of California, San Diego
La Jolla, California

Stephen F. Kry
Department of Radiation Physics
MD Anderson Cancer Center
University of Texas
Houston, Texas

Guillaume Landry
Maastro Clinic
Maastricht, the Netherlands

JinSheng Li
Fox Chase Cancer Center
Philadelphia, Pennsylvania

Michael Ljungberg
Department of Medical Radiation
 Physics
Lund University
Lund, Sweden

C.-M. Charlie Ma
Fox Chase Cancer Center
Philadelphia, Pennsylvania

Harald Paganetti
Department of Radiation
 Oncology
Massachusetts General Hospital
and
Harvard Medical School
Boston, Massachusetts

Hugo Palmans
Division of Acoustics and Ionising
 Radiation
National Physical Laboratory
Teddington, United Kingdom

Katia Parodi
Heidelberg Ion Beam Therapy
 Center
Heidelberg, Germany

Jerimy C. Polf
Department of Physics
Oklahoma State University
Stillwater, Oklahoma

Mark J. Rivard
Department of Radiation Oncology
Tufts University School of Medicine
Boston, Massachusetts

Joao Seco
Department of Radiation Oncology
Harvard Medical School
Boston, Massachusetts

Jan Seuntjens
Medical Physics Unit
McGill University
Montréal, Québec, Canada

Frank Verhaegen
Maastro Clinic
Maastricht, the Netherlands

Jeffrey F. Williamson
VCU Massey Cancer Center
Richmond, Virginia

Monte Carlo
Fundamentals

I

History of Monte Carlo

Alex F. Bielajew
University of Michigan

It is still an unending source of surprise for me to see how a few scribbles on a blackboard or on a sheet of paper could change the course of human affairs.

Stan Ulam
Founder of the modern Monte Carlo method,
in his 1991 autobiography (1991)

1.1 Motivating Monte Carlo

Generally speaking, the Monte Carlo method provides a numerical solution to a problem that can be described as a temporal evolution ("translation/reflection/mutation") of objects ("quantum particles" [photons, electrons, neutrons, protons, charged nuclei, atoms, and molecules], in the case of medical physics) interacting with other objects based upon object–object interaction relationships ("cross sections"). Mimicking nature, the rules of interaction are processed randomly and repeatedly, until numerical results converge usefully to estimated means, moments, and their variances. Monte Carlo represents an attempt to model nature through direct simulation of the essential dynamics of the system in question. In this sense, the Monte Carlo method is, in principle, simple in its approach—a solution to a macroscopic system through simulation of its microscopic interactions. Therein is the advantage of this method. All interactions are microscopic in nature. The geometry of the environment, so critical in the development of macroscopic solutions, plays little role except to define the local environment of objects interacting at a given place at a given time.

The scientific method is dependent on observation (measurement) and hypothesis (theory) to explain nature. The conduit between these two is facilitated by a myriad of mathematical, computational, and simulation techniques. The Monte Carlo method exploits all of them. Monte Carlo is often seen as a "competitor" to other methods of macroscopic calculation, which we will call deterministic and/or analytic methods. Although the proponents of either method sometimes approach a level of fanaticism in their debates, a practitioner of science should first ask, "What do I want to accomplish?" followed by "What is the most efficient way to do it?," and then, "What serves science the best?" Sometimes the correct answer will be "Deterministic," and other times it will be "Monte Carlo." The most successful scientist will avail himself or herself of more than one method of approach.

There are, however, two inescapable realities. The first is that macroscopic theory, particularly transport theory, provides deep insight and allows one to develop sophisticated intuition as to how macroscopic particle fields can be expected to behave. Monte Carlo cannot compete very well with this. In discovering the properties of macroscopic field behavior, Monte Carlo practitioners operate very much like experimentalists. Without theory to provide guidance, discovery is made via trial and error, guided perhaps, by some brilliant intuition.

However complexity is measured, when it comes to developing an understanding of a physical problem, Monte Carlo techniques become, at some point, the most advantageous. A proof is given, in the appendix of this chapter, that the Monte Carlo method is more advantageous in the evolution of five and higher dimensional systems. The dimensionality is just one measure of a problem's "complexity." The problems in RTP (radiotherapy target practice) and dosimetry are typically of dimension $6.\varepsilon$, or $7.\varepsilon$. That is, particles move in Cartesian space, with position $\bar{x}$, that varies continuously, except at particle inception or expiration. They move with momentum, $\bar{p}$, that varies both discretely

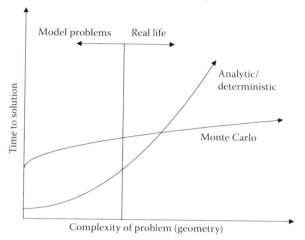

FIGURE 1.1 Time to solution using Monte Carlo versus deterministic/analytic approaches.

and continuously. The dimension of time is usually ignored for static problems, though it cannot be for nonlinear problems, where a particle's evolution can be affected by the presence of other particles in the simulation. (The "space-charge" effect is a good example of this.) Finally, the ε is a discrete dimension that can encompass different particle species, as well as intrinsic spin.

This trade-off, between complexity and time to solution is expressed in Figure 1.1.

Although the name "Monte Carlo method" was coined in 1947, at the start of the computer age, stochastic sampling methods were known long before the advent of computers. The first reference known to this author, is that of Comte de Buffon (1777) who proposed a Monte Carlo-like method to determine the outcome of an "experiment" consisting of repeatedly tossing a needle onto a ruled sheet of paper, to determine the probability of the needle crossing one of the lines. This reference goes back to 1777, well before the contemplation of automatic calculating machines. Buffon further calculated that a needle of length L tossed randomly on a plane ruled with parallel lines, distance d apart, where $d > L$, would have a probability of crossing one of the rules lines of

$$p = \frac{2L}{\pi d}. \tag{1.1}$$

Much later, Laplace (1886) suggested that this procedure could be employed to determine the value of π, albeit slowly. Several other historical uses of Monte Carlo predating computers are cited by Kalos and Whitlock (2008).

The idea of using stochastic sampling methods first occurred to Ulam,[*] who, while convalescing from an illness, played sol-

itaire repeatedly, and then wondered if he could calculate the probability of success by combinatorial analysis. It occurred to him, that it would be possible to do so by playing a large number of games, tallying the number of successful plays (Metropolis, 1987; Eckhart, 1987), and then estimating the probability of success. Ulam communicated this idea to von Neumann who, along with Ulam and Metropolis, were working on theoretical calculations related to the development of thermonuclear weapons. Precise calculations of neutron transport are essential in the design of thermonuclear weapons. The atomic bomb was designed by experiments, mostly, with modest theoretical support. The trigger for a thermonuclear weapon is an atomic bomb, and the instrumentation is destroyed before useful signals can be extracted.[†] Von Neumann was especially intrigued with the idea. The modern Monte Carlo age was ushered in later, when the first documented suggestion of using stochastic sampling methods applied to radiation transport calculations appeared in correspondence between von Neumann and Richtmyer (Metropolis, 1987; Eckhart, 1987), on March 11, 1947. (Richtmyer was the leader of the Theoretical Division at Los Alamos National Laboratories [LANL].) This letter suggested the use of LANL's ENIAC computer to do the repetitive sampling. Shortly afterward, a more complete proposal was written (von Neumann and Richtmyer, 1947). Although this report was declassified as late as 1988, inklings of the method, referred to as a "mix of deterministic and random/stochastic processes," started to appear in the literature, as published abstracts (Ulam and von Neumann, 1945, 1947). Then in 1949, Metropolis and Ulam published their seminal, founding paper, "The Monte Carlo Method" (Metropolis and Ulam, 1949), which was the first unclassified paper on the Monte Carlo methods, and the first to have the name, "Monte Carlo" associated with stochastic sampling.

Already by the 1949, symposia on the Monte Carlo methods were being organized, focusing primarily on mathematical techniques, nuclear physics, quantum mechanics, and general statistical analysis. A later conference, the *Symposium on Monte Carlo Methods*, held at the University of Florida in 1954 (Meyer, 1981) was especially important. There were 70 attendees, many of whom would be recognized as "founding fathers" by Monte Carlo practitioners in the radiological sciences. Twenty papers were presented, including two involving gamma rays, spanning 282 pages in the proceedings. This proceedings also includes a 95-page bibliography, a grand summary of the work-to-date, with many references having their abstracts and descriptions published in the proceedings.

The rest, to quote an overused expression, is history. It is interesting to note the wonderful irony: This mathematical method was created for destruction by means of the most terrible weapon in history, the thermonuclear bomb. Fortunately, this weapon has never been used in conflict. Rather, millions have benefited from the development of Monte Carlo methods

---

[*] The direct quote from Ulam's autobiography (Ulam, 1991). (p. 196, 1991 edition): "The idea for what was later called the Monte Carlo method occurred to me when I was playing solitaire during my illness."

[†] The book Dark Sun, by Richard Rhodes, is an excellent starting point for the history of that topic (Rhodes, 1988).

for medicine. That topic, at least a small subset of it, will occupy the rest of this chapter.

As of this writing, with the data from 2011 still incomplete, we have found that about 3,00,000 papers have been published on the Monte Carlo method. If we restrict this search to only those papers related to medicine, the number of publications is almost 30,000. The 10% contribution to the Monte Carlo method seems to be consistent over time, at least since 1970. That represents an enormous investment in human capital to develop this most useful tool. The temporal evolution of this human effort is shown in Figure 1.2. Before 2005, the growth in both areas appears exponential in nature. The total effort shows three distinct areas of slope, with sudden changes, currently unexplained, though it may be due to the sudden emergence of "vector" and "massively parallel" machines, and the increase in research associated with this fundamentally new computer architecture. The growth in the "Medicine" area has been constant.

Since 2005, both areas are statistically consistent with constant output, with the "Medicine" area leveling out, at greater than 2100 publications/year. It appears that this communication is being written at the pinnacle of this scientific endeavor!

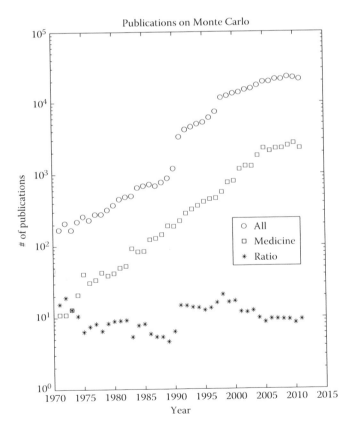

FIGURE 1.2 The number of papers published per year garnered from the Web of Knowledge ("All") and MedLine ("Medicine"). Martin Berger, on a beach near Erice, 1987. (Photograph courtesy of Ralph Nelson.)

1.2 Monte Carlo in Medical Physics

Every historical review has its biases, and the one employed here will restrict the discussion to the applications of radiotherapy and radiation dosimetry. Moreover, the focus will be on the development of electron Monte Carlo, for reasons explained in the following paragraph. There is an abundance of reviews on the use of Monte Carlo in medical physics. A few of the more recent ones that discuss radiotherapy physics and dosimetry are Andreo (1985, 1991), Mackie (1990), Rogers and Bielajew (1990), Ma and Jiang (1999), Verhaegen and Seuntjens (2003), Rogers (2006).

The rise of the use of linear electron accelerators (LINACs) for radiotherapy, also ushered in the need to develop Monte Carlo methods for the purpose of dose prediction and dosimetry. LINACs employed in radiotherapy provide an energetic and penetrating source of photons that enter deep into tissue, sparing the surface and attenuating considerably less rapidly than ^{60}Co or ^{137}Cs beams. Relativistic electrons have a range of about 1 cm for each 2 MeV of kinetic energy in water. At its maximum, starting with a pencil beam of electrons, the diameter of the electron energy deposition, pear-shaped "plume" is also about 1 cm per 2 MeV of initial kinetic energy. These dimensions are commensurate with the organs being treated, as well as the organs at risk. The treatment areas are heterogeneous, with differences in composition and density. Moreover, the instruments used to meter dose are even more diverse. It is true now, as it was then, that the Monte Carlo method provides the only prediction of radiometric quantities that satisfies the accuracy demand of radiotherapy.

Thus, the history of the utility of the Monte Carlo method in medical physics is inextricably tied to the development of Monte Carlo methods of electron transport in complex geometries and in the description of electromagnetic cascades.[*]

The first papers employing the Monte Carlo method using electron transport, were authored by Robert R. Wilson (1950, 1951, 1952, p. 261), who performed his calculations using a "spinning wheel of chance."[†] Although apparently quite tedious, Wilson's method was still an improvement over the analytic methods of the time—particularly in studying the average behavior and fluctuations about the average (Rossi, 1952). Hebbard and P. R. Wilson (1955) used computers to investigate electron straggling and energy loss in thick foils. The first use of an electronic digital computer in simulating high-energy cascades by Monte Carlo methods was reported by Butcher and Messel (1958, 1960), and independently by Varfolomeev and Svetlolobov (1959). These two groups collaborated in a much publicized work (Messel et al., 1962) that eventually led to an extensive set of tables describing the shower distribution functions (Messel and Crawford, 1970)—the so-called "shower book."

---

[*] Certainly there are important applications in brachytherapy and imaging that ignore electron transport. However, we shall leave that description to other authors.

[†] R. R. Wilson is also acknowledged as the founder of proton radiotherapy (Wilson, 1946).

For various reasons, two completely different codes were written in the early-to-mid 1960s to simulate electromagnetic cascades. The first was written by Zerby and Moran (1962a,b, 1963) of the Oak Ridge National Laboratory, motivated by the construction of the Stanford Linear Accelerator Center. Many physics and engineering problems were anticipated as a result of high-energy electron beams showering in various devices and structures at that facility. This code had been used by Alsmiller and others (Alsmiller and Moran, 1966, 1967, 1968, 1969; Alsmiller and Barish, 1969, 1974; Alsmiller et al., 1974) for a number of studies since its development.[*]

The second code was developed by Nagel (Nagel and Schlier, 1963; Nagel, 1964, 1965; Völkel, 1966) and several adaptations have been reported (Völkel, 1966; Nicoli, 1966; Burfeindt, 1967; Ford and Nelson, 1978). The original Nagel version, which Ford and Nelson called SHOWER1, was a FORTRAN code written for high-energy electrons (≤1000 MeV) incident upon lead in cylindrical geometry. Six significant electron and photon interactions (bremsstrahlung, electron–electron scattering, ionization-loss, pair-production, Compton scattering, and the photoelectric effect) plus multiple Coulomb scattering were accounted for. Except for annihilation, positrons and electrons were treated alike and were followed until they reached a cutoff energy of 1.5 MeV (total energy). Photons were followed down to 0.25 MeV. The cutoff energies were as low as or lower than those used by either Messel and Crawford or by Zerby and Moran. The availability of Nagel's dissertation (1964) and a copy of his original shower program provided the incentive for Nicoli (Nicoli, 1966) to extend the dynamic energy range and flexibility of the code in order for it to be made available as a practical tool for the experimental physicist. It was this version of the code that eventually became the progenitor of the EGS (electron gamma shower) code systems (Ford and Nelson, 1978; Nelson et al., 1985; Bielajew and Rogers, 1987; Kawrakow and Rogers, 2000; Hirayama et al., 2005).

On a completely independent track, and apparently independent from the electromagnetic cascade community, was Berger's *e–γ* code. It was eventually released to the public as ETRAN in 1968 (Berger and Seltzer, 1968), though it is clear that internal versions were being worked on at NBS (now NIST) (Seltzer, 1989) since the early 1960s, on the foundations laid by Berger's landmark paper (Berger, 1963). The ETRAN code then found its way, being modified somewhat, into the Sandia codes, EZTRAN (Halbleib and Vandevender, 1971), EZTRAN2 (Halbleib and Vandevender, 1973), SANDYL (Colbert, 1973), TIGER (Halbleib and Vandevender, 1975), CYLTRAN (Halbleib and Vandevender, 1976), CYLTRANNM (Halbleib and Vandevender, 1977), CYLTRANP (unpublished), SPHERE (Halbleib, 1978), TIGERP (Halbleib and Morel, 1979), ACCEPT (Halbleib, 1980), ACCEPTTM (Halbleib et al., 1981), SPHEM (Miller et al., 1981), and finally the all-encompassing ITS (Halbleib and Mehlhorn, 1986; Halbleib et al., 1992) codes. The ITS electron transport code

was incorporated into the MCNP (Monte Carlo N-particle) code at Version 4, in 1990 (Hendricks and Briesmeister, 1991). The MCNP code lays claim to being a direct descendant of the codes written by the originators of the Monte Carlo method, Fermi, von Neumann, Ulam, as well as Metropolis and Richtmyer (Briesmeister, 1986). Quoting directly,[†]

> Much of the early work is summarized in the first book to appear on Monte Carlo by Cashwell and Everett in 1957.[‡] Shortly thereafter the first Monte Carlo neutron transport code MCS was written, followed in 1967 by MCN. The photon codes MCC and MCP were then added and in 1973 MCN and MCC were merged to form MCNG. The above work culminated in Version 1 of MCNP in 1977. The first two large user manuals were published by W. L. Thompson in 1979 and 1981. This manual draws heavily from its predecessors.

The first appearance of electron transport in MCNP occurred with Version 4, in 1990 (Hendricks and Briesmeister, 1991). After that time, MCNP became an important player in medical-related research, to be discussed later.

Berger's contribution (1963) is considered to be the *de facto* founding paper (and Berger the founding father) of the field of Monte Carlo electron and photon transport. That article, 81 pages long, established a framework for the next generation of Monte Carlo computational physicists. It also summarized all the essential theoretical physics for Monte Carlo algorithm development. Moreover, Berger introduced a specialized method for electron transport. Electron transport and scattering, for medical physics, dosimetry, and many other applications, is subject to special treatment. Rather than modeling every discrete electron interaction (of the order of 10^6 for relativistic electrons), cumulative scattering theories, whereby 10^3–10^5 individual elastic and inelastic events are "condensed" into single "virtual" single-scattering events, thereby enabling a speedup by factors of hundreds, typically. Nelson, the originator of the EGS code system, is quoted as saying (W. R. Nelson, personal communication, 2011), "Had I known about Berger's work, I may not have undertaken the work on EGS!"

As for general-purpose uses in medical-related fields, with multi-material, combinatorial geometries, the two historically dominant players in RTP/Dosimetry,[§] are the EGS and MCNP codes, introduced above. In the last decade, GEANT (Brun et al., 1982; Allison et al., 2006) has also made significant contributions as well, presently equal in use to MCNP. A plot of the number of papers published, using these methods is charted

---

[*] According to Alsmiller (R. G. Alsmiller Jr. private communication. (conversation with W. R. Nelson), 1970), the Zerby and Moran source code vanished from ORNL and they were forced to work with an octal version.

[†] The codes and manuals referred to in this quote appear to have vanished.
[‡] 1959, to be exact (Cashwell and Everett, 1959).
[§] There are some very relevant, alternative approaches, that the reader should be aware of, namely FLUKA (Aarnio et al., 1984; Fasso et al., 2005; Battistoni et al., 2007) (that traces its roots to 1964 (Ranft, 1964)), and the Penelope code (Salvat and Parrellada, 1984; Baró et al., 1995). As of this writing, the number of papers produced using these codes in medical areas is about 240, about half that of MCNP.

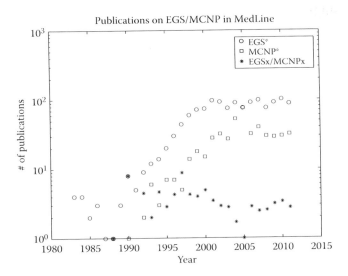

FIGURE 1.3 Papers using EGSx, MCNPx, and GEANTx, as captured on MedLine.

in Figure 1.3. Once MCNP introduced electron transport, we see, from Figure 1.3, that usage of MCNP experienced exponential growth in its use in medical-related areas. That exponential growth ended in about 2000. Since then, both the EGS and MCNP code systems seem to be experiencing steady use, with GEANT still, arguably, on the increase. If one considers all of the nonmedical literature related to Monte Carlo, MCNP is undeniably the most cited Monte Carlo code, by about a factor of 7 over EGS.

It should be emphasized that these two code systems are very different in nature. EGS has specifically targeted the medical area, since 1984, though it has enjoyed some use in other areas of physics as well. Some features are genuinely unique, such as the tracking of separate electron spins, a feature introduced (Namito et al., 1993) in EGS5 (Hirayama et al., 2005), as well as Doppler broadening (Namito et al., 1994), both inclusions of great interest to those doing research in synchrotron radiation light sources. Overall, however, considering Monte Carlo uses over all areas, MNCP's state-of-the-art neutron transport, makes it the world leader in the nuclear and radiological sciences conglomerate. The EGS code systems are supported by practitioners representing a panoply of scientific disciplines: medical physics, radiological scientists, pure and applied physicists. MCNP serves these communities as well, but also enjoys the support of a vibrant nuclear engineering profession, where transport theory of neutrons is a rich and active research area.

1.3 EGSx Code Systems

The history of how EGS, a code developed primarily for high-energy physics shielding and detector simulations, came to be used in medical physics, has never appeared in text, in its entirety. Rogers' (2006) humility probably interfered with its exposition

in his article. However, I was an observer at those early events, and think I may offer some insights. In 1978, SLAC published EGS3 (Ford and Nelson, 1978), and Rogers employed it in several important publications (Rogers, 1982, 1983a,b, 1984a,b). Of particular importance was the publication that offered a patch to the EGS3 algorithms to mimic a technique employed in ETRAN, making electron-dependent calculations reliable, by shortening the steps to virtual interactions. At the time, electron transport step-size artifacts were completely unexplained. Shortening the steps is understood to solve the problem, as explained most eloquently by Larsen (1992), but at the cost of added computational time. These "step-size artifacts," attracted the attention of Nelson, who invited Rogers to participate in authoring the next version of EGS, EGS4 (Nelson et al., 1985), along with Hirayama, a research scientist at the KEK, the High Energy Accelerator Research Organization, Tsukuba, Japan.

Following its release in December 1985, Rogers' institution (a Radiation Standards' Laboratory, and a sister laboratory to Berger's NIST) became a nucleus of medical physics and dosimetry Monte Carlo dissemination. It took over support and distribution of the EGS4 code, and began offering training courses all over the world. Hirayama was engaged in similar efforts in the Asian regions.

Yet, the step-size artifacts in EGS4 remained unexplained. Nahum, who was interested in modeling ionization chamber response, visited Rogers' laboratory in the spring of 1984, to collaborate on this topic. Nahum already had a scholarly past in electron Monte Carlo (Nahum, 1976), producing what would eventually be realized, through a Lewis (1950) moments analysis (Kawrakow and Bielajew, 1998), to be a far superior electron transport algorithm. While using EGS4 to predict ionization chamber response, EGS4 would predict responses that could be 60% too low. Quoting Nahum, "How could a calculation that one could sketch on the back of an envelope, and get correct to within 5%, be 60% wrong using Monte Carlo?" Step-size reduction solved the problem (Bielajew et al., 1985; Rogers et al., 1985), but the search for a resolution to step-size anomalies was commenced, resulting in the PRESTA algorithm (Bielajew and Rogers, 1987, 1989). The release of PRESTA was followed by the demonstration of various small, but important, shortcomings (Rogers, 1993; Foote and Smyth, 1995). There were improvements over the years (Bielajew and Kawrakow, 1997a,b,c, Kawrakow and Bielajew, 1998), eventually resulting in a revision of the EGS code, known as EGSnrc (Kawrakow and Rogers, 2000) and EGS5 (Hirayama et al., 2005). A PRESTA-like improvement of ETRAN (Seltzer, 1991; Kawrakow, 1996) was even developed.

1.4 Application: Ion Chamber Dosimetry

The founding paper for applying Monte Carlo methods to ionization chamber response is attributed to Bond et al. (1978), who, using an in-house Monte Carlo code, calculated ionization chamber response as a function of wall thickness, to [60]Co γ-irradiation. While validating the EGS code for this application, it was found that the EGS code had fundamental algorithmic difficulties with this low-energy regime, as well as this application. The resolution of these difficulties, not patent in other general-purpose Monte Carlo codes, became of great interest to this author. While general improvements to electron transport ensued, the fundamental problem was quite subtle, and was eventually described elegantly by Foote and Smythe (1995). In a nutshell, the underlying algorithmic reason that was identified arose from electron tracks being stopped at material boundaries, where cross sections change. EGS used the partial electron path to model a deflection of the electron, from the accumulated scattering power. The result was a rare, but important effect, the spurious generation of fluence singularities.

The literature on ionization chamber dosimetry is extensive. A partial compilation of very early contributions is (Bielajew et al., 1985; Rogers et al., 1985; Andreo et al., 1986; Bielajew, 1990; Rogers and Bielajew, 1990; Ma and Nahum, 1991).

Presently, the calculation of ionization chamber corrections is a very refined enterprise, with results being calculated to better than 0.1%. The literature on this topic is summarized by Bouchard and Seuntjens in their chapter in this book, "Applications of Monte Carlo to radiation dosimetry." That chapter also summarizes the contribution of Monte Carlo to dosimetry protocol and basic dosimetry data, some of the earliest applications of Monte Carlo to medicine.

1.5 Early Radiotherapy Applications

For brevity, only the earliest papers are cited in this section, and the reader is encouraged to employ the comprehensive reviews already cited earlier in this article. Some of the very early history of radiotherapy applications is rarely mentioned, and I have attempted to gather them here.

The Monte Carlo modeling of Cobalt-60 therapy units was first mentioned in the ICRU Report # 18 (1971). However, a more complete descriptive work followed somewhat later (Rogers et al., 1985; Han et al., 1987).

The modeling of LINAC therapy units was first accomplished by Petti et al. (1983a,b) and then, soon after by Mohan et al. (1985).

Photoneutron contamination from a therapy unit was first described by Ing et al. (1982), although the simulation geometry was simplified.

Mackie et al. pioneered the convolution method (Mackie et al., 1985) and then with other collaborators, generated the first database of "kernels" or "dose-spread arrays" for use in radiotherapy (Mackie et al., 1988). Independently, these efforts were being developed by Ahnesjö et al. (1987). These are still in use today.

Modeling of electron beams from medical LINACs was first accomplished by Teng et al. (1986), Hogstrom et al. (1986), and then Manfredotti et al. (1990).

An original plan to use Monte Carlo calculation for "target button to patient dose" was proposed by Mackie et al. (1990).

That effort became known as "the OMEGA (an acronym for Ottawa Madison Electron Gamma Algorithm) Project." However, early on in that project, a "divide-and-conquer" approach was adopted, whereby the fixed machine outputs ("phase-space files") were used as inputs to a patient-specific target (applicators and patient), to generate a full treatment plan. This bifurcation spawned two industries, treatment head modeling, of which the BEAM/EGSx code is the most refined (Rogers et al., 1995) and is the most cited paper in the "Web of knowledge" with "Monte Carlo" in the title, and "radiotherapy" as a topic. The second industry spawned by the OMEGA project was the development of fast patient-specific Monte Carlo-based dose calculation algorithms (Kawrakow et al., 1995, 1996; Sempau et al., 2000; Gardner et al., 2007). For more discussion on the current fast Monte Carlo methods in use, the reader is encouraged to see the excellent review by Spezi and Lewis (2008).

1.6 The Future of Monte Carlo

The first step in predicting the future is to look where one has been, extrapolate the process, thereby predicting the future. The second step in predicting the future is to realize that the first step involves some very specious, and problematic reasoning!

The progress of time, with the events that it contains, is intrinsically "catastrophic" in nature. A scientific discovery can be so earth-shattering, that new directions of research are spawned, while others dissolve into irrelevance. Yet, we persist in the practice of prediction. Therefore, allow me to be very modest in this effort.

Amdahl's Law (1967): Multiprocessor intercommunication bottlenecks will continue to be a limiting factor, for massively parallel machines. However, gains in traditional, monolayer, single chip speeds do *not* appear to be slowing, following Moore's Law.[*]

Harder to predict is algorithm development, specific to Monte Carlo applications in RTP. There is historical precedent for this in the citation data. In 1991, there was a 2.8 factor increase in productivity in only one year, followed by another in 1998, by a factor of 1.6. These increases are large, and unexplained. Yet, they illustrate the chaotic nature of the field.

There is a strong suggestion that research in Monte Carlo is saturating. Although new areas of research may rise and grow, both the "medical" and "all" data have been flat since about 2005. The EGS and MCNP output has been flat since 2000. Perhaps we are at the pinnacle?

We can safely predict, while the approach to the pinnacle may have been somewhat chaotic, the decline will be gradual. Monte Carlo codes have gotten easier to use, packaged with more

---

[*] In 1965, Gordon E. Moore, co-founder of Intel Corporation, predicted that areal transistor density would double every 18 months. One can infer a commensurate increase in computer speed. Moore changed his prediction to "doubling every two years" in 1975. Current chips designs follow Moore's Law until 2015 at least.

user-friendly interfaces, and developing into "shrink-wrapped," "turn-key" software systems. This is how it should be. The fact that a Monte Carlo method is the "engine" beneath a computational algorithm, should be transparent, but not invisible, to the researcher using the tool.

Paraphrasing the comment made by Martin Berger (Seligman et al., 1991), the founder of our field, during his speech at his retirement *Festschrift* symposium (1991),

> I am not used to so much public attention. Tonight is quite unusual for me. I hope that, after tonight, I can disappear into the anonymity that I so assuredly crave.

And so it may be, for Monte Carlo research—at least, until the next great thing comes along.

A very astute student once said to me, "We no longer have Departments of Railroad Engineering," in an effort to explain the (then) decline of Nuclear Engineering Departments, before the "nuclear renaissance." It may be that the *development* of the Monte Carlo method is bound to decline as it matures, but it will remain an essential component of our scientific infrastructure, forever.

Appendix: Monte Carlo and Numerical Quadrature

In this appendix, we present a mathematical proof that the Monte Carlo method is the most efficient way of estimating tallies in three spatial dimensions when compared to first-order deterministic (analytic, phase-space evolution) methods. Notwithstanding the opinion that the Monte Carlo method is thought of as providing the most accurate calculation, the argument may be made in such a way, that it is independent of the physics content of the underlying algorithm or the quality of the incident radiation field.

A.1 Dimensionality of Deterministic Methods

For the purposes of estimating tallies from initiating electrons, photons, or neutrons, the transport process that describes the trajectories of particles is adequately described by the linear Boltzmann transport equation (Duderstadt and Martin, 1979):

$$\left[\frac{\partial}{\partial s} + \frac{p}{|p|} \cdot \frac{\partial}{\partial x} + \mu(x,p) \right] \psi(x,p,s) = \int dx' \int dp' \mu(x,p,p') \psi(x',p',s),$$

(A.1)

where x is the position, p is the momentum of the particle, $(p/|p|) \cdot \partial/\partial x$ is a directional derivative (in three dimensions $\vec{\Omega} \cdot \vec{\nabla}$, for example) and s is a measure of the particle pathlength. We use the notation that x and p are multi-dimensional variables

of dimensionality N_x and N_p. Conventional applications span the range $1 \le N_{p,x} \le 3$. The macroscopic differential scattering cross section (probability per unit length) $\mu(x,p,p')$ describes scattering from momentum p' to p at location x, and the total macroscopic cross section is defined by

$$\mu(x,p) = \int dp' \mu(x,p,p').$$

(A.2)

$\psi(x,p,s) \, dx \, dp$ is the probability of there being a particle in dx about x, in dp about p and at pathlength s. The boundary condition to be applied is

$$\psi(x,p,0) = \delta(x)\delta(p_0 - p)\delta(s),$$

(A.3)

where p_0 represents the starting momentum of a particle at $s = 0$. The essential feature of Equation A.1, insofar as this proof is concerned, is that the solution involves the computation of a $(N_x + N_p)$-dimensional integral.

A general solution may be stated formally:

$$\psi(x,p,s) = \int dx' \int dp' \, G(x,p,x',p',s) Q(x',p'),$$

(A.4)

where $G(x,p,x',p',s)$ is the Green's function and $Q(x',p')$ is a source. The Green's function encompasses the operations of transport (drift between points of scatter, $x' \rightarrow x$), scattering (i.e., change in momentum) and energy loss, $p' \rightarrow p$. The interpretation of $G(x,p,x',p',s)$ is that it is an operator that moves particles from one point in $(N_x + N_p)$-dimensional phase space, (x',p'), to another, (x,p) and can be computed from the kinematical and scattering laws of physics.

Two forms of Equation A.4 have been employed extensively for general calculation purposes. Convolution methods integrate Equation A.4 with respect to pathlength s and further assume (at least for the calculation of the Green's function) that the medium is effectively infinite. Thus,

$$\psi(x,p) = \int dx' \int dp' \, G\left(|x - x'|, \left[\frac{p}{|p|} \cdot \frac{p'}{|p'|} \right], |p'| \right) Q(x',p'),$$

(A.5)

where the Green's function is a function of the distance between the source point x' and x, the angle between the vector defined by the source p' and p and the magnitude of the momentum of the course, $|p'|$, or equivalently, the energy.

To estimate a tally using Equation A.5 we integrate $\psi(x, p)$ over p, with a response function, $\mathcal{R}(x, p)$ (Shultis and Faw, 1996):

$$T(x) = \int dx' \int dp' \, F(|x - x'|, p') Q(x',p'),$$

(A.6)

where the "kernel," $F(|x - x'|, p')$, is defined by

$$F(|x - x'|, p') = \int dp\, \mathcal{R}(x, p) G\left(|x - x'|, \left[\frac{p}{|p|} \cdot \frac{p'}{|p'|}\right], |p'|\right).$$

(A.7)

$F(|x - x'|, p')$ has the interpretation of a functional relationship that connects particle fluence at phase-space location x', p' to a tally calculated at x. This method has a known difficulty—its treatment of heterogeneities and interfaces. Heterogeneities and interfaces can be treated approximately by scaling $|x - x'|$ by the collision density. This is an exact for the part of the kernel that describes the first scatter contribution but approximate for higher-order scatter contributions. It can also be approximate, to varying degrees, if the scatter produces other particles with different scaling laws, such as the electron set in motion by a first Compton collision of a photon.

For calculation methods that are concerned with primary charged particles, the heterogeneity problem is more severe. The true solution in this case is reached when the path length steps, s in Equation A.4 are made small (Larsen, 1992) and so, an iterative scheme is set up:

$$\psi_1(x, p) = \int dx' \int dp'\, G(x, p, x', p', \Delta s) Q(x', p')$$

$$\psi_2(x, p) = \int dx' \int dp'\, G(x, p, x', p', \Delta s) \psi_1(x', p')$$

$$\psi_3(x, p) = \int dx' \int dp'\, G(x, p, x', p', \Delta s) \psi_2(x', p')$$

.

.

.

(A.8)

$$\psi_N(x, p) = \int dx' \int dp'\, G(x, p, x', p', \Delta s) \psi_{N-1}(x', p')$$

which terminates when the largest energy in $\psi_N(x, p)$ has fallen below an energy threshold or there is no x remaining within the target. The picture is of the phase space represented by $\psi(x, p)$ "evolving" as s accumulates. This technique has come to be known as the "phase-space evolution" model. Heterogeneities are accounted for by forcing Δs to be "small" or of the order of the dimensions of the heterogeneities and using a $G()$ that pertains to the atomic composition of the local environment. The calculation is performed in a manner similar to the one described for convolution. That is,

$$T(x) = \sum_{i=1}^{N} \int dx' \int dp'\, F(x, x', p, p', \Delta s) \psi_i(x', p'),$$

(A.9)

where the "kernel," $F(x, x', p, p', \Delta s)$, is defined by

$$F(x, x', p, p', \Delta s) = \int dp\, \mathcal{R}(x, p) G(x, p, x', p', \Delta s).$$

(A.10)

In the following analysis, we will not consider further any systematic errors associated with the treatment of heterogeneities in the case of the convolution method, nor with the "stepping errors" associated with incrementing s using Δs in the phase space evolution model. Furthermore, we assume that the Green's functions or response kernels can be computed "exactly"—that there is no systematic error associated with them. The important result of this discussion is to demonstrate that the dimensionality of the analytic approach is $N_x + N_p$.

A.2 Convergence of Deterministic Solutions

The discussion of the previous section indicates that deterministic solutions are tantamount to solving a D-dimensional integral of the form:

$$I = \int_D du\, H(u).$$

(A.11)

In D dimensions, the calculation is no more difficult than in two dimensions, only the notation is more cumbersome. One notes that the integral takes the form:

$$I = \int_{u_{1,\min}}^{u_{1,\max}} du_1 \int_{u_{2,\min}}^{u_{2,\max}} du_2 \cdots \int_{u_{D,\min}}^{u_{D,\max}} du_D\, H(u_1, u_2 \cdots u_D)$$

$$= \sum_{i_1=1}^{N_{\text{cell}}^{1/D}} \int_{u_{i_1} - \Delta u_1/2}^{u_{i_1} + \Delta u_1/2} du_1 \sum_{i_2=1}^{N_{\text{cell}}^{1/D}} \int_{u_{i_2} - \Delta u_2/2}^{u_{i_2} + \Delta u_2/2} du_2 \cdots \int_{u_{i_D} - \Delta u_D/2}^{u_{i_D} + \Delta u_D/2} du_D$$

$$\times \sum_{i_D=1}^{N_{\text{cell}}^{1/D}} H(u_1, u_2 \cdots u_D)$$

(A.12)

The Taylor expansion takes the form

$$H(u_1, u_2 \cdots u_D) = H(u_{i_1}, u_{i_2} \cdots u_{i_D}) + \sum_{j=1}^{D} (u_i - u_{i_j})$$

$$\times \partial H(u_{i_1}, u_{i_2} \cdots u_{i_D})/\partial u_j + \sum_{j=1}^{D} \frac{(u_i - u_{i_j})^2}{2}$$

$$\times \partial^2 H(u_{i_1}, u_{i_2} \cdots u_{i_D})/\partial u_j^2 + \sum_{j=1}^{D} \sum_{k \neq j=1}^{D} (u_i - u_{i_j})(u_i - u_{i_k})$$

$$\times \partial^2 H(u_{i_1}, u_{i_2} \cdots u_{i_D})/\partial u_i \partial u_j \cdots$$

(A.13)

The linear terms of the form $(u_i - u_{ij})$ and the bilinear terms of the form $(u_i - u_{ij})(u_i - u_{ik})$ for $k \neq j$ all vanish by symmetry and a relative $N^{-2/D}$ is extracted from the quadratic terms after integration. The result is that

$$\frac{\Delta I}{I} = \frac{1}{24 N_{\text{cell}}^{2/D}} \frac{\left[\sum_{i_1=1}^{N_{\text{cell}}^{1/D}} \sum_{i_2=1}^{N_{\text{cell}}^{1/D}} \cdots \sum_{i_D=1}^{N_{\text{cell}}^{1/D}} \sum_{d=1}^{D} (u_{d,\max} - u_{d,\min})^2 \times \partial^2 H(u_{i_1}, u_{i_2} \cdots u_{i_D})/\partial u_d^2 \right]}{\sum_{i_1=1}^{N_{\text{cell}}^{1/D}} \sum_{i_2=1}^{N_{\text{cell}}^{1/D}} \cdots \sum_{i_D=1}^{N_{\text{cell}}^{1/D}} H(u_{i_1}, u_{i_2} \cdots u_{i_D})}. \quad (A.14)$$

Note that the one- and two-dimensional results can be obtained from the above equation. The critical feature to note is the overall $N_{\text{cell}}^{-2/D}$ convergence rate. The more dimensions in the problem, the slower the convergence for numerical quadrature.

A.3 Convergence of Monte Carlo Solutions

An alternative approach to solving Equation A.1 is the Monte Carlo method whereby N_{hist} particle histories are simulated. In this case, the Monte Carlo method converges to the true answer according to the central limit theorem (Feller, 1967) which is expressed as

$$\frac{\Delta T_{\text{MC}}(x)}{T_{\text{MC}}(x)} = \frac{1}{\sqrt{N_{\text{hist}}}} \frac{\sigma_{\text{MC}}(x)}{T_{\text{MC}}(x)}, \quad (A.15)$$

where $T_{\text{MC}}(x)$ is the tally calculated in a voxel located at x as calculated by the Monte Carlo method and $\sigma_{\text{MC}}^2(x)$ is the variance associated with the *distribution* of $T_{\text{MC}}(x)$. Note that this variance $\sigma_{\text{MC}}^2(x)$ is an intrinsic feature of how the particle trajectories deposit energy in the spatial voxel. It is a "constant" for a given set of initial conditions and is conventionally estimated from the sample variance. It is also assumed, for the purpose of this discussion, that the sample variance exists and is finite.

A.4 Comparison between Monte Carlo and Numerical Quadrature

The deterministic models considered in this discussion precalculate $F(|x - x'|, p')$ of Equation A.7 or $F(x, x', p, p', \Delta s)$ of Equation A.10 storing them in arrays for iterative use. Then, during the iterative calculation phase, a granulated matrix operation is performed. The associated matrix product is mathematically similar to the "mid-point" $N_x + N_p$-multidimensional integration discussed previously:

$$T(x) = \int_{\mathcal{D}} du \, H(u,x), \quad (A.16)$$

where $\mathcal{D} = N_x + N_p$ and $u = (x_1, x_2 \cdots x_{N_x}, p_1, p_2 \cdots p_{N_p})$. That is, u is a multidimensional variable that encompasses both space and momentum. In the case of photon convolution, $H(u, x)$ can be inferred from Equation A.6 and takes the explicit form:

$$H(u,x) = \int dp \, F(|x - x'|, p') Q(x', p'). \quad (A.17)$$

There is a similar expression for the phase space evolution model.

The "mid-point" integration represents a "first-order" deterministic technique and is applied more generally than the convolution or phase space evolution applications. As shown previously, the convergence of this technique obeys the relationship:

$$\frac{\Delta T_{\text{NMC}}(x)}{T_{\text{NMC}}(x)} = \frac{1}{N_{\text{cell}}^{2/D}} \frac{\sigma_{\text{NMC}}(x)}{T_{\text{NMC}}(x)}, \quad (A.18)$$

where $T_{\text{NMC}}(x)$ is the tally in a spatial voxel in an arbitrary N_x-dimensional geometry calculated by a non-Monte Carlo method where N_p momentum components are considered. The D-dimensional phase space has been divided into N_{cell} "cells" equally divided among all the dimensions so that the "mesh-size" of each phase space dimension is $N_{\text{cell}}^{1/D}$. The constant of proportionality as derived previously is

$$\sigma_{\text{NMC}}(x) = \frac{1}{24} \sum_{i_1=1}^{N_{\text{cell}}^{1/D}} \sum_{i_2=1}^{N_{\text{cell}}^{1/D}} \cdots \sum_{i_D=1}^{N_{\text{cell}}^{1/D}}$$
$$\times \sum_{d=1}^{D} (u_{d,\max} - u_{d,\min})^2 \partial^2 H(u_{i_1}, u_{i_2} \cdots u_{i_D})/\partial u_d^2, \quad (A.19)$$

where the u-space of $H(u)$ has been partitioned in the same manner as the phase space described above. $u_{d,\min}$ is the minimum value of u_d while $u_{d,\max}$ is its maximum value. u_{i_j} is the midpoint of the cell in the jth dimension at the i_jth mesh index.

The equation for the proportionality factor is quite complicated. However, the important point to note is, that it depends only on the second derivatives of $H(u)$ with respect to the phase-space variables, u. Moreover, the non-Monte Carlo proportionality factor is quite different from the Monte Carlo proportionality factor. It would be difficult to predict which would be smaller, and almost certainly, would be application dependent.

We now assume that the computation time in either case is proportional to N_{hist} or cell. That is, $T_{\text{MC}} = \alpha_{\text{MC}} N_{\text{hist}}$ and $T_{\text{NMC}} = \alpha_{\text{NMC}} N_{\text{cell}}$. In the Monte Carlo case, the computation time is simply N_{hist} times the average computation time/history. In the non-Monte Carlo case, the matrix operation can potentially attempt to connect every cell in the D-dimensional phase space to the tally at point x. Thus, a certain number of floating-point and integer operations are required for each cell in the problem.

Consider the convergence of the Monte Carlo and non-Monte Carlo method. Using the above relationships, one can show that:

$$\frac{\Delta T_{\mathrm{MC}}(x)/T_{\mathrm{MC}}(x)}{\Delta T_{\mathrm{NMC}}(x)/T_{\mathrm{NMC}}(x)} = \left(\frac{\sigma_{\mathrm{NMC}}(x)}{\sigma_{\mathrm{MC}}(x)}\right)\left(\frac{\alpha_{\mathrm{NMC}}^{D}}{\alpha_{\mathrm{MC}}}\right)^{1/2} t^{(4-D)/2D}, \quad (A.20)$$

where t is the time measuring computational effort for either method. We have assumed that the two calculational techniques are the same. Therefore, given enough time, $D_{\mathrm{MC}}(x) \approx D_{\mathrm{NMC}}(x)$. One sees that, given long enough, the Monte Carlo method is always more advantageous for $D > 4$. We also note, that inefficient programming in the non-Monte Carlo method is severely penalized in this comparison of the two methods.

Assume that one desires to do a calculation to a prescribed $\varepsilon = \Delta T(x)/T(x)$. Using the relations derived so far, we calculate the relative amount time to execute the task to be:

$$\frac{t_{\mathrm{NMC}}}{t_{\mathrm{MC}}} = \left(\frac{\alpha_{\mathrm{MC}}}{\alpha_{\mathrm{NMC}}}\right)\left(\frac{[\sigma_{\mathrm{NMC}}(x)/T_{\mathrm{NMC}}(x)]^{D/2}}{\sigma_{\mathrm{MC}}(x)/T_{\mathrm{MC}}(x)}\right)\varepsilon^{(4-D)/2}, \quad (A.21)$$

which again shows an advantage for the Monte Carlo method for $D > 4$. Of course, this conclusion depends somewhat upon assumptions of the efficiency ratio $\alpha_{\mathrm{MC}}/\alpha_{\mathrm{NMC}}$ which would be dependent on the details of the calculational technique. Our conclusion is also dependent on the ratio $[\{\sigma_{\mathrm{NMC}}(x)/T_{\mathrm{NMC}}(x)\}^{D/2}]/[\sigma_{\mathrm{MC}}(x)/T_{\mathrm{MC}}(x)]$ which relates to the detailed shape of the response functions. For distributions that can vary rapidly, the Monte Carlo method is bound to be favored. When the distributions are flat, non-Monte Carlo techniques may be favored.

Nonetheless, at some level of complexity (large number of N_{cell}'s required) Monte Carlo becomes more advantageous. Whether or not one's application crosses this complexity "threshold," has to be determined on a case-by-case-basis.

Smaller dimensional problems will favor the use of non-Monte Carlo techniques. The degree of the advantage will depend on the details of the application.

References

Aarnio P. A., Ranft J., and Stevenson G. R. A long write-up of the FLUKA82 program. CERN Divisional Report, TIS-RP/106-Rev., 1984.

Ahnesjö A., Andreo P., and Brahme A. Calculation and application of point spread functions for treatment planning with high energy photon beams. *Acta Oncol.*, 26:49–57, 1987.

Allison J. et al. GEANT4 development and applications. *IEEE T. Nucl. Sci.*, 53(1):250–303, 2006.

Alsmiller R. G., Jr and Barish J. High-energy (<18 GeV) muon transport calculations and comparison with experiment. *Nucl. Instr. Meth.*, 71:121–124, 1969.

Alsmiller R. G., Jr and Barish J. Energy deposition by 45 GeV photons in H, Be, Al, Cu, and Ta. Report ORNL-4933, Oak Ridge National Laboratory, Oak Ridge, Tennessee, 1974.

Alsmiller R. G., Jr, Barish J., and Dodge S. R. Energy deposition by high-energy electrons (50 to 200 MeV) in water. *Nucl. Instr. Meth.*, 121:161–167, 1974.

Alsmiller R. G., Jr and Moran H. S. Electron–photon cascade calculations and neutron yields from electrons in thick targets. Report ORNL-TM-1502, Oak Ridge National Laboratory, Oak Ridge, Tennessee, 1966.

Alsmiller R. G., Jr and Moran H. S. Electron–photon cascade calculations and neutron yields from electrons in thick targets. *Nucl. Instr. Methods*, 48:109–116, 1967.

Alsmiller R. G., Jr and Moran H. S. The electron–photon cascade induced in lead by photons in the energy range 15 to 100 MeV. Report ORNL-4192, Oak Ridge National Laboratory, Oak Ridge, Tennessee, 1968.

Alsmiller R. G., Jr and Moran H. S. Calculation of the energy deposited in thick targets by high-energy (1 GeV) electron–photon cascades and comparison with experiment. *Nucl. Sci. Eng.*, 38:131–134, 1969.

Amdahl G. M. Validity of the single processor approach to acheiving large scale computing facilities. *Commun. AFIPS Joint Comp. Conf.*, 1967.

Andreo P. Monte Carlo simulation of electron transport. In *The Computation of Dose Distributions in Electron Beam Radiotherapy*, ed. Nahum A. E., Umea University, Umeå, Sweden, pp. 80–97, 1985.

Andreo P. Monte Carlo techniques in medical radiation physics. *Phys. Med. Biol.*, 36:861–920, 1991.

Andreo P., Nahum A. E., and Brahme A. Chamber-dependent wall correction factors in dosimetry. *Phys. Med. Biol.*, 31:1189–1199, 1986.

Baró J., Sempau J., Fernández-Varea J. M., and Salvat F. PENELOPE: An algorithm for Monte Carlo simulation of the penetration and energy loss of electrons and positrons in matter. *Nucl. Instr. Methods*, B100:31–46, 1995.

Battistoni G., Muraro S., Sala P. R., Cerutti F., Ferrari A., Roestler S., Fasso A., and Ranft J. The FLUKA code: Description and benchmarking. *Proceedings of Hadronic Simulation Workshop*, Fermilab 6–8 September 2006, eds. Albrow M. and Baja R., Melville, New York, pp. 31–49, 2007.

Berger M. J. Monte Carlo calculation of the penetration and diffusion of fast charged particles. *Methods Comput. Phys.*, 1:135–215, 1963.

Berger M. J. and Seltzer S. M. ETRAN Monte Carlo code system for electron and photon transport through extended media. *Radiation Shielding Information Center, Computer Code Collection, CCC-107*, 1968.

Bielajew A. F. Correction factors for thick-walled ionisation chambers in point-source photon beams. *Phys. Med. Biol.*, 35:501–516, 1990.

Bielajew A. F. and Kawrakow I. The EGS4/PRESTA-II electron transport algorithm: Tests of electron step-size stability.

In *Proceedings of the XII'th Conference on the Use of Computers in Radiotherapy*, Medical Physics Publishing, Madison, Wisconsin, pp. 153–154, 1997a.

Bielajew A. F. and Kawrakow I. From "black art" to "black box": Towards a step-size independent electron transport condensed history algorithm using the physics of EGS4/PRESTA-II. In *Proceedings of the Joint International Conference on Mathematical Methods and Supercomputing for Nuclear Applications*, American Nuclear Society Press, La Grange Park, Illinois, USA, pp. 1289–1298, 1997b.

Bielajew A. F. and Kawrakow I. PRESTA-I 25⇒ PRESTA-II: The new physics. In *Proceedings of the First International Workshop on EGS4*, Technical Information and Library, Laboratory for High Energy Physics, Japan, pp. 51–65, 1997c.

Bielajew A. F. and Rogers D. W. O. PRESTA: The parameter reduced electron-step transport algorithm for electron Monte Carlo transport. *Nucl. Instr. Methods*, B18:165–181, 1987.

Bielajew A. F. and Rogers D. W. O. Electron step-size artefacts and PRESTA. In *Monte Carlo Transport of Electrons and Photons*, eds. T. M. Jenkins, W. R. Nelson, A. Rindi, A. E. Nahum, and D. W. O. Rogers. Plenum Press, New York, pp. 115–137, 1989.

Bielajew A. F., Rogers D. W. O., and Nahum A. E. Monte Carlo simulation of ion chamber response to ^{60}Co—Resolution of anomalies associated with interfaces. *Phys. Med. Biol.*, 30:419–428, 1985.

Bond J. E., Nath R., and Schulz R. J. Monte Carlo calculation of the wall correction factors for ionization chambers and A_{eq} for ^{60}Co γ rays. *Med. Phys.*, 5:422–425, 1978.

Briesmeister J. MCNP—A general purpose Monte Carlo code for neutron and photon transport, Version 3A. Los Alamos National Laboratory Report LA-7396-M, Los Alamos, NM, 1986.

Brun R., Hansroul M., and Lassalle J. C. *GEANT User's Guide*. CERN Report DD/EE/82, 1982.

Burfeindt H. Monte-Carlo-Rechnung für 3 GeV-Schauer in Blei. Deutsches Elektronen-Synchrotron Report Number DESY-67/24, 1967.

Butcher J. C. and Messel H. Electron number distribution in electron–photon showers. *Phys. Rev.*, 112:2096–2106, 1958.

Butcher J. C. and Messel H. Electron number distribution in electron–photon showers in air and aluminum absorbers. *Nucl. Phys.*, 20:15–128, 1960.

Cashwell E. D. and Everett C. J. *Monte Carlo Method for Random Walk Problems*. Pergamon Press, New York, 1959.

Colbert H. M. SANDYL: A computer program for calculating combined photon-electron transport in complex systems. Sandia Laboratories, Livermore, Report Number SCL-DR-720109, 1973.

Comte de Buffon G. *Essai d'arithmétique morale*, Vol. 4. Supplément à l'Histoire Naturelle, 1777.

Duderstadt J. J. and Martin W. M. *Transport Theory*. Wiley, New York, 1979.

Eckhart R. Stan Ulam, John von Neumann, and the Monte Carlo method. *Los Alamos Science (Special Issue)*, 131–141, 1987.

Fasso A., Ferrari A., Ranft J., and Sala P. R. FLUKA: A multi-particle transport code. CERN 2005-10, INFN/TC_05/11, SLAC-R-773, 2005.

Feller W. *An Introduction to Probability Theory and Its Applications*, Vol. I, 3rd Edition. Wiley, New York, 1967.

Foote B. J. and Smyth V. G. The modeling of electron multiple-scattering in EGS4/PRESTA and its effect on ionization-chamber response. *Nucl. Inst. Methods*, B100:22–30, 1995.

Ford R. L. and Nelson W. R. The EGS code system—Version 3. Stanford Linear Accelerator Center Report SLAC-210, 1978.

Gardner J., Siebers J., and Kawrakow I. Dose calculation validation of VMC++ for photon beams. *Med. Phys.*, 34:1809–1818, 2007.

Halbleib J. A. SPHERE: A spherical geometry multimaterial electron/photon Monte Carlo transport code. *Nucl. Sci. Eng.*, 66:269, 1978.

Halbleib J. A. ACCEPT: A Three-dimensional multilayer electron/photon Monte Carlo transport code using combinatorial geometry. *Nucl. Sci. Eng.*, 75:200, 1980.

Halbleib J. A., Hamil R., and Patterson E. L. Energy deposition model for the design of REB-driven large-volume gas lasers. *IEEE International Conference on Plasma Science (abstract)*, IEEE Conference Catalogue No. 81CH1640-2 NPS:117, 1981.

Halbleib J. A., Kensek R. P., Mehlhorn T. A., Valdez G. D., Seltzer S. M., and Berger M. J. ITS Version 3.0: The integrated TIGER Series of coupled electron/photon Monte Carlo transport codes. *Sandia report SAND91-1634*, 1992.

Halbleib J. A. and Mehlhorn T. A. ITS: The integrated TIGER series of coupled electron/photon Monte Carlo transport codes. *Nucl. Sci. Eng.*, 92(2):338, 1986.

Halbleib J. A., Sr. and Morel J. E. TIGERP, A one-dimensional multilayer electron/photon Monte Carlo transport code with detailed modeling of atomic shell ionization and relaxation. *Nucl. Sci. Eng.*, 70:219, 1979.

Halbleib J. A. and Vandevender W. H. EZTRAN—A user-oriented version of the ETRAN-15 electron–photon Monte Carlo technique. Sandia National Laboratories Report, SC-RR-71-0598,1971.

Halbleib J. A. and Vandevender W. H. EZTRAN 2: A User-oriented version of the ETRAN-15 electron–photon Monte Carlo technique. Sandia National Laboratories Report, SLA-73-0834, 1973.

Halbleib J. A., Sr. and Vandevender W. H. TIGER, A one-dimensional multilayer electron/photon Monte Carlo transport code. *Nucl. Sci. Eng.*, 57:94, 1975.

Halbleib J. A. and Vandevender W. H. CYLTRAN: A cylindrical-geometry multimaterial electron/photon Monte Carlo transport Code. *Nucl. Sci. Eng.*, 61:288–289, 1976.

Halbleib J. A., Sr. and Vandevender W. H. Coupled electron photon collisional transport in externally applied electromagnetic fields. *J. Appl. Phys.*, 48:2312–2319, 1977.

Han K., Ballon D., Chui C., and Mohan R. Monte Carlo simulation of a cobalt-60 beam. *Med. Phys.*, 14:414–419, 1987.

Hebbard D. F. and Wilson P. R. The effect of multiple scattering on electron energy loss distributions. *Australian J. Phys.*, 1:90–97, 1955.

Hendricks J. S. and Briesmeister J. F. Recent MCNP Developments. *Los Alamos National Laboratory Report LA-UR-91-3456 (Los Alamos, NM)*, 1991.

Hirayama H., Namito Y., Bielajew A. F., Wilderman S. J., and Nelson W. R. The EGS5 Code System. Report KEK 2005-8/SLAC-R-730, *High Energy Accelerator Research Organization/Stanford Linear Accelerator Center*, Tskuba, Japan/Stanford, USA, 2005.

Hogstrom K. R. Evaluation of electron pencil beam dose calculation. *Medical Physics Monograph (AAPM) No. 15*, 532–561, 1986.

ICRU. Specification of high-activity gamma-ray sources. ICRU Report 18, ICRU, Washington, DC, 1971.

Ing H., Nelson W. R., and Shore R. A. Unwanted photon and neutron radiation resulting from collimated photon beams interacting with the body of radiotherapy patients. *Med. Phys.*, 9:27–33, 1982.

Kalos M. H. and Whitlock P. A. *Monte Carlo Methods*, 2nd Edition. John Wiley and Sons-VCH, Weinnheim, Germany, 2008.

Kawrakow I. Electron transport: Longitudinal and lateral correlation algorithm. *Nucl. Instr. Methods*, B114:307–326, 1996.

Kawrakow I. and Bielajew A. F. On the representation of electron multiple elastic-scattering distributions for Monte Carlo calculations. *Nucl. Instr. Methods*, B134:325–336, 1998a.

Kawrakow I. and Bielajew A. F. On the condensed history technique for electron transport. *Nucl. Instr. Methods*, B142:253–280, 1998b.

Kawrakow I. and Rogers D. W. O. The EGSnrc Code System: Monte Carlo simulation of electron and photon transport. Technical Report PIRS–701, National Research Council of Canada, Ottawa, Canada, 2000.

Kawrakow I., Fippel M., and Friedrich K. The high performance Monte Carlo algorithm VMC. *Medizinische Physik, Proceedings*, 256–257, 1995.

Kawrakow I., Fippel M., and Friedrich K. 3D electron dose calculation using a voxel based Monte Carlo algorithm. *Med. Phys.*, 23:445–457, 1996.

Laplace P. S. Theorie analytique des probabilités, Livre 2. In *Oeuvres complétes de Laplace*, Vol. 7, Part 2, pp. 365–366. L'académie des Sciences, Paris, 1886.

Larsen E. W. A theoretical derivation of the condensed history algorithm. *Ann. Nucl. Energy*, 19:701–714, 1992.

Lewis H. W. Multiple scattering in an infinite medium. *Phys. Rev.*, 78:526–529, 1950.

Ma C. -M. and Jiang S. B. Monte Carlo modeling of electron beams for Monte Carlo treatment planning. *Phys. Med. Biol.*, 44:R157–R189, 1999.

Ma C. M. and Nahum A. E. Bragg–Gray theory and ion chamber dosimetry in photon beams. *Phys. Med. Biol.*, 36:413–428, 1991.

Mackie T. R. Applications of the Monte Carlo method in radiotherapy. In *Dosimetry of Ionizing Radiation*, Vol. III, eds.

K. Kase, B. Bjärngard and F. H. Attix. Academic Press, New York, pp. 541–620, 1990.

Mackie T. R., Bielajew A. F., Rogers D. W. O., and Battista J. J. Generation of energy deposition kernels using the EGS Monte Carlo code. *Phys. Med. Biol.*, 33:1–20, 1988.

Mackie T. R., Kubsad S. S., Rogers D. W. O., and Bielajew A. F. The OMEGA project: Electron dose planning using Monte Carlo simulation. *Med. Phys. (abs)*, 17:730, 1990.

Mackie T. R., Scrimger J. W., and Battista J. J. A convolution method of calculating dose for 15 MV x-rays. *Med. Phys.*, 12:188–196, 1985.

Manfredotti C., Nastasi U., Marchisio R., Ongaro C., Gervino G., Ragona R., Anglesio S., and G. Sannazzari. Monte Carlo simulation of dose distribution in electron beam radiotherapy treatment planning. *Nucl. Instr. Methods*, A291:646–654, 1990.

Messel H. and Crawford D. F. *Electron–Photon Shower Distribution Function*. Pergamon Press, Oxford, 1970.

Messel H., Smirnov A. D., Varfolomeev A. A., Crawford D. F., and Butcher J. C. Radial and angular distributions of electrons in electron–photon showers in lead and in emulsion absorbers. *Nucl. Phys.*, 39:1–88, 1962.

Metropolis N. The beginning of the Monte Carlo method. *Los Alamos Science (Special Issue)*, 125–130, 1987.

Metropolis N. and Ulam S. The Monte Carlo method. *Amer. Stat. Assoc.*, 44:335–341, 1949.

Meyer H. A., ed. *Symposium on Monte Carlo Methods*. John Wiley and Sons, New York, 1981.

Miller P. A., Halbleib J. A., and Poukey J. W. SPHEM: A three-dimensional multilayer electron/photon Monte Carlo transport code using combinatorial geometry. *J. Appl. Phys.*, 52:593–598, 1981.

Mohan R., Chui C., and L. Lidofsky. Energy and angular distributions of photons from medical linear accelerators. *Med. Phys.*, 12:592–597, 1985.

Nagel H. H. Die Berechnung von Elektron-Photon-Kaskaden in Blei mit Hilfe der Monte-Carlo Methode. *Inaugural-Dissertation zur Erlangung des Doktorgrades der Hohen Mathematich-Naturwissenschaftlichen Fakultät der Rheinischen Friedrich-Wilhelms-Universtät zu Bonn*, 1964.

Nagel H. H. Elektron-Photon-Kaskaden in Blei: Monte-Carlo-Rechnungen für Primärelektronenergien zwischen 100 und 1000 Me. *Physik V. Z.*, 186:319–346, 1965 (English translation Stanford Linear Accelerator Center Report Number SLAC-TRANS-28, 1965.

Nagel H. H. and Schlier C. Berechnung von Elektron-Photon-Kaskaden in Blei für eine Primärenergie von 200 MeV. *Z. Phys.*, 174:464–471, 1963.

Nahum A. E. Calculations of electron flux spectra in water irradiated with megavoltage electron and photon beams with applications to dosimetry. PhD thesis, University of Edinburgh, UK, 1976.

Namito Y., Ban S., and Hirayama H. Implementation of linearly-polarized photon scattering into the EGS4 code. *Nucl. Instr. Methods*, A322:277–283, 1993.

Namito Y., Ban S., and Hirayama H. Implementation of Doppler broadening of Compton-scattered photons into the EGS4 code. *Nucl. Inst. Meth.*, A349:489–494, 1994.

Nelson W. R., Hirayama H., and Rogers D. W. O. The EGS4 Code System. Report SLAC–265, Stanford Linear Accelerator Center, Stanford, CA, 1985.

Nicoli D. F. The application of Monte Carlo cascade shower generation in lead. Submitted in partial fulfillment of the requirement for the degree of Bachelor of Science at the Massachusetts Institute of Technology, 1966.

Petti P. L., Goodman M. S., Gabriel T. A., and Mohan R. Investigation of buildup dose from electron contamination of clinical photon beams. *Med. Phys.*, 10:18–24, 1983a.

Petti P. L., Goodman M. S., Sisterson J. M., Biggs P. J., Gabriel T. A., and Mohan R. Sources of electron contamination for the Clinac–35 25–MV photon beam. *Med. Phys.*, 10:856–861, 1983b.

Ranft J. Monte Carlo calculation of the nucleon–meson cascade in shielding materials by incoming proton beams with energies between 10 and 1000 GeV. CERN Yellow Report 64–67, 1964.

Rhodes R. *The Making of the Hydrogen Bomb*. Touchstone (Simon & Schuster Inc.), New York, 1988.

Rogers D. W. O. More realistic Monte Carlo calculations of photon detector response functions. *Nucl. Instrum. Meth.*, 199:531–548, 1982.

Rogers D. W. O. The use of Monte Carlo techniques in radiation therapy. *Proceedings of CCPM Course on Computation in Radiation Therapy*, Canadian College of Physicists in Medicine, London, Ontario, 1983a.

Rogers D. W. O. A nearly mono-energetic 6 to 7 MeV photon calibration source. *Health Phys.*, 45:127–137, 1983b.

Rogers D. W. O. Fluence to dose equivalent conversion factors calculated with EGS3 for electrons from 100 keV to 20 GeV and photons from 20 keV to 20 GeV. *Health Phys.*, 46:891–914, 1984a.

Rogers D. W. O. Low energy electron transport with EGS. *Nucl. Inst. Meth.*, 227:535–548, 1984b.

Rogers D. W. O. How accurately can EGS4/PRESTA calculate ion chamber response? *Med. Phys.*, 20:319–323, 1993.

Rogers D. W. O. Fifty years of Monte Carlo simulations for medical physics. *Phys. Med. Biol.*, 51:R287–R301, 2006.

Rogers D. W. O. and Bielajew A. F. Monte Carlo techniques of electron and photon transport for radiation dosimetry. In *The Dosimetry of Ionizing Radiation*, Vol III, eds. K. R. Kase, B. E. Bjärngard, and F. H. Attix. Academic Press, New York and London, pp. 427–539. 1990a.

Rogers D. W. O. and Bielajew A. F. Wall attenuation and scatter corrections for ion chambers: measurements versus calculations. *Phys. Med. Biol.*, 35:1065–1078, 1990b.

Rogers D. W. O., Bielajew A. F., and Nahum A. E. Ion chamber response and A_{wall} correction factors in a ^{60}Co beam by Monte Carlo simulation. *Phys. Med. Biol.*, 30:429–443, 1985.

Rogers D. W. O., Ewart G. M., Bielajew A. F., and G. van Dyk. Calculation of contamination of the ^{60}Co beam from an AECL therapy source. NRC Report PXNR-2710, 1985.

Rogers D. W. O., Faddegon B. A., Ding G. X., Ma C. M., Wei J., and Mackie T. R. BEAM: A Monte Carlo code to simulate radiotherapy treatment units. *Med. Phys.*, 22:503–524, 1995.

Rossi B. B. *High Energy Particles*. Prentice-Hall, New York, 1952.

Salvat F. and Parrellada J. Penetration and energy loss of fast electrons through matter. *J. Appl. Phys. D: Appl. Phys.*, 17(7):1545–1561, 1984.

Seligman H., McLaughlin W. L., Seltzer S. M., and Inokuti M. eds. Proceedings of the Symposium. *Applied Radiation and Isotopes*, 42, DR's file on IAEA, 1991.

Seltzer S. M. An overview of ETRAN Monte Carlo methods. In: *Monte Carlo Transport of Electrons and Photons*, eds. T. M. Jenkins, W. R. Nelson, A. Rindi, A. E. Nahum, and D. W. O. Rogers. Plenum Press, New York, pp. 153–182, 1989.

Seltzer S. M. Electron–photon Monte Carlo calculations: The ETRAN code. *Int. J. Appl. Radiat. Isotopes*, 42:917–941, 1991.

Sempau J., Wilderman S. J., and Bielajew A. F. DPM, a fast, accurate Monte Carlo code optimized for photon and electron radiotherapy treatment planning dose calculations. *Phys. Med. Biol.*, 45:2263–2291, 2000.

Shultis J. K. and Faw R. E. *Radiation Shielding*. Prentice-Hall, Upper Saddle River, 1996.

Spezi E. and Lewis G. An overview of Monte Carlo treatment planning for radiotherapy. *Radiat. Protect. Dosim.*, 131:123–129, 2008.

Teng S. P., Anderson D. W., and Lindstrom D. G. Monte Carlo electron-transport calculations for clinical beams using energy grouping. *Appl. Radiat. Isotopes*, 1189–1194, 1986.

Ulam S. M. *Adventures of a Mathematician*, 2nd Edition. University of California Press, Berkeley, 1991.

Ulam S. M. and J. von Neumann. Random ergodic theorems. *Bull. Amer. Math. Soc. (abstract)*, 51:660, 1945.

Ulam S. M. and J. von Neumann. On combination of stochastic and deterministic processes. *Bull. Amer. Math. Soc. (abstract)*, 53:1120, 1947.

Varfolomeev A. A. and Svetlolobov I. A. Monte Carlo calculations of electromagnetic cascades with account of the influence of the medium on bremsstrahlung. *Sov. Phys. JETP*, 36:1263–1270, 1959.

Verhaegen F. and Seuntjens J. Monte Carlo modelling of external radiotherapy photon beams. *Phys. Med. Biol.*, 48:R107–R164, 2003.

Völkel U. Elektron-Photon-Kaskaden in Blei für Primärteilchen der Energie 6 GeV. Deutsches Elektronen-Synchrotron Report Number DESY-65/6, 1965 (English translation Stanford Linear Accelerator Center Report Number SLAC-TRANS-41, 1966.

von Neumann J. and Richtmyer R. Statistical methods in neutron diffusion. Technical Report LAMS-551, Los Alamos National Laboratory, 1947.

Wilson R. R. Radiological use of fast protons. *Radiology*, 47:487–491, 1946.

Wilson R. R. Monte Carlo calculations of showers in lead. *Phys. Rev. (abstract)*, 79:204, 1950.

Wilson R. R. The range and straggling of high energy electrons. *Phys. Rev.*, 84:100–103, 1951.

Wilson R. R. Monte Carlo study of shower production. *Phys. Rev.*, 86:261–269, 1952.

Zerby C. D. and Moran H. S. A Monte Carlo calculation of the three-dimensional development of high-energy electron–photon cascade showers. Report ORNL-TM-422, Oak Ridge National Laboratory, Oak Ridge, Tennessee, 1962a.

Zerby C. D. and Moran H. S. Studies of the longitudinal development of high-energy electron–photon cascade showers in copper. Report ORNL-3329, Oak Ridge National Laboratory, Oak Ridge, Tennessee, 1962b.

Zerby C. D. and Moran H. S. Studies of the longitudinal development of electron–photon cascade showers. *J. Appl. Phys.*, 34:2445–2457, 1963.

2

Basics of Monte Carlo Simulations

Matthias Fippel
Brainlab AG

2.1 Monte Carlo Method

Monte Carlo (MC) techniques are widely used in natural and social sciences. There are many different "flavors" of how to work with these techniques. As we have seen in the previous chapter, there is a long tradition of using MC methods in different areas, including medical physics. Therefore, it is difficult to provide a general definition of the MC method. Consequently, a literature search results in many definitions. Some examples of introducing literature, useful to read before working with MC methods, are the corresponding textbook chapters in *Numerical Recipes in C* (Press et al., 1992) or in "The Review of Particle Physics" (Nakamura et al., 2010) of the Particle Data Group. A nice introduction is also provided by James in "Monte Carlo Theory and Practice" (James, 1980). More references on MC techniques and random number sampling can be found in these reviews.

For our purposes, we define the MC method in the following short way:

> Monte Carlo is a numerical method to solve equations or to calculate integrals based on random number sampling.

The two aspects of this definition, random number sampling and numerical integration, are outlined in detail in the next two sections.

2.1.1 Random Number Sampling

MC algorithms use a computer program, a procedure or a subroutine, called "random number generator" (RNG). However, computers cannot really generate "random" numbers because the output of any program is—by definition—predictable; hence, it is not truly "random." Therefore, the result of these generators shall correctly be termed "pseudorandom numbers."

A huge sequence of these pseudorandom numbers is required to solve a complex problem. The numbers within a sequence of random numbers shall be uncorrelated, that is, they must not depend on each other. Because this is impossible with a computer program, they should at least appear independent. In other words, any statistical test program shall show that the numbers within the sequence are uncorrelated and any computer code that requires independent random numbers shall produce the same result with different sequences. If this is the case within the uncertainty of the simulation, then these sequences can be called pseudorandom. To keep the notation short throughout this book, we will nevertheless call them just "random numbers." However, we have to keep in mind the real character of these numbers.

A pseudo-RNG must be examined carefully before it can be used for a specific purpose. A useful generator for simulations in radiation therapy must provide two important features:

- The period of the sequence shall be large enough. Otherwise, if the sequence is reused several times, the results of the MC simulation are correlated.
- They must be uniformly distributed in multiple dimensions. That means random vectors created from an *n*-tuple of random numbers must be uniformly distributed in the *n*-dimensional space. Typically, it is not obvious how to detect correlations in higher dimensions.

Most of the generators produce uniformly distributed random numbers in some interval, typically in [0,1]. It is useful to have a look at simple RNGs to understand their operating principle. One class of simple RNGs is called *linear congruential generators*. They generate a sequence of integers $I_1, I_2, I_3, \ldots$, each between 0 and $m - 1$ by the recurrence relation:

$$I_{j+1} = aI_j + c \quad (mod\ m) \tag{2.1}$$

with the parameters

$$a : \text{multiplier}$$
$$c : \text{increment}$$
$$m : \text{modulus}$$

An example is the quick and dirty generator used by old EGS4 (Nelson et al., 1985) implementations:

$$I_{j+1} = aI_j \qquad (2.2)$$

with

$$a = 663608941$$
$$c = 0$$
$$m = 2^{32}$$

On a machine with 32-bit integer representation, the product of two unsigned integers is the low-order 32 bits of the true 64-bit result. Because of $m = 2^{32}$, the modulus is taken into account automatically. Therefore, this RNG is very fast. However, its sequence length of 2^{32} is not enough for MC applications in radiotherapy.

Computer operating system-supplied RNGs are typically of the same *linear congruential* type. Hence, it is not recommended to trust them. For MC applications in radiotherapy, long-sequence RNGs are required. They also should be portable, that is, the same sequence should be produced on different machines. A good source of high-quality RNGs including test programs to examine the created sequences in multiple dimensions is the CERN program library (http://www.cern.ch).

A class of long-sequence RNGs called "subtract-with-borrow" algorithms has been developed by Marsaglia and Zaman (1991). The CERN library function RANMAR for instance has a sequence length of 2^{144}. It is used, for example, by EGS4 (Nelson et al. 1985) and XVMC (Fippel, 1999).

In EGSnrc (Kawrakow, 2000a), the use of RANLUX (Lüscher, 1994) is recommended. It allows different luxury levels between 0 and 4. The quality and also the simulation time increase with increasing luxury level. RANMAR can also be selected in EGSnrc. According to the CERN library documentation, its quality corresponds to RANLUX with luxury level between 1 and 2. According to the EGSnrc manual (Kawrakow et al., 2011), with luxury level 1 or higher, no problems have been discovered in practical EGSnrc calculations.

2.1.2 Numerical Integration

Function $y = f(x)$ shall be integrated in interval $[a,b]$, that is, the area A enclosed by function $f(x)$, the x-axis as well as the interval limits a and b shall be calculated (see Figure 2.1):

$$A = \int_a^b f(x)\,\mathrm{d}x. \qquad (2.3)$$

If this is impossible analytically, some numerical method must be applied. One of the many numerical options is called MC integration because it is based on a sequence of uniformly distributed random numbers. Generally, a computer-generated random number η_i is uniformly distributed in interval $[0,1]$. It can be scaled to interval $[a,b]$ by

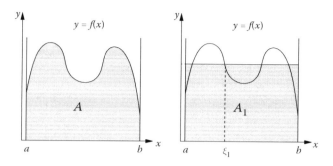

FIGURE 2.1 The left plot shows area A as calculated by integrating function $y = f(x)$ in interval $[a,b]$. The right plot shows a very rough estimate of area A as given by the area of the rectangle $A_1 = (b-a)f(\xi_1)$.

$$\xi_i = (b-a)\eta_i + a, \qquad (2.4)$$

that is, ξ_i is uniformly distributed in $[a,b]$. Some first rough estimate of the real area A is now given by (see Figure 2.1)

$$A_1 = (b-a)f(\xi_1), \qquad (2.5)$$

that is, the rectangle given by the function value at random point ξ_1 and the length of the interval. This estimate is really rough; therefore, we should do this a second time:

$$A_2 = \frac{1}{2}\{(b-a)f(\xi_2) + A_1\} = \frac{b-a}{2}\{f(\xi_1) + f(\xi_2)\}. \quad (2.6)$$

Note that we have averaged the areas from both runs providing in this way a better estimate of the real integral. The generalization is now obvious; after N runs (with N random numbers), we obtain

$$A_N = \frac{b-a}{N}\sum_{i=1}^N f(\xi_i) = (b-a)\langle f(x)\rangle \qquad (2.7)$$

with the average function value for N samples:

$$\langle f(x)\rangle \equiv \frac{1}{N}\sum_{i=1}^N f(\xi_i). \qquad (2.8)$$

The basic theorem of MC integration (Press et al., 1992) also provides information on the uncertainty of the estimate

$$A = A_N \pm (b-a)\sqrt{\frac{\langle f^2(x)\rangle - \langle f(x)\rangle^2}{N}}. \qquad (2.9)$$

The average of the function value squared is defined by

$$\langle f^2(x)\rangle \equiv \frac{1}{N}\sum_{i=1}^N f^2(\xi_i). \qquad (2.10)$$

The estimated area A_N converges to the real integral A in the limit $N \to \infty$. The convergence is slow because of the $1/\sqrt{N}$ behavior, that is, the statistical uncertainty is reduced by a factor of 2 if the number of random points N (and the calculation time) is increased by a factor of 4. Therefore, the MC method should not be used for this simple type of numerical integration. There are better options; see, for example, Press et al. (1992). However, the MC method comes into play if all other methods fail, for example, if the dimensionality of the problem becomes very large, that is, the integral shall be calculated in a space with 10, 100, or even an infinite number of dimensions.

Let us assume that function $f(\vec{x})$ shall be integrated in volume V of a space with D dimensions. Instead of random numbers for MC integration, random points (or vectors) uniformly distributed in the multidimensional volume V are required. Because volume V has D dimensions, we need D random numbers to form one random point. To sample N random points in this volume, we need $D \times N$ random numbers. Therefore, the quality of the RNG should be checked for higher dimensions, not just for the 1D, 2D, or 3D case. Often, correlations can be observed in higher dimensions even if the RNG looks uniform in lower dimensions.

Now, let us assume that we have sampled N random points $\vec{\xi}_1, \dots, \vec{\xi}_N$ uniformly distributed in multidimensional volume V. Then the basic theorem of MC integration is given by (Press et al., 1992)

$$\int dV\, f(\vec{x}) \approx V \langle f(\vec{x}) \rangle \pm \sqrt{\frac{\langle f^2(\vec{x}) \rangle - \langle f(\vec{x}) \rangle^2}{N}} \qquad (2.11)$$

with

$$\langle f(\vec{x}) \rangle \equiv \frac{1}{N} \sum_{i=1}^{N} f(\vec{\xi}_i) \quad \text{and} \quad \langle f^2(\vec{x}) \rangle \equiv \frac{1}{N} \sum_{i=1}^{N} f^2(\vec{\xi}_i). \qquad (2.12)$$

Multidimensional numerical integration is necessary to solve the system of coupled transport equations for problems in radiation therapy, for example, for dose calculation. It is a system of equations because the transport problem for photons and electrons (positrons) must be solved. The system is coupled because electrons influence the photon transport (bremsstrahlung) and vice versa (Compton scatter, photo electric absorption, pair production). Theoretically, the problem has an infinite parameter space because the number of secondary photons and electrons is physically unlimited when we start with a primary particle of definite energy. Therefore, the numerical integration has to be performed in a space with infinite dimensions. Practically, the dimensionality is limited because the region of interest is limited and we usually stop the simulation if the photon or electron energy falls below some minimum energy.

Nevertheless, for numerical integration using MC, a random point in a high-dimensional parameter space must be sampled. This point can be demonstrated by a so-called "particle history," a shower of secondary particles generated by a primary particle including all daughter particles. An example of a particle history

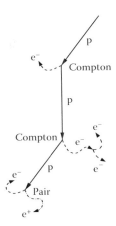

FIGURE 2.2 Example of a particle history starting with a primary photon p (straight line) via Compton interactions and pair production events leading to secondary photons p (straight lines) and secondary electrons e⁻ (dashed lines) and positrons e⁺ (dashed dotted line).

is shown in Figure 2.2. This shows schematically what happens during an MC radiation transport simulation. It starts with a primary particle emitted from a particle source. The simulation takes into account the geometry of the problem, for example, the linac head geometry given by the technical specifications and/or the patient anatomy given by CT images. It also takes into account the material transport properties given by cross-section data. For the primary photon in the example of Figure 2.2, the distance to the first interaction site must be sampled based on the total cross section of the corresponding medium. At the interaction site, the type of interaction (Compton scatter in this case) must be sampled. All secondary particle parameters (secondary particle energies and scattering angles) are determined using differential cross sections and the corresponding probability distribution functions. These steps are repeated until the primary and secondary particles have left the simulation geometry or the particle energy falls below some minimum energy. Dose, for example, is calculated by accumulating the absorbed energy per region. It is obvious that a huge number of particle histories has to be simulated to obtain a result of low noise. Because of Equation 2.11, the noise level and the statistical uncertainty can be reduced by a factor of 2 if the number of histories is increased by a factor of 4. This causes very long calculation times for transport simulation problems in radiation therapy.

2.1.3 Nonuniform Sampling Methods

As we have learned in Section 2.1.1, RNGs usually generate uniformly distributed random numbers. However, for MC transport simulations in radiation therapy, random numbers distributed according to specific probability weight distribution functions $p(x)$ are required many times within the algorithm.

Let us assume that we want to generate a random number ξ in interval $[a,b]$ and distributed according to the nonuniform probability weight function $p(x)$. Available is a RNG generating only uniformly distributed random numbers η in interval

[0,1]. How can ξ be sampled from $p(x)$? This is possible using the cumulative distribution function $P(x)$ defined by the integral

$$P(x) = \int_a^x dx' \, p(x'), \quad a \le x \le b, \quad P(a) = 0, \quad P(b) = 1. \quad (2.13)$$

Function $P(x)$ is monotonically increasing in interval $[a,b]$. The function values $y = P(x)$ are limited to interval $[0,1]$. If formula $y = P(x)$ can be transposed with respect to x, that is, the inverse can be determined as

$$x = P^{-1}(y), \quad (2.14)$$

then it can be shown that

$$\xi = P^{-1}(\eta), \quad (2.15)$$

is distributed according to probability weight function $p(x)$. For a proof, see, for example, Nelson et al. (1985). This method of nonuniform random number sampling is called the *direct* or *transformation* method. It depends on the fact that the inverse of $P(x)$ can be easily and efficiently calculated.

If this is impossible, the *indirect* or *rejection* method (Nelson et al. 1985) has to be applied instead. It takes into account an adequately chosen comparison function $g(x)$ with $g(x) > p(x)$ in interval $[a,b]$. A further condition of $g(x)$ is that a random number ξ_0 can be sampled easily from $g(x)$, for example, using the transformation method. Now, we sample a new uniform random number η_0 from interval $[0,g(\xi_0)]$. We accept ξ_0 as valid random number if $\eta_0 < p(\xi_0)$. We reject ξ_0 if $p(\xi_0) \le \eta_0 \le g(\xi_0)$. In case of rejection, we start with sampling a new random number ξ_0 from $g(x)$ and so on. Again, we do not prove here that ξ_0 sampled in this way is distributed according to $p(x)$. But the rejection method can be demonstrated using the simple example of a constant rejection function $g(x) = p(x_{max}) = const$, with x_{max} as the position of the maximum of $p(x)$ in interval $[a,b]$. In this case, ξ_0 is just sampled from a uniform distribution in $[a,b]$. With the rejection random number $\eta_0 < p(\xi_0)$, it is obvious that ξ_0 is distributed according to $p(x)$.

Of course, the rejection method should be used with care. That is, comparison function $g(x)$ should not differ too much from $p(x)$. If the deviation is too large, the number of rejections can become too large and the method becomes inefficient.

2.2 Monte Carlo Transport in Radiation Therapy

The basic concepts to simulate the transport of photons, electrons, positrons, neutrons, protons, or heavy ions in the energy range of radiation therapy are complex. Therefore, this chapter can only provide the fundamentals of the methodology. To understand all techniques in full detail, the reader is referred to corresponding literature, like the textbook *Monte Carlo Transport of Electrons*

and Photons (Jenkins et al., 1988). A lot of information is available also in the user manuals of frequently used MC packages, like EGS4 (Nelson et al., 1985), EGSnrc (Kawrakow, 2000a; Kawrakow et al., 2011), Penelope (Salvat et al., 2011), MCNP (Briesmeister, 1997), and GEANT4 (Agostinelli et al., 2003). More basic literature is listed in review papers about MC treatment planning (Chetty et al., 2007; Reynaert et al., 2007).

In the next sections, we show how particle transport collision-by-collision is simulated in detail. This so-called analog particle transport simulation scheme is demonstrated using the photon as an example. A further section introduces the condensed history (CH) technique that allows efficient simulations of charged particle transport in radiotherapy.

2.2.1 Analog Particle Transport

Let us assume that a photon of energy E hits the surface of a homogeneous medium. Then, the probability $p(s)$ that this photon interacts after path length s with the medium is given by the attenuation law:

$$p(s) \, ds = \mu(E) \, e^{-\mu(E)s} \, ds. \quad (2.16)$$

The parameter $\mu(E)$ is called the linear attenuation coefficient of the medium for photons of energy E. The mean free path length $\langle s \rangle$ until interaction can be calculated from this distribution function if the medium is extended infinitely below the surface:

$$\langle s \rangle = \int_0^\infty ds \, s \, p(s) = \mu(E) \int_0^\infty ds \, s \, e^{-\mu(E)s} = \frac{1}{\mu(E)}. \quad (2.17)$$

This allows us to express the attenuation law (2.16) in terms of the number of mean free path lengths:

$$\lambda = \frac{s}{\langle s \rangle} = \mu(E) \, s, \quad (2.18)$$

that is

$$p(\lambda) \, d\lambda = e^{-\lambda} \, d\lambda. \quad (2.19)$$

The advantage of this notation is that it also works for heterogeneous geometries if the number of mean free path lengths is defined by

$$\lambda = \sum_{Start}^P \mu_i(E) \, s_i. \quad (2.20)$$

To calculate λ, the photon must be traced on a straight line from the *Start* position on the surface through different regions i containing different materials until the interaction point P. In each region i with linear attenuation coefficient $\mu_i(E)$, the corresponding line segment s_i must be determined. This tracing algorithm to calculate λ is an essential part of MC simulations in radiation therapy. The calculation time can be unnecessarily long if it is implemented inefficiently.

The attenuation law (2.19) provides the probability weight distribution function $p(\lambda)$. The cumulative distribution function is given by

$$P(\lambda) = \int_0^\lambda d\lambda' \, p(\lambda') = \int_0^\lambda d\lambda' \, e^{-\lambda'} = 1 - e^{-\lambda}, \quad P(0) = 0, \quad P(\infty) = 1.$$

$$(2.21)$$

This function is monotonically increasing in $[0,\infty]$. We can now sample λ_1, the distance to the first interaction site, using the transformation method and a uniform random number ξ_1 from the half open interval $[0,1)$:

$$\xi_1 = 1 - e^{-\lambda_1}, \; \Rightarrow \; \lambda_1 = -\ln(1 - \xi_1). \qquad (2.22)$$

Note the special notation of the interval limits above. It means that number 1 must not be included in the sequence of random numbers. If it is included, the logarithm in Equation 2.22 is undefined.

Taking into account the geometric setup of the simulation, the photon is tracked λ_1 mean free path lengths to the first interaction point. Then, the type of the interaction has to be sampled. In the energy range of radiation therapy, four processes are most common, *photoelectric absorption*, *Raleigh scatter*, *Compton scatter*, and *pair production*. They are represented by the corresponding interaction coefficients as material parameters at the interaction site:

$$\mu(E) \equiv \mu_{\text{tot}}(E) = \mu_A(E) + \mu_R(E) + \mu_C(E) + \mu_P(E) \qquad (2.23)$$

and they are used to divide the interval $[0,1]$ into four parts:

$$\begin{array}{lll} [P_0, P_1] & : & \text{photoelectric absorption} \\ [P_1, P_2] & : & \text{Raleigh scatter} \\ [P_2, P_3] & : & \text{Compton scatter} \\ [P_3, P_4] & : & \text{pair production} \end{array} \qquad (2.24)$$

with

$$P_0 = 0, \; P_1 = P_0 + \frac{\mu_A}{\mu_{\text{tot}}}, \; P_2 = P_1 + \frac{\mu_R}{\mu_{\text{tot}}}, \; P_3 = P_2 + \frac{\mu_C}{\mu_{\text{tot}}}, \; P_4 = 1. \qquad (2.25)$$

The interaction type is sampled by using a second uniform random number ξ_2 from interval $[0,1]$ and by checking in which subinterval ξ_2 is located.

With known interaction type, the parameters of all secondary particles can be determined. These parameters, that is, energy and scattering angles, are sampled using the probability distributions given by the corresponding differential cross sections. Furthermore, kinematic conservation laws must be taken into account. In general, for this purpose, the transformation method does not work, that is, the rejection technique is the method of choice.

After that, everything is known to repeat the three steps with the secondary particles. Even electrons and positrons could be simulated in this analog manner. The whole particle history (see an example in Figure 2.2) is simulated including all secondary particles and its daughter particles. The transport simulation of a particle stops if it leaves the geometry of interest or its energy falls below some predefined minimum energy. These cut-off parameters are usually denoted as P_{cut} for photons or E_{cut} for charged particles. During each step of the history, the values of interest are calculated for accumulation. For example, the absorbed energy per region (voxel) is accumulated if the dose is calculated. According to Equation 2.11, the number of simulated particle histories determines the statistical accuracy and thus the calculation time.

2.2.2 Charged Particle Transport

2.2.2.1 Condensed History Technique

Section 2.2.1 outlined the fundamental procedure to simulate the transport of any particle type through matter. In general, this is the standard simulation method for neutral particles because the free path length between two interactions is in the order of the size of the simulation geometry. For example, the mean free path lengths of photons in the therapeutic energy range are in the order of 10 cm in water and human tissue. The region of interest for dose calculation in radiation therapy has a size of about 30 cm. Therefore, on average, very few photon interactions are simulated.

This is completely different for charged particles like electrons or protons. For radiation therapy energies, they undergo a very large number of single interactions. Consequently, the simulation of one electron (positron or proton) history would require a much longer calculation time than the simulation of one photon history. Hence, this approach is impractical for most of the transport problems in radiation therapy.

Fortunately, almost all of these interactions are elastic or semielastic. That means no energy or a small amount of energy is transferred from the charged particle to the surrounding matter. Furthermore, the particle direction changes in general only by small scattering angles. This allows us to group many of these elastic and semielastic events into one CH step. The method has been introduced in 1963 by Berger (1963) and is called CH technique. The majority of MC algorithms, applied in radiotherapy, perform electron, positron, proton, or heavier charged particle transport using the CH technique.

Present-day CH implementations divide all interactions of one charged particle history into hard and soft collisions as well as hard and soft bremsstrahlung production events. The two collision types are distinguished by an arbitrary kinetic energy loss threshold E_c. Hard and soft bremsstrahlung production is distinguished using the parameter k_c. Collision events with an energy transfer lower than E_c to secondary electrons are called soft collisions. These soft collisions are simulated implicitly by continuous energy transfer from the charged particle onto the matter surrounding the particle track. The direction change of the particle due to many small angles is simulated by one large multiple scattering angle. All hard collisions are simulated explicitly as in the case of photons. The minimum energy of secondary particles created during hard collisions is equivalent to E_c. It also provides the maximum energy and consequently the maximum range of charged secondary particles produced during soft collisions. This range has to be smaller than the spatial resolution of the simulation geometry. The meaning of the bremsstrahlung production threshold k_c is similar. Therefore, the arbitrary parameters E_c and k_c must be chosen with care. The simulation result can be influenced negatively if they are too large. On the other hand, the simulation can last too long if these parameters are chosen too small. Please note that the parameter E_c must not be confused with the particle track end energy E_{cut}. However, in many MC simulations, both parameters are chosen to be equal.

Because of the approximate nature of the CH transport, it is furthermore useful to limit the maximum distance traveled in one CH step. This could be realized by another arbitrary user parameter, the global maximum step size s_{max}, or by material- and mass density-dependent parameters s_{max}^i. In many MC algorithms, the maximum step size is determined based on the percentage maximum energy loss E_{step}. This way, the maximum spatial step size automatically depends on the stopping power and the mass density of the present material.

With all of these possibilities, the end of one CH step is determined either by the maximum step size or by the next hard interaction. As a result, a complete electron history may look like the example shown in Figure 2.3. Electrons move in general on straight lines during the CH step. They change the direction due to multiple scattering at the end of the step (see Figure 2.4) or (as in the case of Figures 2.3 and 2.5) between step limits. In

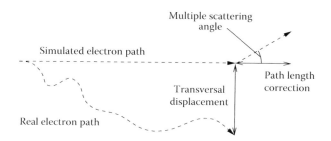

FIGURE 2.4 Example of a simulated electron path using the CH technique and multiple scatter compared to a possible real electron path (or simulated using the analog technique). As a result of the CH transport, the simulated path length must be corrected and a transversal displacement has to be taken into account.

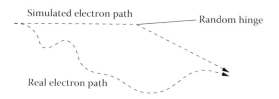

FIGURE 2.5 Same as Figure 2.4, but this time the electron is simulated using the random hinge method. This way, path length corrections and transversal displacements are approximately taken into account.

the second case, the position of the multiple scattering can be determined, for example, randomly using the so-called random hinge method. Figure 2.3 shows that the first CH step is limited by a hard Møller interaction. Møller interactions result in secondary electrons, also called delta electrons. The delta electrons are simulated in the same way using the CH technique until their energy falls below E_{cut} or if they leave the region of interest. Figure 2.3 also shows that the CH steps can be limited by the maximum step length or by hard bremsstrahlung events. Secondary bremsstrahlung photons are simulated like all other photons using the analog technique (see Section 2.2.1).

The various components of the CH technique are outlined in the following subsections using the electron as an example.

2.2.2.2 Continuous Energy Loss

During a CH step the charged particle continuously loses energy due to soft interactions. The average energy loss dE per CH step length ds at point $\vec{r}$ is given by the restricted linear stopping power:

$$L(\vec{r}, E, E_c, k_c) \equiv -\left(\frac{dE}{ds}\right)_{res} = L_{col}(\vec{r}, E, E_c) + L_{rad}(\vec{r}, E, k_c) \quad (2.26)$$

with

$$L_{col}(\vec{r}, E, E_c) \equiv -\left(\frac{dE}{ds}\right)_{res,col} \quad \text{and} \quad L_{rad}(\vec{r}, E, k_c) \equiv -\left(\frac{dE}{ds}\right)_{res,rad}$$

$$(2.27)$$

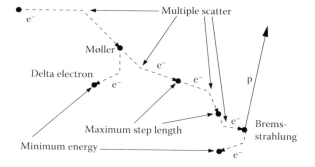

FIGURE 2.3 Example of a particle history starting with a primary electron e⁻ (dashed line) via multiple scatter, Møller interactions, and bremsstrahlung production events leading to secondary (delta) electrons e⁻ (dashed lines) and secondary photons p (straight lines).

defined as restricted linear collision and radiation stopping powers. They can be calculated using the collision cross section $\sigma_{col}(\vec{r}, E, E')$ and the bremsstrahlung production cross section $\sigma_{rad}(\vec{r}, E, k')$ by

$$L_{col}(\vec{r}, E, E_c) = N(\vec{r}) \int_0^{E_c} dE' \, E' \, \sigma_{col}(\vec{r}, E, E')$$

$$L_{rad}(\vec{r}, E, k_c) = N(\vec{r}) \int_0^{k_c} dk' \, k' \, \sigma_{rad}(\vec{r}, E, k'). \tag{2.28}$$

$N(\vec{r})$ is the number of scattering targets per unit volume at point $\vec{r}$. Note the stopping power integration is restricted to energies below E_c and k_c. That means the energy transfer to secondary charged particles is restricted to be below E_c and the energy transfer to secondary photons is restricted to be below k_c. Interactions with higher energy transfer are simulated explicitly within this CH scheme.

The step length s for an electron with initial energy E_0 that loses energy ΔE due to CH transport can be calculated by integrating Equation 2.26:

$$s = -\int_{E_0}^{E_1} \frac{dE}{L(\vec{r}, E, E_c, k_c)} = \int_{E_1}^{E_0} \frac{dE}{L(\vec{r}, E, E_c, k_c)} \tag{2.29}$$

with $E_1 = E_0 - \Delta E$ being the electron energy at the end of the step. The minus sign in the definition (2.26) is used here to exchange the upper and lower integration limits. The integration (2.29) is performed along the simulated electron path and takes into account the spatial dependency of the stopping power on $\vec{r}$. That means function $L(\vec{r}, E, E_c, k_c)$ changes accordingly if a boundary to a region with different material is crossed.

Equation 2.29 provides a unique functional dependency of the CH step length s and the step end energy E_1. Therefore, all electrons with initial energy E_0 and transported along the same step will reach the step end with the same energy E_1. Here, it becomes clear that this approach is an approximation because in reality, only the mean step end energy can be calculated by Equation 2.29. Analog MC simulations show that the real step end energies are random according to a distribution with a mean value of E_1. This effect is known as energy straggling. In the CH simulation scheme defined here, we have to distinguish between soft energy straggling and hard energy straggling. Hard energy straggling is simulated explicitly and the corresponding effects are correctly taken into account. Soft or subthreshold energy straggling is either neglected or it is simulated by an adequate straggling distribution function.

Indeed, subthreshold straggling can be neglected during a CH MC simulation of specific problems if we choose the parameter E_c and the step size small enough. In this case, energy straggling is dominated by the explicitly modeled hard ionization events with energy transfer larger than E_c and the low energy fluctuations have a negligible influence on the final result. Furthermore,

in many cases, subthreshold energy straggling is additionally insignificant compared to the influence of the range straggling (see Section 2.2.2.4).

A potential disadvantage of neglecting soft energy fluctuations is that the calculation time can increase unnecessarily. Therefore, the implementation of some subthreshold straggling theory can help to speed up the simulation or to avoid artifacts. For example, a Gaussian distribution function according to an approach by Bohr (1948) can be used to model energy straggling. More accurate are the theories of Landau (1944) and Vavilov (1957).

2.2.2.3 Multiple Scattering

In contrast to reality and as demonstrated in Figure 2.4, in a CH approach, charged particles move on straight lines during one step. The combined effect of many small-angle elastic and semielastic collisions during one step is simulated by sampling the angular deflection based on a dedicated multiple scattering theory. An example is the theory developed by Fermi and Eyges (Eyges, 1948). This theory models the probability $p(\theta, \varphi) \, d\theta \, d\varphi$ that the electron is scattered within the solid multiple scattering angular section $([\theta, \theta + d\theta], [\varphi, \varphi + d\varphi])$ as 2D Gaussian distribution:

$$p(\theta, \varphi) \, d\theta \, d\varphi = \frac{\theta}{\pi \, \overline{\theta^2}(s)} \exp\left(-\frac{\theta^2}{\overline{\theta^2}(s)}\right) d\theta \, d\varphi, \tag{2.30}$$

where θ is the azimuthal multiple scattering angle, φ is the polar multiple scattering angle, and $\overline{\theta^2}(s)$ is the mean square deflection angle after step length s. The distribution (2.30) is not normalized because θ is limited to the interval $[0, \pi]$ instead of $[0, \infty]$. Equation 2.30 results in two separate cumulative distribution functions:

$$P_\theta(\theta) = 1 - \exp\left(-\frac{\theta^2}{\overline{\theta^2}(s)}\right), \quad P_\varphi(\varphi) = \frac{\varphi}{2\pi}. \tag{2.31}$$

The transformation method from Section 2.1.3 and a uniform random number ξ_θ can be applied to sample θ:

$$\theta = \sqrt{-\overline{\theta^2}(s) \ln(1 - \xi_\theta)}. \tag{2.32}$$

It has to be noted that ξ_θ is uniformly distributed in an interval $[0, \xi_\theta^{max}]$ with $\xi_\theta^{max} < 1$ to ensure $\theta \leq \pi$. Another option is to sample ξ_θ from $[0,1]$ and to reject θ if $\theta > \pi$. Because of the rotational symmetry, the polar scattering angle φ is determined from a uniform distribution in $[0, 2\pi]$.

The quantity $\overline{\theta^2}(s)$ is calculated using the linear scattering power $T_S(\vec{r}, E)$ at point $\vec{r}$:

$$\overline{\theta^2}(s) = \int_0^s ds' \, T_S(s', E). \tag{2.33}$$

The material parameter $T_s(\vec{r}, E)$ depends on the atomic composition at point $\vec{r}$ but also on the electron energy E. This must be taken into account in Equation 2.33 because the electron loses energy between the beginning and the end of the step.

The Gaussian (2.30) is a good approximation of the reality for small cumulative scattering angles θ. However, large scattering angles are underestimated using this distribution. Therefore, very often in the past (e.g., in EGS4 (Nelson et al., 1985)) the multiple scattering distribution of Molière (1948) was used. Nevertheless, this improvement is also based on the small-angle approximation, and especially for large angles, the Molière theory still has limitations.

Most suitable for MC simulations are algorithms based on the exact theory of Goudsmit and Saunderson (1940a,b). Examples are the multiple scattering algorithms as implemented in MCNP (Briesmeister, 1997), Penelope (Salvat et al., 2011), and EGSnrc (Kawrakow et al., 2011). These algorithms sample multiple scattering angles close to reality even for large angles.

2.2.2.4 Transport Mechanics

Figure 2.4 compares a CH electron step to some possible real electron path or to an electron path simulated using analog MC transport. The figure shows that several problems can be expected from a CH simulation. For better comprehension, in this section, we neglect energy loss during the CH step. That means here the electron energy has not changed between the beginning and the end of the step and all interactions during the step are assumed to be elastic. As shown in Figure 2.4, many MC codes sample the multiple scattering angle at the end of the step, that is, the electron moves on a straight line until the final position and then it changes the direction due to multiple scatter. It is obvious that in this case, the electron distance is overestimated. Also shown in Figure 2.4, the real electron path is curved. Therefore, the real electron range is shorter if we assume that both electrons (the real electron and the CH electron) move with the same path length. Furthermore, the real electron range fluctuates around some mean value. This effect is called range straggling and should not be confused with energy straggling as discussed in Section 2.2.2.2. Range straggling is independent of the energy loss of the electron.

Many CH history algorithms employ a path length correction (PLC) algorithm to take into account range overestimation and range straggling. Besides this, a transverse displacement (TD) algorithm is useful to take into account transverse fluctuations of the real electron end position relative to the lateral position as simulated during the CH step. A very simple PLC and TD approach, called random hinge method, is demonstrated in Figure 2.5. It is implemented, for example, in Penelope (Salvat et al., 2011) or XVMC (Kawrakow and Fippel, 2000). In this method, the whole step length s is subdivided using a uniformly distributed random number ξ from interval [0,1] into two substeps ξs and $(1 - \xi)s$. The multiple scattering angle is sampled between the two substeps instead at the end of the step. The average longitudinal (PLC) and lateral (TD) displacements simulated in this manner are very

close to the exact values. More accurate is the parameter-reduced electron-step transport algorithm developed by Kawrakow and Bielajew (1998) (sometimes called PRESTA-II) as implemented in EGSnrc (Kawrakow, 2000a). This algorithm reproduces first- and second-order spatial moments to within 0.1%.

The problems and algorithms discussed in this section could be neglected completely if the charged particle transport steps are small enough. In the limit of infinitesimal small step sizes, any CH algorithm will converge to the correct answer. However, in this way, the simulation becomes extremely inefficient. Therefore, with the implementation of a sophisticated algorithm for the transport mechanics, charged particle transport can be used efficiently for MC simulations in radiation therapy.

2.2.2.5 Boundary Crossing

Theoretically, charged particle MC transport using the CH technique with multiple scattering works only in homogeneous regions containing one definite medium. The problem of conditions with more than one material is shown schematically in Figure 2.6. It shows two regions consisting of different materials as well as two electron tracks, a real electron track calculated using the single scatter scheme and a CH simulated electron path. By definition, the CH technique shall simulate the condensed effect of a huge number of real electron paths, including the real path shown in Figure 2.6. This path is partly located in material II and has to take into account (for this part of the path) the interaction properties of material II. For the rest of the path, it has to consider the properties of material I. Other real electrons may be tracked completely in material I.

Figure 2.6 clearly demonstrates that, for arbitrarily shaped material interfaces, an exact theory of multiple scattering does not exist. Consequently, the CH technique can only be used in regions where the distance to the next material boundary is much larger than the size of the present electron step. Conversely, this means that the maximum step size s_{max} must be decreased if an electron comes close to a material interface and it can be increased again if the electron moves away from the boundary. This kind of step size variation has been introduced, for example, in the original version of PRESTA (Bielajew and Rogers 1987). For geometries with many small regions of different material,

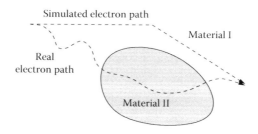

FIGURE 2.6 Example of an electron path simulated using the random hinge method. In contrast to Figures 2.4 and 2.5, two different materials are involved. This can cause simulation artifacts because the electron path simulated using the CH technique is not influenced by material II; however, the real electron path is influenced by material II.

this leads to small step sizes and long calculation times. A further problem is that before an electron crosses the boundary, its CH step size becomes infinitesimally small. To permit boundary crossing anyway, the step size reduction must be switched off if a minimum step size is achieved. However, then the step size becomes similar to the boundary distance and step size artifacts may occur as we have seen in Figure 2.6 and the discussion above.

To avoid step size artifacts, a more accurate algorithm has been implemented in EGSnrc (Kawrakow, 2000a). Here, a minimum distance d_{min} to the next material interface can be defined by the user and CH transport is performed only if the electron is farther away from the boundary than this minimum distance. If the electron comes closer to the boundary than d_{min}, the electron is transported in single scatter mode using the analog technique. This algorithm is free of step size artifacts. However, it also means that electrons are simulated in pure single scatter mode within geometries with very small regions, that is, if the average diameter of one region is smaller than d_{min}. Consequently, calculation times can increase dramatically. Therefore, this technique including an appropriate choice of parameter d_{min} makes sense only for simulations with high accuracy requirements. An example is the simulation of ion chamber responses (Kawrakow, 2000b).

For the majority of MC simulations in radiation therapy, artifacts due to boundary crossing can be ignored because of their negligible influence on the result. A useful approach is, for example, to simulate electron boundary crossing without stopping at the interface using the random hinge technique. Here, the multiple scattering properties of material I are used if the "hinge" is sampled in material I; otherwise, the properties of material II are used. The total electron step length depends on the stopping powers and the corresponding step segments in both materials. According to the PENELOPE user guide (Salvat et al., 2011), this approach provides a "fairly accurate description of interface crossing." It is, for example, an efficient and accurate method for linear accelerator head modeling or an MC-based dose calculation algorithm in radiation therapy treatment planning.

2.2.3 Cross Sections

2.2.3.1 Photon Interaction Coefficients

In Section 2.2.1, the linear attenuation coefficient $\mu(E)$ has been introduced to sample the distance between two photon interactions. For elements, compounds and mixtures, this material parameter can be calculated from the total cross sections $\sigma_i(E)$ of the i-th element in the material as

$$\mu(E) = \sum_i N_i(\vec{r})\, \sigma_i(E). \qquad (2.34)$$

$N_i(\vec{r})$ is the number of atoms of element i per unit volume at point $\vec{r}$ and is calculated by

$$N_i(\vec{r}) = \frac{\rho(\vec{r})\, w_i(\vec{r})}{m_u\, A_i(\vec{r})} = \frac{\rho(\vec{r})\, N_A\, w_i(\vec{r})}{M_i(\vec{r})}. \qquad (2.35)$$

$\rho(\vec{r})$ is the mass density at point $\vec{r}$, $w_i(\vec{r})$ is the weight fraction of the i-th element at point $\vec{r}$, $A_i(\vec{r})$ is the relative atomic mass of the i-th element at point $\vec{r}$, $M_i(\vec{r})$ is molar mass of the i-th element at point $\vec{r}$, $m_u = 1.6605388 \cdot 10^{-27}$ kg is the atomic mass unit, and $N_A = 6.0221418 \cdot 10^{23}$ mol^{-1} is the Avogadro constant (Nakamura et al. 2010).

Corresponding to Equation 2.23, the total atomic cross sections $\sigma_i(E)$ are given by the sum of the cross sections for the different processes:

$$\sigma_i(E) = \sigma_{i,A}(E) + \sigma_{i,R}(E) + \sigma_{i,C}(E) + \sigma_{i,P}(E), \qquad (2.36)$$

that is, photoelectric absorption (A), Raleigh scatter (R), Compton scatter (C), and pair production (P).

If the material composition is exactly known, then the cross sections for the different photon interactions can be calculated using some database, such as XCOM (Berger and Hubbell, 1987) or EPDL97 (Cullen et al., 1997). These databases are available online and they are updated from time to time to include the most recent measurement and calculation results. Most of the general-purpose and specific MC codes are based on one or more of these databases.

Especially in radiation therapy, the material compositions are not always available. Very often, only a CT number (or Hounsfield unit, HU) is known. Therefore, the International Commission on Radiation Units and Measurements (ICRU) has compiled a list of about 100 human tissue types, including material compositions and interaction data (e.g., photon cross sections). These data are published in ICRU Report 46 (*Photon, Electron, Proton and Neutron Interaction Data for Body Tissues*, 1992). With an adequate CT calibration, an HU number can be mapped to a specific body tissue type. The use of a very limited number of materials, for example, six materials such as air, lung, water, soft tissue, soft bone, and hard bone, should be avoided. Linear interpolation between these supporting points can lead to large cross-section inaccuracies.

The interaction probabilities $\mu_k(E,\rho)$ (k = A,R,C,P) of Equation 2.23 for human tissue can also be determined more directly, that is, without knowing the material compositions, using an approach presented for MC dose calculation (Kawrakow et al., 1996; Fippel, 1999):

$$\mu_k(E,\rho) = \frac{\rho}{\rho^w}\, f_k(\rho)\, \mu_k^w(E). \qquad (2.37)$$

The $\mu_k^w(E)$ are the interaction coefficients of water and ρ^w is the mass density of water. As an example, in Figure 2.7, the Compton cross-section ratios $f_C(\rho)$ for all materials of ICRU Report 46 are plotted. It shows that most of the data points (crosses) are located close to the solid line representing the fit function:

$$f_C(\rho) \approx \begin{cases} 0.99 + 0.01\rho/\rho^w, & \rho \le \rho^w \\ 0.85 + 0.15\rho^w/\rho, & \rho \ge \rho^w. \end{cases} \qquad (2.38)$$

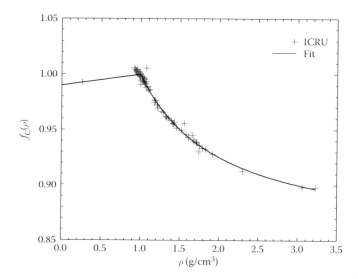

FIGURE 2.7 Ratios of mass Compton interaction coefficients of all materials from ICRU report 46 to the Compton interaction coefficient of water (crosses). The line represents the fit according to Equation 2.38.

The relative uncertainty of function (2.38) is better than 1% for almost all materials of ICRU Report 46. The only exception is gallstone with an error of 1.6%. Similar fits exist for the other interaction types. Thus, it is sufficient to determine only the mass density in a given region and to use equations like (2.38) to calculate the corresponding cross sections.

2.2.3.2 Charged Particle Stopping and Scattering Powers

For the MC simulation of charged particle transport using the CH technique according to Section 2.2.2.2, knowledge of the linear restricted collision and radiation stopping powers $L_{col}(\vec{r}, E, E_c)$ and $L_{rad}(\vec{r}, E, k_c)$ is necessary. These quantities can be calculated if the unrestricted collision and radiation stopping powers $S_{col}(\vec{r}, E)$ and $S_{rad}(\vec{r}, E)$ for the medium at point $\vec{r}$ are known. They provide the average energy loss dE per step ds without restricting the energy loss by some threshold. The total unrestricted linear stopping power is defined by

$$S(\vec{r}, E) \equiv -\left(\frac{dE}{ds}\right) = S_{col}(\vec{r}, E) + S_{rad}(\vec{r}, E). \qquad (2.39)$$

The unrestricted linear collision and radiation stopping powers are calculated using the collision cross section $\sigma_{col}(\vec{r}, E, E')$ and the bremsstrahlung production cross section $\sigma_{rad}(\vec{r}, E, k')$ by

$$S_{col}(\vec{r}, E) = N(\vec{r}) \int_0^E dE' \, E' \, \sigma_{col}(\vec{r}, E, E')$$

$$S_{rad}(\vec{r}, E) = N(\vec{r}) \int_0^E dk' \, k' \, \sigma_{rad}(\vec{r}, E, k'). \qquad (2.40)$$

$N(\vec{r})$ is the number of scattering targets per unit volume at point $\vec{r}$ and can be determined by Equation 2.35.

The mixed CH technique as in Section 2.2.2.2 requires knowledge of the hard inelastic collision and bremsstrahlung cross sections $\sigma_{col}(\vec{r}, E, E')$ and $\sigma_{rad}(\vec{r}, E, k')$ for energy transfer values $E' > E_c$ and $k' > k_c$, respectively. With Equations 2.28 and 2.40, this allows calculation of the restricted stopping powers according to

$$L_{col}(\vec{r}, E, E_c) = S_{col}(\vec{r}, E) - N(\vec{r}) \int_{E_c}^E dE' \, E' \, \sigma_{col}(\vec{r}, E, E')$$

$$(2.41)$$

$$L_{rad}(\vec{r}, E, k_c) = S_{rad}(\vec{r}, E) - N(\vec{r}) \int_{k_c}^E dk' \, k' \, \sigma_{rad}(\vec{r}, E, k').$$

Stopping power tables suitable for MC simulations in radiation therapy have been published, for example, by ICRU for electrons and positrons in Report 37 (*Stopping Powers for Electrons and Positrons*, 1984), for electrons and protons in Report 46 (*Photon, Electron, Proton and Neutron Interaction Data for Body Tissues*, 1992), as well as for protons and alpha particles in Report 49 (*Stopping Powers and Ranges for Protons and Alpha Particles*, 1993). Online databases like ESTAR, PSTAR, and ASTAR (Berger, 1993) can also be used to determine stopping powers for different elements, compounds, and mixtures.

Comparable to Equation 2.37, the stopping powers can also be determined more directly using the mass stopping power ratios $f_k(E, \rho)$ and the mass stopping power of water $S_k^w(E)/\rho^w$ ($k = tot, col, rad, \ldots$):

$$S_k(E, \rho) = \frac{\rho}{\rho^w} f_k(E, \rho) S_k^w(E). \qquad (2.42)$$

This approach has been adopted, for example, to develop an MC dose calculation algorithm for proton therapy (Fippel and Soukup, 2004). Figure 2.8 shows the total proton stopping power ratio $f_S(E, \rho)$ for two proton energies (10 and 100 MeV) and for all materials of ICRU Report 46 (*Photon, Electron, Proton and Neutron Interaction Data for Body Tissues*, 1992). After analyzing the tabulations and calculations of ICRU Report 46 (*Photon, Electron, Proton and Neutron Interaction Data for Body Tissues*, 1992), ICRU Report 49 (*Stopping Powers and Ranges for Protons and Alpha Particles*, 1993), and the online database PSTAR (Berger, 1993) the following fit formula has been published (Fippel and Soukup, 2004):

$$f_S(E, \rho) = \begin{cases} 1.0123 - 3.386 \cdot 10^{-5} E + 0.291(1 + E^{-0.3421})(\rho^{-0.7} - 1) \\ \quad \text{for } \rho \geq 0.9 \\ 0.9925 \quad \text{for } \rho = 0.26 \text{ (lung)} \\ 0.8815 \quad \text{for } \rho = 0.0012 \text{ (air)} \\ \text{interpolate for all other } \rho \leq 0.9 \end{cases}$$

$$(2.43)$$

with the kinetic proton energy E in MeV (rest mass not included) and mass density ρ in g cm^{-3}. Figure 2.8 shows this function

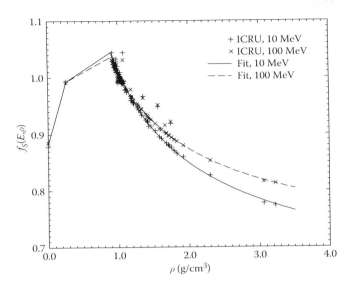

FIGURE 2.8 Mass stopping power ratios of all materials from ICRU report 46 to the mass stopping power of water for 10 and 100 MeV protons (crosses). The lines represent the fit for both energies.

for the two proton energies as solid and dashed lines. Besides some outliers (gallstone, protein, carbohydrate, and urinary stone), Equation 2.43 provides an accuracy of better than 1%, that is, knowledge of the atomic composition in each voxel is not necessary. Equation 2.43 and Figure 2.8 also show that an energy dependence of the stopping power ratio can be taken into account by the fit formulas.

Similar formulas exist for electron and positron collision and radiation stopping powers. The material and mass density dependence of the charged particle multiple scattering distributions can also be modeled this way.

References

Agostinelli, S. et al. 2003. GEANT4—A simulation toolkit, *Nucl. Instrum. Meth. A* **506**: 250–303.

Berger, M. J. 1963. Monte Carlo calculation of the penetration and diffusion of fast charged particles, *Methods in Computational Physics*, Vol. I, Academic Press, New York, pp. 135–215. eds. B. Alder, S. Fernbach, and M. Rotenberg.

Berger, M. J. 1993. *ESTAR, PSTAR, and ASTAR: Computer Programs for Calculating Stopping-Power and Range Tables for Electrons, Protons, and Helium Ions*, Technical Report NBSIR 4999, National Institute of Standards and Technology, Gaithersburg, MD.

Berger, M. J. and Hubbell, J. H. 1987. *XCOM: Photon Cross Sections on a Personal Computer*, Technical Report NBSIR 87-3597, National Institute of Standards and Technology, Gaithersburg, MD.

Bielajew, A. F. and Rogers, D. W. O. 1987. PRESTA: The parameter reduced electron-step transport algorithm for electron Monte Carlo transport, *Nucl. Instrum. Meth. B* **18**: 165–181.

Bohr, N. 1948. The penetration of atomic particles through matter, *Det Kongelige Danske Videnskabernes Selskab Matematisk-Fysiske Meddelelser* **18**(8): 1–144.

Briesmeister, J. F. 1997. *MCNP—A General Monte Carlo N-Particle Transport Code*, Report No. LA-12625-M, Los Alamos National Laboratory.

Chetty, I. J. et al. 2007. Report of the AAPM Task Group No. 105: Issues associated with clinical implementation of Monte Carlo-based photon and electron external beam treatment planning, *Med. Phys.* **34**(12): 4818–4853.

Cullen, D. E., Hubbell, J. H., and Kissel, L. 1997. *EPDL97: The Evaluated Photon Data Library, '97 Version*, Technical Report UCRL-50400, Vol 6, Rev 5, Lawrence Livermore National Laboratory, Livermore, CA.

Eyges, L. 1948. Multiple scattering with energy loss, *Phys. Rev.* **74**: 1534–1535.

Fippel, M. 1999. Fast Monte Carlo dose calculation for photon beams based on the VMC electron algorithm, *Med. Phys.* **26**(8): 1466–1475.

Fippel, M. and Soukup, M. 2004. A Monte Carlo dose calculation algorithm for proton therapy, *Med. Phys.* **31**(8): 2263–2273.

Goudsmit, S. and Saunderson, J. L. 1940a. Multiple scattering of electrons, *Phys. Rev.* **57**: 24–29.

Goudsmit, S. and Saunderson, J. L. 1940b. Multiple scattering of electrons II, *Phys. Rev.* **58**: 36–42.

James, F. 1980. Monte Carlo theory and practice, *Rep. Prog. Phys.* **43**(9): 1145–1189.

Jenkins, T. M., Nelson, W. R., and Rindi, A. (eds) 1988. *Monte Carlo Transport of Electrons and Photons*, Plenum Press, New York and London.

Kawrakow, I. 2000a. Accurate condensed history Monte Carlo simulation of electron transport, I. EGSnrc, the new EGS4 version, *Med. Phys.* **27**: 485–498.

Kawrakow, I. 2000b. Accurate condensed history Monte Carlo simulation of electron transport, II. Application to ion chamber response simulations, *Med. Phys.* **27**: 499–513.

Kawrakow, I. and Bielajew, A. F. 1998. On the condensed history technique for electron transport, *Nucl. Instrum. Meth. B* **142**: 253–280.

Kawrakow, I. and Fippel, M. 2000. Investigation of variance reduction techniques for Monte Carlo photon dose calculation using XVMC, *Phys. Med. Biol.* **45**: 2163–2183.

Kawrakow, I., Fippel, M., and Friedrich, K. 1996. 3D electron dose calculation using a Voxel based Monte Carlo algorithm (VMC), *Med. Phys.* **23**(4): 445–457.

Kawrakow, I., Mainegra-Hing, E., Rogers, D. W. O., Tessier, F., and Walters, B. 2011. *The EGSnrc Code System: Monte Carlo Simulation of Electron and Photon Transport*, NRCC Report PIRS-701, National Research Council Canada, Ottawa.

Landau, L. 1944. On the energy loss of fast particles by ionization, *J. Exp. Phys. USSR* **8**: 201–205.

Lüscher, M. 1994. A portable high-quality random number generator for lattice field theory simulations, *Comp. Phys. Commun.* **79**(1): 100–110.

Marsaglia, G. and Zaman, A. 1991. A new class of random number generators, *Ann. Appl. Probab.* **1**(3): 462–480.

Molière, G. Z. 1948. Theorie der Streuung schneller geladener Teilchen. 2. Mehrfach- und Vielfachstreuung, *Z. Naturforschung A* **3**: 78–97.

Nakamura, K. et al. 2010. (Particle Data Group) The review of particle physics, *J. Phys. G* **37**: 075021. Available on Particle Data Group WWW pages (http://pdg.lbl.gov).

Nelson, W. R., Hirayama, H., and Rogers, D. W. O. 1985. *The EGS4 Code System*, SLAC Report No. SLAC-265, Stanford Linear Accelerator Center.

Photon, Electron, Proton, and Neutron Interaction Data for Body Tissues 1992. ICRU Report 46, International Commission on Radiation Units and Measurements.

Press, W. H., Teukolsky, S. A., Vetterling, W. T., and Flannery, B. P. (eds) 1992. *Numerical Recipes in C*, Cambridge University Press, Cambridge, New York, Port Chester, Melbourne, Sydney.

Reynaert, N. et al. 2007. Monte Carlo treatment planning for photon and electron beams, *Rad. Phys. Chem.* **76**(4): 643–686.

Salvat, F., Fernández-Varea, J.M., and Sempau, J. 2011. PENELOPE-2011: A code system for Monte Carlo simulation of electron and photon transport, nuclear energy agency. NEA/NSC/DOC20115, *Workshop Proceedings*, Barcelona, Spain, 4–7 July 2011.

Stopping Powers and Ranges for Protons and Alpha Particles 1993. ICRU Report 49, International Commission on Radiation Units and Measurements.

Stopping Powers for Electrons and Positrons 1984. ICRU Report 37, International Commission on Radiation Units and Measurements.

Vavilov, P. V. 1957. Ionization losses of high-energy heavy particles, *Soviet Phys. JETP* **5**: 749–751.

3

Variance Reduction Techniques

Matthias Fippel
Brainlab AG

3.1 Introduction

Monte Carlo (MC) calculations can be time consuming, especially for applications in radio therapy (RT). Therefore, algorithmic techniques to speed up the simulations are essential. These techniques are called variance reduction techniques (VRT). In this chapter, VRTs with a special focus on RT are introduced and explained. For clarity, it is written from the point of view of photon–electron interactions. However, the same concepts can be applied for heavy charged and neutral particles (protons, neutrons, etc.).

Additional information about VRTs is available, for example, in book chapters by Bielajew and Rogers (1988) or Sheikh-Bagheri et al. (2006), in an article by Kawrakow and Fippel (2000a), and in two task group reports (Chetty et al., 2007; Reynaert et al., 2007). This literature also contains references of the original publications for most of the VRTs presented in the following sections.

3.1.1 Calculation Efficiency

Depending on the number of histories N, the accuracy of any MC calculated mean value $\langle f(N) \rangle$ of quantity f is limited by its statistical uncertainty. This uncertainty is given by the variance $\sigma(N)$ and provides a measure of the statistical fluctuations of the calculated mean value $\langle f(N) \rangle$ around the true value f of that quantity. It is obvious that $\sigma(N)$ decreases with increasing number of histories N and it becomes zero if N approaches infinity. In general, $\sigma(N)$ cannot be calculated because the true value f is unknown. On the other hand, an estimated variance $s(N)$ can be calculated during an MC simulation by

$$s(N) = \sqrt{\frac{\langle f^2(N) \rangle - \langle f(N) \rangle^2}{N-1}}, \qquad (3.1)$$

with $\langle f^2(N) \rangle$ being the MC calculated mean of f^2. The best estimate of the variance is obtained if $\langle f(N) \rangle$ and $\langle f^2(N) \rangle$ are calculated using the history-by-history method, that is, they are calculated by averaging over all histories (Salvat et al., 2011). $\langle f(N) \rangle$ and $\langle f^2(N) \rangle$ tend to become constant for large numbers of N. Therefore, Equation 3.1 provides a simple method to reduce the variance just by increasing the number of histories N, that is, increasing the calculation time $T(N)$. However, this is not considered to be a VRT. The purpose of variance reduction is to decrease the time of MC simulations by modifying the algorithm while maintaining an unbiased estimate of the variance $s(N)$. Unbiased means that for any realistic history number N, the result of MC, including VRT, must not deviate systematically from the corresponding result without VRT.

Instead of VRT, the methods pointed out in this book should be called efficiency enhancement techniques because they improve the efficiency of MC simulations. The calculation efficiency ε is defined by

$$\varepsilon = \frac{1}{[s(N)]^2 T(N)}. \tag{3.2}$$

From Equation 3.1, it follows that $[s(N)]^2$ becomes proportional to $1/N$ for large N. $T(N)$ is proportional to N. Therefore, the efficiency ε is almost independent of N. The calculation efficiency can be improved by reducing the variance $s(N)$ for a given number of histories N, by decreasing the calculation time $T(N)$ for a given number of histories N, or by doing both. For historical reasons, we will continue to call these techniques VRT throughout this chapter.

3.1.2 Hardware Performance Improvements

The calculation time can be decreased simply by using faster computers or by implementing parallel calculation processes on multicore workstations and computing clusters. However, these methods are not called VRT because they do not make the underlying MC algorithm faster. They just use a given software on a hardware with better performance. Especially, the parallelization of MC calculations is straightforward. Therefore, it is expected that any serious MC algorithm will fully exploit the advantages of present-day computing hardware.

3.1.3 Approximate Methods

In the majority of cases, the calculation time per history is decreased by making approximations. Even if the final result is not affected in a significant way, in fact, the approximate methods do not belong to VRTs. According to its original definition, VRTs must not influence the expected result of an infinitely long MC simulation. However, in literature, approximate methods are often denoted as VRT. Some of them, for example, the condensed history technique (CHT), form the basis of almost all MC calculations in radiation therapy today. Therefore, in this chapter, the approximate methods will also be discussed. It is possible to distinguish between real variance reduction techniques (RVRT) and approximate variance reduction techniques (AVRT).

It is obvious that some techniques are AVRTs, for example, the continuous slowing down approximation (CSDA) of the electron transport. On the other hand, it can be difficult to decide whether a given technique belongs to the RVRTs or to the AVRTs. Sometimes, this decision is purely based on intuition or it is shown numerically that some technique is an RVRT. However, numerical experiments are debatable with this respect because the results of both, RVRT and AVRT, must not be influenced significantly. Only by mathematical proof, it can be shown that a definite technique is an RVRT. These proofs can be complex, that is, they would cover a significant amount of space in this chapter. Hence, they will not be shown here. This book focuses on the use of MC techniques in radiation therapy. For both types of VRTs, this means that they have to speed up the simulations without significant loss of numerical accuracy. Therefore, it makes sense in the following sections not to demonstrate for all cases that a specific VRT is an AVRT or an RVRT.

3.1.4 Condensed History Electron Transport

The majority of MC algorithms in RT perform electron transport using the CHT. The CHT is an AVRT, and it includes approximations and speeds up MC electron transport significantly compared to analog simulation. Because of its fundamental nature, an in-depth explanation of the CHT is provided in Chapter 2.

3.2 Basic Variance Reduction Techniques

In this section, the basic VRTs are outlined. They provide an introduction into the elemental methods of how MC simulation times can be reduced efficiently and what has to be considered to avoid bias of the results. Some of the basic VRTs can be combined to form a more advanced VRT (see Section 3.3).

3.2.1 Uniform Particle Splitting

A common example to explain variance reduction is the particle splitting technique. It can be applied to photons as well as to charged particles. Figure 3.1 demonstrates the splitting technique applied to bremsstrahlung photons. The left drawing represents the MC simulation of an electron hitting the target of a medical linear accelerator (LINAC) and producing one bremsstrahlung photon. For simplicity, we consider the case of only one bremsstrahlung photon because, in general, several photons can be produced by one electron in the target. In a "normal" (also called analog) MC simulation, the photon carries a statistical weight of $w = 1$. This means that one realistic photon is represented by one photon in the simulation. The right drawing represents the case with a particle splitting factor of $N_{split} = 5$, that is, instead of one, five independent bremsstrahlung photons are sampled from the same bremsstrahlung production distribution of the given electron. The VRT is called

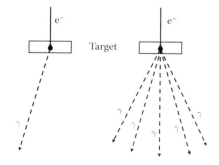

FIGURE 3.1 Schematic representation of an MC simulation of an electron producing one bremsstrahlung photon in the target of a medical linear accelerator (left). On the right-hand side, five bremsstrahlung photons instead of one are created using the splitting technique, each with a statistical weight of $w = 1/5$.

uniform bremsstrahlung splitting (UBS) if all photons are created with the same probability independent of their energy and direction. Then, each of the five photons has to carry a statistical weight of $w = 1/N_{split} = 0.2$, that is, one realistic photon is represented by five photons in the simulation to preserve the total weight.

The reason for bremsstrahlung splitting is given by the purpose of the MC simulation. Two examples are (i) the bremsstrahlung spectrum of the LINAC target shall be calculated or (ii) the target component is part of a full LINAC head simulation with the final goal of dose calculation in a phantom or patient. In both cases, we are interested in a large number of photons to increase the statistical accuracy by lowering the variance. The splitting technique reduces the time required for the creation of five photons in the example of Figure 3.1 because the transport simulation of four additional electrons is saved. In other words, during an analog MC simulation, a lot of time would be "wasted" for the electron transport. However, almost all electrons are absorbed within the target, that is, besides the creation of bremsstrahlung, they do not really contribute to the final result.

Particle splitting is known to be an RVRT, although a proof is not shown here. In the limit of infinite history numbers (number of electrons hitting the target), the results of the analog MC and simulations with photon splitting are identical. This fact is not obvious and it can be wrong if not implemented correctly. An incorrect implementation would arise, for example, if the energy of the electron is reduced by the weighted sum of all split photon energies (five energies in Figure 3.1). Such an approach would bias the shape of the energy spectrum (also known as straggling) of the electrons after the first bremsstrahlung event. Since these electrons can produce secondary and more bremsstrahlung photons, all further results will be influenced by this mistake. To avoid this effect, usually the energy of the electron is reduced just by the energy of one photon arbitrarily selected from all splitting photons. Obviously, the energy is not conserved within one history using this method. On the other hand, the energy conservation law is still fulfilled on an average for large history numbers.

It is important to note, that one should not mix analog MC and splitting or splitting with very different splitting numbers thoughtlessly in one simulation. This could create photons of very different statistical weight, that is, "fat" photons with high weight and "meager" photons with low weight. Only the "fat" photons would contribute to the statistical accuracy of the final result and the time for simulating the "meager" photons would be wasted. A smart implementation of any VRT can produce particles of a very different statistical weight in an intermediate state; however, when the particle properties for the final calculation result are analyzed or processed in a further simulation, it is most efficient if the statistical weight of all the particles is similar.

3.2.2 Russian Roulette

Russian roulette can be considered as the opposite of particle splitting. Very often, both techniques are used in combination.

In a Russian roulette technique, for a definite particle type (photon or charged particle), a survival probability $p_{survive}$ with $p_{survive} \ll 1$ is defined. If a particle of this type is created in an MC simulation, a random number ξ is sampled from a uniform distribution in interval [0,1]. The particle survives if $\xi < p_{survive}$, otherwise it is killed, that is, the simulation of this particle stops. To stay in correspondence with reality, the statistical weight of the surviving particles must be increased by the factor $w = 1/p_{survive}$.

Russian roulette can be applied efficiently, for example, during the simulation of a LINAC head geometry in combination with the splitting technique described in Section 3.2.1. The bremsstrahlung photons with weight $w = 1/N_{split}$ created within the target can hit the secondary collimator of the LINAC. One possible interaction is Compton scattering in the high-Z absorbing material leading to the creation of Compton electrons with weight $w = 1/N_{split}$. Most of these electrons are absorbed without producing any secondary (bremsstrahlung) radiation and the calculation time to transport these electrons would be wasted if all of them are simulated. Therefore, it makes sense to kill these electrons with a probability of $1 - p_{survive} = 1 - 1/N_{split}$. The weight of the surviving electrons must be increased by the factor $1/p_{survive} = N_{split}$, that is, finally they carry a weight of $w = 1$ like the original electrons hitting the bremsstrahlung target. Afterward, it can be useful to again apply splitting if the Compton electrons produce bremsstrahlung. However, in this case, the simulation of these photons makes sense only if they reach the region of interest (ROI). Therefore, some more advanced techniques other than uniform splitting should be the method of choice (see, e.g., Section 3.3.2).

3.2.3 Range Rejection

The Compton electrons created in the example of Section 3.2.2 can also be omitted using range rejection instead of Russian roulette. For this VRT, the shortest distance from the present charged particle position to the region boundary must be calculated and compared to the maximum range of that particle in the regions material. Therefore, this technique can only be applied to particles with a definite maximum range depending on the energy, that is, to charged particles. If this range is smaller than the shortest distance to the region boundary, then the charged particle can never leave the present region and it is useful to stop the simulation here.

Obviously, range rejection is an AVRT because a possible production of bremsstrahlung photons is neglected with this approach and photons have a finite probability of leaving the present region. It is of course possible to correctively take these photons into account by adding them via sampling from an approximate distribution.

3.2.4 Cross-Section Enhancement

For the simulation of an ion chamber or an air cavity in water with the focus on the energy absorbed in air, it is beneficial to artificially increase the total photon cross section by some factor $N_{enhance}$ in a predefined region around the chamber or cavity. Therefore, the number of photon interactions increase by factor $N_{enhance}$ in that region causing a correspondingly increased electron fluence. To maintain an unbiased simulation, the weight of all secondary particles produced in these interactions has to be reduced by multiplication with factor $w = 1/N_{enhance}$.

3.2.5 Interaction Forcing

Comparable to cross-section enhancement and also applicable to photons only is the interaction-forcing method, schematically represented in Figure 3.2. This is a method that can be used for a photon MC dose calculation engine in RT treatment planning. In Figure 3.2, an incoming photon hits the calculation grid surface at the point denoted as **Start**. Then, it is traced along the line denoted with **s** until the interaction point. In a worst-case scenario, this point can be outside the calculation grid, that is, behind the **Stop** position and the photon does not contribute to the tally. Thus, the time spent on sourcing the photon and transporting it through the calculation grid is wasted. To avoid wasting of calculation time, the photon can be forced to interact between **Start** and **Stop**.

Interaction forcing is possible if the number of mean free photon path lengths $\Lambda = \sum_{Start}^{Stop} \mu_i s_i$ between **Start** and **Stop** can be calculated easily and fast for all crossed voxels i. Here, μ_i is the linear attenuation coefficient in voxel i and s_i is the photon step length in voxel i. Then, the number of mean free photon path lengths λ can be selected from the distribution function

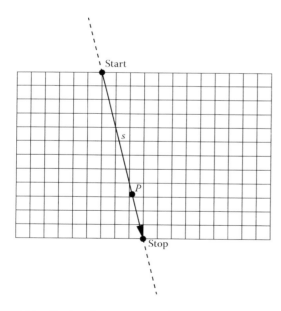

FIGURE 3.2 Forcing the interaction of one simulated photon within the region of interest, that is, between the **Start** and **Stop** positions. The number of mean free photon path lengths $\Lambda = \sum_{Start}^{Stop} \mu_i s_i$ between **Start** and **Stop** must be calculated for all crossed voxels i in advance. This can be time consuming because, in general, each voxel contains a different medium with a different attenuation coefficient μ_i.

$$p(\lambda)d\lambda = \frac{1}{1 - e^{-\Lambda}} e^{-\lambda} d\lambda, \qquad (3.3)$$

with λ restricted to the interval $[0, \Lambda]$. That is, with a uniformly distributed random number ξ from interval $[0,1]$, the number of mean free photon path length λ to the interaction point is calculated by

$$\lambda = -\ln[1 - \xi(1 - e^{-\Lambda})]. \qquad (3.4)$$

The distance to the interaction point P is then determined by tracing the photon along the line s until $\lambda = \sum_{Start}^{P} \mu_i s_i$. Photon forcing requires a weight change of the photon using the factor

$$w = 1 - e^{-\Lambda}, \qquad (3.5)$$

that is, the photon weight decreases.

In general, each voxel contains a different medium with a different attenuation coefficient μ_i. Therefore, interaction forcing is useful only if Λ can be calculated fast enough. Otherwise, the MC calculation efficiency can decrease rather than increase.

3.2.6 Exponential Transform

The exponential transform works in a similar manner as interaction forcing. Here, the exponential depth distribution of photons is stretched or shortened by some factor $F > 0$, that is, the number of mean free photon path lengths λ is sampled from the distribution

$$p(\lambda)d\lambda = \frac{1}{F} e^{-\lambda/F} d\lambda. \qquad (3.6)$$

A uniformly distributed random number ξ from interval $[0,1)$ provides the number of mean free photon path length λ to the interaction point by

$$\lambda = -F\ln(1 - \xi), \qquad (3.7)$$

that is, for $F < 1$, λ is shortened, and for $F > 1$, it is stretched. Exponential transform requires a weight change of the photon using the factor

$$w = Fe^{-\lambda(F-1)/F} = F(1 - \xi)^{(F-1)/F}, \qquad (3.8)$$

that is, the photon weight is changed depending on the result λ or the random number ξ. This VRT is useful to either increase the calculation efficiency close to the surface (photon entrance point) at the cost of a decreased efficiency in larger depth ($F < 1$) or vice versa ($F > 1$).

3.2.7 Woodcock Tracking

The basic idea of Woodcock tracking (also called fictitious interaction method) is to make an inhomogeneous simulation geometry homogeneous by adding a fictitious interaction cross section to the total cross section in each region. The fictitious cross sections are chosen in a way to achieve a constant cross section sum for the whole geometry.

For example, to track photons of energy E in an inhomogeneous material grid, we have to determine the maximum total cross section $\mu_{\max}(E)$ for the whole simulation geometry. Then, in each region or each voxel with index i, a fictitious interaction cross section is determined by

$$\mu_{\text{fict}}^i(E) = \mu_{\max}(E) - \mu_{\text{tot}}^i(E). \tag{3.9}$$

This allows tracking of photons without ray tracing because of the cross section $\mu_{\max}(E)$ being independent of the position within the geometry. As soon as the photon has reached the interaction site, a random number can be used to determine the interaction type, real or fictitious. The probabilities for real and fictitious interactions in each voxel are given by

$$P_{\text{real}}^i(E) = \frac{\mu_{\text{tot}}^i(E)}{\mu_{\max}(E)} \tag{3.10}$$

$$P_{\text{fict}}^i(E) = \frac{\mu_{\text{fict}}^i(E)}{\mu_{\max}(E)}. \tag{3.11}$$

If a real interaction is sampled, the simulation is continued with determining the photon interaction type, for example, Compton scattering or pair production. That is, there is no difference to the conventional MC tracking. However, if a fictitious interaction is sampled, the original photon tracking is continued starting from the present position without changing the photon energy and direction. That is, the fictitious interaction has no effect.

Woodcock tracking implemented in this way does not require weight adjustments of primary and secondary particles. However, it is possible to modify this VRT by introducing weight changes depending on the purpose of the simulation. For example, instead of sampling both real and fictitious interactions, only real interactions are sampled with probability $P = 1$. In this case, the weights of all secondary particles produced during these interactions have to be reduced by the factor

$$w = P_{\text{real}}^i(E) = \frac{\mu_{\text{tot}}^i(E)}{\mu_{\max}(E)}. \tag{3.12}$$

In combination with Russian roulette, further weight modifications are possible to ensure that the statistical weight of all final particles is comparable.

3.2.8 Correlated Sampling

Applied to the problem of dose calculation in a heterogeneous geometry, correlated sampling starts with a known dose distribution $D_{\text{hom}}(\vec{r})$ in a homogeneous phantom, for example, in water. This can be a measured dose distribution, but it can also be a smooth MC dose distribution calculated with very high statistical accuracy. This dose distribution can, for example, be stored in the computer memory. The algorithm then performs a simultaneous calculation of two correlated MC dose distributions, one distribution $D_{\text{hom}}^c(\vec{r})$ in the homogeneous water phantom and another distribution $D_{\text{het}}^c(\vec{r})$ in the heterogeneous geometry. Correlated means that the two simulations are performed using the same sequence of random numbers. To make this calculation fast, the two simulations are performed with low statistical accuracy, that is, with a small number of histories.

The simultaneous simulations are used to determine the distribution of correction factors in each voxel, given by the dose ratios

$$C(\vec{r}) = \frac{D_{\text{hom}}(\vec{r})}{D_{\text{hom}}^c(\vec{r})}. \tag{3.13}$$

The final dose distribution in the heterogeneous geometry is then calculated by

$$D_{\text{het}}(\vec{r}) = C(\vec{r})D_{\text{het}}^c(\vec{r}). \tag{3.14}$$

It has been shown that this technique is useful only if the heterogeneous geometry is not too different from the corresponding homogeneous geometry because strong inhomogeneities destroy the correlation between the two simulations.

This technique can also be generalized to two heterogeneous geometries if an accurate solution for one of the geometries is known. An example is the simultaneous calculation of dose to an air cavity volume in water with and without the wall material present.

3.2.9 Initial Calculation of the Primary Interaction Density

For dose calculations with a simple model of the LINAC head, for example, a point source, the density of primary photon interactions can be precalculated in each voxel by performing an initial photon ray tracing. This is schematically shown in Figure 3.3. The MC simulation then starts after this precalculation step with simulating the corresponding number of secondary photons and charged particles in each voxel. The variance is reduced because the primary photon interaction sites are optimally distributed. Investigations have shown that the efficiency can be improved by a factor of about 2 using this method.

The main disadvantage of this method is that it limits the applicability of the code to relatively simple sources. Advanced MC dose calculation engines in RT, however, model the LINAC head using full simulations, using extended virtual source models, or a combination of both. Therefore, an initial calculation of the primary interaction density is not possible for this type of MC applications.

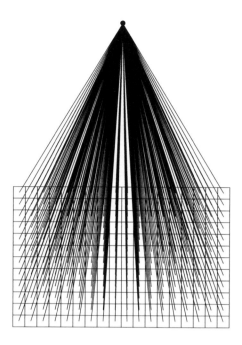

FIGURE 3.3 Precalculation of the primary photon interaction density per voxel of the calculation grid by performing an initial ray tracing between each voxel and the photon source. MC starts with simulating the corresponding number of secondary photons and electrons emitted in each voxel.

3.2.10 Quasi-Random Numbers

The solution of the transport problem in radiation therapy physics corresponds to an integration in a parameter space with an infinite number of dimensions. If this is performed numerically using the MC method, the dimensionality of the problem has to be limited to a finite number of dimensions, for example, n dimensions. Using a random number generator, it is possible to sample random vectors in this n-dimensional space by generating a sequence of n numbers, forming in this way an n-tuple. In general, one n-tuple corresponds to one particle history of the MC simulation. The simulation of N particle histories corresponds to the generation of a sequence of N n-tuples. If a normal pseudorandom number generator of good quality is used for that purpose, the distribution of the n-tuples is suboptimal, that is, the statistical error of the result decreases as $1/\sqrt{N}$.

Sequences of n-tuples that fill the n-dimensional space more uniformly than uncorrelated pseudorandom numbers are called quasi-random sequences. These sequences are known to substantially improve the integration efficiency in certain types of problems. Using quasi-random numbers, the statistical error of the result can decrease asymptotically as $1/N$ if the quasi-random n-tuples are optimally distributed. There are different types of quasi-random numbers; examples are Halton's sequence or the Sobol sequence (Press et al., 1992).

Similar to pseudorandom numbers, quasi-random numbers are not really random. However, in contrast to pseudorandom numbers, quasi-random numbers cannot be considered as uncorrelated. For example, the Sobol sequence in one dimension starts with 0.5, 0.75, 0.25, 0.375, …, that is, it tries to fill the interval [0,1] in an optimum way. The method is obvious for the one-dimensional case, but it becomes complex for higher dimensions (Press et al. 1992).

3.3 Advanced Variance Reduction Techniques

Advanced VRTs are formed in general by combining different basic VRTs. Some of them have demonstrated to be very efficient for MC calculations in RT, for example, for dose calculations. They are outlined shortly in this section.

3.3.1 Selective Bremsstrahlung Splitting

With UBS (Section 3.2.1), the splitting factor N_{split} is a constant throughout the whole simulation, that is, it is independent of the direction of the electron producing the bremsstrahlung photons. With selective bremsstrahlung splitting (SBS) (Sheikh-Bagheri et al., 2006), $N_{\text{split}}(\theta)$ becomes a function of the variable θ, the angle between the present direction of the electron and the central beam axis. Because bremsstrahlung photons are forward peaked into a direction that differs only slightly from the direction of the initial electron, it makes sense to use smaller splitting factors $N_{\text{split}}(\theta)$ with increasing θ. The maximum should be at $\theta = 0$ and the minimum at $\theta = 180°$. Most of the photons transported with large angles are just absorbed within the primary or secondary collimators of the LINAC head and their calculation time would be wasted. SBS reduces this time waste. Photons with a better chance of reaching the ROI are created with a higher probability using SBS.

There is, however, a serious drawback of this method; it introduces a nonuniform distribution of statistical weights

$$w(\theta) = \frac{1}{N_{\text{split}}(\theta)}, \tag{3.15}$$

whereas θ is *not* the angle of the photon; it is the angle of the original electron. This means that photons in a specific small scoring location can have different weights. This influences the variance in this scoring location and the final efficiency gain of the method is much smaller than expected. SBS improves the efficiency by a factor of 2–3 compared to UBS (Kawrakow et al., 2004).

3.3.2 Directional Bremsstrahlung Splitting

The disadvantages of SBS are eliminated using the directional bremsstrahlung splitting (DBS) method. Unlike UBS and SBS, DBS is a complex algorithm. Hence, the original and systematic publication of Kawrakow et al. (2004) is referred if the reader is interested in details of the method. Here, only a summary will be provided.

DBS uses (as UBS) a constant user-defined splitting factor N_{split} for the production of bremsstrahlung. Then, the algorithm analyzes the direction of the photons produced. They are transported always, if they are directed into the ROI. If not, Russian roulette (Section 3.2.2) with the survival probability $p_{\text{survive}} = 1/N_{\text{split}}$ is played. Therefore, photons directed into the ROI carry a weight of $w = 1/N_{\text{split}}$. Photons aiming away from the ROI carry a weight of $w = 1$. Additional calculation time can be saved if the directional dependence of the photons is calculated in advance by a smart modification of the bremsstrahlung production cross section (Kawrakow et al., 2004).

By further combining splitting and Russian roulette, by processing "fat" and "meager" particles differently throughout the rest of the simulation as well as by treating photons in gas and higher-density materials differently, it be ensured that all photons inside the ROI will be "meager," that is, have a weight of $w = 1/N_{\text{split}}$ and those outside the ROI will be "fat" with a weight of $w = 1$.

A drawback of this default DBS implementation is that all electrons are "fat." When they contribute to the result, for example, for dose calculation, the electrons would influence the variance in a negative manner. For these purposes, a DBS with electron splitting is available. This technique also ensures that the electrons reaching the ROI are "meager"; however, the overall calculation efficiency decreases compared to the default DBS.

3.3.3 Macro Monte Carlo

Macro Monte Carlo (MMC) is based on the premise that the simulation time can be decreased by the use of precomputed results. One of these precomputing techniques for the purpose of dose calculation in RT treatment planning of electron beams is called MMC (Neuenschwander and Born, 1992). As shown in Figure 3.4, this technique is based on precalculated fluence distributions on the surface of spheres for electrons hitting the spheres. A general-purpose MC algorithm can be used to calculate the distributions for incoming electrons of different energy as well as spheres with different diameters and consisting of different materials. The left drawing in Figure 3.4 shows schematically a "full" MC simulation of an electron path within such a homogeneous sphere. With these presimulations, the distributions of exit points, angles, and energies are calculated and stored as a function of sphere radius, material, and initial electron energy.

Dose calculation using the patient geometry starts with analyzing the computed tomography (CT) information and processing the calculation grid. Each voxel is assigned with a definite material and a definite sphere diameter, corresponding with the distance till the material changes. In homogeneous regions, spheres with larger diameters can be used. In regions with a strong density fluctuation, smaller diameters must be assigned to the voxels. When an electron hits the calculation grid, the sphere radius and material are determined based on the preprocessing result. The parameters of the electron leaving the sphere

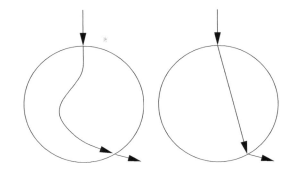

FIGURE 3.4 Principle of macro Monte Carlo. The left drawing shows a "full" MC simulation of an electron path in a homogeneous sphere with a definite diameter. A huge number of electrons hitting those spheres are precalculated and the distributions of exit points, angles, and energies are stored as a function of sphere radius, material, and initial electron energy. During an MC dose calculation, using the patient geometry, these macroscopic distributions are used to directly jump from the entry to the exit point (right drawing).

are randomly selected from the precomputed macroscopic fluence distributions, that is, the electron directly jumps from the entry to the exit point of the sphere (see the right drawing of Figure 3.4). The energy loss within the sphere is used to calculate the absorbed dose in each voxel.

The process continues with determining the new sphere parameters for the electron leaving the old sphere. It is repeated until the whole electron history is simulated and the energy of the initial electron is consumed.

Compared to conventional MC, MMC is faster by about one order of magnitude. One disadvantage is, however, only the electron transport speed can be increased. This means that MMC cannot be applied to photon beams efficiently.

3.3.4 History Repetition

Another technique with recycling of simulation results is called history repetition. This VRT, schematically represented in Figure 3.5, has been developed originally to increase the speed of MC dose calculations for electron beams in RT. Electron history repetition is based on complete electron histories precalculated in a homogeneous water phantom of infinite size. An example of such a precalculated history is shown in Figure 3.5 left from the box. Precalculated means that all parameters of this history (condensed history step lengths, multiple scattering angles, energy losses, secondary particle parameters) are saved for reuse.

On the right-hand side of Figure 3.5, the history is applied to the actual patient geometry $N_{\text{repeat}} = 2$ times, starting at different positions and with different directions at the surface of the calculation matrix. In reality, the optimal N_{repeat} depends on the size of the electron field. The optimum efficiency gain (approximately a factor of 2) is achieved with about one repetition per 5 cm² of the field, that is, for a 10×10 cm² electron field, the repetition number should be about $N_{\text{repeat}} = 20$. If N_{repeat} is smaller, the potential of history repetition is not fully exploited. If N_{repeat}

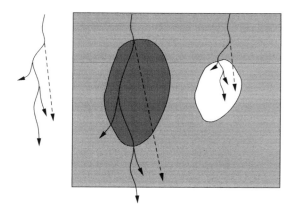

FIGURE 3.5 Electron history repetition. An electron history is simulated in a homogeneous water phantom of infinite size and all parameters (step lengths, scattering angles, energy losses, and secondary particle parameters) are saved for reuse within the actual simulation (left). In the right drawing, this history is applied to the present patient geometry several times by scaling all parameters depending on the medium and mass density in each voxel.

is too large, the average distance between two neighbor histories becomes small and the variance is influenced by correlations between these neighbor histories.

The precalculations can be performed in advance (e.g., using a general-purpose MC algorithm) for a given large number of electron histories in water. Then, all parameters of all histories have to be stored in a database for use during the dose calculation. In a second approach, one history in water is simulated during the dose calculation immediately before it is applied to the patient geometry N_{repeat} times. After the repetition of this history, the corresponding memory is cleared and the next history in water can be simulated and applied to the patient. This second method allows the simulation of an arbitrary number of particle histories and the final variance of the dose can become as small as desired. The first method, on the other hand, is limited by the number of precalculated histories in the database.

Bremsstrahlung photons are generally not included in the repetition. They are simulated on the fly and they are killed with Russian roulette or they are transported using kinetic energy released per unit mass (KERMA) approximation.

To take the heterogeneities correctly into account, when applied to the patient geometry, the histories in water must be adjusted by scaling the stored parameters depending on the medium, the stopping power, the scattering power, and the mass density in each voxel. Mainly the electron step lengths and multiple scattering angles are scaled.

An important requirement of history repetition is that the histories must be scalable. Scalability can be assured, for example, if all transport parameters, such as stopping and scattering powers, are approximated as

$$S^T(M,E) = f_c^T(M) f_0^T(E). \quad (3.16)$$

This means that each transport parameter function of type T that depends on the medium M and the electron energy E is approximated as a product of a correction function depending only on the medium and another function depending only on the energy. It has been shown that this is possible for the different types of human tissue, including some phantom materials such as water with an accuracy of 1–2%. With this factorization and without further loss of generality, a reference medium can be selected. In most cases, this is water, that is, for $f_0^T(E)$, the corresponding transport parameter in water is selected:

$$f_0^T(E) = S^T(H_2O, E). \quad (3.17)$$

Thus, we get

$$S^T(M,E) = f_c^T(M) S^T(H_2O, E). \quad (3.18)$$

If, for example, S^T is the linear collision stopping power $S^{\mathrm{coll}} = dE/dx$, then for a given infinitesimal small energy loss dE, the step length dx according to collision loss is scaled by

$$dx(M) = \frac{dE}{f_c^{\mathrm{coll}}(M) S^{\mathrm{coll}}(H_2O, E)} = \frac{dx(H_2O)}{f_c^{\mathrm{coll}}(M)}. \quad (3.19)$$

The factorization of Equation 3.16 is not a necessary condition for electron histories to be scalable. The correction $f_c^T(M)$ can be extended by including a slight energy dependence. However, electron history repetition does not work for arbitrary materials, for example, metals. Therefore, this technique is an AVRT.

3.3.5 Simultaneous Transport of Particle Sets

The limitations of history repetition can be avoided using the simultaneous transport of particle sets (STOPS) technique. With STOPS, several particles of the same type (electron, positron, or photon) and with the same energy are transported simultaneously as a set. As with history repetition, the initial positions, directions, and weights are different. In contrast to history repetition, the histories in one set are not transformed into each other by path length and scattering angle scaling. However, a variety of parameters are independent of the material, so they are sampled once for all particles in the set.

STOPS can be applied for photons as well as for charged particles. Here, electrons are used to explain the technique because STOPS has been developed originally in VMC++ (Kawrakow, 2000; Kawrakow and Fippel, 2000b) to replace electron history repetition. Electrons are transported in most of the MC algorithms using a class II CHT (see Chapter 2 for more details) with continuous energy loss to model semielastic collisions as well as discrete Møller (Bhabha for positrons) and bremsstrahlung interactions to model hard inelastic events. The CHT allows one to express distances between discrete interactions as energy losses. For the material of type M, they are sampled using the total discrete interaction cross section (number of interactions) per unit energy loss:

$$\Sigma_E(M,E) = \frac{\Sigma(M,E)}{L(M,E)}. \qquad (3.20)$$

$\Sigma(M,E)$ is the total cross section per unit length and $L(M,E)$ is the restricted stopping power. An advantage of $\Sigma_E(M,E)$ is that it depends only weakly on the material type M as well as on the energy E and a global maximum Σ_E^{max} for all M and E can be efficiently used to perform Woodcock tracking (Section 3.2.7). This means that the geometry becomes homogeneous in terms of number of interactions (fictitious or real) and the energy loss ΔE can be sampled once for all electrons within the set. Thus, at the end of the step, all electrons have the same energy.

The geometric step lengths between the initial and final points depend on ΔE and on the stopping power $L(M,E)$. They are calculated separately for each electron in the set because $L(M,E)$ is different for different materials. Furthermore, the multiple scattering properties are material dependent, that is, the multiple scattering angles are also sampled individually for each particle in the set.

At the end of the step, the interaction type (fictitious, Møller, or bremsstrahlung for electrons) must be determined for each particle separately; however, the same random number can be used for all particles. This causes a very high chance of identical interactions, especially for human tissue because of the weak material dependence of the interaction properties. If the interaction type is identical for all particles, the whole set stays alive. If there are particles with a different interaction type, the set is split into subsets and henceforward the new subsets are transported independently.

If a specific (nonfictitious) discrete interaction is sampled, secondary particle parameters such as the δ-electron energy for Møller interactions or the photon energy for bremsstrahlung production must be determined using the differential cross section. The differential Møller cross section is material independent and if we assume the same for bremsstrahlung, the secondary particle energies can be sampled once for the entire set. The polar scattering angles are determined uniquely by the kinematics of the processes. Azimuthal scattering angles always follow a uniform distribution in the interval $[0, 360°]$. This distribution is independent of the material type, that is, the azimuthal scattering angle can also be selected once for the particle set. Because identical energies are sampled for equivalent secondary particles within the set, the simulation continues with determining the energy loss to the next interaction by again using the Woodcock scheme. All steps of the simulation are repeated until the entire energy of the set is absorbed or the particles have left the simulation geometry.

The efficiency gain of STOPS is approximately a factor of 2; that is, it is comparable to history repetition. In contrast to history repetition, STOPS can be used for arbitrary media.

3.3.6 Continuous Boundary Crossing

General-purpose MC algorithms usually stop the simulation of charged particles at material interfaces. This is necessary because the underlying physics is valid only for a homogeneous region consisting of one specific material. Especially, the multiple scattering properties are material dependent. Therefore, close to a material boundary, an accurate MC algorithm should switch into a single scattering mode.

However, this is extremely time consuming and for many applications, this is not necessary. For dose calculation in RT treatment planning, history repetition (Section 3.3.4) can be used without loss of accuracy. The multiple scattering angles and step lengths are precalculated in water and scaled depending on the material properties in each voxel. Therefore, it is possible for charged particles to cross boundaries continuously and to trace them efficiently like photons.

3.3.7 Multiple Photon Transport

In combination with photon splitting and Russian roulette, history repetition (Section 3.3.4) and STOPS (Section 3.3.5) can also be used for photon Monte Carlo. To avoid confusion with the various splitting techniques outlined in Sections 3.2.1, 3.3.1, and 3.3.2, here, we call this technique multiple photon transport (MPT). The method has been introduced in Kawrakow and Fippel (2000a) to speed up MC dose calculation for photon beams in RT treatment planning.

Figure 3.6 schematically represents this VRT. The basic idea comes from interaction forcing (Section 3.2.5); the photon is forced to interact within the calculation grid. However, the

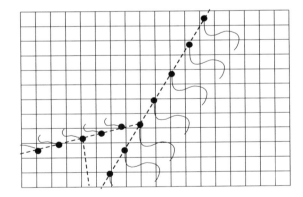

FIGURE 3.6 Multiple photon transport is a combination of photon splitting, Russian roulette, and electron history repetition or STOPS. Instead of one photon, a multiple number of identical photons (dashed lines) are simulated in one ray tracing through the actual geometry. Only the free path length to the interaction point is sampled differently for these photons. Because the interactions are assumed to be identical, for all secondary electrons (solid lines), the history repetition technique (or STOPS) can be applied. Secondary photons such as Compton photons are killed using Russian roulette. The surviving secondary photons are simulated like the primary photons by using the same algorithm.

time-consuming precalculation of the number of mean free photon path lengths Λ between the entry and exit points of the grid is avoided with MPT.

With MPT, instead of one photon, a multiple number of N_{split} identical photons are simulated simultaneously in one ray tracing through the actual geometry. Each of the photons is carrying a weight of $w = 1/N_{\text{split}}$. Identical means that the photons have the same energy, direction, and entry position at the surface of the calculation grid. Only the free path lengths λ_i to the first interaction points are sampled differently for the photons. In units of the mean free photon path length, they are calculated using just one random number ξ uniformly distributed in [0,1) by

$$\lambda_i = -\ln\left[1 - \frac{\xi + i}{N_{\text{split}}}\right], \quad i = 0, \dots, N_{\text{split}} - 1. \quad (3.21)$$

This means that ξ is used to generate the random numbers for the path length determination of N_{split} photons, whereas these numbers are distributed with optimum uniformity in interval [0,1) (see Figure 3.7).

Combined with electron history repetition, the MC simulation is performed by repeating the following steps:

1. If $i = 0$, trace primary photons from the cube surface to λ_0 or trace scattered photons from the interaction site to λ_0.
2. If $i = 0$, sample one interaction type (Compton scatter, pair production, photoelectric absorption, etc) for all N_{split} subphotons of the same ray.
3. If $i = 0$, sample secondary particle parameters and charged particle histories in water:
   a. For photoelectric absorption, sample the photoelectron direction, create a corresponding history in water, and store the parameters.
   b. For pair production, sample the electron's and positron's energy and direction, create the two histories in water, and store the parameters.
   c. For Compton scattering, sample the scattered particle's energy and direction, create the Compton electron history in water, and store the parameters.
4. If $i > 0$, trace photons from λ_{i-1} to λ_i through the inhomogeneous voxel geometry.
5. Apply the charged particle histories to the patient geometry starting at the i-th interaction point.
6. In case of Compton scatter, play Russian roulette with the secondary photon using the survival probability $p_{\text{survive}} = 1/N_{\text{split}}$ and store its parameters if it survives.

7. Increase i by 1, that is, take the next subphoton from the ray and go to step 4.

The procedure for the N_{split} primary photons stops when the cube boundary is reached. Then, the remaining Compton photons are simulated in the same way by executing the MPT function recursively. The procedure with STOPS instead of history repetition works correspondingly.

MPT is fast because N_{split} photons are transported with only one single ray tracing through the geometry. Furthermore, many quantities are reused with history repetition or STOPS. MPT also reduces the variance because the photon interaction sites are distributed optimally along the ray. With a splitting number of about $N_{\text{split}} = 40$, the MC dose calculation efficiency is increased by a factor of 5–10 using MPT.

3.4 Pure Approximate Variance Reduction Techniques

Some of the AVRTs should not really be considered as VRT if the corresponding technique is dominated by one or more approximations. Here, some of these techniques, called pure AVRTs, are presented.

3.4.1 KERMA Approximation

An approximation based on KERMA is motivated by the fact that the energy released by low-energy photons is absorbed in the direct neighborhood of the photon ray. For these photons, the MC transport of secondary electrons can be switched off. That is, for all voxels along the photon ray, the deposited energy is calculated by

$$\Delta E_{\text{dep}} = E \mu_{\text{en}}^M(E) \Delta s. \quad (3.22)$$

Here, Δs is the length of the photon path within the given voxel, E is the energy of the photon, and $\mu_{\text{en}}^M(E)$ is the linear energy absorption coefficient for photons of energy E and the material M in the corresponding voxel.

Systematic deviations of the dose distributions can be minimized with this approach if the maximum range of the secondary electrons is smaller than the spatial resolution of the calculation grid. This is possible if the KERMA approximation is applied only to photons with an energy E smaller than some predefined maximum energy K_{cut}. Furthermore, it is useful to apply this technique only for secondary and higher-order scattered photons, but not to primary photons coming directly from the LINAC head.

3.4.2 Continuous Slowing Down Approximation

In CSDA, electrons (or other charged particles) are transported in one step without the creation of secondary particles. The length of the step is calculated from the energy of the electron

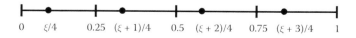

FIGURE 3.7 For the multiple photon transport technique, one uniform random number ξ is used to generate $N_{\text{split}} = 4$ (in this example) random numbers, distributed with optimum uniformity in interval [0,1].

and the unrestricted total stopping power. This continuous slowing down range is a unique function of the energy of the electron. Therefore, energy straggling is neglected using this approach.

There are versions of CSDA with and without multiple scattering. Without multiple scattering, the charged particle is transported on a straight line. Otherwise, a multiple scattering angle is sampled and included in the simulation.

CSDA should be used only for low-energy charged particles, for example, for electrons within an MC simulation, when their energy drops below some user-defined threshold energy E_{cut}. However, CSDA for subthreshold electrons is more accurate and produces less statistical fluctuations than local absorption of the track-end energy.

3.4.3 Transport Parameter Optimization

Both the accuracy and speed of coupled electron–photon transport MC simulations in RT depend on the selection of various transport parameters such as particle transport cut-off energies, particle production threshold energies, and condensed history step sizes.

MC algorithms usually employ a photon energy cut-off parameter P_{cut}, that is, photons are not transported if they are generated within the simulation with an energy below P_{cut}. The remaining energy can be neglected or it can be deposited locally. It is obvious that the accuracy of the result increases with decreasing P_{cut}, however, the calculation time also increases. On the other hand, if the remaining photon energy is absorbed locally, a large value for P_{cut} can cause additional fluctuations. This can decrease the calculation efficiency, although the calculation time has also decreased.

Another important parameter is the photon production threshold energy P_{min}, for example, for bremsstrahlung. This means that only bremsstrahlung photons with an energy larger than P_{min} can be generated during the MC simulation. The effect of electron stopping due to bremsstrahlung photons below P_{min} is taken into account by the restricted radiative stopping power. The question here is how to deal with the radiative energy released by these electrons? Both local absorption and total disappearance of this energy are approximations. To avoid a significant influence on the result on the one hand and to ensure short calculation times on the other, P_{min} should be selected with care depending on the type of the calculation.

The problem is less complex for the cut-off and production threshold energies of charged particles, E_{cut} and E_{min}. Because of the strong correlation between energy and range, both parameters can be selected depending on the spatial resolution of the calculation geometry. On the other hand, efficiency can be improved by using a higher value of E_{cut} and by simulating track-end electrons with energy below E_{cut} in CSDA (Section 3.4.2).

Another important parameter in condensed history MC electron algorithms is the step size (or the procedure used to determine the actual step size depending on energy, material, geometry, etc.). The step size must be restricted by some maximum if an approximate multiple scattering theory, for example,

a small-angle approximation is implemented. In widespread use are approaches with a user-defined maximum energy loss per step parameter E_{step}. The value of E_{step} should be optimized depending on the algorithm and the type of the application.

References

Bielajew, A. F. and Rogers, D. W. O. 1988. Variance-reduction techniques, In: T. M. Jenkins, W. R. Nelson, and A. Rindi (eds), *Proceedings of the International School of Radiation Damage and Protection Eighth Course: Monte Carlo Transport of Electrons and Photons below 50 MeV*, Plenum Press, New York, Chapter 18, pp. 407–419.

Chetty, I. J. et al. 2007. Report of the AAPM Task Group No. 105: Issues associated with clinical implementation of Monte Carlo-based photon and electron external beam treatment planning, *Med. Phys.* **34**(12): 4818–4853.

Kawrakow, I. 2000. VMC++, electron and photon Monte Carlo calculations optimized for radiation treatment planning, In: A. Kling, F. Barao, M. Nakagawa, L. Tavora and P. Vaz (eds), *Advanced Monte Carlo for Radiation Physics, Particle Transport Simulation and Applications, Proceedings of the Monte Carlo 2000 Conference*, Springer-Verlag, Berlin, pp. 229–236.

Kawrakow, I. and Fippel, M. 2000a. Investigation of variance reduction techniques for Monte Carlo photon dose calculation using XVMC, *Phys. Med. Biol.* **45**: 2163–2183.

Kawrakow, I. and Fippel, M. 2000b. VMC++, a MC algorithm optimized for electron and photon beam dose calculations for RTP, *World Congress on Medical Physics and Biomedical Engineering, Med. Phys. 27, Meeting Issue*, Chicago, Illinois, USA.

Kawrakow, I., Rogers, D. W. O., and Walters, B. R. B. 2004. Large efficiency improvements in BEAMnrc using directional bremsstrahlung splitting, *Med. Phys.* **31**(10): 2883–2898.

Neuenschwander, H. and Born, E. 1992. A macro Monte Carlo method for electron beam dose calculations, *Phys. Med. Biol.* **37**: 107–125.

Press, W. H., Teukolsky, S. A., Vetterling, W. T., and Flannery, B. P. (eds) 1992. *Numerical Recipes in C*, Cambridge University Press, Cambridge, New York, Port Chester, Melbourne, Sydney.

Reynaert, N. et al. 2007. Monte Carlo treatment planning for photon and electron beams, *Rad. Phys. Chem.* **76**(4): 643–686.

Salvat, F., Fernández-Varea, J. M., and Sempau, J. 2011. *PENELOPE-2011: A Code System for Monte Carlo Simulation of Electron and Photon Transport*, Nuclear Energy Agency. NEA/NSC/DOC(2011)5, Workshop Proceedings, Barcelona, Spain, July 4–7, 2011.

Sheikh-Bagheri, D., Kawrakow, I., Walters, B., and Rogers, D. W. O. 2006. Monte Carlo simulations: Efficiency improvement techniques and statistical considerations, *Integrating New Technologies into the Clinic: Monte Carlo and Image-Guided Radiation Therapy, Proceedings of the 2006 AAPM Summer School*, Medical Physics Publishing, Madison, WI, pp. 71–91.

II

Application of Monte Carlo Techniques in Radiation Therapy

4

Applications of Monte Carlo to Radiation Dosimetry

Hugo Bouchard
Centre hospitalier de l'Université de Montréal

Jan Seuntjens
McGill University

4.1 Introduction and Scope

Over the last decades, developments in Monte Carlo radiation transport algorithms have had tremendous impact in different areas of radiation dosimetry. During the experimental determination of absorbed dose, several quantities are often difficult (or impossible) to estimate accurately without numerical models. In the context of radiation dosimetry, detectors are most of the time constituted of several components, whose material differs substantially from the medium where absorbed dose is to be known. This situation induces a well-known problem that can be characterized in terms of perturbation factors and changes in response to energy (or energy dependence). Over the different generations of radiation dosimetry protocols, progress in Monte Carlo techniques has affected the improvement in accuracy during the determination of absorbed dose. While air kerma-based protocols of the 1980s, for example AAPM's TG-21 (AAPM, 1983), were the first ones to consider the ionization chamber design in detail, some quantities such as the replacement and the wall perturbation factors (i.e., P_{repl} and P_{wall}, respectively) were estimated with approximate measurements, analytical models, or simply taken as unity based on judicious assumptions. Nowadays, such factors can be calculated with Monte Carlo methods without the need for such approximations, improving the accuracy of absorbed dose determination with radiation detectors.

This chapter presents an overview of the different applications of Monte Carlo to photon and electron beam radiation dosimetry for the energy range used in radiation therapy. Following the introduction, the basics of measurement dosimetry is defined, including physical quantities and units as well as the formalism of reference absorbed dose determination. In the third section, the calculation of quantities related to the standard formalism of absorbed dose measurement is described. The fourth section summarizes several studies using detector response simulations in a context relevant to clinical dosimetry and describes the tools currently available to perform such studies. The fifth section treats the topic of Monte Carlo simulation accuracy in measurement dosimetry, describing the different limitations contributing to that issue. In the sixth section, the evolution of nonstandard beam dosimetry techniques is treated with emphasis on the role of Monte Carlo in the acquisition of dosimetry-related data. Finally, a conclusion is made describing the expected role of the Monte Carlo method in the future of radiation dosimetry.

4.2 Basics of Measurement Dosimetry of Photon and Electron Beams

4.2.1 Historical Context of Cavity Theory

In 1936, Gray published one of the first papers dealing with the fundamental problem of converting absorbed dose in an air-filled cavity to absorbed dose in water (Gray, 1936), including earlier work by W. L. Bragg (1912) and himself (Gray, 1929). In their approach, an idealistic situation is defined where a volume of interest achieves two artificial conditions implying that the amount of energy lost by electrons through electronic collisions is equal to the energy absorbed in that cavity. As a consequence, the expression of absorbed dose in photon (or electron) beams is reduced to absorbed dose to gas in the cavity times the ratio of stopping powers, which depends on the beam energy (Laurence, 1937). Spencer and Attix (1955) took a closer look at the Bragg–Gray formulation in an attempt to solve discrepancies between experiments and predictions due to a breakdown of the conditions required by the principle. To account for "fast" secondary electrons (delta rays) that could exit the cavity, Spencer and Attix defined a two-group energy deposition schematization where electronic inelastic collisions are considered dissipative only if their kinetic energy is below a threshold value (also known as cut-off energy), denoted Δ. The electron slowing down fluence spectrum was derived by solving the Boltzmann transport equation for infinite media (see, e.g., McGinnies, 1959) and based on that a new formulation of stopping-power ratio was defined, providing a better agreement to experiments than the Bragg–Gray formulation.

With technological developments, Monte Carlo methods emerged as a solution to the radiation transport problem allowing accurate estimation of the electron spectra within media (Berger, 1963). Despite the treatment by Spencer and Attix as well as Burch (1955, 1957), the conditions of the Bragg–Gray theory formed the basis of radiation dosimetry protocols in the 1960s (e.g., the ICRU Report no. 14, 1969) with Spencer–Attix stopping-power ratios, adopted in the 1970s (e.g., the ICRU Report no. 21, 1972). Significant theoretical progress in ionization chamber dosimetry was made by Almond and Svensson (1977), whereas progress in Monte Carlo techniques allowed for practical calculation of stopping-power ratios. Nahum (1978) published a modified formulation of stopping-power ratios suitable for use with Monte Carlo methods. By adding a track-end term to the Spencer–Attix's formulation, as reported earlier by Burch (1957), stopping-power ratio data were further calculated extensively based on this modified formulation (e.g., Andreo and Nahum, 1985). Despite the fact that other improved approaches to cavity theory were also proposed (e.g., Burlin, 1966; Janssens et al., 1974; Zheng-Ming, 1980; Janssens, 1981; Bielajew, 1986), the radiation dosimetry community widely adopted the Spencer–Attix–Nahum formulation of stopping-power ratios, being applicable to megavoltage photon and electron beams (e.g., the ICRU Report no. 35, 1984). For several decades, the Spencer–Attix–Nahum formulation has been extensively included in dosimetry protocols (e.g., AAPM's TG-21, 1983; IAEA's TRS-277, 1987), air kerma standards, electron beam dosimetry (e.g., AAPM's TG-25, 1991), and beam quality correction factors (e.g., AAPM's TG-51, 1999; IAEA's TRS-398, 2001).

4.2.2 Theory of Measurement Dosimetry

The determination of absorbed dose is performed using a radiation detector in a reference dosimetry setup, defined by the radiation source and geometry of interest. By performing an experimental measurement of the detector reading or signal in these conditions, one can obtain the absorbed dose to the medium of interest by converting the detector signal, being induced by a physical process specific to the device, to absorbed dose to the medium at the point of measurement. In the context of this review, we will consider that appropriate corrections have been applied to the detector signal, so as to (1) refer to standard reference conditions (e.g., environmental conditions in ionization chambers) and (2) correct for effects that influence the measurement of the physical process itself (e.g., ionization chamber recombination and polarity effects). We call the result the *detector signal*. Note that Monte Carlo calculations can also play a role in the determination of these corrections (see, e.g., Abdel-Rahman et al., 2006), for the polarity effect, but we will not review these studies further as part of this review.

The detector signal is first converted to absorbed dose in the sensitive volume of the detector, also called the *detector cavity*. For a given detector signal M_{det}, the absorbed dose in the detector cavity, denoted D_{det}, can be expressed as follows:

$$D_{\mathrm{det}} = C_{\mathrm{Q}} M_{\mathrm{det}}, \qquad (4.1)$$

where C_{Q} is a global factor representing the conversion of detector signal to absorbed dose in the detector cavity for a specific beam quality Q. The factor depends on the physical detection mechanism and incorporates the intrinsic energy dependence of the detector, its linearity with dose and its dose rate dependence. Note that C_{Q} can also be expressed as the product of a series of factors, each accounting for these effects (see, e.g., Rogers, 2009). The intrinsic energy dependence describes the link between the physical effect and the dose to the detector sensitive volume. The determination of the intrinsic energy dependence thus requires models of the physical or chemical behavior of the mechanism that links the observed reading to the absorbed dose to the sensitive volume. For example, Klassen et al. (1999) described the energy dependence of the G-value of Fricke dosimeters in megavoltage beams. In practice, the intrinsic energy dependence is usually determined indirectly, by calibration against a standard at a specific beam quality (e.g., ^{60}Co for linac radiotherapy) and after extracting the extrinsic energy dependence as calculated by condensed history (CH) Monte Carlo methods (see below). The importance of differentiating intrinsic and extrinsic energy dependence lies in the ability to determine the impact of beam quality change on the response of a detector arising from fundamental effects (i.e., physical signal to detector dose conversion) versus dosimetric effects, arising from detector embedding, walls, and so on, that can be modeled through CH Monte Carlo techniques.

Typical point-dose detectors are constituted of a homogeneous detection cavity (the sensitive volume) as well as other structural components being designed to both assure the physical integrity of the detector and ensure basic operation. As the radiation field interacts with these nonsensitive components, the result from these interactions is particular to the detector geometry and materials where such components are constituted differently from the medium in which the detector lies. Therefore, since the measured cavity response differs from what would be measured without these components being present, the assessed quantity is perturbed. This ensemble of phenomena can be termed the *component-specific* perturbation effects.

For a detection cavity freed of nonsensitive components, there remains some perturbation effect during radiation detection, since the cavity is also (most of the time) constituted of a material different from the medium. As physical properties such as mass density, atomic number, and the mean ionization potential (*I*-value) affect the charged particle fluence differential in energy and direction, which is a result of interactions in the geometry, the energy fluence in the cavity differs from the one at the point of measurement without the detector being present. This overall phenomenon is defined as the detector replacement perturbation effect. This means that if, hypothetically, one replaces the detection cavity, so far assumed freed of nonsensitive components, with an infinitely small detection cavity (or small enough within the limits of the definition of absorbed dose), the charged particle fluence at the point of measurement would be the same as the one in the medium without any detector in place. Similarly, as the detector perturbation effect, the replacement perturbation effect depends on the cavity size and shape, the nature of the material constituting the cavity, as well as the irradiation conditions.

Note that historically, radiation dosimetry was based on Bragg–Gray principles and Fano's theorem (Fano, 1954), which implies that there is no replacement perturbation effect under charged particle equilibrium (CPE) conditions. For practical situations, Fano's theorem is not strictly applicable (Bjärngard and Kase, 1985; Bouchard et al., 2012) and therefore a replacement perturbation effect is always present, especially for low energies where electron scattering power is high.

While the perturbation effects are also energy dependent, the overall conversion of absorbed dose in the detection cavity to absorbed dose to medium, considering the detector in its entire physical integrity, is written as follows (Rogers, 2009):

$$D_{\mathrm{med}} = f\, D_{\mathrm{det}},\qquad(4.2)$$

where D_{med} is the absorbed dose to the medium at the point of measurement and *f* is a factor representing the overall absorbed-dose energy dependence of the detector (also called the extrinsic energy dependence). This factor encompasses the component-specific and replacement perturbation effects as well as the absorbed-dose energy dependence of the detection material.

Although the component-specific and replacement perturbation effects must be evaluated to accurately determine absorbed dose to the medium in which the detector is placed in, the absorbed-dose energy dependence of the detection material relative to the medium must also be assessed. During radiation transport, the density of ionizations in charged particle tracks is much higher than for photons, which in principle interact only once, and therefore the contribution of charged particles to dose is dominant, whether the primary beam is a photon or electron beam. Furthermore, as electrons are set in motion by ionization, their fluence is higher than for positrons, typically by at least a few orders of magnitude for radiotherapy beams. Therefore, collision mass stopping power averaged over the electron spectrum is a major indicator of the absorbed-dose energy dependence of the detection material, with respect to the overall solution of Boltzmann radiation transport equations achieved by the Monte Carlo method (see, e.g., Bouchard, 2012). This means that even for an infinitely small detection cavity, freed of the effect of components, the conversion of absorbed dose to the detector to absorbed dose to medium must consider the energy dependence of the mean mass stopping power, which is defined by the electron fluence spectrum and the physical properties of the media involved.

The detector response (sometimes termed detector sensitivity) can be defined as the ratio of its reading, M_{det}, to the quantity of interest, which could be air kerma or dose to a medium. For example, the detector absorbed dose response or sensitivity is defined as

$$R_{\mathrm{D}} \equiv \frac{M_{\mathrm{det}}}{D_{\mathrm{med}}},\qquad(4.3)$$

where M_{det} is the reading of the detector placed in the medium med and D_{med} the absorbed dose at the effective point of measurement (EPOM) in the medium *med*. By combining Equations 4.1 and 4.2 into Equation 4.3, the absorbed dose response can be written as

$$R_{\mathrm{D}} = \frac{1}{f\, C_{\mathrm{Q}}}.\qquad(4.4)$$

As discussed above, Monte Carlo simulations of detector response are involved with the extrinsic component of the response, that is, the calculation of *f*.

4.2.3 In-Phantom Dosimetry

Absorbed dose measurement techniques are specific to a given detector, geometry, and beam quality. In the context of radiation dosimetry, the beam quality should be viewed as a beam-specific and geometry-dependent concept, rather than a descriptive of the beam only. It is defined as the overall condition induced by a given radiation beam in a specific region of the geometry of interest. While this condition is multidimensional, quality can be related to dosimetric factors being detector specific and position dependent. In practice, beam quality is encapsulated in a single quantity known as the in-water beam quality specifier (e.g., AAPM TG-51's %$dd(10)_x$ (Almond et al., 1999)). This quantity is used in specific applications where dosimetric factors need to be characterized as a function of beam quality. The function C_Q and the factor f, defined in Equations 4.1 and 4.2, respectively, depend on the nature of the detector and the beam quality. While the standard formalism is valid for any detector, it was historically developed for ionization chambers, the gold standard in clinical reference dosimetry. For a given beam quality, the standard formalism to convert detection signal to absorbed dose in the detection cavity is as follows:

$$C_Q = N_{gas}, \tag{4.5}$$

where N_{gas} is a signal-to-dose calibration coefficient that can be expressed as

$$N_{gas} \equiv \frac{(W/e)_{gas}}{m_{gas}}. \tag{4.6}$$

Here, $(W/e)_{gas}$ is the energy necessary to create an ion pair in the gas filling the chamber for a cobalt-60 beam (e.g., 33.97 eV/pair for sea-level dry air (Boutillon and Perroche-Roux, 1987)) and m_{gas} is the mass of gas (air) in the effective sensitive volume of the chamber. The absorbed-dose energy dependence factor f, used to convert detector dose to dose to medium, is expressed as

$$f = \left(\frac{\bar{L}}{\rho}\right)_{gas}^{med} P_{repl} P_{wall} P_{cel} P_{stem}, \tag{4.7}$$

where $(\bar{L}/\rho)_{gas}^{med}$ is the medium-to-gas Spencer–Attix stopping-power ratio, representing the absorbed-dose energy dependence of the detection material, P_{repl} is the replacement perturbation factor, and P_{wall}, P_{cel}, and P_{stem} are the wall, central electrode, and stem perturbation factors, respectively.

In a more general context, the standard formalism can be generalized to any detector. While the approach to convert detector signal to absorbed dose is linear for ionization chambers, it can be stated differently and specifically to the detection mechanism. As for the absorbed-dose energy dependence factor stated in Equation 4.7, the component-specific perturbation factors (i.e., P_{wall}, P_{cel}, and P_{stem}) are clearly defined by the presence of these components. Generally, there could be a series of perturbation factors that require attention to account for all nonsensitive components constituting the detector. That is, one can define the following for a more general context:

$$f = \left(\frac{\bar{L}}{\rho}\right)_{det}^{med} P_{repl} \cdot \prod_i P_i \tag{4.8}$$

where P_i are the component-specific perturbation factors of the detector.

4.2.4 Air Kerma Dosimetry

Air kerma measurements with ionization chambers have been, for several decades, fundamental to clinical radiation dosimetry (e.g., AAPM's TG21, 1983; IAEA's TRS-277, 1987). The development of calorimetry techniques (Ross and Klassen, 1996; Seuntjens and DuSautoy, 2003; Seuntjens and Duane, 2009) allowed the development of absorbed-dose-to-water-based dosimetry protocols (TG-51, TRS-398). This approach has become, over the last decade, the standard method for megavoltage beam dosimetry. Despite this change of practice, air kerma standards remain essential for the calibration of ionization chambers for kilovoltage photon beam dosimetry (e.g., AAPM's TG-61 (Ma et al., 1999)).

The rationale behind air kerma measurements is based on Fano's theorem (Fano, 1954), stating that under the condition of CPE, the charged particle fluence differential in energy and direction is independent of point-to-point density variations in the medium. If one defines a geometry in which an air-filled cavity is irradiated by a photon beam, Fano's theorem implies that if the cavity is surrounded by air-equivalent dense material and that CPE is achieved around and in the cavity, the electron fluence is unperturbed by the presence of the cavity. In practice, photon attenuation makes it impossible to achieve exact CPE. However, CPE can be achieved in a virtual situation, where photon interactions would set secondary electrons in motion but would artificially be regenerated (see, e.g., Bouchard et al., 2012). In such a situation, the electron fluence is characterized by the beam quality and the nature of the material, and absorbed dose equals collision kerma (see illustration in Johns and Cunningham's book, 1983).

In practice, a thick wall of air-equivalent material is chosen to enclose the air cavity, such that the perturbation of the electron spectrum between the wall and the air is minimal. The beam quality-specific wall thickness is chosen larger than the electron range (in the wall material), such that electron equilibrium (strictly speaking, a transient equilibrium) is established between the photon beam and secondary electrons in the wall. Using Fano's theorem, one can assume that the electron fluence perturbation by the presence of the cavity is negligible, this statement corresponding to the first condition required for Bragg–Gray dosimetry (which is also required in the Spencer–Attix cavity theory). The second required condition, that is, that photon interactions in the cavity are negligible, can also be assumed. This way, one can state that the absorbed dose to a mass of wall occupying the chamber sensitive volume is expressed as

$$D_{cav,wall} = \overset{Fano}{\left(\frac{\bar{L}}{\rho}\right)_{air}^{wall}} D_{cav,air}, \tag{4.9}$$

where $D_{cav,air}$ is the absorbed dose in the detection cavity filled with air.

The next step of the rationale is to convert absorbed dose in the detection cavity filled with wall material into absorbed dose to the medium consisting the geometry. Assuming that the attenuation of the photon fluence in the wall and the medium are identical, this is achieved by simply taking the ratio of collision kerma wall-to-medium, as follows:

$$K_{\text{coll,med}} \overset{\text{no att. or scat.}}{=} \left(\overline{\frac{\mu_{\text{en}}}{\rho}}\right)^{\text{med}}_{\text{wall}} D_{\text{cav,wall}},\qquad(4.10)$$

where $\left(\overline{\mu_{\text{en}}/\rho}\right)^{\text{med}}_{\text{wall}}$ is the ratio of mass–energy absorption coefficients, medium-to-wall. In Equation 4.10, it is implicitly assumed that the photon energy fluence in medium (air) and wall is identical, and that there is perfect CPE in the wall (i.e., $K_{\text{coll,wall}} = D_{\text{wall}}$). In the context of realistic detectors, one admits that (1) CPE is not perfectly achieved and (2) the presence of the wall causes perturbation in photon fluence. In a more general manner, one can state the absorbed-dose energy dependence factor f as follows:

$$f = \left(\overline{\frac{L}{\rho}}\right)^{\text{wall}}_{\text{air}} \left(\overline{\frac{\mu_{en}}{\rho}}\right)^{\text{med}}_{\text{wall}} K_{\text{wall}} K,\qquad(4.11)$$

where K_{wall} is a correction factor accounting for differences in photon and electron attenuation and scatter between the wall and the medium, and K is a general correction factor accounting for the fact that Fano's theorem is not entirely applicable (i.e., CPE is not established or the wall is not perfectly air-equivalent). The factor K also accounts for other component-specific effects (e.g., a build-up cap, a central electrode), as well as the fact that the chamber does not determine air kerma at a point, while this is what is meant to be measured, due to the anisotropy of the source and the nonlinear fluence-to-distance relation.

The final step of the method consists of choosing the medium as air, and to convert absorbed dose in free air to air kerma. The result yields the following expression:

$$K_{\text{air}} = \frac{1}{(1 - g_{\text{air}})} f \cdot D_{\text{cav,air}},\qquad(4.12)$$

where $(1 - g_{\text{air}})$ is the ratio between collision kerma and kerma, accounting for the kinetic energy lost in radiative interactions (bremsstrahlung, annihilation-in-flight).

4.3 Calculation of Radiation Dosimetry-Related Quantities

4.3.1 Stopping-Power Ratios

As described in the above section, Spencer–Attix cavity theory is fundamental to radiation dosimetry of photons and electron beams. The expression of medium-to-detector stopping-power ratio is given by Spencer–Attix–Nahum's formulation as follows (Spencer and Attix, 1955; Nahum, 1978):

$$\left(\frac{L}{\rho}\right)^{\text{med}}_{\text{det}} \equiv \frac{\int_{\Delta}^{E_{\max}} \varphi_{e-}(T)\left(L_{\text{med}}(T)/\rho_{\text{med}}\right)dT + TE_{\text{med}}}{\int_{\Delta}^{E_{\max}} \varphi_{e-}(T)\left(L_{\text{det}}(T)/\rho_{\text{det}}\right)dT + TE_{\text{det}}},\qquad(4.13)$$

where $\varphi_{e-}(T)$ is the electron fluence differential in energy (in MeV^{-1} cm^{-2}). Here, for either material (i.e., the medium or the detector), $L(T)$ is the restricted collision stopping power, ρ is the mass density, det is the medium of the sensitive volume of the detector, med is the medium the detector is placed in, and TE is the track-end term, discussed shortly. Note here that as described by Spencer–Attix cavity theory (which assumes a detector satisfying Bragg–Gray conditions), the electron fluence $\varphi_{e-}(T)$ is by definition unperturbed by the presence of the detector, and therefore corresponds to the electron fluence in the medium without any detector being present.

There are two ways in which stopping-power ratios are calculated. The first method consists of calculating stopping-power ratios analytically, using Equation 4.13. In such an approach, a Monte Carlo simulation of the transport through the medium of interest is used to estimate the electron fluence $\varphi_{e-}(T)$, differential in energy, by calculating the average path length in the cavity over its volume as a function of kinetic energy. The track-end term in the analytical method is defined as follows (Nahum, 1978):

$$TE_{\text{med}} \equiv \varphi_{e-}(\Delta)\left(\frac{S_{\text{coll,med}}(\Delta)}{\rho_{\text{med}}}\right)\Delta,\qquad(4.14)$$

where S_{coll} is the collision electron stopping power in the medium in question. Obviously for such approach, the choice of the simulation energy threshold, which separates discrete interactions and condensed steps, must be smaller than Δ.

The second and more modern approach to stopping-power ratio calculations (see SPRRZNRC's manual (Rogers et al., 2011)) is a direct implementation of Equation 4.13 in the code, avoiding an offline calculation and the computation and storage of electron fluence spectra. This is particularly relevant if stopping-power ratios are calculated for several positions in the geometry. The radiation transport is performed in a single medium. During the simulation, the electron condensed step and electron track ends are scored in two ways: (1) in the medium, where the transport is determined from physical properties of the geometry, and (2) in the detection material, where the absorbed dose is scaled by the ratio of restricted or unrestricted stopping powers. During a CH step, the absorbed dose in the detection material is given by

$$\text{DDEP}_{\text{det}} = \text{DDEP}_{\text{med}} \frac{\left(\overline{L_{\text{det}}/\rho_{\text{det}}}\right)}{\left(\overline{L_{\text{med}}/\rho_{\text{med}}}\right)},\qquad(4.15)$$

where DDEP represents the absorbed dose during an electron step. Here, the numerator and denominator are the restricted collision stopping power averaged over the energy of the step, in the detection material and the medium, respectively. Note that since electron steps are usually small, the averaged stopping power can be approximated by the stopping power of the mean kinetic energy in the step. The electron track-ends are calculated as follows:

$$TE_{det} = TE_{med} \frac{(S_{coll,det}(\Delta)/\rho_{det})}{(S_{coll,med}(\Delta)/\rho_{med})}. \qquad (4.16)$$

Note that this scaling of the track-end, represented by the ratio of stopping powers, comes from the analytic definition of track-ends at Equation 4.14. This accounts for the fact that the number of electrons with kinetic energy falling below Δ and being absorbed is scaled by that factor. Also note that this approach to stopping-power ratio calculations requires that the choice of the simulation energy threshold must be equal to Δ.

In each algorithm, the transport calculation is only performed in a single medium (the medium in which the secondary electron fluence has been established) and in which the stopping-power ratio is meant to be calculated. As it is defined by Bragg–Gray cavity theory, the electron fluence is only calculated once and is thus assumed unperturbed by the presence of the detector, and this condition is achieved in both approaches. It is clear that the calculation of stopping-power ratios is based on Spencer–Attix cavity theory and Bragg–Gray conditions, and therefore can only represent realistic detector response in specific conditions. In reality, both conditions of Bragg–Gray are violated. Indeed, the electron fluence is perturbed by the presence of the cavity, and photons interact in the volume of interest, filled with either material. Therefore, perturbation factors must be evaluated to convert the detection cavity response to the medium response accurately (as stated by Equation 4.8). Another matter to consider is the choice of the parameter Δ, which can lead to contradictions (see, e.g., Bouchard, 2012). Nonetheless, stopping-power ratios are useful to determine the absorbed-dose energy dependence of a given detection material.

4.3.2 Interaction Coefficients

Other important quantities of interest that can be calculated with Monte Carlo simulations are mass–energy absorption coefficients. In the context of dosimeter response for in-air and in-phantom measurements, several studies have addressed kilovoltage photon beams (Seuntjens et al., 1987, 1988; Ma and Seuntjens, 1999) as well as high-energy photon beams (Cunningham et al., 1986; Furhang et al., 1995). Calculated data have been used to obtain correction factors in respect to dosimetry protocols (AAPM's TG-21, 1983; IAEA's TRS-277, 1987; AAPM'S TG-51, 1999; IAEA's TRS-398, 1999; AAPM's TG-61, 2001) as well as detector response scaling. The average mass-absorption coefficients in a given medium is defined as follows:

$$\left(\frac{\overline{\mu_{en}}}{\rho}\right)_{med} \equiv \frac{\int_0^{E_{max}} \varphi_\gamma(h\nu)\left(\mu(h\nu)/\rho\right)_{med} E_{ab}(h\nu)dh\nu}{\int_0^{E_{max}} \varphi_\gamma(h\nu)h\nu\,dh\nu}, \qquad (4.17)$$

where $\varphi_\gamma(h\nu)$ is the photon fluence differential in energy (in $MeV^{-1}\,cm^{-2}$), μ is the energy-dependent linear attenuation coefficient in the medium (in cm^{-1}), and $E_{ab}(h\nu)$ is the average energy being absorbed by the medium from photon interactions with initial energy $h\nu$. Note that one needs Monte Carlo simulations to evaluate $E_{ab}(h\nu)$. The evaluation of such coefficients requires linear attenuation data and a proper knowledge of the photon fluence spectrum, either measured or calculated with Monte Carlo simulations under specific conditions. Systematic compilations of mass–energy absorption coefficients for elements and materials of dosimetric interest were performed by Hubbell (1982) and Seltzer (1993) for monoenergetic photons. To avoid storage of a large number of spectra, direct scoring techniques with Monte Carlo can be used to calculate average ratios under specific conditions.

4.4 Modeling Detector Response

4.4.1 Ionization Chamber Correction Factors

Over the last decades, considerable improvements in particle transport accuracy (e.g., Bielajew et al., 1985; Bielajew and Rogers, 1987; Andreo et al., 1993; Bielajew et al., 1994; Baro et al., 1995; Sempau et al., 1997; Kawrakow and Bielajew, 1998a, 1998b; Kawrakow, 2000a; Bote and Salvat, 2008) have allowed the use of Monte Carlo methods for detector response simulation to an accuracy of 0.1% or better (Kawrakow, 2000b; Seuntjens et al., 2002; Poon et al., 2005; Sempau and Andreo, 2006). Such improvements have allowed studies to be performed in dosimetric situations that could not be done by other means with the same level of accuracy. A recent example is the behavior of ionization chamber response in nonstandard beam configurations (small beams or modulated beams), which was extensively studied using Monte Carlo models and compared with experimental measurements (Paskalev et al., 2003; Bouchard and Seuntjens, 2004; Capote et al., 2004; Sánchez-Doblado et al., 2005). These studies demonstrated the need for Monte Carlo calculations in the evaluation of correction factors for such dosimetric situations.

More recent publications used Monte Carlo modeling to perform studies or to generate accurate data related to ionization chamber response to dose in context of standards or reference dosimetry protocols (Buckley and Rogers, 2006a,b; Burns, 2006; Kawrakow, 2006; Mainegra-Hing et al., 2008; McEwen et al., 2008; Wang and Rogers, 2008, 2009a,b,c; Wulff et al., 2008a; Zink and Wulff, 2008, 2011; Bouchard et al., 2009; Crop et al., 2009; Tessier and Kawrakow, 2010).

In general, detector sensitivity (or response) calculations cover three areas of importance: (1) in-phantom reference dosimetry, (2) in-phantom relative dosimetry, and (3) in-air chamber correction factors.

In the context of reference dosimetry, calculation of component-specific and replacement perturbation factors can be performed by taking the ratio of absorbed doses calculated in a series of detector geometries, only differing in their components or in the composition of the cavity. In this manner, perturbation correction factors can be calculated as a ratio of doses scored in geometries connected to each other by chain of correlated geometries. The chain of correlated geometries can consist of a fully-modeled chamber having components gradually removed from it to calculate the effect of a specific perturbation. It can also contain a virtual volume in which dose is scored, such as, for example, a bare chamber scoring volume in which stopping-power ratios are calculated or a small scoring sphere serving to evaluate dose at a point. The ratio of absorbed doses in the user-defined scoring volumes corresponds to a perturbation factor related to the differences in characteristics of the volumes (i.e., their geometry or composition). For example, the ratio between calculated absorbed doses to the chamber without wall and the fully modeled chamber is, by definition in this case, the wall perturbation factor. From this starting point, the ratio between calculated absorbed doses to the chamber without wall and central electrode and the chamber without wall is, also by definition in this case, the central electrode perturbation factor. The calculation chain goes on until the all-scoring volumes are defined to represent the factors in Equation 4.7. This method represents a significant improvement in consistency from approximate calculations or measurements. Figure 4.1 illustrates the principle chain to determine perturbation factors used by Wulff et al. (2008b).

For relative dosimetry, Monte Carlo modeling of ionization chamber response has been applied to aid an accurate conversion of depth ionization into depth dose for both photon and electron beams. In electron beams, this entailed calculation of stopping-power ratios (Burns et al., 1995) and perturbation correction factors as a function of depth (Buckley and Rogers, 2006a,b; Verhaegen et al., 2006).

In photon beams, stopping-power ratios have been shown to be quasi-depth independent, except for the build-up region (Andreo and Brahme, 1986) but recent efforts have readdressed the issue of the EPOM in the determination of percent depth dose. Kawrakow (2006) proposed a method to reevaluate the EPOM proposed by dosimetry protocols (AAPM's TG-21, 1983; IAEA's TRS-277, 1987;

AAPM's TG-51, 1999; IAEA's TRS-398, 2001) using Monte Carlo simulations. Calculating the ionization curve (IC), which is the chamber response to dose as a function of depth normalized to the maximum response, and comparing the results with calculated percent depth dose (PDD) in water allows finding the best shift to be applied on the IC, such that the mean square difference between the shifted IC and the PDD is minimal (Kawrakow, 2006; McEwen et al., 2008; Tessier and Kawrakow, 2010). This work has demonstrated that the standard prescription provided by reference dosimetry protocols needs to be fine-tuned.

Other detectors have also been modeled in the context of relative dosimetry. Diode detectors are very efficient for electron beam PDD measurements, since at clinically relevant depths when placed with the center of the sensitive volume at the appropriate depth, measured reading does not require any correction. However, because electrons reach sub-MeV kinetic energy in the high-gradient region of the PDD curve beyond d_{max} and because the relative contribution of bremsstrahlung photons to dose increases for depths beyond d_{max}, one might expect that the response of the diode must be corrected in this dose fall-off region and especially the bremsstrahlung tail. With the effective atomic number of water roughly between $Z = 7$ and $Z = 8$, and the one of amorphous silicon $Z = 14$, the large differences in electron scattering power at low kinetic energies might lead to an appreciable shift in the position of the depth of 50% maximum dose (R_{50}) and the practical range (R_p). However, Wang and Rogers (2007) extensively studied diode response to electron beam with depth based on a complete Monte Carlo model and conclude that the effect on the position of R_{50} is well below 1 mm when using a diode to measure electron depth dose curves. However, they showed that corrections on the depth dose curve in the bremsstrahlung tail amounts to 5–10% of the maximum dose. This has been demonstrated experimentally for 6–21 MeV clinical electron beams, where the dose in the bremsstrahlung tail relative to the maximum dose measured with the TG-51 dosimetry protocol was compared to charge collection ratios with a restricted water-to-detector mass stopping-power ratio correction and highly constrained Monte Carlo simulation, with the parallel plate chamber underresponding in the bremsstrahlung tail by 10%, the silicon diode overresponding by up to 5% (see Faddegon et al., 2005, 2009).

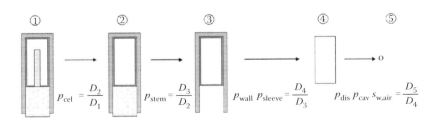

FIGURE 4.1 Illustration of the chain technique of Wulff et al. (2008b) for calculating ionization chamber perturbation factors. The various perturbation factors are given by the dose ratios from one step to another in the ionization chambers' cavity (1–4) and the dose to a small portion of water (5). The step from models 3 and 4 can be further subdivided into separate calculation of p_{wall} and p_{sleeve}. (From Wulff, J., Heverhagen, J. T., and Zink K. 2008b. Monte-Carlo-based perturbation and beam quality correction factors for thimble ionization chambers in high-energy photon beams. *Phys Med Biol* 53: 2823–2836. With permission.)

An important early area of application for chamber simulations was in the context of air kerma radiation standards. A historical application has been the calculation of the wall correction factors in air kerma standards for ^{60}Co and for the dissemination of the air kerma standards through protocols such as TG-21 and TRS-277. In-air ionization chamber wall corrections are calculated in a process of correlated scoring of the energy deposited by (1) the total energy deposition, that is, all secondary electrons and their progeny from primary and scattered photons, (2) the "first-collision" energy deposition by all secondary electrons and their progeny from only primary photons, and (3) the "unattenuated first-collision," which is the first-collision energy deposition, corrected for attenuation of the primary photon. The wall correction is calculated as a product of two factors, the attenuation correction and a scatter correction. The attenuation correction is defined as a ratio of doses scored in (3) by those in (2), whereas the scatter correction is defined as a ratio of doses scored in (2) by those in (1). All scoring is done simultaneously for each history, making the scoring highly correlated, which greatly reduces the statistical uncertainty in the ratios. Despite rigorous proof of accuracy by Bielajew and colleagues and despite the low uncertainties to which wall corrections could be calculated, it was not until detailed experiments by McCaffrey et al. (2004) and Buërmann et al. (2003) confirmed the Monte Carlo correction factors that the standards dosimetry community accepted Monte Carlo-based wall correction factors.

A final application that has received recent attention is the reevaluation of free-air chamber correction factors by Monte Carlo techniques as recently reviewed in detail by Burns and Buërmann (2009).

4.4.2 Simulation Geometries

In general, common Monte Carlo systems provide a geometry package in which radiation transport is simulated. In the context of photon and electron beam radiation therapy, accurate detector response calculations are performed using mainly two systems: (1) EGSnrc (Kawrakow, 2000a) and (2) PENELOPE (Salvat et al., 2009). Both systems have powerful tools to describe

detector geometries. As for EGSnrc, the user code *cavity* and *egs_chamber* (Wulff et al., 2008b), being specialized in detector response simulation, allow for the use of several geometries based on the *egs++* library (Kawrakow, 2005). Typical user-defined geometry can be done using simple solids such as boxes, spheres, or cylinders along with the use of Boolean operators (union, intersection, etc.) on these objects to combine them logically and make complicated geometries. Also, geometries can be defined by the surfaces surrounding them, this using Boolean operators to define volumes from a combination of simple surfaces, for example, planes or infinite cylinders, and solid objects. The structure of the library can also allow the definition of new geometry classes. In PENELOPE, the main idea is the use of quadratic surfaces, defined by a second-order equation with x, y, and z, to define the geometries. Solid bodies are constructed with a combination of surfaces in which the space is filled with a homogeneous medium. The priorities of the bodies are specified by the user, allowing embedding bodies into other volumes to obtain complex geometries. In both systems, rotations and translations of the geometries are allowed, making it possible to simulate any detector with very good precision in detail. Figure 4.2 shows the chamber modeled by Crop et al. (2009) using the *egs++* library.

4.4.3 Efficiency Improvements

To reach a precision level comparable to the calculation accuracy (i.e., 0.1% (Kawrakow, 2000a)), simulations must be highly efficient. The first matter being considered in such types of application is the determination of region indexing, which without any proper consideration can become time consuming when the geometry is complex. The design of algorithms allowing efficient transport through boundaries has been given a high priority in the design and implementation of some codes, like the EGSnrc geometry package *egs++* (Kawrakow, 2005). Calculations performed in geometries having a single scoring volume allow for significant efficiency improvements as compared to brute force calculations. One common variance reduction technique (VRT) used in practice for detector response simulation is range rejection (RR), being used in both EGSnrc and PENELOPE. The idea in this technique is that any charged particle having a range

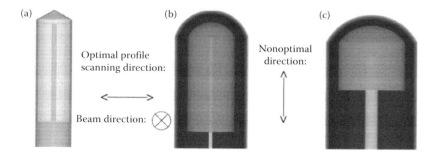

FIGURE 4.2 Geometrical models of the following chambers achieved by Crop et al. (2009): (a) NE2571, (b) PinPoint 31006, and (c) PinPoint 31016 (images are not on the same scale). (From Crop, F. et al., 2009. The influence of small field sizes, penumbra, spot size and measurement depth on perturbation factors for microionization chambers. *Phys Med Biol* 54: 2951–2969. With permission.)

smaller than its distance to the scoring volume will be discarded. This avoids simulating particle transport that will yield no effective result. One approximation in this technique is in the amount of bremsstrahlung radiation that will reach the scoring volume. To control this approximation, the user can define a threshold below which RR is considered.

Correlated sampling (CS) is also a VRT having good affinities with detector simulation. For applications where relative detector response is assessed, some information on particle transport being common to several geometries can be recycled. For example, if one is interested in calculating the wall correction factor, two geometries are defined at the same position: (1) a fully modeled chamber and (2) a chamber modeled without wall. It is clear that for any particle being transported from the beam in either geometry, the transport will be identical until the particle reaches the region where the wall is located. Defining a CS region enclosing both detectors, one can save a significant amount of computational time simulating both geometries using the same information gathered on the surface defining the CS region. Another example is for detector response relative to another location, to mimic IC measurements. In such application, the CS region is made large enough to enclose all detector positions. When scoring with the CS technique, the energies and variances are not only scored but the covariances as scored energies are also positively correlated between each other. Mathematically, this implies that the uncertainty on the ratio is smaller than the one obtained for independent simulations using the same number of histories. The CS technique was implemented in CSnrc by Buckley et al. (2004), a modification of the user code CAVRZnrc, and is illustrated in Figure 4.3. The user code *egs_chamber*, being a modification of the code *cavity*, was designed by Wulff et al. (2008a), defining the intermediate phase-space scoring (IPSS) technique to use CS in the subgeometries. Finally, the user code CScavity, also being a modification of the code *cavity*, was implemented by Crop et al. (2009). While in early studies, typical gains in efficiency with the CS technique were reported between 10 and 100 (Ma and Nahum, 1993), they are roughly between 20 and 40 for CSnrc and *CScavity*, respectively.

Another powerful VRT relevant to detector response simulation in photon beams is photon cross section enhancement (XCSE). Defining a region around the scoring volume in which the photon interaction cross sections are enhanced, the shower of particles is amplified. The resulting information extracted from this shower, initiated from single particles entering that region, is such that the overall simulation efficiency is optimized. There are strict rules to obey to assure an unbiased result and equal particle weighting. For that matter, particle splitting and Russian roulette are played in such a way that each particle type carry a unique statistical weight, given a XCSE factor (see Wulff et al., 2008b). Note that electrons surviving the range rejection game make an exception to this rule. Figure 4.4 illustrates the combination of CS technique, through the IPSS method, and XCSE in the user code *egs_chamber* (Wulff et al., 2008b). Note that during the validation of this code, the gain in efficiency by the use of XCSE itself was reported to be 130 for the cases simulated (Wulff et al., 2008b). When comparing

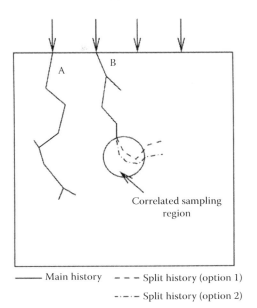

FIGURE 4.3 Illustration of the correlated sampling method of Buckley et al. (2004). The small region within the larger phantom is the correlated sampling region and changes material with each new geometry option. Particle A shows a main history that never enters the correlated sampling region and is transported only once during the entire simulation. Particle B shows a particle for which the main history (solid line) is transported only once, and the split history (dashed lines) transport is repeated for each geometry option. For positive correlations, the trajectories will be similar for each of the geometry options. (From Buckley L. A., Kawrakow I., and Rogers D. W. O. 2004. CSnrc: Correlated sampling Monte Carlo calculations using EGSnrc. *Med Phys* 31: 3425–3435. With permission.)

a simulation without any VRT with an optimal simulation (i.e., using RR, CS, and XCSE), total gains in efficiency are between 650 and nearly 10^4, depending on the nature of the simulation.

4.4.4 Positioning Uncertainties in Detector Response Calculations

In radiation dosimetry, another quantity of interest is experimental uncertainty, which is typically evaluated either by a statistical analysis of observations or by other nonstatistical methods (ISO, 1995). Among several sources of uncertainty, one that could potentially be evaluated with Monte Carlo calculations is the uncertainty on detector response induced by setup positioning errors. This topic is relevant particularly in situations where high accuracy is required and dose gradients are significant, such as nonstandard photon beams (Chung et al., 2010) or electron beam dosimetry (Zink and Wulff, 2009). For these applications, the effect of positioning errors on detector response can be considerable and have a substantial effect on the confidence of the measured data.

In a recent paper, Bouchard et al. (2011) constructed and implemented a method in the user code *egs_chamber* to evaluate these effects in a systematic manner. The method is based

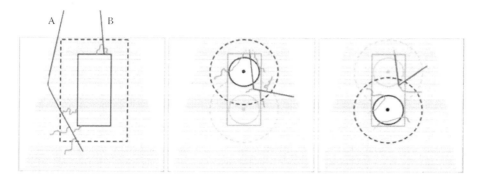

FIGURE 4.4 Illustration of the method of Wulff et al. (2008a) using intermediate phase-space scoring (IPSS) and photon cross section enhancement (XCSE). In the left base geometry photon, tracks A and B are started from the simulation source. The IPSS volume (left) surrounds both ion chamber positions (middle, right). The IPSS volume (solid line box) in the base geometry is surrounded by an XCSE region (dashed line box). Each of the two chamber geometries has its own XCSE region that only partially overlaps with the XCSE region of the base geometry. The phase space of photons and electrons is stored and the transport is terminated as soon as the particles enter the IPSS region (left). The phase space is used for both ion chamber positions as a particle source. Electrons must survive a Russian roulette game, if they do not start inside the XCSE region of the respective ion chamber geometry. In this illustration, no IPSS electrons originating from photon A survive the game for the first chamber position (middle) and no IPSS electrons from photon B survive for the second (right). (From Wulff, J., Zink, K., and Kawrakow, I. 2008b. Efficiency improvements for ion chamber calculations in high energy photon beams. *Med Phys* 35: 1328–1336. With permission.)

on the principle that Monte Carlo simulations of radiation transport can be used to score other quantities than typical energy deposition. In fact, any Monte Carlo method consists of scoring statistical estimators along with their respective uncertainties. In the present context, the quantity of interest is the standard deviation of the detector response, each individual response calculated in a randomly generated geometry, based on a user-defined statistical behavior of geometry positioning. Intuitively, the simplest approach to calculate such quantity would be to proceed as follows. First, one generates random detectors or geometry positions (including translations and rotations), this in respect to what is defined as probable positioning errors (according to certain probability distributions). Second, one calculates the dose response of the detector at each position. Finally, one estimates the standard deviation of the calculated responses and normalizes it to the mean response. While this seems to be a trivial way to proceed, the direct result from this intuitive approach is statistically biased, since results from Monte Carlo simulations are estimated and contain statistical uncertainties (Bouchard et al., 2011).

In the method proposed by Bouchard et al. (2011), an unbiased estimator of dose uncertainty induced by positioning errors is constructed for accurate calculations. An unbiased estimator of dose ratios, as is used in the context of CS, is also built taking into account the statistical correlations. The method is implemented in *egs_chamber* and allows for positioning errors on (1) the phantom according to the beam and (2) the detector according to the phantom. Several spatial operations (i.e., translation and rotation) as well as probability distributions (i.e., Gaussian and uniform) are allowed to characterize the random nature of positioning errors. Although the method requires an optimization procedure to obtain efficient calculations, it is shown that

with optimal parameters one can achieve the same statistics as for standard dose calculations with only twice the number of histories. In typical situations, since the statistics on the uncertainty estimators need to be typically of the order of a few tenth of a percent, calculations can be time consuming. Nonetheless, such simulations find uses in situations where no other means to evaluate these effects exist, especially in nonstandard fields. Figure 4.5 shows results obtained by the method by Bouchard et al. (2011) for small field output factor measurements using a diode.

4.5 Accuracy of Detector Response Calculations

4.5.1 General Considerations

In assessing the accuracy of Monte Carlo calculated detector response, there are four main sources of systematic uncertainties:

1. The accuracy of the CH transport system
2. The accuracy of the cross-section data in the system, including both fundamental properties of elements (*I*-values, etc.) and how well the true composition of the materials of the real detector are modeled (impurities, etc.)
3. The accuracy of the geometry of the model of the detector
4. The accuracy of the radiation source model (i.e., small field, collimation system, spot size, etc.)

4.5.2 Accurate Condensed History Technique

During the simulation of charged particle transport, the analog simulation of discrete interactions is limited to a certain

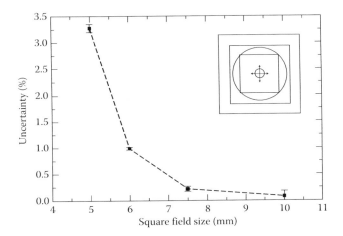

FIGURE 4.5 Calculation results by Bouchard et al. (2011) of setup positioning-induced detector output factors uncertainty in small beam measurements using a PTW 60012 diode model with both isocenter and detector uniform displacement of ±1 mm (in directions perpendicular to the beam axis only). Error bars correspond to the type A uncertainty on the estimated type B uncertainty. The sketch in the top right corner shows a beam's eye view of the diode placed in the fields, with the circles representing the sensitive volume and the total volume of the diode, the squares representing the projection of the fields on isocenter, and the cross representing a ±1 mm range of displacement in both directions. (From Bouchard, H., Seuntjens, J., and Kawrakow, I. 2011. A Monte Carlo method to evaluate the impact of positioning errors on detector response and quality correction factors in nonstandard beams. *Phys Med Biol* 56: 2617–2634. With permission.)

energy threshold below which the interactions are condensed into a single step, or CH (Berger, 1963). The calculation of electron (and positron) condensed steps is performed based on the continuous slowing down approximation (CSDA) and the multiple scattering theory. It is worth noting that both approaches must, on average, converge to an accurate result. During electron transport with the CH technique, the two main algorithms involved are (1) the electron-step algorithm, where the electron nonlinear trajectory is determined, and (2) the boundary-crossing algorithm. The approach by Kawrakow (2000a), implemented in the EGSnrc system, uses several improvements of the techniques behind these algorithms, as compared to earlier versions of the system (Nelson et al., 1985; Bielajew et al., 1994). The constraints of the Molière multiple scattering theory were resolved by the implementation of the Goudsmit–Saunderson formulation (Kawrakow and Bielajew, 1998a), diminishing the truncation error by two orders of magnitudes (Kawrakow and Bielajew, 1998b). The accuracy of the boundary-crossing algorithm was also improved by the use of a single-scattering mode near the boundaries, resolving a known singularity (Kawrakow, 2000a). To evaluate the potential accuracy of Monte Carlo simulations in radiation dosimetry and detector simulation applications, the convergence of these improved algorithms was extensively studied.

4.5.3 Fano Test

Accurate Monte Carlo calculation of the extrinsic response of gas-filled ionization chambers or cavities is one of the most stringent tests of a CH Monte Carlo code. The electron transport physics, the boundary-crossing algorithm, the evaluation of energy loss along the CH step, as well as the cross sections must be accurate. During the 1980s and 1990s, the ability of Monte Carlo to accurately calculate chamber response was assessed by comparing results to a calculation of f using the Spencer–Attix cavity theory at ^{60}Co for graphite-walled chambers (Rogers, 1993). However, this test of accuracy of a Monte Carlo code depends on (1) the accuracy of the theory against which it is assessed and (2) the accuracy of the cross sections involved in the calculation. Therefore, modern CH Monte Carlo algorithms are tested for consistency (not accuracy) using the so-called Fano test. This test relies on the validity of the Fano theorem (Fano, 1954), which states that under conditions of CPE, the medium electron fluence differential in energy and angle is independent of density variations from point to point. The theorem can be used to test the extent to which the electron step algorithm, path length correction, boundary-crossing algorithm, and energy loss evaluation have been implemented adequately in a Monte Carlo simulation program when dealing with transport in a situation consisting of multiple regions and densities.

The Fano test is usually carried out on a cavity filled with the gaseous form of the phantom material in which the cavity is embedded. The idea is that the gaseous form of the wall material has exactly the same cross sections as the wall material itself but has a 1000-fold lower density. This means that density effect as part of the stopping power of the cavity gas and wall must be taken as the same or artificially removed from the calculation. The Fano test first requires to setup, artificially, perfect CPE. There are two vintages of the test, differing in the way this is achieved.

One way is close to the physical reality of a photon beam interacting in a material and effectively present itself as a distributed source of charged particles set in motion. In a typical cavity (ionization chamber) setup, the beam is interacting in the phantom (or wall) of the simulated ionization chamber and either restore the primary photon once it has interacted, discard the scattered photon, and follow the secondary electron slowing down (the regeneration technique) or, alternatively, by dividing the cavity dose by A_{wall}, the in-air correction factor for photon attenuation and scatter. Effectively, all this means that while the photon beam interacts, it is not allowed to be scattered and scattering and attenuation effects are "undone." This clearly represents a situation where energy is not conserved; in fact, the radiant energy inserted in the form of a photon beam is recovered after the secondary electrons are generated and the photon beam is used for no more than generating a distributed electron source in the medium. The advantage of performing the test in this manner, however, is that the energy inserted in the medium can be retrieved from photon mass energy transfer coefficients.

The Fano test entails the comparison of collision kerma in the wall, under CPE with dose to the cavity, as follows:

$$D_{cav}^{unw} = \frac{D_{cav}}{A_{wall}} = K_{coll,wall}. \qquad (4.18)$$

This technique has been used by several authors during the investigation of accuracy of the EGSnrc code Kawrakow (2000b), Seuntjens et al. (2002), the PENELOPE code (Yi et al., 2006), and the GEANT4 code (Poon et al., 2005).

The Fano test requires the calculation of collision kerma to the wall (or build-up medium), which requires the estimation of energy fraction expended in bremsstrahlung in the wall (g_{wall}). This is usually done in auxiliary calculations and should be based on the same photon interaction data as used in the cavity dose calculation so that the evaluation of Equation 4.18 becomes cross-section independent.

A second and more direct way, which avoids these auxiliary calculations, is to realize Fano conditions by directly initiating electron fluence with uniform intensity per unit mass in an infinite medium with a gap of infinite extent. In this case, the dose is just the energy of the electrons times their number per unit mass. Sempau and Andreo (2006) used this method to compare dose calculations with the CH Monte Carlo simulations with event-by-event calculations and arrived at a consistency of 0.2% with PENELOPE.

4.5.4 Accuracy of Physical Data

The Fano test does not assess one of the components in an accurate detector response calculation in realistic situations, that is, the accuracy of the interaction data. There are several sources of cross-section data in literature, as well as several models behind them, and some physical data depend on few parameters, such as electron and positron stopping powers, which depend on the mean excitation energy (*I*-value) (ICRU37, 2007). A recent study by Wulff et al. (2010) studied systematic variations of ionization chamber quality correction factors k_Q induced by systematic changes in cross-section data and *I*-values. It was found in general that the effect on k_Q factors by uncertainties in photon cross-section data, being estimated to be of the order of 1–2%, is only up to 0.4% for typical thimble chambers. As for the effect of uncertainties on the *I*-value, the effect is remarkable only for $Z = 6$, assuming an uncertainty on the *I*-value of 6.1% for that material, yielding an effect of up to 0.37% on k_Q factor calculations. It is worth noting that another similar study by Muir and Rogers (2010) comes to slightly different conclusions, obtaining systematic errors up to 0.55% for k_Q from uncertainties in photon cross sections, and up to 0.19% on k_Q from uncertainties on *I*-values (here, an uncertainty on the *I*-value for $Z = 6$ was taken to be 4.5%).

The value of the energy necessary to create an ion pair in air, W_{air}, has been the subject of considerable recent research and controversy. The complication is that most sources of values of W_{air} involve graphite cavity ionization chambers in conjunction with graphite calorimeters and the value of *I* for graphite enters in the determination of W_{air} through the restricted collision stopping-power ratio graphite to air. Büermann and Hilgers (2007) compared exposure measurements at 300 and 400 keV using a large free-air and a graphite cavity chamber. This allowed a determination of the graphite-to-air mass stopping power ratio to a stated uncertainty of 0.3% and the *I*-value for graphite consistent with this is close to the Bichsel and Hiraoka (1992) *I*-value. Büermann et al. (2007) did a second experiment, where they use a graphite cavity chamber in the physikalisch technische bundesanstalt (PTB) water calorimeter in a ^{60}Co beam, and used a Monte Carlo calculated ratio of dose to water to dose to cavity to compare absorbed dose from the ionometric to the calorimetric method. Reconciling both methods for absorbed dose to water determination requires either an increase in *I*-value to 87 eV or a 1.5% reduction in the W_{air} value (or a combination of both). A result qualitatively similar, albeit only amounting to an effect of 0.7%, is obtained by considering the key comparison data of worldwide absorbed dose standards at ^{60}Co, maintained by the bureau international des poids et mesures (BIPM).

Thomson and Rogers (2010) reanalyzed the calorimeter experiments that led to the determination of the product of the energy necessary to create an ion pair in air per charge, $(W/e)_{air}$, with the graphite-to-air mass stopping power ratio. This was done by reanalyzing the gap corrections in the graphite calorimeter experiment and by analyzing the sensitivity of the product to secondary particle production thresholds and *I*-values. As suggested, extracted $(W/e)_{air}$ value from this analysis is 33.68 J/C ± 0.2%, which is 0.8% lower than the currently accepted 33.97 J/C ± 0.1% (Boutillon and Perroche-Roux, 1987). The impact of the uncertainty in $(W/e)_{air}$ on k_Q was studied by Muir and Rogers (2010). They demonstrated an effect of 0.5% on k_Q factor calculations based on the uncertainty of 0.75% on *W/e* for air. In a more recent paper, Muir et al. (2011) showed an excellent agreement between measured and Monte Carlo calculated k_Q factors. From their study, they concluded that an upper limit of variation of 0.4% exists on $(W/e)_{air}$ in the energy range from cobalt-60 to 25 MV.

4.5.5 Effects of Machine Modeling

In the context of detector response calculation, machine modeling is a crucial step of the procedure. While it can be understood from Equation 4.8 that detector response is dependent on the electron spectrum, stopping-power ratios are not quite sensitive to the beam lateral fluence (Sánchez-Doblado et al., 2003; Bouchard et al., 2009). Among the parameters determining the model, the details in the geometry and materials and details on the electron beam, such as the radial distribution and the energy, are important parameters to consider. A paper by Sheikh-Bagheri and Rogers (2002) is a good starting point to systematically determine these parameters from experimental measurements. Based on a proper model of the machine geometry from the manufacturer's specifications, one can estimate the incident electron beam energy and spatial distribution, calculated off-axis factors and central-axis depth dose curves can be compared with measurements. Central-axis depth dose data is useful to determine the mean energy of the electrons. Once the mean energy is determined, the in-air off-axis factors are used to derive the spot size of the electron beam spatial distribution.

Recent studies reported that the detector response to linac beams is significantly sensitive to the spot size of the incident electron (Francescon et al., 2008; Crop et al., 2009). During the simulation of small beam output factors, Francescon et al. (2008) reported that a change of 1.2 mm in the full width half maximum (FWHM) can have an effect up to 6% on the calculation of output factors with several detectors, such as ionization chambers, diodes, and diamond detectors. Another study by Crop et al. (2009) concludes that the strong influence of the spot size on ionization chamber response for small field sizes limit their use. The difficulties related to the estimation of the spot size depend on how the machine is modeled from the start. If one adjusts the FWHM based on open field measurements, such as recommended by Sheikh-Bagheri and Rogers (2002), small errors in the shape of the flattening filter can potentially yield erroneous spot sizes. One way to improve this method is to determine the FWHM from small field output factor measurements and explicitly accounting for the detector in the model (Francescon et al., 2008; Kawrakow, 2011). Then the shape of the flattening filter and other influencing components of the linac are adjusted with large field profiles. This method has demonstrated to improve the agreement between the calculations from the model and experimental measurements (Francescon et al., 2011). Another approach is to measure the spot size and other simulation details directly, putting narrow constraints on the parameter adjustment (Sawkey and Faddegon, 2009).

4.5.6 Size of Volume of Interest

The size of the volume in which absorbed dose is calculated influences the calculation accuracy. While the Monte Carlo method intrinsically defines energy thresholds below which particles deposit their energy locally, this approximation can break down for volumes with dimensions comparable to the range of electrons having sub-threshold energies. As energy is being conserved during particle transport simulation, what contributes to the error between the dose calculated with model and the true dose is the fraction of energy being locally absorbed that could be transported across the boundaries of the volume of interest. Therefore, the range (of mean free path) of the particles having kinetic energy (or energy) below the threshold determines the critical region of the volume where the absorbed energy is approximated. If this region is small compared to the volume, the error is also small. Therefore, an appropriate choice of energy thresholds is necessary to calculate dose accurately in small volumes.

In typical cases, the accuracy of the method related to the size of the scoring volume can be evaluated by showing the convergence of the calculations as a function of the energy thresholds. However, it can be argued that there are some exceptions in which the particle range or mean free path do not strictly decrease with decreasing energy. One simple example is the behavior of photon cross sections with energies around different atomic shells, where discontinuities are observed due to a drastic change in the availability of electrons. However, in general, the particle range (or mean free path) decreases with decreasing energy threshold. Therefore it is expected that the smaller is the scoring volume, the smaller the energy threshold needs to be for accurate calculation. However, it is also worth noting that since the Monte Carlo method is a semiclassical approach, it is limited to the transport of particles with kinetic energy (or energy) equal or larger than 1 keV, by Heisenberg's uncertainty principle (Thomson and Kawrakow 2011). This implies that the accuracy of the method is seriously compromised in the case where the range (or mean free path) of particles having a kinetic energy (or energy) of 1 keV is comparable to the size of the cavity.

4.5.7 Effects of Geometry Accuracy

The accuracy to which geometrical details are described in detector models also has an effect on the calculation accuracy. The achievement of such detailed modeling depends on the tools available, which have considerably evolved over the past decade. While earlier studies modeled ionization chambers with simplistic geometries (Bouchard and Seuntjens, 2004; Capote et al., 2004; Sánchez-Doblado et al., 2005), the accuracy to which they generate ionization chamber response-related data was limited. With the new *egs++* library (Kawrakow, 2005) and the geometry package in PENELOPE (Salvat et al., 2009), detector geometry can be fully modeled. A study by Ubrich et al. (2008) demonstrated that ionization chamber components can have a crucial effect on their response, specifically for photon beams in the kilovoltage range. Indeed, they showed that a small difference in the shape of the central electrode results in a deviation of 8.5% for beam quality correction factor k_Q. However, for megavoltage photon beams, recent studies demonstrated that the effects are small (i.e., a few tenth of a percent) in reference conditions (Muir and Rogers, 2010; Wulff et al., 2010). Figure 4.6 illustrates the method of Muir and Rogers (2010) that showed maximum differences in k_Q factors of 0.2% for the geometries used: (a) fully-modeled chamber, (b) chamber with simplistic stem, and (c) cylindrical wall with central electrode. From these reported studies, it can be concluded that although for typical situations the geometrical accuracy has a small impact on calculations, they remain an important matter to be considered, especially for nonstandard beam configurations. This is relevant especially for kilovoltage beams, metallic detector components, and nonstandard megavoltage beams. In fact, photoelectric effects can affect detector response significantly, and beams where lateral disequilibrium exists can have an important effect on the modeling of volume averaging effects and the displacement of the water volume by the air cavity.

4.6 Small and Nonstandard Field Dosimetry

4.6.1 Small Fields

Small fields and combinations of small fields are increasingly used in radiation therapy and can be generally termed nonstandard

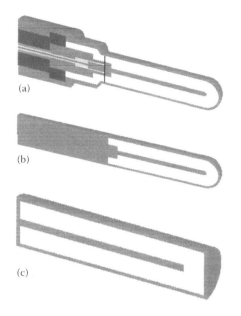

FIGURE 4.6 (See color insert.) Illustration of the method by Muir and Rogers (2010) to evaluate the effect of the geometry modeling on ionization chamber response calculations. Three models of the Exradin A12 ionization chamber (scales differ): (a) blueprint model; the solid line shows where the active cavity begins; (b) user manual, with a spherical top and a 2 cm long stem; (c) user manual, purely cylindrical geometry. (From Muir, B. R. and Rogers, D. W. O. 2010. Monte Carlo calculations of k_Q, the beam quality conversion factor. *Med Phys* 37: 5939–5950. With permission.)

fields as they are different from conventional flat radiation therapy beams. For the purpose of measurement dosimetry, there are three characteristics that determine the presence of small field conditions in external photon beams. These are: (i) loss of lateral CPE, (ii) partial occlusion of the primary photon source by the collimating devices, and (iii) size of the detector compared to the beam dimensions. The first two characteristics are beam related, while the third is detector related and depends on the relative size of the detector versus the size of the radiation field.

Loss of lateral CPE occurs in photon beams if the beam radius becomes smaller than the maximum range of secondary electrons. Monte Carlo calculations by Li et al. (1995) have aided in quantitatively establishing a criterion for small field conditions by determining the minimal radius of a circular photon field for which collision kerma, corrected for center of electron production, becomes equal to absorbed dose. The second characteristic is related to the finite size of the primary photon beam source, the extended focal spot, which is usually determined by the FWHM of the bremsstrahlung photon fluence distribution exiting the target. Shielding of the primary source will strongly affect the output of the beam. Monte Carlo calculations involving detailed modeling of the treatment head with special collimators confirm these characteristics (Paskalev et al., 2003; Sham et al., 2008). The third feature that characterizes a small field is the relative size of the detector in comparison to the size of the radiation field. Finite-size detectors will produce a signal

that is proportional to the *average* dose over the sensitive volume of the detector and this is affected by homogeneity (or lack thereof) of dose over the detection volume (volume averaging) in a small field; from this signal, a deconvolution process is required to derive the dose at a point. The ideal detector for small fields samples the dose at a point, is water-equivalent, and has a linear dose response, which is energy independent and dose rate independent. The minimal chamber size, however, is determined by technical considerations related to the magnitude of the extracameral signal as compared to the signal from the cavity. Since detector perturbations such as volume averaging and density effects are dramatically influenced by the presence of high gradients and charged particle disequilibrium, small field conditions are also assumed to exist when the sensitive volume of the detector is less than a lateral equilibrium length away from the edge of the field.

In a recent proposal by a consultancy group of the International Atomic Energy Agency (IAEA) in collaboration with the American Association of Physicists in Medicine (AAPM), Alfonso et al. (2008) proposed the concept of an intermediate machine-specific reference (*msr*) field to link reference dosimetry in small fields to the calibration chain in conventional reference fields. The reason for this proposal is that many of the specialized external beam machines are unable to meet field size and reference dosimetry setup requirements of traditional absorbed dose protocols and one could, therefore, not follow the reference dosimetry protocol recommendations of TG-51 or TRS-398, leading to clinical dosimetry uncertainties. Alfonso's proposal gives a framework for calibration in *msr* fields and output measurements in other fields relative to *msr* fields. The proposal also defines correction factors on output factor measurements taking into account the nature of the detector used for the output factor measurements. One can easily show that the ratio of correction factors for a pair of detectors of different type is equal to the ratio of the readings of that pair of detectors. Sets of ratios of these correction factors for distinct detector types can thus be extracted from measurements for varying field size. Monte Carlo calculations have been used extensively to estimate correction factors for reference dosimetry or to correct measured output factors in small field radiation therapy and can, thus, be used to verify the above-mentioned ratios. Two recent example studies that have followed the Alfonso nomenclature are the extensive small field dosimetry work by the Francescon group (Francescon et al., 2008, 2011) and Cranmer-Sargison et al. (2011). Francescon et al. (2011) performed a detailed analysis on the correction factors, $k_{Q_{clin},Q_{msr}}^{f_{clin},f_{msr}}$ for different small field detectors and field sizes on two accelerators. What makes the work specifically useful is that a sensitivity investigation of the effect of source parameters on the uncertainties is included. Similarly, Cranmer-Sargison et al. (2011) recently published values of $k_{Q_{clin},Q_{msr}}^{f_{clin},f_{msr}}$ for a set of diode detectors and different field sizes. They discuss in detail the Monte Carlo commissioning of the source and the coupling of this to the detailed modeling of the detectors and conclude that the work provides a basis for correcting measured output ratios for diodes.

4.6.2 Intensity-Modulated Beams

Intensity modulated radiation therapy (IMRT) photon beams, a second form of a nonstandard field, is another type of delivery not complying with absorbed dose reference dosimetry protocols. Earlier studies have demonstrated that additional correction factors may have to be considered for accurate IMRT field calibration (Bouchard and Seuntjens, 2004; Capote et al., 2004). While stopping-power ratios were demonstrated to have a small sensitivity to such configurations (Sánchez-Doblado et al., 2003), ionization chamber gradient effects were shown to be the major factor responsible for the significance of these corrections (Bouchard et al., 2009). In a second part of the Alfonso et al. (2008) proposal, the term *plan class specific reference field (pcsr)* is coined to define a calibration field representative of a modulated delivery for a class of IMRT treatments. The formalism defines the *pcsr* correction factor to correct the calibration coefficient obtained in conventional reference conditions for conditions occurring in dynamic nonstandard fields. Alfonso et al. (2008) recommend that the factors are calculated with Monte Carlo simulation alone or measured using a suitable detector applying corrections obtained from a Monte Carlo simulation. With tools such as EGSnrc (including BEAMnrc (Rogers et al., 2009)) or PENELOPE, detailed models of modulated fields by dynamic multileaf collimators and detector correction factors have already been established for composite deliveries using step-and-shoot (Sánchez-Doblado et al., 2005) and dynamic techniques (Chung et al., 2010; De Ost et al., 2011) and will allow detailed quantification of *pcsr* correction factors under different delivery scenarios. One of the issues with the implementation of the *pcsr* concept is the criterion imposed on such fields with respect to reproducibility, modulation, and homogeneity of the dose distribution. Monte Carlo calculations and measurements such as by Chung et al. (2012) will help establish suitable criteria for defining and clinically implementing these fields.

4.7 Conclusions

Monte Carlo calculations have been part of measurement dosimetry developments since the late 1970s with applications ranging from stopping-power ratios, wall correction factors to absolute calculation of detector response from blueprints of radiation source and detector. The accuracy requirements imposed by these stringent detector dosimetry applications have promoted the development and self-consistency of the modern CH codes. In parallel, VRTs, new flexible geometry packages, and the development in computing capabilities now enable the calculation of detector response in extreme detail. One can arguably state that the developments of accurate CH Monte Carlo codes driven by the need for more accurate detector response have also been key in the development and success of other applications in medical physics such as Monte Carlo treatment planning dose calculations.

The developments in Monte Carlo code accuracy and speed have also opened the door to new applications including even more detailed simulations of detectors, including details such as electric field in ionization chamber dosimetry, the *in silico* design of novel detectors with superior dosimetric characteristics, or the on-the-fly Monte Carlo simulation of detector response as part of a Monte Carlo treatment planning system commissioning, to name a few. Finally, Monte Carlo calculations of ionization chambers have enabled complex studies of modulated field dosimetry, which will also lead to more accurate, better tuned, and conceptually simpler clinical reference dosimetry protocols.

References

AAPM Task Group 21 Radiation Therapy Committee 1983. A protocol for the determination of absorbed dose from high-energy photon and electron beams. *Med Phys* 10: 741–771.

AAPM Task Group 25 Radiation Therapy Committee 1991. Clinical electron-beam dosimetry. *Med Phys* 18: 73–107.

Abdel-Rahman, W., Seuntjens, J. P., Verhaegen, F., and Podgorsak, E. B. 2006. Radiation induced currents in parallel plate ionization chambers: Measurement and Monte Carlo simulation for megavoltage photon and electron beams. *Med Phys* 33: 3094–3104.

Alfonso, R., Andreo, P., Capote, R. et al. 2008. A new formalism for reference dosimetry of small and nonstandard fields. *Med Phys* 35: 5179–5186.

Almond, P. R. and Svensson, H. 1977. Ionization chamber dosimetry for photon and electron beams. *Acta Radiol Ther Phys Biol* 16: 177–186.

Almond, P. R., Biggs, P. J., Coursey, B. M., Hanson, W. F., Huq, M. S., Nath, R., and Rogers, D. W. O. 1999. AAPM's TG-51 protocol for clinical reference dosimetry of high-energy photon and electron beams. *Med Phys* 26: 1847–1869.

Andreo, P. and Brahme, A. 1986. Stopping power data for high-energy photon beams. *Phys Med Biol* 31: 839–858.

Andreo, P. and Nahum, A. E. 1985. Stopping-power ratio for a photon spectrum as a weighted sum of the values for monoenergetic photon beams. *Phys Med Biol* 30: 1055–1065.

Andreo, P., Medin, J., and Bielajew, A. F. 1993. Constraints of the multiple-scattering theory of Molière in Monte Carlo simulations of the transport of charged particles. *Med Phys* 20: 1315–1325.

Baro, J., Sempau, J., Fernandez-Varea, J. M., and Salvat, F. 1995. PENELOPE: An algorithm for Monte Carlo simulation of penetration and energy loss of electrons and positrons in matter. *Nucl Inst Meth B* 100: 31–46.

Berger, M. J. 1963. Monte Carlo calculation of the penetration and diffusion of fast charged particles. In: *Methods of Computational Physics Vol 1*, B. Alder, S. Fernbach, and M. Rotenberg (eds), Academic Press, New York, pp. 135–215.

Bichsel, H. and Hiraoka, T. 1992. Energy loss of 70MeV protons in elements. *Nucl Instrum Methods Phys Res B* 66: 345–351.

Bielajew, A. F., Rogers, D. W. O., and Nahum, A. E. 1985. The Monte Carlo simulation of ion chamber response to 60-Co-resolution of anomalies associated with interfaces. *Phys Med Biol* 30: 419–427.

Bielajew, A. F. 1986. Ionisation cavity theory: A formal derivation of perturbation factors for thick-walled ion chambers in photon beams. *Phys Med Biol* 31: 161–170.

Bielajew, A. F. and Rogers, D. W. O. 1987. PRESTA: The parameter reduced electron-step transport algorithm for electron Monte Carlo transport. *Nucl Inst Meth B* 18: 165–171.

Bielajew, A. F. 1994. Plural and multiple small-angle scattering from a screened Rutherford cross section. *Nucl Inst Meth B* 86: 257–269.

Bjärngard, B. E. and Kase, K. R. 1985. Replacement correction factors for photon and electron dose measurements. *Med Phys* 12: 785–787.

Bote, D. and Salvat, F. 2008. Calculations of inner-shell ionization by electron impact with the distorted-wave and plane-wave Born approximations. *Phys Rev A* 77 042701: 1–24.

Bouchard, H. and Seuntjens, J. 2004. Ionization chamber-based reference dosimetry of intensity modulated radiation beams. *Med Phys* 31: 2454–2465.

Bouchard, H., Seuntjens, J., Carrier, J., and Kawrakow, I. 2009. Ionization chamber gradient effects in nonstandard beam configurations. *Med Phys* 36: 4654–4663.

Bouchard, H., Seuntjens, J., and Kawrakow, I. 2011. A Monte Carlo method to evaluate the impact of positioning errors on detector response and quality correction factors in nonstandard beams. *Phys Med Biol* 56: 2617–2634.

Bouchard, H., Seuntjens, J., and Palmans, H. 2012. On charged particle equilibrium violation in external photon fields. *Med Phys* 39: 1473–1480.

Bouchard, H. 2012. A theoretical re-examination of Spencer Attix cavity theory. *Phys Med Biol* 57: 3333–3358.

Boutillon, M. and Perroche-Roux, A. M. 1987. Re-evaluation of the W value for electrons in dry air. *Phys Med Biol* 32: 213–219.

Bragg, W. H. 1912. *Studies in Radioactivity.* First edition. Macmillan and Co., Ltd., London, pp. 91–99, 161–169.

Buckley L. A., Kawrakow I., and Rogers, D. W. O. 2004. CSnrc: Correlated sampling Monte Carlo calculations using EGSnrc. *Med Phys* 31: 3425–3435.

Buckley, L. A. and Rogers, D. W. O. 2006a. Wall correction factors, Pwall, for thimble ionization chambers. *Med Phys* 33: 455–464.

Buckley, L. A. and Rogers, D. W. O. 2006b. Wall correction factors, Pwall, for parallel-plate ionization chambers. *Med Phys* 33: 1788–1796.

Büermann, L., Kramer, H-M., and Csete, I. 2003. Results supporting calculated wall correction factors for cavity chambers. *Phys Med Biol* 48: 3581–3594.

Büermann, L. and Hilgers, G. 2007. Significant discrepancies in air kerma rates measured with free-air and cavity ionization chambers. *Nucl Instrum Methods* A 580: 477–480.

Büermann, L., Gargioni, E., Hilgers, G., and Krauss, A. 2007. Comparison of ionometric and calorimetric determination of absorbed dose to water for cobalt-60 gamma rays. Workshop on Absorbed Dose and Air Kerma Primary Standards (LNE, Paris, May 9–11, 2007), Available online at http://www.nucleide.org/ADAKPSWS/Presentations2007.htm.

Burch, P. R. J. 1955. Cavity ion chamber theory. *Radiat Res* 3: 361–378.

Burch, P. R. J. 1957. Comment on recent cavity ionization theories. *Radiat Res* 6: 79–84.

Burlin, T. E. 1966. A general theory of cavity ionisation. *Br J Radiol* 39: 727–734.

Burns, D. T. 2006. A new approach to the determination of air kerma using primary-standard cavity ionization chambers. *Phys Med Biol* 51: 929–942.

Burns, D. T. and Büermann, L. 2009. Free air ionization chambers. *Metrologia* 46: S9–S23.

Burns, D. T., Duane, S., and McEwen, M. R. 1995. A new method to determine ratios of electron stopping powers to an improved accuracy. *Phys Med Biol* 40: 733–739.

Capote, R., Sánchez-Doblado, F., Leal, A., Lagares, J. I., Arráns, R., and Hartmann, G. H. 2004. An EGSnrc Monte Carlo study of the microionization chamber for reference dosimetry of narrow irregular IMRT beamlets. *Med Phys* 31: 2416–2422.

Chung, E., Bouchard, H., and Seuntjens, J. 2010. Investigation of three radiation detectors for accurate measurement of absorbed dose in nonstandard fields. *Med Phys* 37: 2402–2413.

Chung, E., Soisson, E., and Seuntjens, J. 2012. Dose homogeneity specification for reference dosimetry of nonstandard fields. *Med Phys* 39: 407–414.

Cranmer-Sargison, G., Weston, S., Evans, J. A., Sidhu, N. P., and Thwaites, D. I. 2011. Implementing a newly proposed Monte Carlo based small field dosimetry formalism for a comprehensive set of diode detectors. *Med Phys* 38: 6592–6602.

Crop, F., Reynaert, N., Pittomvils, G., Paelinck, L., De Wagter, C., Vakaet, L., and Thierens, H. 2009. The influence of small field sizes, penumbra, spot size and measurement depth on perturbation factors for microionization chambers. *Phys Med Biol* 54: 2951–2969.

Cunningham, J. R., Woo, M., Rogers, D. W. O., and Bielajew, A. F. 1986. The dependence of mass energy absorption coefficient ratios on beam size and depth in a phantom. *Med Phys* 13: 496–502.

De Ost, B., Schaeken, B., Vynckier, S., Sterpin, E., and Van den Weyngaert, D. 2011. Reference dosimetry for helical tomotherapy: Practical implementation and a multicenter validation. *Med Phys* 38: 6021–6026.

Faddegon, B., Schreiber, E., and Ding, X. 2005. Monte Carlo simulation of large electron fields, *Phys Med Biol* 50: 741–753.

Faddegon, B. A., Sawkey, D., O'Shea, T., McEwen, M., and Ross, C. 2009. Treatment head disassembly to improve the accuracy of large electron field simulation. *Med Phys* 36: 4577–4591.

Fano, U. 1954. Note on the Bragg–Gray cavity principle for measuring energy dissipation. *Rad Res* 1: 237–240.

Francescon, P., Cora, S., and Cavedon, C. 2008. Total scatter factors of small beams: A multidetector and Monte Carlo study. *Med Phys* 35:504–513.

Francescon, P., Cora, S., and Satariano, N. 2011. Calculation of k(Q(clin),Q(msr)) (f(clin),f(msr)) for several small detectors and for two linear accelerators using Monte Carlo simulations. *Med Phys* 38: 6513–6527.

Furhang, E. E., Chui, C. S., and Lovelock, M. 1995. Mean mass energy absorption coefficient ratios for megavoltage x-ray beams. *Med Phys* 22: 525–530.

Gray, L. H. 1929. The absorption of penetrating radiation. *Proc Roy Soc* A 122: 647–668.

Gray, L. H. 1936. An ionization method for the absolute measurement of gamma ray energy. *Proc R Soc Lond A* 156: 578–596.

Hubbell, J. H. 1982. Photon mass attenuation and energy-absorption coefficients. *Int J Appl Radiat Isot* 33: 1269–1290.

IAEA 1987. Absorbed Dose Determination in Photon and Electron Beams: An International Code of Practice. Volume 277 of Technical Report Series. IAEA, Vienna.

IAEA 2001. Absorbed Dose Determination in External Beam Radiotherapy: An International Code of Practice for Dosimetry Based on Standards of Absorbed Dose to Water, Volume 398 of Technical Report Series. IAEA, Vienna.

ICRU 1969. Radiation Dosimetry: X-Rays and Gamma Rays with Maximum Photon Energies between 0.6 and 50 MeV. Technical Report no. 14, International Commission on Radiation Units and Measurements, Washington D.C.

ICRU 1972. Radiation Dosimetry: Electrons with Initial Energies between 1 and 50 MeV. Technical Report no. 21, International Commission on Radiation Units and Measurements, Washington D.C.

ICRU 1984. Radiation Dosimetry: Electron Beams with Energies between 1 and 50 MeV. Technical Report no. 35, International Commission on Radiation Units and Measurements, Washington D.C.

ICRU 2007. Radiation Dosimetry: Stopping Powers for Electrons and Positions. Report no. 37, International Commission on Radiation Units and Measurements, Washington D.C.

ISO 1995. Guide to Expression of Uncertainty in Measurement. Technical Report, International Organization on Standardization, Geneva, Switzerland.

Janssens, A., Eggermont, G., Jacobs R., and Thielens, G. 1974. Spectrum perturbation and energy deposition models for stopping power ratio calculations in general cavity theory. *Phys Med Biol* 19: 619–630.

Janssens, A. 1981. Modified energy-deposition model, for the computation of the stopping-power ratio for small cavity sizes. *Phys Rev A* 23: 1164–1173.

Johns, H. E. and Cunningham, J. R. 1983. *The Physics of Radiology*, 4th ed., Charles C. Thomas, Springfield, IL.

Kawrakow, I. 2006. On the effective point of measurement in megavoltage photon beams. *Med Phys* 33: 1829–1839.

Kawrakow, I. and Bielajew, A. F. 1998a. On the representation of electron multiple elastic-scattering distributions for Monte Carlo calculations. *Nucl Inst Meth B* 134: 325–336.

Kawrakow I. and Bielajew A. F. 1998b. On the condensed history technique for electron transport. *Nucl Inst Meth B* 142: 253–280.

Kawrakow, I. 2000a. Accurate condensed history Monte Carlo simulation of electron transport. I: EGSnrc, the new EGS4 version. *Med Phys* 27: 485–498.

Kawrakow, I. 2000b. Accurate condensed history Monte Carlo simulation of electron transport. II: Application to ion chamber response simulations. *Med Phys* 27: 499–513.

Kawrakow, I. 2005. EGSpp: The EGSnrc C ++ Class Library. Report PIRS-899. National Research Council, Ottawa, Canada.

Kawrakow, I. 2006. On the effective point of measurement in megavoltage photon beams. *Med Phys* 33: 1829–1839.

Kawrakow, I. 2011. Beam modeling. Paper presented at the International Workshop on Recent Advances in Monte Carlo Techniques for Radiation Therapy, Montréal, Québec, Canada, June 8–10, 2011.

Klassen, N. V., Shortt, K. R., Seuntjens, J., and Ross, C. K. 1999. Fricke dosimetry: The difference between G(Fe^{3+}) for ^{60}Co γ-rays and high-energy x-rays. *Phys Med Biol* 44: 1609–1624.

Laurence, G. C. 1937. The measurement of extra hard X-rays and gamma-rays in roentgens. *Can J Res A* 15: 67–78.

Li, X. A., Soubra, M., Szanto, J., and Gerig, L. H. 1995. Lateral electron equilibrium and electron contamination in measurements of head-scatter factors using mini phantoms and brass caps. *Med Phys* 22:1167–1170.

Ma, C. M. and Nahum, A. E. 1993. Calculation of absorbed dose ratios using correlated Monte Carlo sampling. *Med Phys* 20: 1189–1199.

Ma, C. M. and Seuntjens, J. P. 1999. Mass energy-absorption coefficient and backscatter factor ratios for kilovoltage x-ray beams. *Phys Med Biol* 44: 131–143.

Ma, C. M., Coffey, C. W., DeWerd, L. A., Liu, C., Nath, R., Seltzer, S. M., and Seuntjens J. P. 1999. AAPM protocol for 40–300 kV x-ray beam dosimetry in radiotherapy and radiobiology. *Med Phys* 28: 868–893.

Mainegra-Hing, E., Reynaert, N., and Kawrakow, I. 2008. Novel approach for the Monte Carlo calculation of free-air chamber correction factors. *Med Phys* 35: 3650–3660.

McCaffrey, J. P., Mainegra-Hing, E., Kawrakow, I., Shortt, K. R., and Rogers, D. W. O. 2004. Evidence for using Monte Carlo calculated wall attenuation and scatter correction factors for three styles of graphite-walled ion chamber. *Phys Med Biol* 49: 2491–2501.

McEwen, M. R., Kawrakow, I., and Ross, C. K. 2008. The effective point of measurement of ionization chambers and the build-up anomaly in MV x-ray beams. *Med Phys* 35: 950–958.

McGinnies, R. T. 1959. Energy Spectrum Resulting from Electron Slowing Down, NBS Circular 597.

Muir, B. R. and Rogers, D. W. O. 2010. Monte Carlo calculations of k$_Q$, the beam quality conversion factor. *Med Phys* 37: 5939–5950.

Muir, B. R., McEwen, M. R., and Rogers, D. W. O. 2011. Measured and Monte Carlo calculated k$_Q$ factors: Accuracy and comparison. *Med Phys* 38: 4600–4609.

Nahum, A. E. 1978. Water/air mass stopping power ratios for megavoltage photons and electrons beams. *Phys Med Biol* 23: 24–38.

Nelson, W. R., Hirayama, H., and Rogers D. W. O. 1985. The EGS4 Code System. Report no. SLAC-265, National Research Council, Ottawa, Canada.

Paskalev, K. A., Seuntjens, J. P., Patrocinio, H. J., and Podgorsak, E. B. 2003. Physical aspects of dynamic stereotactic radiosurgery with very small photon beams (1.5 and 3 mm in diameter). *Med Phys* 30: 111–118.

Poon, E., Seuntjens, J., and Verhaegen, F. 2005. Consistency test of the electron transport algorithm in the GEANT4 Monte Carlo code. *Phys Med Biol* 50:681–694.

Rogers, D. W. O. 1993. How accurately can EGS4/PRESTA calculate ion-chamber response? *Med Phys* 20: 319–323.

Rogers, D. W. O. 2009. General characteristics of radiation dosimeters and a terminology to describe them. In: *Clinical Dosimetry Measurements in Radiotherapy*, D. W. O. Rogers and J. E. Cygler (eds), Medical Physics Publishing, Madison, pp. 137–145.

Rogers, D. W. O., Walters, B. R. B., and Kawrakow I. 2009. BEAMnrc Users Manual. Report no. PIRS-0509a, National Research Council, Ottawa, Canada.

Rogers, D. W. O., Kawrakow, I., Seuntjens, J. P., Walters, B.R.B., and Mainegra-Hing, E. 2011. NRC User Codes for EGSnrc. Report PIRS-702. National Research Council, Ottawa, Canada, pp. 69–73.

Ross, C. K. and Klassen, N. V. 1996. Water calorimetry for radiation dosimetry. *Phys Med Biol* 41: 1–29.

Salvat, F., Fernández-Varea, J. M., and Sempau, J. 2009. PENELOPE-2008: A Code System for Monte Carlo Simulation of Electron and Photon Transport. Workshop Proceedings, Barcelona, Spain 30 June–3 July 2008, Nuclear Energy Agency no. 6416, OECD, Paris, France.

Sempau, J., Acosta, E., Baro, J., Fernandez-Varea, J. M., and Salvat, F. 1997. An algorithm for Monte Carlo simulation of coupled electron-photon transport. *Nucl Inst Meth B* 132: 377–390.

Sánchez-Doblado, F., Andreo, P., Capote, R. et al. 2003. Ionization chamber dosimetry of small photon fields: A Monte Carlo study on stopping-power ratios for radiosurgery and IMRT beams. *Phys Med Biol* 48: 2081–2099.

Sánchez-Doblado, F., Capote, R., Rosello, J. V. et al. 2005. Micro ionization chamber dosimetry in IMRT verification: Clinical implications of dosimetric errors in the PTV. *Radiother Oncol* 75: 342–348.

Sawkey, D. and Faddegon, B. A. 2009. Simulation of large x-ray fields using independently measured source and geometry details. *Med Phys* 36: 5622–5632.

Seltzer, S. M. 1993. Calculation of photon mass energy-transfer and mass energy-absorption coefficients. *Rad Res* 136: 147–170.

Sempau, J. and Andreo, P. 2006. Configuration of the electron transport algorithm of PENELOPE to simulate ion chambers. *Phys Med Biol* 51: 3533–3548.

Seuntjens, J., Thierens, H., Van der Plaetsen, A., and Segaert, O. 1987. Conversion factor f for X-ray beam qualities, specified by peak tube potential and HVL value. *Phys Med Biol* 32: 595–603.

Seuntjens, J., Thierens, H., Van der Plaetsen, A., and Segaert, O. 1988. Determination of absorbed dose to water with ionisation chambers calibrated in free air for medium-energy x-rays. *Phys Med Biol* 33: 1171–1185.

Seuntjens, J., Kawrakow, I., Borg, J., Hobeila, F., and Rogers, D. W. O. 2002. Calculated and measured air-kerma response of ionization chambers in low and medium energy photon beams in recent developments in accurate radiation dosimetry. In: *Proceedings of an International Workshop*, J. P. Seuntjens and P. Mobit (eds), Medical Physics Publishing, Madison, WI, pp. 69–84.

Seuntjens, J. P. and DuSautoy, A. R. 2003. Review of calorimeter based absorbed dose to water standards. Standards and Codes of Practice in Medical Radiation Dosimetry (Proc. Int. Symp. Vienna, 2002), IAEA, Vienna. 1: 37–66.

Seuntjens, J. and Duane, S. 2009. Photon absorbed dose standards. *Metrologia* 46: S39–S58.

Sham, E., Seuntjens, J., Devic, S., and Podgorsak, E. B. 2008. Influence of focal spot on characteristics of very small diameter radiosurgical beams. *Med Phys* 35: 3317–3330.

Sheikh-Bagheri, D. and Rogers, D. W. O. 2002. Monte Carlo calculation of nine megavoltage photon beam spectra using the BEAM code. *Med Phys* 29: 391–402.

Spencer, L. V. and Attix, F. H. 1955. A theory of cavity ionization. *Radiat Res* 3: 239–254.

Tessier, F. and Kawrakow, I. 2010. Effective point of measurement of thimble ion chambers in megavoltage photon beams. *Med Phys* 37: 96–107.

Thomson, R. M. and Rogers, D. W. O. 2010. Re-evaluation of the product of (W/e) air and the graphite to air stopping-power ratio for ^{60}Co air kerma standards. *Phys Med Biol* 55: 3577–3595.

Thomson, R. M. and Kawrakow, I. 2011. On the Monte Carlo simulation of electron transport in the sub-1 keV energy range. *Med Phys* 38: 4531–4534.

Ubrich, F., Wulff, J., Kranzer, R., and Zink, K. 2008. Thimble ionization chambers in medium-energy x-ray beams and the role of constructive details of the central electrode: Monte Carlo simulations and measurements. *Phys Med Biol* 53: 4893–4906.

Verhaegen, F., Zakikhani, R., Dusautoy, A. et al. 2006. Perturbation correction factors for the NACP-02 plane-parallel ionization chamber in water in high-energy electron beams. *Phys Med Biol* 51:1221–1235.

Wang, L. L. W. and Rogers, D. W. O. 2007. Monte Carlo study of Si diode response in electron beams. *Med Phys* 34: 1734–1742.

Wang, L. L. W. and Rogers, D. W. O. 2008. Calculation of the replacement correction factors for ion chambers in megavoltage beams by Monte Carlo simulation. *Med Phys* 35: 1747–1755.

Wang, L. L. W. and Rogers, D. W. O. 2009a. The replacement correction factors for cylindrical chambers in high-energy photon beams. *Phys Med Biol* 54:1609–1620.

Wang, L. L. W. and Rogers, D. W. O. 2009b. Study of the effective point of measurement for ion chambers in electron beams by Monte Carlo simulation. *Med Phys* 36: 2034–2042.

Wang, L. L. W. and Rogers, D. W. O. 2009c. Replacement correction factors for cylindrical ion chambers in electron beams. *Med Phys* 36: 4600–4608.

Wulff, J., Heverhagen, J. T., and Zink, K. 2008a. Monte-Carlo-based perturbation and beam quality correction factors for thimble ionization chambers in high-energy photon beams. *Phys Med Biol* 53: 2823–2836.

Wulff, J., Zink, K., and Kawrakow, I. 2008b. Efficiency improvements for ion chamber calculations in high energy photon beams. *Med Phys* 35: 1328–1336.

Wulff, J., Heverhagen, J. T., Zink, K., and Kawrakow, I. 2010. Investigation of systematic uncertainties in Monte Carlo-calculated beam quality correction factors. *Phys Med Biol* 55: 4481–4493.

Yi, C. Y., Hah, S. H., and Yeom, M. S. 2006. Monte Carlo calculation of the ionization chamber response to ^{60}Co beam using PENELOPE. *Med Phys* 33:1213–1221.

Zheng-Ming, L. 1980. An electron transport theory of cavity ionization. *Rad Res* 84: 1–15.

Zink, K. and Wulff, J. 2008. Monte Carlo calculations of beam quality correction factors k_Q for electron dosimetry with a parallel-plate Roos chamber. *Phys Med Biol* 53: 1595–1607.

Zink, K. and Wulff, J. 2009. Positioning of a plane-parallel ionization chamber in clinical electron beams and the impact on perturbation factors. *Phys Med Biol* 54: 2421–2435.

Zink, K. and Wulff, J. 2011. On the wall perturbation correction for a parallel-plate NACP-02 chamber in clinical electron beams. *Med Phys* 38: 1045–1054.

5

Monte Carlo Modeling of External Photon Beams in Radiotherapy

Frank Verhaegen
Maastro Clinic

5.1 Introduction

It is well accepted that Monte Carlo simulation methods offer the most powerful tool for modeling and analyzing radiation transport for radiotherapy applications. One of the most frequent and important uses of Monte Carlo modeling in external beam radiotherapy is the creation of a virtual model of the radiation source. Applications are source design and optimization, studying radiation detector response, and treatment planning. Since the vast majority of radiotherapy is performed with megavolt (MV) photon beams from a linear accelerator (linac), this is also reflected in the research efforts. To a lesser extent, external beam radiotherapy is performed with electron beams, lower-energy photon beams emanating from an x-ray tube, or hadron beams. This chapter will cover the Monte Carlo modeling of photon beams and Chapter 6 will cover the Monte Carlo modeling of electron beams, while other chapters will deal with hadron beams.

Historically, Monte Carlo methods provided a simple alternative to measurements to derive photon spectra from radiation sources (McCall et al., 1978; Patau et al., 1978; Faddegon et al., 1990, 1991). Monte Carlo techniques were used in the early days of linac design in the 1970s to aid in the optimization of the photon beams (McCall et al., 1978). The first simplified models of photon beams from linacs were reported from the late 1970s till the early 1990s (Patau et al., 1978; Nilsson and Brahme, 1981; Mohan et al., 1985). For linac electron beams, an important early effort was the work by Udale (1988, 1992) in the late 1980s. They introduced the modular approach for Monte Carlo modeling of electron beams, which is nowadays very popular since the advent of the BEAM Monte Carlo user interface (Rogers et al., 1995). Extensive literature reviews of Monte Carlo modeling of electron beams were provided by Ma and Jiang (1999) and of photon beams by Verhaegen and Seuntjens (2003). They cover many details and are in many respects still current.

In this chapter, we will provide an overview of the work performed in the field of Monte Carlo modeling of clinical photon beams over the last 30 years or so. We will discuss what is required to derive a Monte Carlo model of such beams, and show some applications. Arguably the most important clinical application, radiotherapy treatment planning, will be covered in other chapters. Monte Carlo models of radiation sources discussed in this chapter will start from the primary electron beam and, hence, we do not cover the extensive work done in the area of accelerator design on, for example, the beam generation and transport in a linac flight tube.

5.2 Photon Beams from Clinical Linacs

Nowadays, clinically used photon beams in radiotherapy usually are within the energy range of 4–25 MeV. A linac has an essential modular construction as shown generically in Figure 5.1. Different manufacturers may have the components of a linac in a different order but the ones indicated in Figure 5.1 are usually those that are required in a Monte Carlo model. For an overview of linac technology, the reader is referred to several excellent texts available (Greene and Williams, 1997; Van Dyk, 2005). The Monte Carlo codes employed in this chapter are mostly EGS4 (Nelson et al., 1985) or EGSnrc (Kawrakow, 2000) (often in conjunction with the user interface BEAM (Rogers et al., 1995)), GEANT4 (Agostinelli et al., 2003), MCNP (Briesmeister, 2000), and PENELOPE (Sempau et al., 1997). They are all discussed in other chapters. Techniques to speed up simulations, variance reduction techniques, are also discussed elsewhere in this volume. They are heavily relied upon in linac simulations, which can be prohibitively long, even with today's powerful computers. A discussion of variance reduction techniques as they pertain to photon linac modeling can be found in the literature (Verhaegen and Seuntjens, 2003).

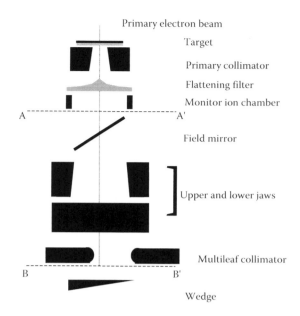

Primary electron beam

Target

Primary collimator

Flattening filter

Monitor ion chamber

A ---------------------------- A'

Field mirror

Upper and lower jaws

Multileaf collimator

B ---------------------------- B'

Wedge

FIGURE 5.1 Schematic representation of linac components in a Monte Carlo model of a photon linac. Different linac manufacturers may have the linac components in a different order.

5.2.1 Components of a Monte Carlo Model of a Linac Photon Beam

The components depicted in Figure 5.1 must be known in detail to build a faithful model of a linac. The composition of materials and alloys; their mass densities; the position, dimensions, and shape of defining surfaces of the components; and their motion must all be known in great detail to build an accurate Monte Carlo model. Knowledge of tolerances may help to determine uncertainties in the calculated results. While seemingly trivial, this is a task that has daunted many workers. It usually requires interaction with linac manufacturers to disclose constructional

details. Needless to say, errors made at this stage will often translate into systematic errors in the calculated output of the linac. Therefore, it is of utmost importance that linac blueprints are verified as much as possible and that Monte Carlo linac models are validated against an extensive set of dose measurements of, as a minimum, depth and lateral dose profiles in water, and output factor in water (ratio of dose at a depth in water at the central axis for a certain field size divided by the same for a reference field size). The validation procedure may comprise comparisons against measurements at various source-to-surface distances of a phantom, should treatment planning be the main application. The literature contains many examples of extensive comparisons of Monte Carlo linac models against measurements (Chaney et al., 1994; Mazurier et al., 1999; Deng et al., 2000; Hartmann Siantar et al., 2001; Ding, 2002a).

In 1995, the BEAM/EGS4 user interface was released (Rogers et al., 1995), since then upgraded to BEAM/EGSnrc (or BEAMnrc). This user interface allows easy modeling of radiotherapy linacs and led to a spate of papers presenting photon and electron beam models. It was, and still is, seen as a major step forward in the field. It is currently the most widely used linac simulation package. Figure 5.2 shows a typical example of a simple photon linac model where linac components and particle tracks can be discerned (Figure 5.2a) and where interactions in the flattening filter can be observed in detail (Figure 5.2b). The BEAM code allows easy assembly of a linac model (it also enables building models for x-ray tubes and a few other geometries) with a wide choice of building blocks consisting of geometrical shapes such as disks, cones, parallelepipeds, trapezoids, and so on. The code relies heavily on the fact that linac components do not overlap in the beam direction, which is mostly the case. Separate parts of a linac can, therefore, be built, validated, and studied separately. In addition to linac components that may be defined with any material (the BEAM package provides an extensive cross-section dataset), a wide choice of primary source

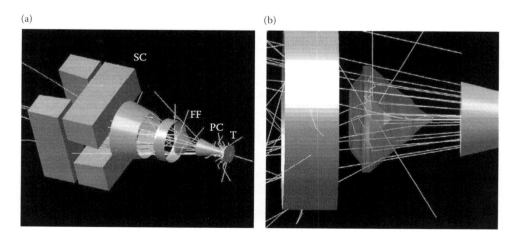

(a)   (b)

FIGURE 5.2 **(See color insert.)** (a) Simplified BEAM Monte Carlo model of an 18 MV linac photon beam. From right to left, we encounter the primary electron beam hitting the target (T), the primary collimator (PC), the flattening filter (FF), and some secondary collimators (SC). Photon tracks are shown in yellow, electron tracks in blue, and positron tracks in red. In (b), a close-up of the particle interactions in the FF is shown in exquisite detail. By depicting the FF in a semitransparent fashion, the details of the geometry (notice the cone inside a cone structure) and the interactions are revealed.

geometries is available. Particle transport can be done for electrons, photons, and positrons. Particles can be tagged according to interaction types, interaction sites, and so on, which provide a powerful beam analysis method.

Other user interfaces similarly intended to enable easy assembly of complex geometries have been introduced, but these have yet to equal the success of BEAM. An example is PENLINAC for the PENELOPE Monte Carlo code (Rodriguez, 2008). The GATE user interface (Jan et al., 2004) for the GEANT4 Monte Carlo code also seems to hold some promise.

Around the same time BEAM was first released, Lovelock et al. (1995) also presented a modular Monte Carlo approach to simulate linac photon beams. Both are predated by papers by Udale (1988, 1992) who introduced modular Monte Carlo models for electron beams. In Lovelock et al. (1995), the radius of a uniform primary electron beam hitting the target and its energy distribution (a truncated Gaussian) could be taken into account. Information on particles crossing a predefined plane could be stored in a phase-space file. Simulations were done in stages, that is, phase-space information from one stage was used as input source for a subsequent stage. Particles could be tagged according to different types of interactions they underwent. They noticed that calculated depth dose profiles in water are relatively insensitive to the primary electron energy, whereas the horns in the lateral dose profiles are a good indicator for the primary electron energy. A lower primary electron energy will result in more pronounced horns. As such, these authors were among the first to introduce a tuning procedure of the Monte Carlo primary electron source characteristics to derive the primary electron energy. This is often needed in case the manufacturer does not provide the required information or where simulations of the electron trajectories in the flight tube are not available. This information is rarely at hand.

The photon linac components that are the most important in the model in Figure 5.1 are the target, flattening filter, secondary collimators such as jaws and multileaf collimator (T, FF, SC), and wedges. After the accelerated primary electron beam exits the flight tube, it possesses a narrow-energy, angular and spatial distribution. This electron beam will hit the target, commonly consisting of a high-Z metal in which the electrons will produce bremsstrahlung photons. It has to be pointed out that in the Monte Carlo models of clinical linacs, usually no modeling is done of the electron beam, prior to exiting the flight tube. This will inevitably entail making assumptions about the primary electron beam, which will be discussed later. The bremsstrahlung photons are then collimated initially by a primary collimator and the photon fluence is differentially attenuated by the flattening filter to produce a reasonably flat dose distribution in water at a certain depth. Target and flattening filters are the most important sources of contaminant electrons, unless a wedge is present. In the generic model of Figure 5.1, the next components to be considered are the monitor ion chamber and the field mirror. Both present only a small attenuation to the photon beam and are often omitted from Monte Carlo models. However, when backscatter to the monitor chamber is the subject of study

(Section 5.2.1.3), some model of the ion chamber should be present. The photon beam is then finally shaped and modulated by secondary collimators and beam modifiers such as jaws, blocks, multileaf collimators (MLC), and wedges. In the following sections, Monte Carlo modeling of the various linac components will be discussed in more detail.

5.2.1.1 Primary Electron Beam Distribution and Photon Target

5.2.1.1.1 *Photon Target Simulation for Radiotherapy Treatment*

The bremsstrahlung photons produced in the target by the primary electrons in the energy range 4–25 MeV are mostly derived from a relatively thin layer on the upstream side. This is because of the quick energy degradation of the electron energy in a high-Z material, the almost linear dependence of the bremsstrahlung cross section on the electron kinetic energy for high energies and electron scattering in the target. At high electron energies, the average bremsstrahlung photon emission angle is given approximately by m_0c^2/E_0 (m_0c^2 is the electron's rest energy; E_0 its total energy) yielding a strongly forward-peaked angular distribution. Targets in clinical linacs are usually thick enough to stop the primary electrons completely and are for that reason often referred to as "thick targets." In such targets, the bremsstrahlung photon angular distribution will be spread out because of electrons undergoing multiple scattering, but the result will nevertheless be a strongly anisotropic photon fluence, while the photon spectrum is fairly isotropic. Owing to this complex picture, it is very difficult to obtain the spectral and angular bremsstrahlung photon distribution from thick targets using analytical methods such as the Schiff theory (Koch and Motz, 1959; Desobry and Boyer, 1991). More complex analytical calculation schemes have been developed but the most comprehensive method to generate bremsstrahlung photon distributions from targets in clinical linacs is Monte Carlo simulation. In what follows, we will discuss a few studies related to this topic.

Patau et al. (1978) were among the first to present a simple Monte Carlo model, using an in-house code, of a complete photon beam linac. It consisted of a 5.7 MeV electron pencil beam hitting a W/Cu target, followed by a flattening filter of Pb and an unspecified collimator. The authors calculated photon energy fluences, photon transmission through diverse materials, and electron spectra in water. McCall et al. (1978) were among the first to attempt to improve the design of a linac by performing Monte Carlo simulations. They used the EGS3 code (Ford and Nelson, 1978) for the calculation of bremsstrahlung spectra and for the generation of secondary electrons in water. For several combinations of target materials and flattening filters, they noticed a linear correlation between the depth of the dose maximum in water and the average energy of the photon spectrum exiting the linac for 10–25 MV photon beams. They concluded that Monte Carlo simulations are a powerful tool for designing target and flattening filter.

Starting in the late 1980s, workers from the National Research Council of Canada (NRCC) in Ottawa devoted several studies to Monte Carlo modeling of bremsstrahlung production in linac targets. Bielajew et al. (1989) implemented bremsstrahlung angular sampling from the Koch and Motz distribution (1959) in EGS4. The Koch and Motz (KM) angular distribution was found to lead to a significant difference in the degree of self-absorption in the target compared to using a fixed angle of m_0c^2/E_0 (a common approach in many Monte Carlo codes, at one time). This work was complemented by several careful experimental studies (mostly by Faddegon) to determine the energy and angular distribution of emitted photons by targets (Faddegon et al., 1990, 1991). They found reasonable, but not perfect, agreement between simulated and measured spectra. The accuracy of cross sections for bremsstrahlung production remains a study subject today (Poon and Verhaegen, 2005).

5.2.1.1.2 Photon Beam Spot Size in Radiotherapy Accelerators

A point that deserves special attention is the focal spot size of the photon beam or the primary electron beam. The latter is a crucial parameter in Monte Carlo simulations, which influences calculated dose and fluence distributions, as will be demonstrated later. Munro and Rawlinson (1988) were among the first to estimate the size of the x-ray source in linacs. A slit camera, in combination with tomography techniques, was used to project an image of the photons that are emitted under small angles with the central beam axis on a diode detector. Determining the size of the photon source this way is also a direct measure for the spot size of the primary electron beam hitting the target. The presence of a flattening filter does not disturb the spot size measurement since only those photons that go through it without interacting can contribute to the image. For 6–25 MV linacs, they found that the size and shape of the x-ray source can differ from machine to machine, with mostly an elliptical shape with the long axis perpendicular to the gantry axis. The full width half maximum (FWHM) of the measured spots varied between 0.7 and 3.3 mm. Similar studies have been reported (Lutz et al., 1988; Loewenthal et al., 1992; Treuer et al., 1993; Schach von Wittenau et al., 2002). Jaffray et al. (1993) conducted extensive measurements of focal spot sizes for 6–25 MV photon beams from seven different linacs, using Munro and Rawlinson's tomographic technique (Munro et al., 1988). They reported measured elliptical spot sizes with an FWHM of 0.5–3.4 mm with eccentricities of 1.0–3.1. They concluded from repeated measurements that the long-term stability of the spot size is mostly determined by the linac design and not so much by linac tuning. They noticed that different energies on the same linac can have different source spot centers: a shift of 0.8 mm was noted for 6 and 18 MV beams on a Varian Clinac 2100C. Focal spots have also been reported to move during the start-up phase of irradiations by up to 0.7 mm in the gun-target direction for Elekta linacs (Sonke et al., 2003). Recently, a method was proposed to estimate the size and the shape of the focal spot from measured dose profile data (Wang and Leszczynski, 2007).

5.2.1.1.3 Photon Target Simulation for Radiotherapy Imaging

A final topic we will discuss briefly in this section is the use of Monte Carlo simulations for the optimization of radiotherapy accelerator targets for portal imaging, which is also relevant for megavolt cone beam CT imaging. Simulation of the imaging detector is of relevance here but detailed discussion of this is not within the scope of this review. Figure 5.3 shows the complete simulation geometry, including the patient and the portal imager downstream. The figure also shows a simulated versus a recorded portal image of a contrast phantom. A linac target for imaging would typically consist of a lower atomic number material than a therapy target. This results in a photon spectrum that is considerably softer, which is needed for imaging purposes due to the imaging contrast at low photon energies.

In some of the older studies (Galbraith, 1989; Mah et al., 1993), detailed experimental work and simulations led to advocating a thin low-Z target (Be or graphite) in a linac with the flattening filter removed to enhance the fraction of photons below 150 keV for the purpose of improving the quality of portal imaging. These studies also demonstrated that a large fraction of the low-energy photons coming of the special target are absorbed in the phantom, thereby reducing the sought improved imaging contrast. Several others (Ostapiak et al., 1998; Tsechanski et al., 1998) used EGS4 simulations of 2–10 MeV electron beam interacting with a range of target materials (Be, C, Al, Ti, Cu). For 4 and 10 MeV electron beams, the optimum Cu target thickness to maximize the integrated photon fluence was found to be 1.5 and 4 mm (Tsechanski et al., 1998), respectively (these targets cannot be considered "thick" as defined above). Somewhat at variance with the previously cited studies (Galbraith, 1989; Ostapiak et al., 1998), Tsechanski et al. (1998) find that there is not much motivation to use imaging targets with an atomic number below Al. This is because in thin targets indeed, more low-energy photons are produced for lower-Z materials but these will mostly be attenuated in the imaged patient and, therefore, contribute to an increased dose. Thin Al or even Cu targets were recommended by these workers.

Building on these works, Flampouri et al. (2002, 2005) presented a comprehensive experimental and simulation study of an optimized portal imaging beam, which was later refined (Roberts et al., 2008). They used the BEAM/EGS4 MC code for detailed simulations of a complete linac treatment head, including a specially designed contrast phantom and several imaging systems. Optical contrast derived from simulations and experiments was found to be in good agreement (Figure 5.3). Recently, Monte Carlo techniques were used for the study of low-atomic-number targets for MV cone beam CT (Connell and Robar, 2010; Robar et al., 2009; Faddegon et al., 2010).

5.2.1.2 Flattening Filter

In this section, we turn our attention to the next component in a linac downstream from the target that has a major influence on the beam: the flattening filter. Flattening filters are designed to generate flat dose distributions at a certain depth in water. We will highlight a number of Monte Carlo simulation studies that provided insight in the influence of the usually complex-shaped flattening filter on photon fluence distributions (see Figure 5.2). It has to be pointed out that there are linacs that do not use

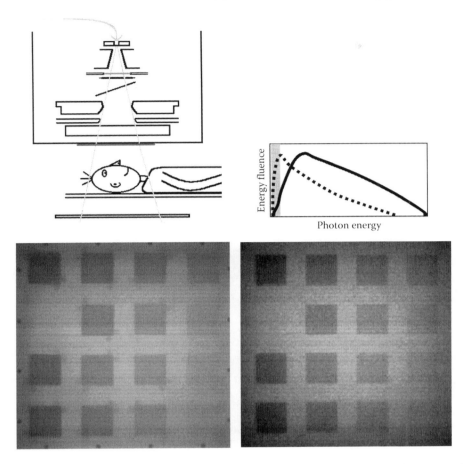

FIGURE 5.3 Schematic representation of a linac simulation, including a patient and a portal imager below the patient. The lower-energy photon spectrum (dotted line) used for imaging purposes is compared to the therapy photon spectrum (full line) in the graph. The bottom panels compare a recorded portal image (left) of a contrast phantom to a simulated one (right). (Work by S. Flampouri, University of Florida Proton Therapy Institute.)

flattening filters. The Microtron MM50 accelerator, tomotherapy machines, and some modern volumetric arc therapy devices do not employ flattening filters (for a recent example of a Monte Carlo study on one of these machines, see (Kry et al. 2010)).

McCall et al. (1978) were probably among the first to report the now well-known off-axis differential softening of the photon spectrum, based on Monte Carlo simulations of simplified flattener models made of Al, Ni, and W for a 25 MV photon beam. Since Al was found to cause the largest softening, they recommended that flattening filters for high-energy photon beams should be made of medium-Z materials such as Cu or steel. Mohan et al. (1985) simulated 4–24 MV photon beams from a number of different linacs. They were the first to build accurate models of the flattening filters (presumably made of steel) by developing a special geometry package that allowed easy modeling of the complex geometry (another forerunner of BEAM). They demonstrated that flattening filters cause significant spectral hardening both on and off the beam axis. For example, a 15 MV beam with no flattening filter present yielded average photon energies at 100 cm from the source of 2.8 and 2.5 MeV, respectively, at the central axis and in an annular scoring region between 10 and 25 cm. These values changed

to 4.1 and 3.3 MeV when a flattener was added, increasing the differential off-axis spectral softening significantly. It can be seen from their work that central axis and off-axis spectra differ mostly for energies below 1 MeV. They also reported that the average photon energy from a linac is lower than the one-third of the nominal energy as then was—and today still is—commonly assumed.

A generic model of a linac ("McRad") based on the EGS4 code was presented by Lovelock et al. (1995) at around the same time BEAM was released. For a 6 MV beam, these authors observed a 10% off-axis softening at 100 cm for an unflattened beam whereas adding a flattening filter augmented this to about 30%. The fact that all these studies report a relatively small off-axis softening for unflattened beams is related to the observation that the bremsstrahlung spectrum from a thick target is fairly isotropic within the angular range of the photons that can reach the patient, whereas the bremsstrahlung intensity is highly anisotropic, as discussed before (Section 5.2.1.1.1). Faddegon et al. (1999) used BEAM simulations to successfully redesign a flattening filter of a clinical accelerator to obtain larger flat 6 MV fields. This involved changing the material of the flattener from steel to brass.

5.2.1.3 Monitor Ion Chamber Backscatter

In some clinical linacs, the signal from the beam monitor ion chamber is affected by the position of the secondary movable beam collimators (jaws). This only occurs in linacs where the distal monitor chamber window is sufficiently thin, where no backscatter plate is present and where the distance between chamber and the upper surface of the collimators is small enough. In that case, particles backscattering from the movable collimators can deposit charge in the monitor chamber, in addition to particles moving in the forward direction in the beam. In a small field, more backscatter from the collimators will occur than in a large field. This means that the monitor chamber will reach its preset number of monitor units (MU) quicker and terminate the beam. This will cause the linac output to decrease with decreasing field size. The magnitude of this effect is usually limited to a few percent but for some linac types, larger effects have been reported. The decreased output is automatically included in output factor measurements, but when Monte Carlo simulations are used to determine output factors, the effect has to be taken into account separately. This applies to all cases where outputs from fields of different sizes are combined, so it also plays a role in modeling dynamic (or virtual) wedges and possibly even in intensity-modulated radiation therapy (IMRT) modeling. This only is possible in certain types of linacs, but one should be aware of this effect that may cause simulations to disagree with measurements.

A few studies have used Monte Carlo techniques to investigate the backscatter effect. Liu et al. (1997a, 2000) performed several detailed simulations of Varian linacs to study the differential effects of the different collimators on monitor backscatter. They concluded that the width of the jaw opening is important to determine the backscatter correction, whereas the actual off-axis position of the field is not important. Verhaegen et al. (2000) modeled Varian 6 and 10 MV photon beams, including a detailed model of the monitor ion chamber. By tagging particles and selectively transporting photons and electrons, it was found that electrons cause most of the backscatter effect. A spectral analysis of the forward and backward moving particles was shown. In this study it was also found that in this linac the backscatter from MLC is negligible at the level of the monitor chamber. In the two studies cited where MC simulations and measurements were compared (Liu et al., 2000; Verhaegen et al., 2000), good agreement was observed.

5.2.1.4 Wedges

Inserting a physical wedge in a photon beam alters the beam characteristics significantly. Not only is the dose distribution modified by attenuation and scatter from the wedge, but the photon spectrum is also affected. Usually, spectral hardening is seen below some parts of the wedge. When a moving linac jaw is used to form a dynamic (or virtual) wedge, the photon spectrum is much more similar to the open-field spectrum. For very-high-energy beams (>20 MV), beam wedges may cause beam softening due to, for example, pair production and annihilation.

Monte Carlo simulations can be used to model physical or dynamic wedges. When modeling a physical wedge, great care has to be taken that the model for the wedge is exact. Not only the exact shape of the wedge has to be implemented, but the composition and density also have to be known exactly. In particular, steel wedges have been known to be difficult to model because of uncertainties in the manufacturer's specified composition.

Liu et al. (1997b) extended their dual-source model to include a wedge model for a 6–18 MV photon beams of a Varian 2100C linac. Steel and lead wedges with angles from 15° to 45° were modeled. The wedge was seen as part of the patient-dose calculation. Special photon dose kernels to take into account the effect of the wedge were added in their convolution/superposition dose calculation method. These kernels were calculated using MC techniques in bimaterial spheres (lead and water). The beam hardening by the wedge was included in the model. Their dose calculations were in agreement with full Monte Carlo simulations, except in the build-up region because of lack of electrons coming from the wedge in their source model. It was found that the wedge-generated photon fluence contributed significantly to the total dose: for example, a 45° wedge in a 20×20 cm^2 10 MV photon beam caused 8.5% of the total dose at the central axis at a depth of 5 cm. By distinguishing between annihilation, bremsstrahlung, and Compton-scattered photons from the wedge, it was found that the latter was responsible for the majority of the wedge-generated dose. The wedge was found to produce a near-Gaussian-shaped lateral dose profile.

Schach von Wittenau et al. (2000) used crude Monte Carlo wedge models and found that wedge-generated bremsstrahlung photons carried about 20% of the outgoing energy in a 10 MV beam. Li et al. (2000) presented their Monte Carlo code MCDOSE, which takes wedges into account as a part of the patient-dose calculation. Spezi et al. (2001) modified DOSXYZ (now updated to DOSXYZnrc (Walters et al., 2009)) to include a rectilinear voxel geometry module and the ability to collect phase-space information behind the wedge. They built a model for the lead wedges of their 6 MV photon beam of a Varian 2100CD linac. A step-resolution of 1 mm in the wedged and beam direction was found to be sufficient for the dose calculations.

Van der Zee and Welleweerd (2002) introduced a new component module in the BEAM code that allows modeling complex wedges such as the Elekta internal wedge. Figure 5.4 shows the CM WEDGE that was used by the authors to model the internal wedge in their 6–10 MV photon beams of an Elekta SL linac. They found that the presence of the wedge altered the primary and scattered photon components from the linac significantly: beam hardening by 0.3 and 0.7 MeV was observed for the two components, respectively. They also noted that the wedge-generated photons were mostly generated close to the distal edge of the wedge. Figure 5.4 also shows the photon energy fluence originating from various sources in the linac, including scattered photons from the wedge. Monte Carlo techniques are unique in extracting such information.

In building Monte Carlo models for dynamic wedges, the time-dependent movement of the jaw has to be included. The

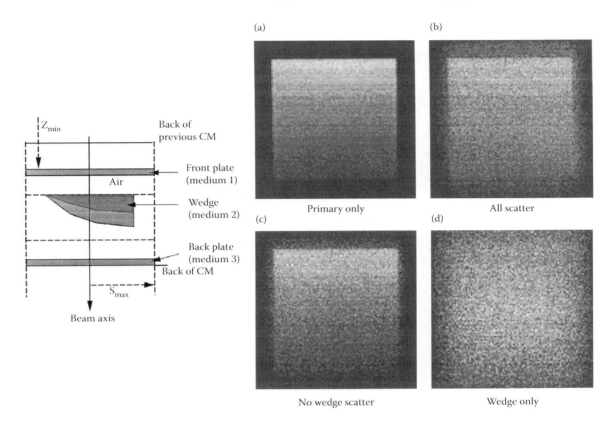

FIGURE 5.4 The left panel shows the component module for the BEAM Monte Carlo code for modeling physical wedges. (From van der Zee, W., and J. Welleweerd. 2002. *Med Phys* 29:876–885. With permission.) This module is not part of the standard BEAM package. The right panel shows the relative distribution of the energy fluence of primary and scattered photons for a 30 × 30 cm² wedged field for a 10 MV photon beam: (a) primary photons only, (b) scattered photons from all scatter sources, (c) scattered photons excluding wedge scatter, and (d) scattered photons from the wedge only. The thicker part of the wedge (heel) is at the lower part of the figures.

studies we will discuss here laid the foundations for modeling motion of MLC in IMRT, to be discussed in the next section. Verhaegen and Das (1999) built a BEAM MC model for the Virtual Wedge of 6–10 MV photon beams of a Siemens MD2 linac. The wedge was modeled by a discrete sum of a large number of open fields. By comparing calculated photon spectra from the heel, toe, and central regions, they found that the photon spectrum of the Virtual Wedge was only slightly hardened, with no significant difference for the three regions. In contrast, 60° physical wedges of tungsten introduced a significant beam hardening for the 6 MV beam across the whole wedge. Both the transmitted and wedge-generated photons were found to be harder for a tungsten wedge than for an open field. Steel wedges were found not to alter the spectrum significantly.

A full dynamic simulation technique to model a Varian enhanced dynamic wedge was introduced by Verhaegen and Liu (2001). This is the first report of a fully dynamic Monte Carlo simulation of a linac component. They used the BEAM Monte Carlo code to model 6–10 MV dynamically wedged photon beams of a 2100C Varian linac by sampling the position of the moving jaw from the segmented treatment table. More information is given in Chapter 7. They termed their technique the "position-probability-sampling method," which is implemented

in a BEAM CM, DMLCQ. The varying backscatter to the monitor chamber during the jaw movement was included in the model. The authors obtained excellent agreement between measured and calculated wedged dose profiles. Shih et al. (2001) presented detailed information on photon and electron spectra and angular distributions produced by dynamic and physical wedges.

Monte Carlo models for other beam modifiers such as blocks, compensators, and so on were discussed in the literature (Jiang and Ayyangar, 1998; Li et al., 2000; Spezi et al., 2001; Ma et al., 2002).

5.2.1.5 Multileaf Collimators and Intensity-Modulated Radiation Therapy

MLC, consisting of banks of tungsten leaves that allow shaping of conformal or intensity-modulated fields in IMRT, are among the most challenging geometrical structures in a linac to model. The MLC leaves can have very complex designs for the leaf ends and the leaf edges. An additional complication is that MLC leaves can move during beam delivery in step-and-shoot and dynamic treatments. The only way that these complex beam-shaping devices can be fully taken into account is by Monte Carlo simulation, but this is a far from trivial task. Furthermore, significant differences in radiation spectrum in different sections of IMRT fields could have significant effects in radiobiology or for film measurements often used in IMRT verification (Liu and Verhaegen, 2002).

The Monte Carlo research for modeling MLCs has mainly been focused on two aspects: exactness of the geometrical models and methods to improve calculation efficiency. Because different linac manufacturers have implemented different designs of MLCs, there is no generic MLC model available in codes like BEAM. There are also differences in how IMRT beams are delivered in different linacs. All of this has to be taken into account in a Monte Carlo model for IMRT. We would also like to point out that the mass density of the tungsten used in the MLC leaves usually has to be ascertained for every linac, due to significant variations in different linacs. Backscatter from the MLC leaves to the monitor chamber does not play a significant role due to the large distance of the MLC to the chamber (Verhaegen et al., 2000), the presence of a backscatter plate (Hounsell, 1998), or the thickness of the monitor chamber downstream window thickness (Verhaegen and Das, 1999).

An early Monte Carlo study of MLC leaf geometries was Küster (1999). She used the GEANT3 Monte Carlo code to investigate the characteristics of two experimental designs of MLCs for a Siemens Primus linac with the purpose of determining the leaf end shape required to optimize dose penumbras. In this feasibility study, leaf leakage was also investigated but good agreement between calculations and measurements was not obtained.

One of the first to show good agreement between measured and simulated MLC dose distributions was De Vlamynck et al. (1999). They built a model for a 6 MV MLC–shaped photon beam from an Elekta SL25 linac. The rounded shape of the MLC leaf ends was approximated by using a stack of MLC modules in BEAM simulations. Interleaf leakage was not modeled in this study. Figure 5.5 gives an example of the good agreement obtained between calculated and measured dose profiles and output factors for small fields with an off-axis position of up to

12.5 cm. No significant spectral differences were noted when large and small fields were compared. In the same paper, a new component module for the BEAM code, MLCQ, was introduced. This allowed modeling of MLC leaves with curved leaf ends.

Ma et al. (2000) modified BEAM/DOSXYZ to allow simulation of multiple-beam fixed-gantry IMRT treatments. The information in the linac MLC leaf sequence file was used to derive a fluence map from which particles were sampled during the simulations. The statistical weight of the particles was altered using the information in the fluence map. The MLC leakage was incorporated by reducing the particle statistical weight for the MLC blocked sections. Their model did not include the effects of the leaf shape, tongue-and-groove design, and scattering in the MLC material, but still they reached an agreement within 2% for measured and calculated dose distributions. For complex IMRT plans, they showed differences between their MC system and a conventional pencil-beam algorithm of up to 20%. The same group of workers (Deng et al., 2001) later studied the MLC tongue-and-groove effect on IMRT dose distributions. They used a ray-tracing approach to modify the fluence maps so that the presence of the tongue-and-groove geometry was accounted for. For single-field IMRT dose differences due to the tongue-and-groove effect of 4.5% were noted; when more than five fields were used the differences fell below 1.6%. They remark that the effect of the tongue-and-groove geometry is probably insignificant in IMRT, especially when organ/patient movement is considered. The same workers (Kapur et al., 2000) introduced a new BEAM component module, VARMLC, that allows modeling the tongue-and-groove effect of some MLCs.

Work by another group (Keall et al., 2001; Kim et al., 2001; Siebers et al., 2002) introduced an elegant model to take the MLC geometry into account in Monte Carlo simulations without actually having to model any particle transport in the MLC material. In Keall et al. (2001) the MLC was compressed in a thin layer in the beam direction. Geometrical paths of photons passing through the MLC were determined for particles sampled

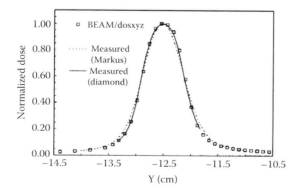
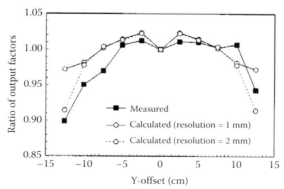

FIGURE 5.5 Comparison between measured and Monte Carlo calculated MLC dose profiles ($10 \times 1\ cm^2$ field, 12.5 cm off-axis, left panel) and output factors ($10 \times 1\ cm^2$ field at various off-axis distances, right panel). The agreement is good in general, especially for relative doses but it also shows that absolute dose output for these small fields is hard to calculate correctly. (From De Vlamynck, K. et al. 1999. *Med Phys* 26:1874–1882. With permission.)

from a phase-space file above the MLC. The statistical weights of the particles are modified according to the probability that the particles reach the bottom of the MLC geometry, where a new phase-space file is constructed. The model took into account intraleaf thickness variations and the curved shape of the leaf end but omitted interleaf leakage and charged particle transport; hence, production of bremsstrahlung photons and annihilation photons. First-scattered Compton photons were included but transport across leaves was not considered. This model was extended (Siebers et al., 2002) to take the leaf edge effect into account. Cross-leaf photon transport and charged particle transport in the leaves was still ignored. They make the interesting observation that it may suffice to model only first Compton scatter and omit electron transport in other linac components upstream from the MLC. The same group also fully simulated MLC geometries (Kim et al., 2001). The leaf ends, leaf edges, mounting slots, and holes were included in the study. They investigated radiation leaking through the MLC. Depending on the field size defined by the jaws above the MLC and the photon energy they found that fully blocked MLC fields had a radiation leakage dose of about 1.5–2%, compared to the dose in an open field. It was observed that the MLC generated photons add a broad background dose to the transmitted radiation (Figure 5.6). Electrons emitted from the MLC cause up to 35% of the surface dose (Kim et al., 2001) in an 18 MV MLC blocked photon beam. For 6 MV photons, significant hardening of the photon spectrum behind the MLC was observed, resulting in a significant shift of the depth dose curve. No such effect was noted for 18 MV photons.

Recently, a method to greatly speed up MLC simulations was reported (Brualla et al., 2009) in which sophisticated variance reduction techniques are used to automatically determine geometrical regions that require detailed transport and others

where transport may be performed in an approximate fashion. They also show that modeling only single Compton scatter in the MLCs does not suffice.

Liu et al. (2001) were the first to implement fully dynamic MLC motion to allow modeling step-and-shoot or dynamic IMRT delivery. This resulted in a new BEAM component module, DMLCQ. During sampling of particles for transport, the leaf positions were randomized according to the number of MU in each field segment. This information was derived from the linac leaf sequence file. A correction for the difference between nominal and actual leaf position was required. Figure 5.7 compares Monte Carlo calculations and measurements for an IMRT field delivered in step-and-shoot mode. One of the most detailed MLC models ever published (Heath and Seuntjens, 2003), based on Liu et al. (2001), included leaf leakage, mounting screws, leaf divergence, support railing groove, leaf tips, and dynamic motion.

Li et al. (2001) built a model for intensity-modulated arc therapy (IMAT) delivery, which involves an Elekta linac SL20 arcing around the patient while the MLCs are moving. An additional complication is that the SL20 uses its internal wedge for some fractions of the delivery. They obtained good agreement with measurements but found up to 10% difference between Monte Carlo and conventional dose calculation techniques for clinical IMAT fields. Many more papers on Monte Carlo simulations for arc therapy have been published since (Bush et al., 2008; Teke et al., 2009); more on this topic will be discussed in Chapter 7.

The effect of the MLC on portal dose distributions was studied using Monte Carlo techniques (Fix et al., 2001; Spezi and Lewis, 2002).

5.2.2 Full Linac Modeling

In the following two sections, we will describe complete linac models for photon beams. The photon beam models can be basically divided into two categories. The first approach is to do complete simulations of linacs and either use the obtained particles directly in calculations of dose, fluence, and so on in phantoms, or store phase-space information about these particles at possibly several planar levels in the linac for further use. The phase-space can be sampled for further particle transport in the rest of the geometry (linac or phantom). If the phase-space is situated at the bottom of the linac, this then effectively replaces the linac and becomes the virtual linac. We will call this the phase-space approach. In Figure 5.1, two levels where phase-space information might be collected are indicated (levels AA' and BB'). A particularly elegant approach couples BEAMnrc directly to DOSXYZnrc simulations, thereby eliminating the need to store intermediate phase-space files.

The second approach calculates particle distributions differential in energy, position, or angle in any combination of these in one-, two-, or three-dimensional histograms. This is usually done for several subsources of particles in the linac and is termed a Monte Carlo source model. The linac is then replaced by these subsources that constitute the virtual linac. The source model approach often starts from information

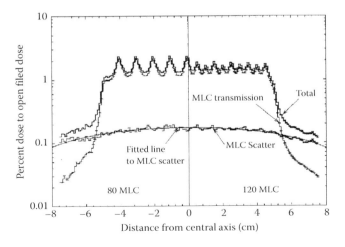

FIGURE 5.6 A dose profile (% of open field dose) perpendicular to the direction of leaf motion of the Monte Carlo computed radiation leakage from an MLC blocked 6 MV 10 × 10 cm² field at 5 cm depth in a water phantom for the 80-leaf (left-hand side) and the 120-leaf (right-hand side) MLCs (From Kim, J. O. et al. 2001. *Med Phys* 28:2497–2506. With permission.)

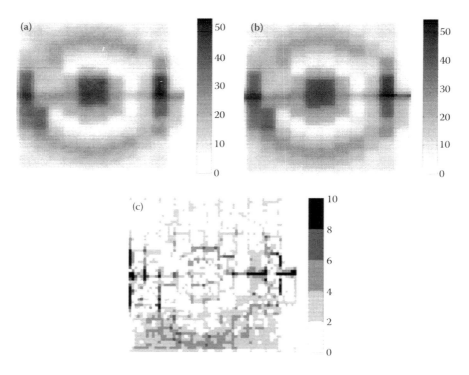

FIGURE 5.7 Monte Carlo calculated (a) and film measured (b) dose distributions at 1.5 cm depth for a 6 MV step-and-shoot IMRT field. Panel (c) shows the voxel-to-voxel differences in obtained dose (|calculated—measured|/max(measured)). (From Liu, H. H., F. Verhaegen, and L. Dong. 2001. A method of simulating dynamic multileaf collimators using Monte Carlo techniques for intensity-modulated radiation therapy. *Phys Med Biol* 46:2283–2298. With permission.)

collected in phase-spaces in the linac. Source models inevitably have approximations since they condense individual particle information in histograms, whereas all the information on individual particles is preserved in the phase-space approach. The correlation between angle, energy, and position of particles might be partially or completely lost depending on the degree of complexity of the source model. The disadvantage of working with phase-spaces is the large amount of information to be stored and the slower sampling speed during retrieval of all this information. Also, approximations in the linac model will result in uncertainties in the phase-space approach. Source models often involve data smoothing, so they usually result in less statistical noise.

5.2.2.1 Complete Linac Simulations

As mentioned before, Patau et al. (1978) pioneered the Monte Carlo simulation of a more or less complete model of a photon linac. They presented energy and angular distributions of photons before and after a flattening filter but attempted no further analysis of the influence of the linac components on the photon beam. In another early study (Nilsson and Brahme, 1981), linac components were sequentially made transparent (i.e., no interactions were allowed in them) to study the influence of the components on the particle beam. They found that in a 21 MV beam, the thicker flattening filter, combined with the more forward-peaked bremsstrahlung angular distribution, caused most of the photon scatter. In a 6 MV beam, the primary collimator contributed

most to the scatter, corresponding to the larger angular bremsstrahlung distribution for lower primary electron energies.

Mohan et al. (1985) presented a much-quoted EGS3 simulation study of a set of Varian linacs producing 4–24 MV photon beams. The model included a target/backing, a primary collimator, a flattening filter, and a secondary collimator system. They noted that off-axis photon spectra were softer than at the central axis. The collimating jaws had no significant effect on the energy and angular photon distributions. The beam was mostly determined by the target and the flattening filter. For 15 MV photons, they found that when photons reached a plane at 100 cm from the source, 93.5% had never scattered whereas 2.8%, 3.5%, and 0.2% suffered scattering interactions in the primary collimator, flattening filter, and secondary collimator system, respectively. The photon spectra presented in this study are a reasonable approximation; they are therefore still frequently used today, for example, for patient dose calculations. Their angular distributions, however, are only crude approximations.

A model of a 6 MV Siemens linac photon beam including the exit window, target, primary collimator, flattening filter, monitor chamber, and collimating jaws was published by Chaney et al. (1994). They focused on head scatter contributions. Despite the model being very approximate, the calculated scatter contributions are nevertheless found to be in fair agreement with more detailed studies. They found that about 9% of the photons in a large field were derived from scatter. Figure 5.8 shows their scatter-plot of the sites of origin of scattered particles. From this

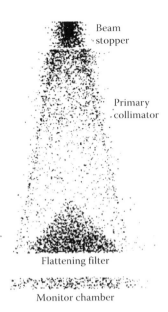

Beam stopper

Primary collimator

Flattening filter

Monitor chamber

FIGURE 5.8 Distribution of sites of origin of photons for a 6 MV photon beam. Clearly visible are the target/beam stopper, the primary collimator, the flattening filter, and the monitor ion chamber. The center of mass of all the origin sites was found to be inside the flattening filter at 6.2 cm from target. (From Chaney, E. L., T. J. Cullip, and T. A. Gabriel. 1994. *Med Phys* 21:1383–1390. With permission.)

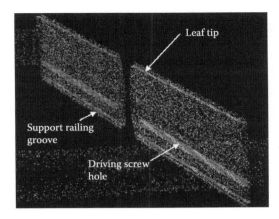

Leaf tip

Support railing groove

Driving screw hole

FIGURE 5.9 Photon ray-tracing an MLC leaf geometry in Monte Carlo simulations. (From Heath, E., and J. Seuntjens. 2003. *Phys Med Biol* 48:4045–4063. With permission.)

figure, it is evident that Monte Carlo simulations—besides producing aesthetically pleasing pictures—can provide clear insight into which parts of the linac play an important role as a source of scatter or particle generation.

Libby et al. (1999) give detailed information on how to validate Monte Carlo linac models. As an example of verification of linac geometry, they show how the complex shape of the flattening filter can be reproduced and compared to manufacturer's data by bombarding it with monoenergetic, monodirectional photons and calculating the attenuation profile. Another approach (Heath and Seuntjens, 2003) for verifying complex geometries is ray-tracing photon tracks in such a way that photon coordinates are stored every time a photon crosses a boundary between MLC leaf material and air. Figure 5.9 gives an example of the geometry that can be visualized in this fashion.

Detailed models of photon linacs were presented by many others (Mazurier et al., 1999; van der Zee and Welleweerd, 1999; Lin et al., 2001; Ding, 2002a; Reynaert et al., 2005; Titt et al., 2006; Vassiliev et al., 2006; Mesbahi et al., 2007, 2005; Bednarz and Xu, 2009). Highly detailed studies on the influence of linac components and the characteristics of the primary electron beam were done by Sheikh-Bagheri and Rogers, (2002a,b) and Sheikh-Bagheri et al. (2000). They advocated using in-air dose profiles measured with an ion chamber and a build-up cap to derive the energy and spatial characteristics of the primary electron beam. Figure 5.10 illustrates the sensitivity of the off-axis factor (defined as the ratio of measured dose to water with an ion chamber in air at an off-axis position and at the central axis) to the primary

electron energy and the width of the radial intensity distribution of the electron beam. By avoiding phantom scatter as a confounding factor, the authors clearly demonstrate the differential effects of the components in their model. Interestingly, the insensitivity of dose calculations to the energy distribution of the primary electron beam for photon beam simulations is in contrast to electron beams where both build-up and build-down regions are highly sensitive to the primary electron energy distribution.

A general finding is that the average photon energy is below one-third of the nominal energy at the central axis (more so for the highest-energy beam) and that the average energy 20 cm off-axis is decreased by 0.5–2.0 MeV (the greatest decrease usually corresponds to the highest nominal energy). Ding (2002a) noted that the mean energy of the charged particles in a photon beam varies less with off-axis distance than the photon mean energy. A procedure to tune a Monte Carlo linac model was outlined by Verhaegen and Seuntjens (2003).

5.2.2.2 Source Models

Early Monte Carlo-based source models were first introduced for electron beams and then later for photon beams (Liu et al., 1997a,b; Ma, 1998). Liu et al. (1997a,b) introduced a dual-photon source model: one source for the primary photons coming directly from the target and a second extra-focal source to describe scatter from mainly the primary collimator and the flattening filter. A third source took the electron contamination into account. Derivation of the source model started by performing a full linac Monte Carlo simulation and phase-space analysis. The obtained distribution of points of origin of photons for a 10 MV beam is illustrated in Figure 5.11. Obviously, it is very important in such a model that the geometrical position of the scatter sources is estimated correctly. Inevitably, replacing the real three-dimensional source distributions with two-dimensional ones is an approximation. The primary photon source at the target was modeled with a radius of 0.1 cm. The effect of the jaws was modeled as eclipsing collimators, thereby ignoring the small amount of scattered radiation the jaws contribute. Backscatter

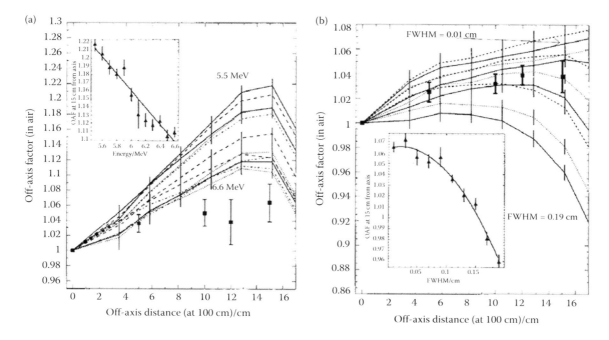

FIGURE 5.10   Linear dependence of off-axis factors at 15 cm off axis distance (see inset) to the primary electron energy in a 6 MV Siemens photon beam (Figure 5.9a). Quadratic dependence of off-axis factors at 15 cm off axis distance (see inset) to the Gaussian width of the primary electron intensity distribution in an 18 MV Varian photon beam (Figure 5.9b). (From Sheikh-Bagheri, D., and D. W. Rogers. 2002. *Med Phys* 29:391–402. With permission.)

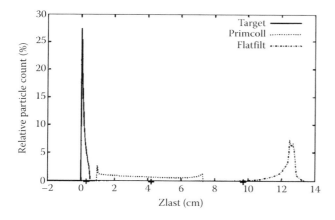

FIGURE 5.11   Distribution of the points of origin of photons in the target, primary collimator, and flattening filter in a 10 MV photon beam. The derived estimates for the source positions are indicated by the crosses on the x-axis. (From Liu, H. H., T. R. Mackie, and E. C. McCullough. 1997. *Med Phys* 24:1975–1985. With permission.)

toward the monitor chamber was added explicitly, which is often omitted in later works. Despite the approximations in the model (Liu et al., 1997a,b), good agreement was reached between calculated and measured dose distributions and output factors.

Other complex source models can be found in the literature (Schach von Wittenau et al., 1999; Deng et al., 2000; Fix et al., 2000, 2001; Hartmann Siantar et al., 2001; Sikora et al., 2007; Tillikainen et al., 2007). The PEREGRINE group (Schach von

Wittenau et al., 1999) performed a very detailed study to investigate the influence of linac components on photon fluence, energy spectra, angular distributions, and so on. They introduced the concept of *correlated histograms*, which contains information extracted from phase-space files in such a way that correlations between energy, angles, and so on are preserved to some degree. Target photons were found to originate from a small disk source with a radial fluence distribution, the primary collimator was found to generate photons in a ring source close to its upper edge and the flattening filter was found to produce photons uniformly throughout the filter geometry. Figure 5.12 shows the correlated histogram used for the two subsources for a 6 MV beam: target and scattered photons. The authors reached very good agreement between full phase-space-based simulations and their model. A subsource was added to model the electron contamination (Hartmann Siantar et al., 2001) in the PEREGRINE photon Monte Carlo treatment planning system. The jaws were modeled as a masking collimator, which causes the dose outside the penumbra to be up to 15% too low. Backscatter to the monitor chamber was included as an empirical correction.

One of the most complex source models published to date was presented by Fix et al. (2001). Using GEANT3 simulations, they derived a model for a 6 MV linac beam with 12 subsources: one each for the target, flattening filter, primary collimator, and mirror, and two each for all four of the secondary collimators. The two subsources for each secondary collimator represent radiation emanating from the inner side and the lower side of the collimators. It was found that the inner-side subsource generates most particles close to a 1 cm

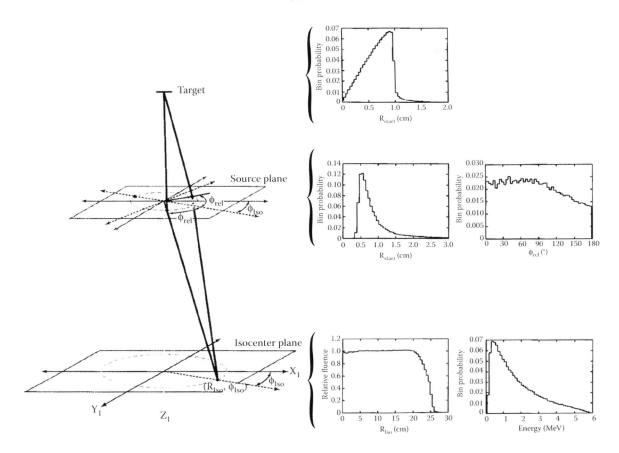

FIGURE 5.12 Correlated histograms of starting position and starting angle for the two subsources in a virtual source model of a 6 MV photon beam. The energy fluence and energy spectrum are described at the isocenter. The origin of photons from the primary collimator or the flattening filter is sampled from the radial and angular distributions. The starting position of the target photons is sampled from a radial distribution, with assumed angular isotropy. (From Schach von Wittenau, A. E. et al. 1999. *Med Phys* 26:1196–1211. With permission.)

distance from the top of the collimator, whereas the bottom-side subsource generated most particles close to the field edge. The total contribution from the eight collimator sources was very small and is usually ignored in most linac source models. Data was generated for eight field sizes, which allowed interpolation for arbitrary fields.

The source model approach has to be used with care due to its inherent approximations and potential for systematic error. Its strength lies in the fact that the requirements on data storage are greatly reduced with respect to a full phase-space approach; figures of 400–10,000 have been quoted (Chetty et al., 2000; Deng et al., 2000; Fix et al., 2000). Because of the reduced phase-space reading/writing and because source models inherently involve some degree of fluence smoothing, the calculation time required to achieve a certain specified statistical variation in a dose calculation is also reduced. Monte Carlo-based treatment planning systems often resort to virtual source models.

5.2.2.3 Absolute Dose Calculations (Monitor Unit Calculations)

Dose calculations based on Monte Carlo simulations of linacs (or any other device) usually result in doses expressed in Gy/

particle. In the BEAMnrc package, the term "particle" refers to the initially simulated particles, so in a linac simulation, the calculated doses are in Gy per initial primary electron hitting the target. Even in simulations that are done in several parts, involving multiple phase-space files at different levels in the linac, this information is preserved in the phase-space files. This ensures correct normalization of the doses per initial particle. Other Monte Carlo codes use different schemes or leave this up to the user. This absolute dose can be related easily to the absolute dose in the real world. By running a simulation that exactly matches an experimental setup to determine absolute dose, one can obtain the ratio of the measured and calculated doses, which are expressed in units of [Gy/MU]/[Gy/particle], where MU stands for monitor unit. The latter corresponds to the reading of the monitor ionization chamber in the linac. This ratio may serve as conversion factor for all beams produced by the linac for that particular photon energy. By obtaining these conversion factors for all linac energies, Monte Carlo calculated doses may be converted to absolute doses as used in radiotherapy practice. This approach is also valid for electron beams (see Chapter 6), and is employed by most users. The reference setup that is frequently modeled is a dose determination in an open 10×10 cm^2

field in water, but in principle any setup, not necessarily using open fields, can be modeled to determine the conversion factors for each photon energy. This approach was described in more detail in the literature (Leal et al., 2003; Ma et al., 2004) with special attention to IMRT treatments. A different approach that employs the relation between the simulated monitor chamber reading, the initial number of particles impinging on the target, and the field size was introduced by Popescu et al. (2005). This elegant method requires attention to the issue of backscatter of particles to the monitor chamber (Section 5.2.1.3). It does not rely on renormalization to measurements and relative output factors. Although in principle a superior approach, it is not often used in practice, as far as we know.

5.2.3 Stereotactic Beams

Linac-based stereotactic irradiation of intracranial and extracranial lesions is a frequently employed radiotherapy technique. Monte Carlo simulation of the collimated, narrow photon beams (10–30 mm at the isocenter) is therefore important in two areas. First, knowledge of detailed beam data allows a proper understanding of detector response that is often affected by the absence of lateral electronic equilibrium. Second, treatment planning systems may employ Monte Carlo techniques to reliably take into account tissue heterogeneities near small targets. This implies that the basic properties of the beam should be accurately calculated using geometric treatment head information.

Several Monte Carlo studies on stereotactic beams (Kubsad et al., 1990; Sixel and Podgorsak, 1993, 1994; Sixel and Faddegon, 1995; Verhaegen et al., 1998; Chaves et al., 2004; Belec et al., 2005; Jones and Das, 2005; Ding et al., 2006; Scott et al., 2008, 2009; García-Pareja et al., 2010) and a recent review paper (Taylor et al., 2010) have been published. Studies have focused on the accuracy of dose calculations in small fields, photon scatter, photon and electron spectra, and correction factors for measurements. A full simulation (Verhaegen et al., 1998) of a dedicated 6 MV stereotactic linac with field sizes ranging from 5 to 50 mm showed that photon spectra are influenced by the small beams; mean photon energies at d_{max} in water were found to vary between 2.05 MeV for a 5 mm cone to 1.65 MeV for a 5 cm cone. They also observed that Spencer–Attix stopping-power ratios at the central axis were depth independent, which is an important information for measurements. Another study simulated a small-field dynamic radiosurgery unit capable of producing 1.5 mm pencil beams (Paskalev et al., 2002, 2003) and focused on aspects of lateral electronic disequilibrium. Differences of up to 60% were found between Spencer–Attix stopping-power ratios and dose-to-water and dose-to-cavity ratios; a correction procedure was proposed. It has been demonstrated (Scott et al., 2008, 2009) that small-field output factors are very sensitive to the dimensions of the primary electron beam, which suggests a way to commission this part of the linac geometry.

Some have suggested that Monte Carlo dose calculations for treatment planning in stereotactic beams may not be needed (Ayyangar and Jiang, 1998), while others (Solberg et al., 1995) have argued that Monte Carlo simulation is required for reliable small-field dose calculation in regions of tissue heterogeneity for which, in conventional planning systems, often only the primary photon beam component (and not the scatter component) is considered. Full Monte Carlo simulations are preferred for small fields but virtual source models have also been proposed (Chaves et al., 2004).

As an example, Figure 5.13 shows the level of detail that is needed to model a linac with a micro-MLC for stereotactic irradiation. The geometric accuracy of the leaf ends and sides, and the composition of the leaves are very important. The figure also shows the level of agreement with measurements that can be achieved by an accurate Monte Carlo model. Conventional dose calculation was shown to be considerably less accurate.

Monte Carlo studies of a CyberKnife robot-mounted small-field linac have also been published (Yamamoto et al., 2002; Araki, 2006; Sharma et al., 2010; Wilcox et al., 2010). A recent study introduced an exotic technique to enhance the simulation efficiency of small fields. The "ant colony" method (García-Pareja et al., 2010) was employed whereby geometric cells are overlaid with an importance map that is derived in an analogous fashion in which ants leave pheromone trails to food sources. In their approach, photons that reach regions of interest determine the importance maps so that subsequent photons may be simulated more efficiently.

5.2.4 Contaminant Particles

5.2.4.1 Electrons

Electron contamination in photon beams has been studied by numerous authors (Padikal and Deye, 1978; Nilsson and Brahme, 1979; Ling et al., 1982; Mackie and Scrimger, 1982; Petti et al., 1983a; Rustgi et al., 1983; Yorke et al., 1985; Gerbi and Khan, 1990; Lamb and Blake, 1998; Zhu and Palta, 1998; Malataras et al., 2001; Ding, 2002b; Ding et al., 2002; Yang et al., 2004; Abdel-Rahman et al., 2005; Lopez Medina et al., 2005; Mesbahi, 2009; Sikora and Alber, 2009; Ververs et al., 2009). Electrons in the treatment beam are generated either in the high-Z components of the linac or in the air. It is known that electrons in the photon beam are responsible for a significant fraction of the surface dose in a phantom or patient. The dose from contaminant electrons drops off rapidly with depth and is usually insignificant at the depth of maximum dose of the photon beam, d_{max} (for an overview of experimental findings, see Verhaegen and Seuntjens, 2003).

Petti et al. (1983a,b) were the first to dedicate a Monte Carlo study to electron contamination in photon beams. They observed that 70% of the electrons at the phantom surface were generated in the flattening filter and monitor ion chamber, 13% from the collimating jaws, and 17% in the air for a source to surface distance (SSD) of 80 cm. For an SSD of 400 cm, these fractions changed to 34%, 5%, and 61%, respectively. They concluded from this that at all SSDs the contribution from the flattening filter and monitor chamber is significant whereas the contribution from air-generated electrons is only important at very large SSD. The former was also found to cause the shift of d_{max} with field size because the collimating jaws act as an eclipsing absorber. At an SSD of 80 cm, these electrons had a most probable energy of

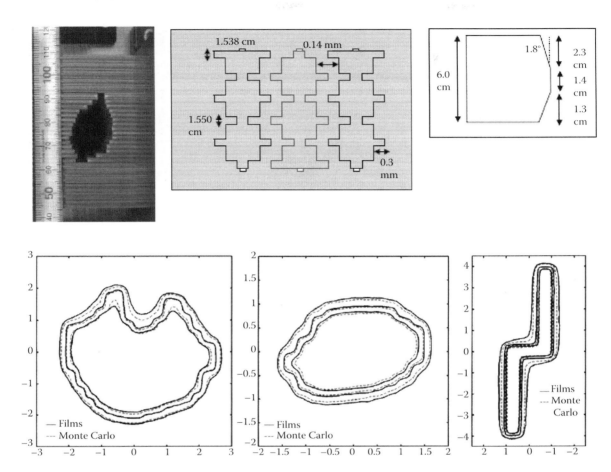

FIGURE 5.13 Top panels: a micro-MLC for stereotactic irradiation (left) and its complex geometry Monte Carlo model of the leaf ends (middle) and sides (right). Bottom panels: a few examples of Monte Carlo calculations and film measurements of the isodose distribution in small fields. (From Belec, J., H. Patrocinio, and F. Verhaegen. 2005. Development of a Monte Carlo model for the Brainlab microMLC. *Phys Med Biol* 50:787–799. With permission.)

1.5 MeV (with 40% of them having an energy exceeding 5 MeV) and a forward-peaked angular distribution with 60% of the electrons moving at an angle with respect to the beam axis of less than 16°. These findings explain why these electrons still play a significant role at very large SSDs. The electrons generated in the collimating jaws, which are commonly assumed to be important, were found to play a minor role.

Some Monte Carlo studies (Fix et al., 2001; Malataras et al., 2001) showed energy spectra, energy fluence distributions, and mean energy distributions for the electrons reaching the phantom or bottom of the linac. They found that the energy spectra of electrons generated in the flattening filter and in air are very different: the former has a broad energy spectrum with an average electron energy comparable to the average photon energy. The latter has a narrow energy distribution, peaked at low energies. These results may be very linac dependent. It was pointed out (Malataras et al., 2001) that the increasing probability of pair production causes the flattening filter to become the dominant electron source at high photon energies (see Figure 5.14).

For quite a long time, Monte Carlo studies of surface dose in photon beams showed significant differences with measurements

(van der Zee and Welleweerd, 1999; Hartmann Siantar et al., 2001; Ding, 2002b; Sheikh-Bagheri and Rogers, 2002a,b). Only with very detailed linac models and taking into account that radiation detectors may need various response corrections close to an air/phantom interface, a few workers (Sheikh-Bagheri et al., 2000; Abdel-Rahman et al., 2005) succeeded in achieving good agreement. These studies showed that Monte Carlo dose calculations near surfaces are often more reliable than measurements, especially if corrections are not taken into account.

5.2.4.2 Neutrons

It is known that photons with an energy exceeding about 8 MeV can produce neutrons in the target and collimating structures. For an overview of experimental techniques, we refer to AAPM Report 19 (1986). Several authors used Monte Carlo techniques to simulate 15–50 MV photon beams from a variety of linacs (Ing et al., 1982; Agosteo et al., 1993; Gudowska et al., 1999; Ongaro et al., 2000). They included methods to estimate (γ,n) production in linac components (mostly the target) and phantom geometries. Monte Carlo codes that can handle photo-neutron production are, for example, MCNP(X). It has been determined

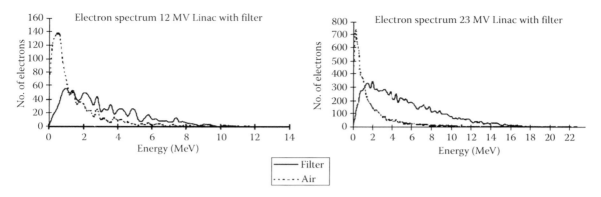

FIGURE 5.14 Monte Carlo generated energy spectra of secondary electrons at 100 cm from the target in a Saturne linac. The electrons originating in flattening filter and in the air below it are depicted. They exhibit a clear difference in spectral shape. (From Malataras, G., C. Kappas, and D. M. Lovelock. 2001. A Monte Carlo approach to electron contamination sources in the Saturne-25 and -41. *Phys Med Biol* 46:2435–2446. With permission.)

that photo-neutron production in soft tissue phantoms leads to a neutron dose of about 0.003% of the photon dose inside a 25 MV photon beam (Agosteo et al., 1993). In studies for linac-generated neutrons from 15–18 MV photon beams, maximum equivalent neutron doses at the central axis of about 2–5 mSv per Gy absorbed photon dose were found (d'Errico et al., 1998; Ongaro et al., 2000; Ding et al., 2002). Outside the photon field, the linac-generated neutron dose has been observed to decrease by about a factor of 10, meaning the ratio of absorbed dose by the neutrons is higher outside the field compared to inside the field. Most of the linac-generated neutrons depositing energy in a tissue-equivalent phantom had an energy >1 MeV (d'Errico et al., 1998). Recently, the photo-neutron spectrum from a wedged 18 MV photon beam was studied (Ghavami et al., 2010). They found that the neutron fluence decreases with increasing field size for open beams, and increases with increasing field size for wedged beams. Another issue of interest is calculation of neutron dose in linac mazes for radioprotection (Falcao et al., 2007; Facure et al., 2008).

5.2.5 Radiotherapy Kilovolt X-Ray Units

Kilovolt x-ray tubes have been used in radiotherapy for many decades but are now being used less frequently for cancer treatment. Linac electron beams and brachytherapy sources have largely replaced the use of kV x-rays for superficial or skin lesions, respectively. This by no means implies that there is no role for Monte Carlo modeling of kV x-ray tubes. In diagnostic imaging, Monte Carlo methods are commonly used to model x-ray tubes for planar or CT imaging. A recent example of CT scanner modeling versus Compton scatter spectroscopy is Bazalova and Verhaegen (2007). The literature contains many examples of Monte Carlo modeling of x-ray tubes for diagnostic and mammography applications (Boone et al., 2000; Verhaegen and Castellano, 2002; Ay et al., 2004) for example, dosimetric purposes. Detailed discussion of these mostly diagnostic applications is outside the scope of this chapter. There are some specialized applications of kV x-rays in radiotherapy that are under investigation and where Monte Carlo techniques play an important role, such as contrast-enhanced

radiotherapy, in which kV x-rays are used to bombard targets laden with dose enhancers such as high atomic number contrast media (Verhaegen et al., 2005) or gold nanoparticles (Jones et al., 2010). To model the interaction of kV x-rays with high atomic number materials accurately, a number of physical processes must be implemented, which makes photon transport in the kV energy range harder than in the MV energy range where Compton scatter is the dominant interaction. Examples of things that need to be implemented in the transport models and which largely can be ignored at MV energies are electron impact ionization, Rayleigh scattering, photo-electron angular distributions, and fluorescent decay. Another specialized area is the modeling of miniature x-ray tubes that are small enough to be implanted in lesions to irradiate them from within. These devices often consist of very thin transmission x-ray targets made of a composite of various high-atomic-number materials, stressing the importance of the accuracy of bremsstrahlung cross sections and the modeling of electron impact ionization. Examples are the work of Yanch and Harte (1996) on brain applications and Liu et al. (2008) on electronic brachytherapy. Another special application is small animal irradiation for preclinical studies that is usually done with kV x-rays to avoid the build-up regions of MV photon beams in small animals such as mice (Tryggestad et al., 2009). Monte Carlo modeling also plays a role in beam simulation and dose calculation. Since these applications are outside the field of external beam radiotherapy, they will not be discussed in detail here.

X-ray spectra are used in many different applications, for example, in studies in radioprotection, radiobiology, radiation quality, and response of radiation detectors. Therefore, several authors created methods to generate x-ray spectra as a function of target angle (most efforts focused on tungsten targets), accelerating voltage and filtration. Whereas earlier work (Birch and Marshall, 1979; Iles, 1987) employed analytical techniques, the most recent work relies on Monte Carlo methods for detailed electron penetration in the target (Poludniowski, 2007; Poludniowski and Evans, 2007). The SpekCalc code (Poludniowski et al., 2009) uses a graphical user interface to perform a very fast, yet accurate, x-ray spectrum calculation for tungsten targets bombarded by electrons in the range 50–150 keV.

In the late 1990s, the BEAM code was used to simulate entire radiotherapy x-ray tubes (50–300 kV), including effects such as tube voltage ripple (Verhaegen et al., 1999). The results were compared to spectral measurements and another Monte Carlo code, MCNP. Both codes had problems with the characteristic x-ray lines due to limitations in the modeling of electron impact ionization. Compared to the measurements, there was a clear overestimation of the low-energy component in the photon spectrum at the expense of contributions in the high-energy part of the spectrum. Other workers modeled x-ray tubes for radiotherapy or diagnostic applications (Omrane et al., 2003; Ay et al., 2004; Ay and Zaidi, 2005; Kakonyi et al., 2009; Taleei and Shahriari, 2009) and special attention was paid to variance reduction techniques (Mainegra-Hing and Kawrakow, 2006). A recent study modeled in great detail the backscatter from electrons at x-ray targets, and the subsequent reentry of the electrons due to interaction with the electric field inside the x-ray tube (Ali and Rogers, 2008). This effect causes the well-known extrafocal radiation of which the authors demonstrate that it may influence the spectral shape (therefore, the half-value layer) and the air kerma. Figure 5.15 shows the magnitude of the backscatter coefficients for first- and higher-generation electrons obtained from Monte Carlo simulation. For applications with a hard filtration and tight beam collimation, the effect of the extrafocal radiation appears minimized.

A final application that deserves brief mention in this section is the simulation of x-ray units for kV cone beam CT imaging, which is a popular technique used in image-guided radiotherapy (Ay and Zaidi, 2005; Jarry et al., 2006; Mainegra-Hing and Kawrakow, 2008; Downes et al., 2009; Spezi et al., 2009). Figure 5.16 gives an overview of a simulated cone beam CT and the distribution of the last interaction sites of the photons in the geometry (Spezi et al., 2009). These simulation studies may serve to remove scatter from the cone beam images, to optimize the imaging chain, or to derive the patient imaging dose.

5.2.6 ^{60}Co Teletherapy Units

As the relative importance of traditional ^{60}Co teletherapy in comparison with accelerator therapy has dwindled in industrialized countries, the number of Monte Carlo studies on these modalities is also limited. Although ^{60}Co decay gives rise to only two gamma-ray lines of 1.17 and 1.33 MeV, the spectrum of photons in a teletherapy unit is affected by the source encapsulation as well as by the collimating system. Contaminant electrons are also an issue, especially in calibration setups. Monte Carlo calculations have been used to characterize ^{60}Co beams in clinical as well as standard dosimetry laboratory setups.

Han et al. (1987) modeled the Theratron 780 teletherapy machine with a simplified source capsule. Only about 72% of the photons were found to be either 1.17 or 1.33 MeV; the other 28% was single or multiple scattered photons mixed with a small number of unscattered bremsstrahlung and positron annihilation photons. They studied photon fluence spectra, the effect of field size on output, and dose calculations. It was found that an increase in output of the ^{60}Co unit with increasing field size is caused by scattered photons from the primary tungsten collimator. The electron contamination of an AECL ^{60}Co teletherapy unit was studied (Rogers et al., 1988) and it was found that electrons originate from the source capsule (most effective at close SSD), the collimators, and the air (most effective at large SSD). At 80 cm SSD, for a field size equivalent to 35×35 cm^2, about 45% of the dose at d_{max} is due to contaminant electrons

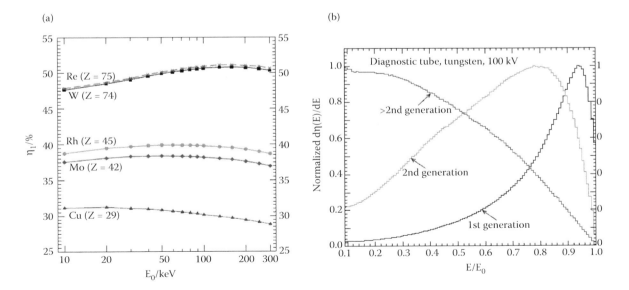

(a)

(b)

FIGURE 5.15 First-generation electron backscatter coefficient η_1 defined as the ratio of the number of electrons backscattering of an x-ray target, over the number of total primary electrons impinging on it (a). Energy spectra from electrons backscattered at all angles from a tungsten x-ray target (b). (From Ali, E. S., and D. W. Rogers. 2008. *Med Phys* 35:4149–4160. With permission.)

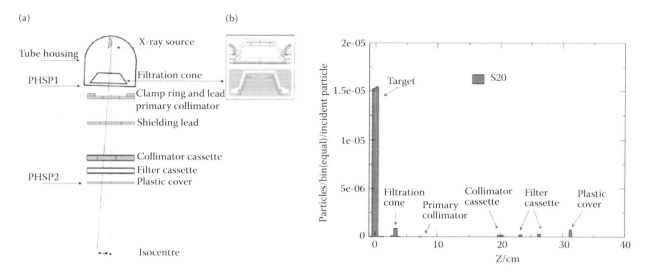

FIGURE 5.16 Left: (a) schematic representation of an Elekta Synergy cone beam CT device. (b) The tube filtration components are shown in the inset. Right: distribution of last interaction sites. (From Spezi, E. et al. 2009. *Med Phys* 36:127–136. With permission.)

of which 22% come from the source capsule and air-generated electrons; 10% and 13% are from electrons emerging from the outer and inner collimators, respectively. The major source of scattered photons was the source capsule itself and scattered photons contribute 18% of the maximum dose. The inner and outer collimators were found to increase the effect of photon scatter by a few percent. More recent studies (Mora et al., 1999; Carlsson Tedgren et al., 2010) modeled ^{60}Co units from calibration laboratories in great detail and focused on small but important spectral effects.

Besides its use in traditional teletherapy, ^{60}Co is implicated in special treatment techniques such as the GammaKnife. Of special interest for this treatment modality is the study of dose distributions and output factors for the very small field sizes used for which detailed source modeling is important. Detailed Monte Carlo studies can be found in the literature (Moskvin et al., 2002; Cheung and Yu, 2006).

5.3 Summary

Monte Carlo simulations are an extremely powerful tool in modern radiotherapy. That the Monte Carlo community is alive and well can be seen from successful workshops that were held in recent years in, for example, in Montreal, Ghent, Cardiff, and Sevilla.

Most of the studies cited in this chapter used EGS/EGSnrc-based codes (including the user interface BEAM) but other codes such as GEANT4 and PENELOPE are also gaining popularity. We highlighted the capabilities of these codes for modeling and analyzing external radiotherapy photon beams, but they also have proven to be of great value for kV x-ray systems, brachytherapy sources, and patient dose calculations. In addition, Monte Carlo simulations play an increasingly important role in imaging for diagnostics and patient treatment verification. In this and other

chapters, the 4D modeling of linac beam delivery in volumetric arc therapy was mentioned. Elsewhere in this book, the capabilities of Monte Carlo codes to incorporate 4D motion, such as in patient/organ movement and deformation, were discussed.

Simulation time and storage of large amounts of data, once prohibitive for complex Monte Carlo modeling, are now less of an issue. Technology is aiding here, for example, with the advent of ultrafast GPU processors.

In the near future, Monte Carlo-based treatment planning will become the workhorse of patient planning. Speed and accuracy will be of the essence, and clear guidelines on commissioning and model generation will be needed.

It is clear that Monte Carlo methods will remain very important in the study of new technology in radiotherapy. To name just one example, Monte Carlo methods are facing new challenges in, for example, the design and dosimetry in hybrid MRI–linac systems (Kirkby et al., 2008; Lagendijk et al., 2008) where electrons are perturbed by the strong magnetic fields employed in MRI imaging.

References

AAPM. 1986. Neutron measurements around high energy x-ray radiotherapy machines, AAPM Report 19.

Abdel-Rahman, W., J. P. Seuntjens, F. Verhaegen, F. Deblois, and E. B. Podgorsak. 2005. Validation of Monte Carlo calculated surface doses for megavoltage photon beams. *Med Phys* 32:286–298.

Agosteo, S., A. Para, F. Gerardi, M. Silari, A. Torresin, and G. Tosi. 1993. Photoneutron dose in soft tissue phantoms irradiated by 25 MV x-rays. *Phys Med Biol* 38:1509–1528.

Agostinelli, S. et. al. 2003. Geant4-a Simulation Toolkit. *Nucl Instrum Methods Phys Res A, Accel Spectrom Detect Assoc Equip* 506:250–303.

Ali, E. S., and D. W. Rogers. 2008. Quantifying the effect of off-focal radiation on the output of kilovoltage x-ray systems. *Med Phys* 35:4149–4160.

Araki, F. 2006. Monte Carlo study of a Cyberknife stereotactic radiosurgery system. *Med Phys* 33:2955–2963.

Ay, M. R., M. Shahriari, S. Sarkar, M. Adib, and H. Zaidi. 2004. Monte Carlo simulation of x-ray spectra in diagnostic radiology and mammography using MCNP4C. *Phys Med Biol* 49:4897–4917.

Ay, M. R., and H. Zaidi. 2005. Development and validation of MCNP4C-based Monte Carlo simulator for fan- and cone-beam x-ray CT. *Phys Med Biol* 50:4863–4885.

Ayyangar, K. M., and S. B. Jiang. 1998. Do we need Monte Carlo treatment planning for linac based radiosurgery? A case study. *Med Dosim* 23:161–168.

Bazalova, M., and F. Verhaegen. 2007. Monte Carlo simulation of a computed tomography x-ray tube. *Phys Med Biol* 52:5945–5955.

Bednarz, B., and X. G. Xu. 2009. Monte Carlo modeling of a 6 and 18 MV Varian Clinac medical accelerator for in-field and out-of-field dose calculations: Development and validation. *Phys Med Biol* 54:N43–N57.

Belec, J., H. Patrocinio, and F. Verhaegen. 2005. Development of a Monte Carlo model for the Brainlab microMLC. *Phys Med Biol* 50:787–799.

Bielajew, A., R. Mohan, and C.-S. Chui. 1989. Improved bremsstrahlung photon angular sampling in the EGS4 code system In NRCC Report 1-22.

Birch, R., and M. Marshall. 1979. Computation of bremsstrahlung X-ray spectra and comparison with spectra measured with a Ge(Li) detector. *Phys Med Biol* 24:505–517.

Boone, J. M., M. H. Buonocore, and V. N. Cooper, 3rd. 2000. Monte Carlo validation in diagnostic radiological imaging. *Med Phys* 27:1294–1304.

Briesmeister, J. 2000. MCNPTM-A general Monte Carlo N-Particle transport code, Version 4C, LA-13709-M.

Brualla, L., F. Salvat, and R. Palanco-Zamora. 2009. Efficient Monte Carlo simulation of multileaf collimators using geometry-related variance-reduction techniques. *Phys Med Biol* 54:4131–4149.

Bush, K., R. Townson, and S. Zavgorodni. 2008. Monte Carlo simulation of RapidArc radiotherapy delivery. *Phys Med Biol* 53:N359–N370.

Carlsson Tedgren, Å., S. de Luelmo, and J.-E. Jan-Erik Grindborg. 2010. Characterization of a ⁶⁰Co unit at a secondary standard dosimetry laboratory: Monte Carlo simulations compared to measurements and results from the literature. *Med Phys* 37:2777–2786.

Chaney, E. L., T. J. Cullip, and T. A. Gabriel. 1994. A Monte Carlo study of accelerator head scatter. *Med Phys* 21:1383–1390.

Chaves, A., M. C. Lopes, C. C. Alves, C. Oliveira, L. Peralta, P. Rodrigues, and A. Trindade. 2004. A Monte Carlo multiple source model applied to radiosurgery narrow photon beams. *Med Phys* 31:2192–2204.

Chetty, I., J. J. DeMarco, and T. D. Solberg. 2000. A virtual source model for Monte Carlo modeling of arbitrary intensity distributions. *Med Phys* 27:166–172.

Cheung, J. Y., and K. N. Yu. 2006. Study of scattered photons from the collimator system of Leksell Gamma Knife using the EGS4 Monte Carlo code. *Med Phys* 33:41–45.

Connell, T., and J. L. Robar. 2010. Low-Z target optimization for spatial resolution improvement in megavoltage imaging. *Med Phys* 37:124–131.

d'Errico, F., R. Nath, L. Tana, G. Curzio, and W. G. Alberts. 1998. In-phantom dosimetry and spectrometry of photoneutrons from an 18 MV linear accelerator. *Med Phys* 25:1717–1724.

De Vlamynck, K., H. Palmans, F. Verhaegen, C. De Wagter, W. De Neve, and H. Thierens. 1999. Dose measurements compared with Monte Carlo simulations of narrow 6 MV multileaf collimator shaped photon beams. *Med Phys* 26:1874–1882.

Deng, J., S. B. Jiang, A. Kapur, J. Li, T. Pawlicki, and C. M. Ma. 2000. Photon beam characterization and modelling for Monte Carlo treatment planning. *Phys Med Biol* 45:411–427.

Deng, J., T. Pawlicki, Y. Chen, J. Li, S. B. Jiang, and C. M. Ma. 2001. The MLC tongue-and-groove effect on IMRT dose distributions. *Phys Med Biol* 46:1039–1060.

Desobry, G. E., and A. L. Boyer. 1991. Bremsstrahlung review: An analysis of the Schiff spectrum. *Med Phys* 18:497–505.

Ding, G. X. 2002a. Energy spectra, angular spread, fluence profiles and dose distributions of 6 and 18 MV photon beams: Results of Monte Carlo simulations for a varian 2100EX accelerator. *Phys Med Biol* 47:1025–1046.

Ding, G. X. 2002b. Dose discrepancies between Monte Carlo calculations and measurements in the buildup region for a high-energy photon beam. *Med Phys* 29:2459–2463.

Ding, G. X., C. Duzenli, and N. I. Kalach. 2002. Are neutrons responsible for the dose discrepancies between Monte Carlo calculations and measurements in the build-up region for a high-energy photon beam? *Phys Med Biol* 47:3251–3261.

Ding, G. X., D. M. Duggan, and C. W. Coffey. 2006. Commissioning stereotactic radiosurgery beams using both experimental and theoretical methods. *Phys Med Biol* 51:2549–2566.

Downes, P., R. Jarvis, E. Radu, I. Kawrakow, and E. Spezi. 2009. Monte Carlo simulation and patient dosimetry for a kilovoltage cone-beam CT unit. *Med Phys* 36:4156–4167.

Facure, A., A. X. da Silva, L. A. da Rosa, S. C. Cardoso, and G. F. Rezende. 2008. On the production of neutrons in laminated barriers for 10 MV medical accelerator rooms. *Med Phys* 35:3285–3292.

Faddegon, B. A., M. Aubin, A. Bani-Hashemi, B. Gangadharan, A. R. Gottschalk, O. Morin, D. Sawkey, V. Wu, and S. S. Yom. 2010. Comparison of patient megavoltage cone beam CT images acquired with an unflattened beam from a carbon target and a flattened treatment beam. *Med Phys* 37:1737–1741.

Faddegon, B. A., P. O'Brien, and D. L. Mason. 1999. The flatness of Siemens linear accelerator x-ray fields. *Med Phys* 26:220–228.

Faddegon, B. A., C. K. Ross, and D. W. Rogers. 1990. Forward-directed bremsstrahlung of 10- to 30-MeV electrons incident on thick targets of Al and Pb. *Med Phys* 17:773–785.

Faddegon, B. A., C. K. Ross, and D. W. Rogers. 1991. Angular distribution of bremsstrahlung from 15-MeV electrons incident on thick targets of Be, Al, and Pb. *Med Phys* 18:727–739.

Falcao, R., A. Facure, and A. Silva. 2007. Neutron dose calculation at the maze entrance of medical linear accelerator rooms. *Rad Prot Dosim* 123:283–287.

Fix, M. K., H. Keller, P. Ruegsegger, and E. J. Born. 2000. Simple beam models for Monte Carlo photon beam dose calculations in radiotherapy. *Med Phys* 27:2739–2747.

Fix, M. K., P. Manser, E. J. Born, R. Mini, and P. Ruegsegger. 2001. Monte Carlo simulation of a dynamic MLC based on a multiple source model. *Phys Med Biol* 46:3241–3257.

Fix, M. K., M. Stampanoni, P. Manser, E. J. Born, R. Mini, and P. Ruegsegger. 2001. A multiple source model for 6 MV photon beam dose calculations using Monte Carlo. *Phys Med Biol* 46:1407–1427.

Flampouri, S., P. M. Evans, F. Verhaegen, A. E. Nahum, E. Spezi, and M. Partridge. 2002. Optimization of accelerator target and detector for portal imaging using Monte Carlo simulation and experiment. *Phys Med Biol* 47:3331–3349.

Flampouri, S., H. A. McNair, E. M. Donovan, P. M. Evans, M. Partridge, F. Verhaegen, and C. M. Nutting. 2005. Initial patient imaging with an optimised radiotherapy beam for portal imaging. *Radiother Oncol* 76:63–71.

Ford, R., and W. Nelson. 1978. The EGS Code System: Computer programs for the Monte Carlo simulation of electromagnetic cascade showers (Version 3). In Stanford Linear Accelerator Center Report SLAC-210.

Galbraith, D. M. 1989. Low-energy imaging with high-energy bremsstrahlung beams. *Med Phys* 16:734–746.

García-Pareja, S., P. Galán, F. Manzano, L. Brualla, and A. M. Lallena. 2010. Ant colony algorithm implementation in electron and photon Monte Carlo transport: Application to the commissioning of radiosurgery photon beams. *Med Phys* 37:3782–3790.

Gerbi, B. J., and F. M. Khan. 1990. Measurement of dose in the buildup region using fixed-separation plane-parallel ionization chambers. *Med Phys* 17:17–26.

Ghavami, S. M., A. Mesbahi, and E. Mohammadi. 2010. The impact of automatic wedge filter on photoneutron and photon spectra of an 18-MV photon beam. *Radiat Prot Dosimetry* 138:123–128.

Greene, D., and P. Williams. 1997. *Linear Accelerators for Radiation Therapy*. Institute of Physics Publishing, Bristol and Philadelphia.

Gudowska, I., A. Brahme, P. Andreo, W. Gudowski, and J. Kierkegaard. 1999. Calculation of absorbed dose and biological effectiveness from photonuclear reactions in a bremsstrahlung beam of end point 50 MeV. *Phys Med Biol* 44:2099–2125.

Han, K., D. Ballon, C. Chui, and R. Mohan. 1987. Monte Carlo simulation of a cobalt-60 beam. *Med Phys* 14:414–419.

Hartmann Siantar, C. L. et al. 2001. Description and dosimetric verification of the PERE GRINE Monte Carlo dose calculation system for photon beams incident on a water phantom. *Med Phys* 28: 1322–1337.

Heath, E., and J. Seuntjens. 2003. Development and validation of a BEAMnrc component module for accurate Monte Carlo modelling of the Varian dynamic Millennium multileaf collimator. *Phys Med Biol* 48:4045–4063.

Hounsell, A. R. 1998. Monitor chamber backscatter for intensity modulated radiation therapy using multileaf collimators. *Phys Med Biol* 43:445–454.

Iles, W. 1987. Computation of X-ray bremsstrahlung spectra over an energy range 15 keV–300 keV National Radiological Protection Board Report R204.

Ing, H., W. R. Nelson, and R. A. Shore. 1982. Unwanted photon and neutron radiation resulting from collimated photon beams interacting with the body of radiotherapy patients. *Med Phys* 9:27–33.

Jaffray, D. A., J. J. Battista, A. Fenster, and P. Munro. 1993. X-ray sources of medical linear accelerators: Focal and extra-focal radiation. *Med Phys* 20:1417–1427.

Jan, S. et al. 2004. GATE: A simulation toolkit for PET and SPECT. *Phys Med Biol* 49:4543–4561.

Jarry, G., S. A. Graham, D. J. Moseley, D. J. Jaffray, J. H. Siewerdsen, and F. Verhaegen. 2006. Characterization of scattered radiation in kV CBCT images using Monte Carlo simulations. *Med Phys* 33:4320–4329.

Jiang, S. B., and K. M. Ayyangar. 1998. On compensator design for photon beam intensity-modulated conformal therapy. *Med Phys* 25:668–675.

Jones, A. O., and I. J. Das. 2005. Comparison of inhomogeneity correction algorithms in small photon fields. *Med Phys* 32:766–776.

Jones, B., S. Krishnan, and S. Cho. 2010. Estimation of microscopic dose enhancement factor around gold nanoparticles by Monte Carlo calculations B. *Med Phys* 37:3809–3816.

Kakonyi, R., M. Erdelyi, and G. Szabo. 2009. Monte Carlo analysis of energy dependent anisotropy of bremsstrahlung x-ray spectra. *Med Phys* 36:3897–3905.

Kapur, A., C. Ma, and A. Boyer. 2000. Monte Carlo simulations for multileaf-collimator leaves: design and dosimetry. In *World Congress on Medical Physics and Biomedical Engineering*, Chicago, Illinois, p. 1410.

Kawrakow, I. 2000. Accurate condensed history Monte Carlo simulation of electron transport. II. Application to ion chamber response simulations. *Med Phys* 27:499–513.

Keall, P. J., J. V. Siebers, M. Arnfield, J. O. Kim, and R. Mohan. 2001. Monte Carlo dose calculations for dynamic IMRT treatments. *Phys Med Biol* 46:929–941.

Kim, J. O., J. V. Siebers, P. J. Keall, M. R. Arnfield, and R. Mohan. 2001. A Monte Carlo study of radiation transport through multileaf collimators. *Med Phys* 28:2497–2506.

Kirkby, C., T. Stanescu, S. Rathee, M. Carlone, B. Murray, and B. G. Fallone. 2008. Patient dosimetry for hybrid MRI-radiotherapy systems. *Med Phys* 35:1019–1027.

Koch, H., and J. Motz. 1959. Bremsstrahlung cross section formulas and related data Rev. *Mod Phys* 31:920–955.

Kry, S. F., O. N. Vassiliev, and R. Mohan. 2010. Out-of-field photon dose following removal of the flattening filter from a medical accelerator. *Phys Med Biol* 55:2155–2166.

Kubsad, S. S., T. R. Mackie, M. A. Gehring, D. J. Misisco, B. R. Paliwal, M. P. Mehta, and T. J. Kinsella. 1990. Monte Carlo and convolution dosimetry for stereotactic radiosurgery. *Int J Radiat Oncol Biol Phys* 19:1027–1035.

Küster, G. 1999. Monte Carlo Studies for the Optimisation of Hardware Used in Conformal Radiation Therapy. PhD Thesis, University of Heidelberg, Heidelberg.

Lagendijk, J. J., B. W. Raaymakers, A. J. Raaijmakers, J. Overweg, K. J. Brown, E. M. Kerkhof, R. W. van der Put, B. Hardemark, M. van Vulpen, and U. A. van der Heide. 2008. MRI/linac integration. *Radiother Oncol* 86:25–29.

Lamb, A., and S. Blake. 1998. Investigation and modelling of the surface dose from linear accelerator produced 6 and 10 MV photon beams. *Phys Med Biol* 43:1133–1146.

Leal, A., F. Sanchez-Doblado, R. Arrans, J. Rosello, E. C. Pavon, and J. I. Lagares. 2003. Routine IMRT verification by means of an automated Monte Carlo simulation system. *Int J Radiat Oncol Biol Phys* 56:58–68.

Li, J. S., T. Pawlicki, J. Deng, S. B. Jiang, E. Mok, and C. M. Ma. 2000. Validation of a Monte Carlo dose calculation tool for radiotherapy treatment planning. *Phys Med Biol* 45:2969–2985.

Li, X. A., L. Ma, S. Naqvi, R. Shih, and C. Yu. 2001. Monte Carlo dose verification for intensity-modulated arc therapy. *Phys Med Biol* 46:2269–2282.

Libby, B., J. Siebers, and R. Mohan. 1999. Validation of Monte Carlo generated phase-space descriptions of medical linear accelerators. *Med Phys* 26:1476–1483.

Lin, S. Y., T. C. Chu, and J. P. Lin. 2001. Monte Carlo simulation of a clinical linear accelerator. *Appl Radiat Isot* 55:759–765.

Ling, C. C., M. C. Schell, and S. N. Rustgi. 1982. Magnetic analysis of the radiation components of a 10 MV photon beam. *Med Phys* 9:20–26.

Liu, H. H., T. R. Mackie, and E. C. McCullough. 1997a. Calculating output factors for photon beam radiotherapy using a convolution/superposition method based on a dual source photon beam model. *Med Phys* 24:1975–1985.

Liu, H. H., T. R. Mackie, and E. C. McCullough. 1997b. A dual source photon beam model used in convolution/superposition dose calculations for clinical megavoltage x-ray beams. *Med Phys* 24:1960–1974.

Liu, H. H., T. R. Mackie, and E. C. McCullough. 2000. Modeling photon output caused by backscattered radiation into the monitor chamber from collimator jaws using a Monte Carlo technique. *Med Phys* 27:737–744.

Liu, D., E. Poon, M. Bazalova, B. Reniers, M. Evans, T. Rusch, and F. Verhaegen. 2008. Spectroscopic characterization of a novel electronic brachytherapy system. *Phys Med Biol* 53:61–75.

Liu, H. H., and F. Verhaegen. 2002. An investigation of energy spectrum and lineal energy variations in mega-voltage photon beams used for radiotherapy. *Radiat Prot Dosimetry* 99:425–427.

Liu, H. H., F. Verhaegen, and L. Dong. 2001. A method of simulating dynamic multileaf collimators using Monte Carlo techniques for intensity-modulated radiation therapy. *Phys Med Biol* 46:2283–2298.

Loewenthal, E., E. Loewinger, E. Bar-Avraham, and G. Barnea. 1992. Measurement of the source size of a 6- and 18-MV radiotherapy linac. *Med Phys* 19:687–690.

Lopez Medina, A., A. Teijeiro, J. Garcia, J. Esperon, J. A. Terron, D. P. Ruiz, and M. C. Carrion. 2005. Characterization of electron contamination in megavoltage photon beams. *Med Phys* 32:1281–1292.

Lovelock, D. M., C. S. Chui, and R. Mohan. 1995. A Monte Carlo model of photon beams used in radiation therapy. *Med Phys* 22:1387–1394.

Lutz, W. R., N. Maleki, and B. E. Bjarngard. 1988. Evaluation of a beam-spot camera for megavoltage x rays. *Med Phys* 15:614–617.

Ma, C.-M. 1998. Characterisation of computer simulated radiotherapy beams for Monte Carlo treatment planning. *Radiat Phys Chem* 35:329–344.

Ma, C. M., and S. B. Jiang. 1999. Monte Carlo modelling of electron beams from medical accelerators. *Phys Med Biol* 44:R157–R189.

Ma, C. M., J. S. Li, T. Pawlicki, S. B. Jiang, J. Deng, M. C. Lee, T. Koumrian, M. Luxton, and S. Brain. 2002. A Monte Carlo dose calculation tool for radiotherapy treatment planning. *Phys Med Biol* 47:1671–1689.

Ma, C. M., T. Pawlicki, S. B. Jiang, J. S. Li, J. Deng, E. Mok, A. Kapur, L. Xing, L. Ma, and A. L. Boyer. 2000. Monte Carlo verification of IMRT dose distributions from a commercial treatment planning optimization system. *Phys Med Biol* 45:2483–2495.

Ma, C. M., R. A. Price, Jr., J. S. Li, L. Chen, L. Wang, E. Fourkal, L. Qin, and J. Yang. 2004. Monitor unit calculation for Monte Carlo treatment planning. *Phys Med Biol* 49:1671–1687.

Mackie, T. R., and J. W. Scrimger. 1982. Contamination of a 15-MV photon beam by electrons and scattered photons. *Radiology* 144:403–409.

Mah, D. W., D. M. Galbraith, and J. A. Rawlinson. 1993. Low-energy imaging with high-energy bremsstrahlung beams: Analysis and scatter reduction. *Med Phys* 20:653–665.

Mainegra-Hing, E., and I. Kawrakow. 2006. Efficient x-ray tube simulations. *Med Phys* 33:2683–2690.

Mainegra-Hing, E., and I. Kawrakow. 2008. Fast Monte Carlo calculation of scatter corrections for CBCT images. *J Phys Conf Ser* 10.1088/1742–6596/102/1/012017

Malataras, G., C. Kappas, and D. M. Lovelock. 2001. A Monte Carlo approach to electron contamination sources in the Saturne-25 and -41. *Phys Med Biol* 46:2435–2446.

Mazurier, J., F. Salvat, B. Chauvenet, and J. Barthe. 1999. Simulation of photon beams from a Saturne 43 accelerator using the code PENELOPE. *Phys Med* XV:101–110.

McCall, R. C., R. D. McIntyre, and W. G. Turnbull. 1978. Improvement of linear accelerator depth-dose curves. *Med Phys* 5:518–524.

Mesbahi, A. 2007. Dosimetric characteristics of unflattened 6 MV photon beams of a clinical linear accelerator: A Monte Carlo study. *Appl Radiat Isot* 65:1029–1036.

Mesbahi, A. 2009. A Monte Carlo study on neutron and electron contamination of an unflattened 18-MV photon beam. *Appl Radiat Isot* 67:55–60.

Mesbahi, A., M. Fix, M. Allahverdi, E. Grein, and H. Garaati. 2005. Monte Carlo calculation of Varian 2300C/D Linac photon beam characteristics: A comparison between MCNP4C, GEANT3 and measurements. *Appl Radiat Isot* 62:469–477.

Mesbahi, A., P. Mehnati, A. Keshtkar, and A. Farajollahi. 2007. Dosimetric properties of a flattening filter-free 6-MV photon beam: A Monte Carlo study. *Radiat Med* 25:315–324.

Mohan, R., C. Chui, and L. Lidofsky. 1985. Energy and angular distributions of photons from medical linear accelerators. *Med Phys* 12:592–597.

Mora, G. M., A. Maio, and D. W. Rogers. 1999. Monte Carlo simulation of a typical ^{60}Co therapy source. *Med Phys* 26:2494–2502.

Moskvin, V., C. DesRosiers, L. Papiez, R. Timmerman, M. Randall, and P. DesRosiers. 2002. Monte Carlo simulation of the Leksell Gamma Knife: I. Source modelling and calculations in homogeneous media. *Phys Med Biol* 47:1995–2011.

Munro, P., J. A. Rawlinson, and A. Fenster. 1988. Therapy imaging: Source sizes of radiotherapy beams. *Med Phys* 15:517–524.

Nelson, W., H. Hirayama, and D. Rogers. 1985. The EGS4 code system In Stanford Linear Accelerator Center Report SLAC-265.

Nilsson, B., and A. Brahme. 1979. Absorbed dose from secondary electrons in high energy photon beams. *Phys Med Biol* 24:901–912.

Nilsson, B., and A. Brahme. 1981. Contamination of high-energy photon beams by scattered photons. *Strahlentherapie* 157:181–186.

Omrane, L. B., F. Verhaegen, N. Chahed, and S. Mtimet. 2003. An investigation of entrance surface dose calculations for diagnostic radiology using Monte Carlo simulations and radiotherapy dosimetry formalisms. *Phys Med Biol* 48:1809–1824.

Ongaro, C., A. Zanini, U. Nastasi, J. Rodenas, G. Ottaviano, C. Manfredotti, and K. W. Burn. 2000. Analysis of photoneutron spectra produced in medical accelerators. *Phys Med Biol* 45:L55–61.

Ostapiak, O. Z., P. F. O'Brien, and B. A. Faddegon. 1998. Megavoltage imaging with low Z targets: Implementation and characterization of an investigational system. *Med Phys* 25:1910–1918.

Padikal, T. N., and J. A. Deye. 1978. Electron contamination of a high-energy X-ray beam. *Phys Med Biol* 23:1086–1092.

Paskalev, K., J. Seuntjens, and E. Podgorsak. 2002. Dosimetry of Ultra Small Photon Fields. In *Recent Developments in Accurate Radiation Dosimetry*, Medical Physics Publishing, Madison, WI, pp. 298–318.

Paskalev, K. A., J. P. Seuntjens, H. J. Patrocinio, and E. B. Podgorsak. 2003. Physical aspects of dynamic stereotactic radiosurgery

with very small photon beams (1.5 and 3 mm in diameter). *Med Phys* 30:111–118.

Patau, J., C. Vernes, M. Terrissol, and M. Malbert. 1978. Calcul des caracteristiques qualitatives (TEL, F.Q., equivalent de dose) d'un faisceau de photons de freinage a usage medical, par simulation de sa creation et de son transport. In *Sixth Symposium on Microdosimetry*. J. Booz, and H. Ebert, editors. Harwood Academic Publishing, London, pp. 579–588.

Petti, P. L., M. S. Goodman, J. M. Sisterson, P. J. Biggs, T. A. Gabriel, and R. Mohan. 1983a. Sources of electron contamination for the Clinac-35 25-MV photon beam. *Med Phys* 10:856–861.

Petti, P. L., M. S. Goodman, T. A. Gabriel, and R. Mohan. 1983b. Investigation of buildup dose from electron contamination of clinical photon beams. *Med Phys* 10:18-24.

Poludniowski, G. G. 2007. Calculation of x-ray spectra emerging from an x-ray tube. Part II. X-ray production and filtration in x-ray targets. *Med Phys* 34:2175–2186.

Poludniowski, G. G., and P. M. Evans. 2007. Calculation of x-ray spectra emerging from an x-ray tube. Part I. electron penetration characteristics in x-ray targets. *Med Phys* 34:2164–2174.

Poludniowski, G., G. Landry, F. DeBlois, P. M. Evans, and F. Verhaegen. 2009. SpekCalc: A program to calculate photon spectra from tungsten anode x-ray tubes. *Phys Med Biol* 54: N433–N438.

Poon, E., and F. Verhaegen. 2005. Accuracy of the photon and electron physics in GEANT4 for radiotherapy applications. *Med Phys* 32:1696–1711.

Popescu, I. A., C. P. Shaw, S. F. Zavgorodni, and W. A. Beckham. 2005. Absolute dose calculations for Monte Carlo simulations of radiotherapy beams. *Phys Med Biol* 50:3375–3392.

Reynaert, N., M. Coghe, B. De Smedt, L. Paelinck, B. Vanderstraeten, W. De Gersem, B. Van Duyse, C. De Wagter, W. De Neve, and H. Thierens. 2005. The importance of accurate linear accelerator head modelling for IMRT Monte Carlo calculations. *Phys Med Biol* 50:831–846.

Robar, J. L., T. Connell, W. Huang, and R. G. Kelly. 2009. Megavoltage planar and cone-beam imaging with low-Z targets: Dependence of image quality improvement on beam energy and patient separation. *Med Phys* 36: 3955–3963.

Roberts, D. A., V. N. Hansen, A. C. Niven, M. G. Thompson, J. Seco, and P. M. Evans. 2008. A low Z linac and flat panel imager: Comparison with the conventional imaging approach. *Phys Med Biol* 53:6305–6319.

Rodriguez, M. L. 2008. PENLINAC: Extending the capabilities of the Monte Carlo code PENELOPE for the simulation of therapeutic beams. *Phys Med Biol* 53:4573–4593.

Rogers, D., G. Ewart, A. Bielajew, and G. van Dyk. 1988. Calculation of electron contamination in a ^{60}Co therapy beam. In *Dosimetry in Radiotherapy*. International Atomic Energy Agency, Vienna, pp. 303–312.

Rogers, D. W., B. A. Faddegon, G. X. Ding, C. M. Ma, J. We, and T. R. Mackie. 1995. BEAM: A Monte Carlo code to simulate radiotherapy treatment units. *Med Phys* 22:503–524.

Rustgi, S. N., Z. C. Gromadzki, C. C. Ling, and E. D. Yorke. 1983. Contaminant electrons in the build-up region of a 4 MV photon beam. *Phys Med Biol* 28:659–665.

Schach von Wittenau, A. E., P. M. Bergstrom, Jr., and L. J. Cox. 2000. Patient-dependent beam-modifier physics in Monte Carlo photon dose calculations. *Med Phys* 27:935–947.

Schach von Wittenau, A. E., L. J. Cox, P. M. Bergstrom, Jr., W. P. Chandler, C. L. Hartmann Siantar, and R. Mohan. 1999. Correlated histogram representation of Monte Carlo derived medical accelerator photon-output phase space. *Med Phys* 26:1196–1211.

Schach von Wittenau, A. E., C. M. Logan, and R. D. Rikard. 2002. Using a tungsten rollbar to characterize the source spot of a megavoltage bremsstrahlung linac. *Med Phys* 29:1797–1806.

Scott, A. J., A. E. Nahum, and J. D. Fenwick. 2008. Using a Monte Carlo model to predict dosimetric properties of small radiotherapy photon fields. *Med Phys* 35:4671–4684.

Scott, A. J., A. E. Nahum, and J. D. Fenwick. 2009. Monte Carlo modeling of small photon fields: Quantifying the impact of focal spot size on source occlusion and output factors, and exploring miniphantom design for small-field measurements. *Med Phys* 36:3132–3144.

Sempau, J., E. Acosta, J. Baro, J. Fernandez-Varea, and F. Salvat. 1997. An algorithm for Monte Carlo simulation of coupled electron-photon transport. *Nucl Instr Meth Phys Res B* 132:377–390.

Sharma, S. C., J. T. Ott, J. B. Williams, and D. Dickow. 2010. Clinical implications of adopting Monte Carlo treatment planning for CyberKnife. *J Appl Clin Med Phys* 11:3142.

Sheikh-Bagheri, D., and D. W. Rogers. 2002a. Monte Carlo calculation of nine megavoltage photon beam spectra using the BEAM code. *Med Phys* 29:391–402.

Sheikh-Bagheri, D., and D. W. Rogers. 2002b. Sensitivity of megavoltage photon beam Monte Carlo simulations to electron beam and other parameters. *Med Phys* 29:379–390.

Sheikh-Bagheri, D., D. W. Rogers, C. K. Ross, and J. P. Seuntjens. 2000. Comparison of measured and Monte Carlo calculated dose distributions from the NRC linac. *Med Phys* 27:2256–2266.

Shih, R., X. Li, and J. Chu. 2001. Dosimetric characteristics of dynamic wedged fields: A Monte Carlo study. *Phys Med Biol* 46:N281–292.

Siebers, J. V., P. J. Keall, J. O. Kim, and R. Mohan. 2002. A method for photon beam Monte Carlo multileaf collimator particle transport. *Phys Med Biol* 47:3225–3249.

Sikora, M., and M. Alber. 2009. A virtual source model of electron contamination of a therapeutic photon beam. *Phys Med Biol* 54:7329–7344.

Sikora, M., O. Dohm, and M. Alber. 2007. A virtual photon source model of an Elekta linear accelerator with integrated mini MLC for Monte Carlo based IMRT dose calculation. *Phys Med Biol* 52:4449–4463.

Sixel, K. E., and B. A. Faddegon. 1995. Calculation of x-ray spectra for radiosurgical beams. *Med Phys* 22:1657–1661.

Sixel, K. E., and E. B. Podgorsak. 1993. Buildup region of high-energy x-ray beams in radiosurgery. *Med Phys* 20: 761–764.

Sixel, K. E., and E. B. Podgorsak. 1994. Buildup region and depth of dose maximum of megavoltage x-ray beams. *Med Phys* 21:411–416.

Solberg, T. D., F. E. Holly, A. A. De Salles, R. E. Wallace, and J. B. Smathers. 1995. Implications of tissue heterogeneity for radiosurgery in head and neck tumors. *Int J Radiat Oncol Biol Phys* 32:235–239.

Sonke, J. J., B. Brand, and M. van Herk. 2003. Focal spot motion of linear accelerators and its effect on portal image analysis. *Med Phys* 30:1067–1075.

Spezi, E., P. Downes, E. Radu, and R. Jarvis. 2009. Monte Carlo simulation of an x-ray volume imaging cone beam CT unit. *Med Phys* 36:127–136.

Spezi, E., and D. G. Lewis. 2002. Full forward Monte Carlo calculation of portal dose from MLC collimated treatment beams. *Phys Med Biol* 47:377–390.

Spezi, E., D. G. Lewis, and C. W. Smith. 2001. Monte Carlo simulation and dosimetric verification of radiotherapy beam modifiers. *Phys Med Biol* 46:3007–3029.

Taleei, R., and M. Shahriari. 2009. Monte Carlo simulation of X-ray spectra and evaluation of filter effect using MCNP4C and FLUKA code. *Appl Radiat Isot* 67:266–271.

Taylor, M., T. Kron, and R. Franich. 2010, pending. A contemporary review of stereotactic radiotherapy: Inherent dosimetric complexities and the potential for detriment. *Acta Oncol* 50:483–508.

Teke, T., A. M. Bergman, W. Kwa, B. Gill, C. Duzenli, and I. A. Popescu. 2009. Monte Carlo based, patient-specific RapidArc QA using Linac log files. *Med Phys* 37:116–123.

Tillikainen, L., S. Siljamaki, H. Helminen, J. Alakuijala, and J. Pyyry. 2007. Determination of parameters for a multiple-source model of megavoltage photon beams using optimization methods. *Phys Med Biol* 52:1441–1467.

Titt, U., O. N. Vassiliev, F. Ponisch, S. F. Kry, and R. Mohan. 2006. Monte Carlo study of backscatter in a flattening filter free clinical accelerator. *Med Phys* 33:3270–3273.

Treuer, H., R. Boesecke, W. Schlegel, G. Hartmann, R. Müller, and V. Sturm. 1993. The source-density function: Determination from measured lateral dose distributions and use for convolution dosimetry. *Phys Med Biol* 38:1895–1909.

Tryggestad, E., M. Armour, I. Iordachita, F. Verhaegen, and J. W. Wong. 2009. A comprehensive system for dosimetric commissioning and Monte Carlo validation for the small animal radiation research platform. *Phys Med Biol* 54:5341–5357.

Tsechanski, A., A. F. Bielajew, S. Faermann, and Y. Krutman. 1998. A thin target approach for portal imaging in medical accelerators. *Phys Med Biol* 43:2221–2236.

Udale, M. 1988. A Monte Carlo investigation of surface doses for broad electron beams. *Phys Med Biol* 33:939–953.

Udale-Smith, M. 1992. Monte Carlo calculations of electron beam parameters for three Philips linear accelerators. *Phys Med Biol* 37:85–105.

van der Zee, W., and J. Welleweerd. 1999. Calculating photon beam characteristics with Monte Carlo techniques. *Med Phys* 26:1883–1892.

van der Zee, W., and J. Welleweerd. 2002. A Monte Carlo study on internal wedges using BEAM. *Med Phys* 29:876–885.

Van Dyk, J. 2005. *Modern Technology of Radiation Oncology*, Volumes 1 and 2 (hardcover) Medical Physics Publishing, Madison, WI.

Vassiliev, O. N., U. Titt, S. F. Kry, F. Ponisch, M. T. Gillin, and R. Mohan. 2006. Monte Carlo study of photon fields from a flattening filter-free clinical accelerator. *Med Phys* 33: 820–827.

Verhaegen, F., and I. A. Castellano. 2002. Microdosimetric characterisation of 28 kVp Mo/Mo, Rh/Rh, Rh/Al, W/Rh and Mo/Rh mammography X ray spectra. *Radiat Prot Dosimetry* 99:393–396.

Verhaegen, F., and I. J. Das. 1999. Monte Carlo modelling of a virtual wedge. *Phys Med Biol* 44:N251–N259.

Verhaegen, F., I. J. Das, and H. Palmans. 1998. Monte Carlo dosimetry study of a 6 MV stereotactic radiosurgery unit. *Phys Med Biol* 43:2755–2768.

Verhaegen, F., and H. H. Liu. 2001. Incorporating dynamic collimator motion in Monte Carlo simulations: An application in modelling a dynamic wedge. *Phys Med Biol* 46:287–296.

Verhaegen, F., A. E. Nahum, S. Van de Putte, and Y. Namito. 1999. Monte Carlo modelling of radiotherapy kV x-ray units. *Phys Med Biol* 44:1767–1789.

Verhaegen, F., B. Reniers, F. Deblois, S. Devic, J. Seuntjens, and D. Hristov. 2005. Dosimetric and microdosimetric study of contrast-enhanced radiotherapy with kilovolt x-rays. *Phys Med Biol* 50:3555–3569.

Verhaegen, F., and J. Seuntjens. 2003. Monte Carlo modelling of external radiotherapy photon beams. *Phys Med Biol* 48:R107–164.

Verhaegen, F., R. Symonds-Tayler, H. H. Liu, and A. E. Nahum. 2000. Backscatter towards the monitor ion chamber in high-energy photon and electron beams: Charge integration versus Monte Carlo simulation. *Phys Med Biol* 45:3159–3170.

Ververs, J. D., M. J. Schaefer, I. Kawrakow, and J. V. Siebers. 2009. A method to improve accuracy and precision of water surface identification for photon depth dose measurements. *Med Phys* 36:1410–1420.

Walters, B., I. Kawrakow, and D. Rogers. 2009. *DOSXYZnrc Users Manual*. NRCC, Ottawa.

Wang, L. L., and K. Leszczynski. 2007. Estimation of the focal spot size and shape for a medical linear accelerator by Monte Carlo simulation. *Med Phys* 34:485–488.

Wilcox, E. E., G. M. Daskalov, H. Lincoln, R. C. Shumway, B. M. Kaplan, and J. M. Colasanto. 2010. Comparison of planned dose distributions calculated by Monte Carlo and Ray-Trace algorithms for the treatment of lung tumors with cyberknife: A preliminary study in 33 patients. *Int J Radiat Oncol Biol Phys* 77:277–284.

Yamamoto, T., T. Teshima, S. Miyajima, M. Matsumoto, H. Shiomi, T. Inoue, and H. Hirayama. 2002. Monte Carlo calculation of depth doses for small field of CyberKnife. *Radiat Med* 20:305–310.

Yanch, J. C., and K. J. Harte. 1996. Monte Carlo simulation of a miniature, radiosurgery x-ray tube using the ITS 3.0 coupled electron-photon transport code. *Med Phys* 23:1551–1558.

Yang, J., J. S. Li, L. Qin, W. Xiong, and C. M. Ma. 2004. Modelling of electron contamination in clinical photon beams for Monte Carlo dose calculation. *Phys Med Biol* 49:2657–2673.

Yorke, E. D., C. C. Ling, and S. Rustgi. 1985. Air-generated electron contamination of 4 and 10 MV photon beams: A comparison of theory and experiment. *Phys Med Biol* 30:1305–1314.

Zhu, T. C., and J. R. Palta. 1998. Electron contamination in 8 and 18 MV photon beams. *Med Phys* 25:12–19.

6

Monte Carlo Modeling of External Electron Beams in Radiotherapy

Frank Verhaegen
Maastro Clinic

6.1 Introduction

The introduction given in Chapter 5 also serves as an introduction to this chapter. In this chapter, we will give an overview of the work performed in the field of Monte Carlo modeling of clinical electron beams over the last 30 years or so. It is widely believed that to calculate a patient dose correctly in electron beams, Monte Carlo simulations are needed, more so than for radiotherapy photon beams where there exist alternative dose calculation methods that rival Monte Carlo simulations in accuracy and performance. For electron beams, the situation is different, with many papers exposing flaws in analytical dose calculation methods, even in modern treatment planning systems. It is, therefore, not surprising that Monte Carlo simulations are seeing a widespread acceptation in the radiotherapy community.

6.2 Electron Beams from Clinical Linacs

Electron beam radiotherapy is administered much less frequently than photon beam treatment, most commonly as a boost to a small superficial target. The energies used nowadays mostly range from 4 to 20 MeV. The schematics of an electron beam linac for Monte Carlo modeling are represented in Figure 6.1. The most important differences with a photon beam linac are the absence of the photon target, the presence of a scattering foil to broaden the beam, and a multistage collimator to shape the beam close to the irradiated volume. The latter is needed because electrons scatter much more in air than photons and they therefore require collimation close to the surface of the patient. Monte Carlo modeling of electron beams will be treated more concisely in this chapter than photon beam modeling because of its less frequent clinical use. The BEAM Monte Carlo code is again the most heavily used tool in many studies.

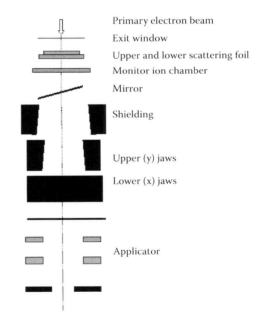

FIGURE 6.1 Schematic representation of an electron beam linac.

6.2.1 Early Work in Electron Beam Modeling

Berger and Seltzer (1978) were among the first to model the interaction of electrons with lead scattering foils, which is the linac component that influences the electron beam most significantly and where contaminant bremsstrahlung photon production takes place. They found that the intervening air causes a significant energy degradation of the electron beam, while the effect of the air can be ignored in high-energy photon beams. It needs to be pointed out that the same group that pioneered Monte Carlo photon beam modeling was also one of

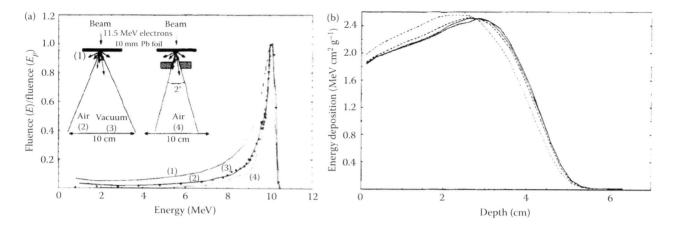

FIGURE 6.2   (a) The influence of the simulation geometry in electron beams on the electron energy spectrum. (b) The influence of the spectral variation on water depth dose. (From Andreo, P. et al. 1989. *Phys Med Biol* 34:751–768. With permission.)

the first to present a Monte Carlo model for a clinical electron beam. Borrell-Carbonell et al. (1980) published in 1980 simplified models of several linacs; they modeled beam collimators as apertures with no interactions in their walls; therefore, no secondary particles were produced. This approximation does not result in realistic particle fluence for electron beams. Rogers and Bielajew (1986) compared calculated and measured depth dose curves in water for monoenergetic electrons. They noted that the simulations predicted a less steep dose gradient near the surface and a too steep dose fall-off beyond the depth of maximum dose. When the electrons were made to pass through simulated exit window, scattering foils, and air, the differences were reduced. They pointed out that electron energy straggling has an important influence on the depth dose. Andreo and Fransson (1989) studied stopping-power ratios in water derived from a variety of simple and more complex electron beam models. They showed that stopping powers are relatively insensitive to the details of the electron spectra. They also indicated that it is important to preserve the correlation between energy and angle, which has implications for virtual source models. In another paper (Andreo et al., 1989), they studied the influence of the electron energy and angular distribution on the depth dependence of stopping-power ratios. Figure 6.2 shows that for a variety of simulation geometries, from simple to more complex (depicted in the insert in (a)), the electron energy spectra are more or less broadened. This causes small differences in water depth doses and, in turn, in water to air stopping-power ratios. Ebert and coworkers (Ebert and Hoban, 1995a,b, 1996) studied simple models of applicators and cerrobend cutouts. They identified two main processes by which the applicator/cutout may influence the dose distributions: by electron scatter of the inner edges of the aperture and by bremsstrahlung production. Studies like these may lead to optimized collimator design. Other early studies of simplified Monte Carlo models have been published (Manfredotti et al., 1987; Keall and Hoban, 1994).

6.2.2 Monte Carlo Modeling of Complete Electron Linac Beams

6.2.2.1 First Full Linac Models

In general, it is accepted that modeling electron beams is more difficult than photon beams because of the greater sensitivity of particle fluences and absorbed dose distributions to the details of the primary electron beam (energy, spatial, and angular distribution) and the geometry of the linac, in particular, the scattering foils that are present in most linacs, and the applicators/collimators. The pioneering efforts to model complete electron beam geometries with the EGS4 code were made by Udale/Udale-Smith (1988, 1992). Figure 6.3 shows her models for Philips linacs, which included the exit window, the primary collimator, the scattering foils, the monitor chamber, the mirror, the movable photon jaws, the accessory ring, and the applicator.

The modular approach of building linacs paved the way for the later BEAM code with its "component module" approach. To quantify the effects of various parts of the treatment head, she simulated five cases, from a monoenergetic pencil beam in vacuum to the full linac geometry. She used measured depth dose distributions to tune the monoenergetic primary electron source energy by attempting to match the depth of 50% of the maximum dose, R_{50}, and the practical electron range, R_p. Electron range rejection was employed as variance reduction technique to avoid transporting electrons that could not reach the linac exit at great computational cost. Computers have since then become many orders of magnitude faster but it is still good advice to invest time in selecting variance reduction techniques wisely. Udale scored phase-space files at the bottom of the linac, and in a second step these were used for phantom dose calculations, still a common approach today. She also used a second approach, namely, by extracting energy and angular distributions from the phase-space files and using them in a virtual linac model. Correlations between particle position, energy, and angle are ignored this way. She demonstrated that some degree of correlation must be maintained to avoid loss of

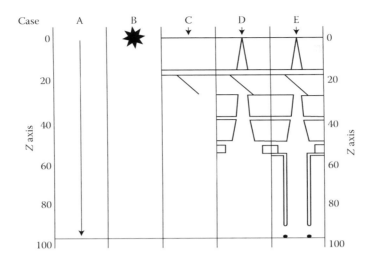

FIGURE 6.3 Five simulated electron beam geometries with increasing complexity from case A through E. In case A, a monoenergetic pencil beam in vacuum impinges directly on the phantom. In case B, a monoenergetic isotropic point source sends electrons through 95 cm of air. Case C adds complexity by modeling the interactions of a monoenergetic pencil beam with an electron window, a scattering foil, mirror, and air. In case D movable (photon), jaws were added, and in case E, the electron applicator is present as well. (From Udale, M. 1988. *Phys Med Biol* 33:939–953. With permission.)

simulation accuracy. Another approach to reduce calculation time was to calculate dose in water separately for primary electrons, secondary electrons, and photons and to scale their contributions.

Udale-Smith (1992) compared models of several linacs and established that some had superior designs, that is, lead to improved dose distributions, in that they produced less contaminant photons, less energetic contaminant photons, less scattered electrons, and narrower electron angular distributions. Monte Carlo simulation is the ideal tool for these kinds of studies.

Other early work was done by Kassaee et al. (1994) on applicator modeling and by Burns et al. (1995) on electron stopping powers and practical ranges in clinical electron beams.

6.2.2.2 Applications of the BEAM Code

As stated before in the sections on photon beam modeling, the advent of the BEAM code in 1995 (Rogers et al., 1995) was a major step forward for Monte Carlo modeling of linacs, including electron beams. In fact, most early results reported with the BEAM code were on electron beams. Also, for electron beams, a wide variety of geometry modules ("component modules"), source geometries, variance reduction techniques, scoring techniques, and tagging methods are available. It remains a very useful and popular tool nowadays.

Figure 6.4, taken from the original BEAM paper (Rogers et al., 1995), demonstrates the excellent agreement that can be obtained for dose distributions with BEAM simulations, provided all the necessary details of the linac are known. This is often a problem such that the user has to resort to "tuning" the model, but in this case it concerned a research linac with very well-known characteristics. A more complete report including medical linacs was given by Ding and Rogers (1995). An example is given in Figure 6.5, including again a representation of particle tracks interacting with the linac.

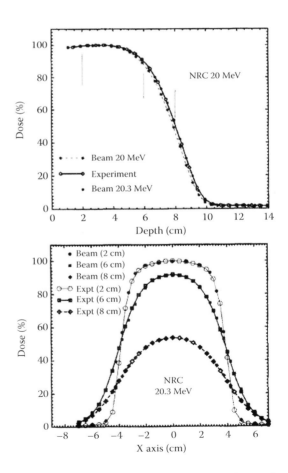

FIGURE 6.4 Measured and Monte Carlo calculated electron dose distributions in water for a 20 MeV electron beam from a research linac with very well-known characteristics. (From Rogers, D. W. et al. 1995. *Med Phys* 22:503–524. With permission.)

In a series of reports, Ma et al. (1997) and Ma and Rogers (1995) investigated the concept of multiple source models that made use of virtual point source positions due to the diffusivity of electron scatter. This makes deriving source models for electron beams more complicated than for photon beams, which exhibit less scatter. Using realistic linac models, they calculated mean energy in water and stopping-power ratios in water (Ding and Rogers, 1996). With a similar approach, electron fluence correction factors used in the conversion of dose measured in plastic to dose in water were also reported (Ding et al., 1997).

Various authors focused on studying components of electron linacs. Verhaegen et al. (2000) studied backscatter to the monitor chamber, an effect potentially present in both electron and photon beams in some linacs. The relative increase in backscatter (increased signal of monitor chamber) was found to be 2% when the photon jaws decrease from a square field size of 40 to 0 cm in a 6 MeV beam. For higher energies, the effect was smaller. The magnitude of the effect is comparable to photon beams. The contribution to the backscattered fluence from downstream linac components other than the jaws (applicator scrapers, field-defining cut-out) was found to be negligible. Figure 6.6 shows the spectra of forward- and backward-moving particles at the monitor chamber. In contrast to photon beams, in electron beams, we obtain a significant difference in spectral shape between the forward and backscattered electrons. As in photon beams, the backscattered photons were found to contribute insignificantly to the energy deposited in the chamber.

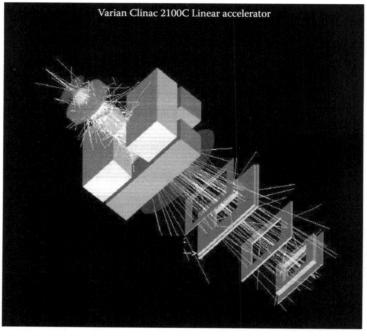

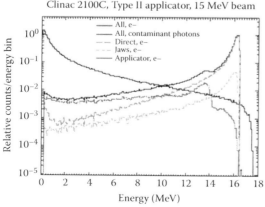

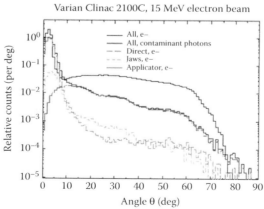

FIGURE 6.5 (**See color insert.**) Monte Carlo model for Varian electron beam (top). The bottom shows particle energy and angular spectra for a 15 MeV electron beam, differentiated according to their origin or interaction site. (Reproduced from Ding, G., and D. Rogers. 1995. Energy spectra, angular spread, and dose distributions of electron beams from various accelerators used in radiotherapy. Report PIRS-0439 (Ottawa: NRCC). With permission.)

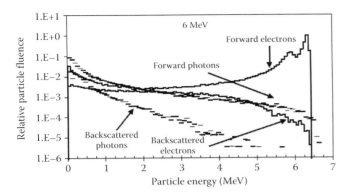

FIGURE 6.6 Monte Carlo simulated forward and backward particles reaching the monitor chamber in a 6 MeV electron beam. (From Verhaegen, F. et al. 2000. *Phys Med Biol* 45:3159–3170. With permission.)

Dose calculations in electron beams are sensitive to the characteristics of the electron spot size hitting the scattering foils. Information on the photon focal spot size in photon beam mode is not helpful for the electron mode, so it must be determined separately. An experimental technique based on a slit camera has been proposed (Huang et al., 2005) to derive the size of the source of bremsstrahlung photons emerging from the scattering foils, which is equivalent to the size of the electron beam impinging on them. They derived a full width at half maximum (FWHM) size of the elliptically shaped electron primary beams as 1.7–2.2 mm for 6–16 MeV electron beams. They also noted a shift of the primary beam up to 8 mm away from the center of the linac. These authors also studied the influence of primary electron beam divergence on dosimetry in large fields. Introducing a 1.6° primary electron beam divergence changed in air measured beam profiles drastically. Primary electron beam information is among the hardest to estimate in clinical electron beams.

It is known that electron beam models are sensitive to the details of the scattering foils, especially for large fields of high-energy electrons. A study (Bieda et al., 2001) considered four parameters of the scattering foils and found that the distance between the foils is critical. This study emphasizes the importance of communication with the manufacturer or verification of the geometry. Large electron fields were also studied in great detail by Faddegon and coworkers (Faddegon et al., 2005, 2008, 2009; Schreiber and Faddegon, 2005; Shea et al., 2010).

6.2.2.3 Studies for Electron Treatment Planning

It is known that conventional treatment planning systems for electron beams have errors in irregular fields and heterogeneous targets (Ding et al., 2005). It is generally accepted that Monte Carlo algorithms have unparalleled accuracy for electron dose calculations. Recently, several Monte Carlo-based treatment planning systems became available. These have been evaluated extensively (Cygler et al., 2004, 2005; Ding et al., 2006; Popple

et al., 2006), which is covered in another chapter in this book. Much of the present-day work is based on the pioneering efforts of a few groups. Fast Monte Carlo codes with electron treatment planning as their main application were introduced by Neuenschwander et al. (1995) and Kawrakow et al. (1996) in the mid-1990s.

Ma and coworkers considered various aspects of Monte Carlo electron treatment planning. They simulated clinical electron beams (Kapur et al., 1998; Ma et al., 1999) and studied beam characterization and modeling for dose calculations (Ma, 1998; Jiang et al., 2000), air gap factors for treatment at extended distances, the commissioning procedure for electron beam treatment planning, stopping-power ratios for dose conversions (Kapur and Ma, 1999), and output factors (Kapur et al., 1998). The latter were also investigated by others (Zhang et al., 1999; Verhaegen et al., 2001). In most of these studies, the BEAM code was used. Virtual source models for electron beams remain an active field of research (Wieslander and Knoos, 2006, 2007).

6.2.2.4 Other Studies with Electron Beams

Several workers developed intensity-modulated electron therapy, in analogy with photon beams. This involves designing or modifying an electron collimator close to the patient and developing optimization methods for the beam delivery. Al-Yahya et al. (2005a,b, 2007) designed a few-leaf electron collimator, consisting of four motor-driven trimmer bars. Monte Carlo modeling was heavily relied upon during the design stage. The full range of rectangular fields that can be delivered with the device, in combination with the available electron energies, were used as input for an inverse planning algorithm based on simulated annealing (Al-Yahya et al., 2005a). This system optimizes the beam delivery based on precalculated patient-specific dose kernels. They demonstrated that highly conformal treatments can be planned this way.

Modulated electron radiation therapy (MERT) whereby both intensity and energy of the beams are modulated was studied intensively by Ma and coworkers (Ma et al., 2000; Lee et al., 2001, 2000; Deng et al., 2002; Jin et al., 2008; Klein et al., 2008). They first did this for a dedicated electron multileaf collimator (Ma et al., 2000; Lee et al., 2001, 2000; Deng et al., 2002), and later by employing the standard photon multileaf collimator to shape the electron beams at short treatment distances (Jin et al., 2008; Klein et al., 2008). They also used inverse Monte Carlo treatment planning.

A final application of electron Monte Carlo modeling to be mentioned in this chapter is the use of portal imaging in electron beams. The bremsstrahlung photon contamination may be used to obtain an image of the patient as was demonstrated by relying on Monte Carlo techniques (Jarry and Verhaegen, 2005) to study the response of the portal imager to the bremsstrahlung photon fluence. Figure 6.7 compares calculated with recorded portal images in an electron beam. It is clear that also here Monte Carlo techniques play an important role in the optimization of the imaging chain.

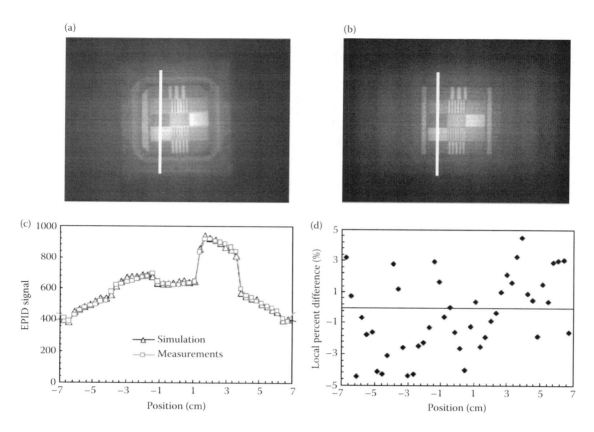

FIGURE 6.7 Measured (a) and simulated (b) images of a test phantom obtained with the bremsstrahlung photons in a 12 MeV electron beam. Profiles (c) and local differences (d) are also shown. The position of the profile is indicated by the white line in (a) and (b). (Reproduced from Jarry, G., and F. Verhaegen. 2005. *Phys Med Biol* 50:4977–4994. With permission.)

6.3 Summary

As in the Chapter 5 on Monte Carlo simulations for photon beams, these techniques are a very useful powerful tool in modern radiotherapy with electron beams. This is even more the case here due to known deficiencies of analytical electron dose calculation techniques. In electron beams, Monte Carlo simulations also play a decisive role in designing complex beam delivery techniques such as in modulated electron therapy. Also in this field, studies with the EGS/EGSnrc-based codes (including the user interface BEAM) dominate, but other codes such as GEANT4 are experiencing an increasing use. Since Monte Carlo codes usually differ more in their cross sections and transport methods for electrons than for photons, care has to be taken to properly benchmark simulations for electron beams. Variance reduction techniques for electrons are another point of interest. Many clinical treatment planning systems are offering Monte Carlo modules for electrons, which will be covered elsewhere in this book.

References

Al-Yahya, K., D. Hristov, F. Verhaegen, and J. Seuntjens. 2005a. Monte Carlo based modulated electron beam treatment planning using a few-leaf electron collimator—Feasibility study. *Phys Med Biol* 50:847–857.

Al-Yahya, K., M. Schwartz, G. Shenouda, F. Verhaegen, C. Freeman, and J. Seuntjens. 2005b. Energy modulated electron therapy using a few leaf electron collimator in combination with IMRT and 3D-CRT: Monte Carlo-based planning and dosimetric evaluation. *Med Phys* 32:2976–2986.

Al-Yahya, K., F. Verhaegen, and J. Seuntjens. 2007. Design and dosimetry of a few leaf electron collimator for energy modulated electron therapy. *Med Phys* 34:4782–4791.

Andreo, P., A. Brahme, A. Nahum, and O. Mattsson. 1989. Influence of energy and angular spread on stopping-power ratios for electron beams. *Phys Med Biol* 34:751–768.

Andreo, P., and A. Fransson. 1989. Stopping-power ratios and their uncertainties for clinical electron beam dosimetry. *Phys Med Biol* 34:1847–1861.

Berger, M., and S. Seltzer. 1978. The influence of scattering foils on absorbed dose distributions from electron beams. Report NBSIR 78-1552 (Gaitherburg: NBS).

Bieda, M. R., J. A. Antolak, and K. R. Hogstrom. 2001. The effect of scattering foil parameters on electron-beam Monte Carlo calculations. *Med Phys* 28:2527–2534.

Borrell-Carbonell, A., J. P. Patau, M. Terrissol, and D. Tronc. 1980. Comparison between experimental measurements and calculated transport simulation for electron dose distributions inside homogeneous phantoms. *Strahlentherapie* 156:186–191.

Burns, D. T., S. Duane, and M. R. McEwen. 1995. A new method to determine ratios of electron stopping powers to an improved accuracy. *Phys Med Biol* 40:733–739.

Cygler, J. E., C. Lochrin, G. M. Daskalov, M. Howard, R. Zohr, B. Esche, L. Eapen, L. Grimard, and J. M. Caudrelier. 2005. Clinical use of a commercial Monte Carlo treatment planning system for electron beams. *Phys Med Biol* 50:1029–1034.

Cygler, J. E., G. M. Daskalov, G. H. Chan, and G. X. Ding. 2004. Evaluation of the first commercial Monte Carlo dose calculation engine for electron beam treatment planning. *Med Phys* 31:142–153.

Deng, J., M. C. Lee, and C. M. Ma. 2002. A Monte Carlo investigation of fluence profiles collimated by an electron specific MLC during beam delivery for modulated electron radiation therapy. *Med Phys* 29:2472–2483.

Ding, G. X., and D. W. Rogers. 1996. Mean energy, energy-range relationships and depth-scaling factors for clinical electron beams. *Med Phys* 23:361–376.

Ding, G. X., J. E. Cygler, C. W. Yu, N. I. Kalach, and G. Daskalov. 2005. A comparison of electron beam dose calculation accuracy between treatment planning systems using either a pencil beam or a Monte Carlo algorithm. *Int J Radiat Oncol Biol Phys* 63:622–633.

Ding, G. X., D. M. Duggan, C. W. Coffey, P. Shokrani, and J. E. Cygler. 2006. First macro Monte Carlo based commercial dose calculation module for electron beam treatment planning—New issues for clinical consideration. *Phys Med Biol* 51:2781–2799.

Ding, G., and D. Rogers. 1995. Energy spectra, angular spread, and dose distributions of electron beams from various accelerators used in radiotherapy. Report PIRS-0439 (Ottawa: NRCC).

Ding, G. X., D. W. Rogers, J. E. Cygler, and T. R. Mackie. 1997. Electron fluence correction factors for conversion of dose in plastic to dose in water. *Med Phys* 24:161–176.

Ebert, M. A., and P. W. Hoban. 1995a. A model for electron-beam applicator scatter. *Med Phys* 22:1419–1429.

Ebert, M. A., and P. W. Hoban. 1995b. A Monte Carlo investigation of electron-beam applicator scatter. *Med Phys* 22:1431–1435.

Ebert, M. A., and P. W. Hoban. 1996. The energy and angular characteristics of the applicator scattered component of an electron beam. *Australas Phys Eng Sci Med* 19:151–159.

Faddegon, B. A., J. Perl, and M. Asai. 2008. Monte Carlo simulation of large electron fields. *Phys Med Biol* 53:1497–1510.

Faddegon, B. A., D. Sawkey, T. O'Shea, M. McEwen, and C. Ross. 2009. Treatment head disassembly to improve the accuracy of large electron field simulation. *Med Phys* 36:4577–4591.

Faddegon, B., E. Schreiber, and X. Ding. 2005. Monte Carlo simulation of large electron fields. *Phys Med Biol* 50:741–753.

Huang, V. W., J. Seuntjens, S. Devic, and F. Verhaegen. 2005. Experimental determination of electron source parameters for accurate Monte Carlo calculation of large field electron therapy. *Phys Med Biol* 50:779–786.

Jarry, G., and F. Verhaegen. 2005. Electron beam treatment verification using measured and Monte Carlo predicted portal images. *Phys Med Biol* 50:4977–4994.

Jiang, S. B., A. Kapur, and C. M. Ma. 2000. Electron beam modeling and commissioning for Monte Carlo treatment planning. *Med Phys* 27:180–191.

Jin, L., C. M. Ma, J. Fan, A. Eldib, R. A. Price, L. Chen, L. Wang et al. 2008. Dosimetric verification of modulated electron radiotherapy delivered using a photon multileaf collimator for intact breasts. *Phys Med Biol* 53:6009–6025.

Kapur, A., and C. M. Ma. 1999. Stopping-power ratios for clinical electron beams from a scatter-foil linear accelerator. *Phys Med Biol* 44:2321–2341.

Kapur, A., C. M. Ma, E. C. Mok, D. O. Findley, and A. L. Boyer. 1998. Monte Carlo calculations of electron beam output factors for a medical linear accelerator. *Phys Med Biol* 43:3479–3494.

Kassaee, A., M. D. Altschuler, S. Ayyalsomayajula, and P. Bloch. 1994. Influence of cone design on the electron beam characteristics on clinical accelerators. *Med Phys* 21:1671–1676.

Kawrakow, I., M. Fippel, and K. Friedrich. 1996. 3D electron dose calculation using a Voxel based Monte Carlo algorithm (VMC). *Med Phys* 23:445–457.

Keall, P., and P. Hoban. 1994. The angular and energy distribution of the primary electron beam Australas. *Phys Eng Sci Med* 17:116–123.

Klein, E. E., M. Vicic, C. M. Ma, D. A. Low, and R. E. Drzymala. 2008. Validation of calculations for electrons modulated with conventional photon multileaf collimators. *Phys Med Biol* 53:1183–1208.

Lee, M. C., J. Deng, J. Li, S. B. Jiang, and C. M. Ma. 2001. Monte Carlo based treatment planning for modulated electron beam radiation therapy. *Phys Med Biol* 46:2177–2199.

Lee, M. C., S. B. Jiang, and C. M. Ma. 2000. Monte Carlo and experimental investigations of multileaf collimated electron beams for modulated electron radiation therapy. *Med Phys* 27:2708–2718.

Ma, C.-M. 1998. Characterization of computer simulated radiotherapy beams for Monte Carlo treatment planning Radiat. *Phys Chem* 53 329–344.

Ma, C. M., B. A. Faddegon, D. W. Rogers, and T. R. Mackie. 1997. Accurate characterization of Monte Carlo calculated electron beams for radiotherapy. *Med Phys* 24:401–416.

Ma, C. M., E. Mok, A. Kapur, T. Pawlicki, D. Findley, S. Brain, K. Forster, and A. L. Boyer. 1999. Clinical implementation of a Monte Carlo treatment planning system. *Med Phys* 26:2133–2143.

Ma, C. M., T. Pawlicki, M. C. Lee, S. B. Jiang, J. S. Li, J. Deng, B. Yi, E. Mok, and A. L. Boyer. 2000. Energy- and intensity-modulated electron beams for radiotherapy. *Phys Med Biol* 45:2293–2311.

Ma, C.-M., and D. Rogers. 1995. Beam characterization: A multiple source model. Report PIRS-0509(D) (Ottawa: NRCC).

Manfredotti, C., U. Nastasi, R. Ragona, and S. Anglesio. 1987. Comparison of three dimensional Monte Carlo simulation and the pencil beam algorithm for an electron beam from a linear accelerator. *Nucl Instrum. Methods A* 255:355.

Neuenschwander, H., T. R. Mackie, and P. J. Reckwerdt. 1995. MMC—A high-performance Monte Carlo code for electron beam treatment planning. *Phys Med Biol* 40:543–574.

O'Shea, T. P., D. L. Sawkey, M. J. Foley, and B. A. Faddegon. 2010. Monte Carlo commissioning of clinical electron beams using large field measurements. *Phys Med Biol* 55:4083–4105.

Popple, R. A., R. Weinber, J. A. Antolak, S. J. Ye, P. N. Pareek, J. Duan, S. Shen, and I. A. Brezovich. 2006. Comprehensive evaluation of a commercial macro Monte Carlo electron dose calculation implementation using a standard verification data set. *Med Phys* 33:1540–1551.

Rogers, D. W., and A. F. Bielajew. 1986. Differences in electron depth-dose curves calculated with EGS and ETRAN and improved energy-range relationships. *Med Phys* 13:687–694.

Rogers, D. W., B. A. Faddegon, G. X. Ding, C. M. Ma, J. We, and T. R. Mackie. 1995. BEAM: A Monte Carlo code to simulate radiotherapy treatment units. *Med Phys* 22:503–524.

Schreiber, E. C., and B. A. Faddegon. 2005. Sensitivity of large-field electron beams to variations in a Monte Carlo accelerator model. *Phys Med Biol* 50:769–778.

Udale, M. 1988. A Monte Carlo investigation of surface doses for broad electron beams. *Phys Med Biol* 33:939–953.

Udale-Smith, M. 1992. Monte Carlo calculations of electron beam parameters for three Philips linear accelerators. *Phys Med Biol* 37:85–105.

Verhaegen, F., C. Mubata, J. Pettingell, A. M. Bidmead, I. Rosenberg, D. Mockridge, and A. E. Nahum. 2001. Monte Carlo calculation of output factors for circular, rectangular, and square fields of electron accelerators (6–20 MeV). *Med Phys* 28:938–949.

Verhaegen, F., R. Symonds-Tayler, H. H. Liu, and A. E. Nahum. 2000. Backscatter towards the monitor ion chamber in high-energy photon and electron beams: Charge integration versus Monte Carlo simulation. *Phys Med Biol* 45:3159–3170.

Wieslander, E., and T. Knoos. 2006. A virtual-accelerator-based verification of a Monte Carlo dose calculation algorithm for electron beam treatment planning in homogeneous phantoms. *Phys Med Biol* 51:1533–1544.

Wieslander, E., and T. Knoos. 2007. A virtual-accelerator-based verification of a Monte Carlo dose calculation algorithm for electron beam treatment planning in clinical situations. *Radiother Oncol* 82:208–217.

Zhang, G. G., D. W. Rogers, J. E. Cygler, and T. R. Mackie. 1999. Monte Carlo investigation of electron beam output factors versus size of square cutout. *Med Phys* 26:743–750.

Dynamic Beam Delivery and 4D Monte Carlo

Emily Heath
Ryerson University

Joao Seco
Harvard Medical School

7.1 Introduction

Dynamic delivery techniques such as dynamic wedges and intensity-modulated radiation therapy (IMRT) employing dynamic multileaf collimators (MLCs), tomotherapy, and intensity-modulated arc therapy (IMAT) are being used on a routine basis in many radiotherapy clinics. In a dynamic beam delivery, the particle fluence is modulated by a beam modifier whose position within the beam is varied as a function of time. In some techniques, not only the fluence is modified but also the incident direction and energy of the beam may be changed, for example, in an arc therapy delivery or a scanned proton beam. In the absence of Monte Carlo (MC) simulations, the influence of the beam modifiers on the spatial and energy distribution of incident particles is often approximated by the treatment planning system with poorly defined dosimetric consequences, especially when multiple field segments or beam angles are delivered. Accurate dose calculation methods are required to characterize the dynamic beam modifiers, to test the accuracy of the treatment planning system dose distribution, and to perform independent monitor unit calculations as well as to reconstruct the patient dose delivery. If properly validated, a dynamic MC model of the beam can serve as a commissioning tool to replace extensive complicated measurements, especially if measurement resolution or accuracy is questionable. This topic comprises the first part of this chapter. The methods for simulation of dynamic beam delivery devices will be discussed followed by examples of specific applications to dynamic radiotherapy techniques.

The simulation of dynamic patient geometries has only been investigated recently as the research interest in four-dimensional (4D) radiotherapy techniques for management of respiratory motion has increased. The use of MC-based dose calculation algorithms is a natural choice for such applications due to their excellent dosimetric accuracy in low-density lung tissue compared to the more standard analytical dose calculation algorithms. Different approaches to dynamic patient dose accumulation using MC methods will be discussed in the second part of this chapter as well as issues that may limit their practical application.

Finally, the interaction of dynamic beam delivery methods with tumor motion, referred to as the "interplay effect" is a concern in intensity-modulated photon and proton therapy, which employ moving MLCs or scanned beams. Numerical methods are best suited for such studies, including approaches that use MC algorithms. This topic comprises the final section of this chapter.

7.2 Simulations of Dynamic Beam Delivery

7.2.1 Strategies for Simulating Time-Dependent Beam Geometries

In developing MC algorithms to model the IMRT delivery, a very accurate "blueprint" (geometry, atomic composition, etc.) is required of the beam-modifying devices used to generate the intensity profile. The approaches for simulating dynamic geometries in MC simulations range, in the order of increasing complexity, from modification of particle weights to account for attenuation, performing multiple static simulations of "discrete" geometrical states, to treating the beam motion as a probabilistic problem where the beam geometry is sampled on a particle-by-particle basis from probability distributions describing the fraction of the total delivery time for which each geometrical configuration exists.

The simplest approach, particle weighting, arose from techniques used in analytical dose calculation algorithms to model the influence of beam modifiers. This approach has been favored for its relative computational efficiency compared to simulating particle transport in the beam modifiers. Weighting factors may be determined from linear attenuation based on a ray tracing through the beam modifier geometry. Temporal variations of the beam modifier positions are accounted by scaling the weighting factors by the fraction of the total delivery time that the modifier blocks the beam path. For example, Ma et al. (2000) converted MLC leaf sequence files to intensity maps by accumulating monitor units (MUs) for unblocked areas and for the blocked areas, where the latter contribution is weighted by the average leaf leakage. This approach ignores the effect of the leaf tongue and groove as well as the scatter contributions from the leaves. To model more accurately the MLC leaf geometry, a ray-tracing algorithm can be used to calculate intensity maps (Deng et al., 2001) to more accurately model the influence of the leaf geometry. An example of the fluence maps calculated with and without consideration of the leaf geometry is shown in Figure 7.1.

Keall et al. (2001) also included the scatter contributions by calculating probability maps for leaves being open, closed, or a leaf tip passing through a grid point. The weighting factors for blocking leaves are calculated from linear attenuation along the path length of the particle through the MLC considering the energy of the incident photon. The scatter contribution is obtained by integration of attenuation factors obtained for successive positions of the leaf tips by subsampling the intensity map grid. These MLC simulations should be distinguished from the other previously mentioned approaches because they also include modeling of first-Compton scattering of each incident photon; thus, a simplified photon transport in the MLC was performed. The weights of these scattered photons are modified by the probability of a leaf blocking the particular grid point. A limitation of the particle weighting approach is that the nonuniformity of particle weights leads to greater statistical variance (Ma et al., 2000). To improve the particle weight uniformity, particle splitting may be applied to particles with large weights, while Russian roulette can be used to reduce the number of low-weight particles.

Simulations that model in detail the different geometrical states of a dynamic beam delivery can be divided into two main approaches. The first uses multiple discrete simulations and will be referred to as the "static component simulation" (SCS) method. This approach is logical when geometry changes occur in discrete steps (e.g., the step-and-shoot IMRT); however, for a continuously variable geometry, a reasonable limit must be imposed on the number of geometry samples and thus the temporal resolution. It should be pointed out that simulating more geometries does not necessarily lengthen the calculation time

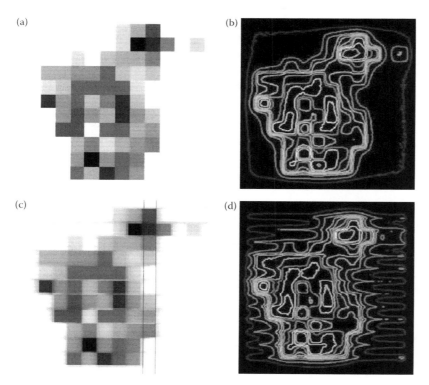

FIGURE 7.1 (**See color insert**.) (a) and (c) Intensity maps used for particle weighting calculated for one field of a step-and-shoot IMRT plan. On the right, (b and d) are shown as the corresponding Monte Carlo dose distributions. The intensity map in (c) is calculated using a ray tracking algorithm, which includes the tongue-and-groove geometry of the MLC leaves. (From Deng J. et al. The MLC tongue-and-groove effect on IMRT dose distributions. *Phys. Med. Biol.* 46, 1039–1060, 2001. With permission.)

as the total number of histories to be simulated can be distributed between the individual geometries. Rather, the overhead comes from the input file preparation, initialization, and postprocessing steps. These SCS simulations can be run separately and the results can be recombined later or by restarting a previous simulation with an updated geometry (Shih et al., 2001). Furthermore, the simulations on different geometries can be performed for equal numbers of particle histories and the results can be weighted by the fractional MUs to determine the dose. Alternatively, the number of histories to be simulated for each geometry can be calculated based on the fractional MUs. With this latter approach, the statistical variance will be higher for segments that deliver fewer MUs.

The time dependence of the geometry can be incorporated into a fully dynamic MC simulation by continuously updating the geometry after a predefined number of initial particles, corresponding to the time resolution. In GEANT4, this is possible through the use of the G4UI messenger class, which uses the method SetNewValue to update the geometry parameters (Paganetti, 2004). This approach updates the geometry in a linear time progression. Another approach is to randomize the sampling of the different geometries. This "position-probability sampling" (PPS) method (Liu et al., 2001) requires cumulative probability functions (cpdf) for each geometrical parameter that varies. The probabilities are calculated from the fraction of the total delivery time that the particular geometrical element (i.e., MLC leaf and jaw) spends at a certain location or configuration. From an operational overhead point of view, the PPS approach may be more efficient than the SCS method. Both approaches should result in the same statistical variance for an equal number of incident particles if the SCS method calculates the number of histories to run for each geometry based on the same cpdf used for the PPS method.

7.2.2 Applications of Monte Carlo to Model Dynamic Radiotherapy Techniques

7.2.2.1 Dynamic Wedge

Dynamic or virtual wedges are an intensity-modulated delivery approach where a wedge-shaped dose distribution is delivered by the dynamic motion of the collimating jaws. The jaw motion is specified as a function of fractional MUs in the so-called segmented treatment table (STT). The spectrum of particles emerging from a virtual wedge may vary significantly from a static solid wedge, which differentially hardens the beam.

The first reported simulation of a dynamic wedge was performed by Verhaegen and Das (1999). They used the EGS4/BEAM code to simulate delivery with a dynamic wedge on a Siemens linac and to compare the energy spectrum with that obtained with a physical wedge. In the first step, the simulation of transport in the upper section of the treatment head is performed and a phase space file is scored before the upper jaws. In the next step, 20 discrete simulations of transport through the treatment jaws are performed between which one of the upper jaws moved in 1 cm steps. The resulting 20 phase space files are combined by taking a precalculated number of particles from each phase space file based on a formula for the ratio of dose, or MUs, that must be delivered at each jaw position to obtain a certain wedge angle. This assumption of a one-to-one relationship between the number of delivered MUs and the number of simulated histories is not entirely accurate as it ignores the effect of particles that are backscattered from the jaws into the monitor unit chamber. A correction for this can be applied if the amount of backscatter as a function of jaw position can be characterized for the linac geometry (Liu et al., 2000). In the study of Verhaegen and Das (1999), the dose in the monitor unit chamber resulting from backscattered radiation was less than 1% for all simulated jaw positions. The authors modeled both virtual and physical wedges from 15° to 60° for energies 6–10 MV and obtained a good agreement with measurements, except for the penumbra region of the toe end of the wedge where the discrepancies were up to 4%. They also compared the virtual wedge with the physical wedge in the heel, center, and toe end of the wedges for beam spectrum variations. In comparing physical wedges with virtual wedges for beam hardening effects, no major differences were observed except that the 60° physical wedges produce significantly harder beams across the whole field due to higher absorption by the tungsten.

In a similar fashion, Shih et al. (2001) reproduced the dynamic wedge delivery of a Varian linac, as specified in an STT file, by calculating the number of particles to be simulated for each jaw setting based on the weights in the STT. After the simulation of the first STT entry, the simulation is restarted using the IRESTART feature in BEAM with the updated jaw position and incident histories. In this way, the phase spaces scored for each jaw setting are appended to the main phase space file and no additional postprocessing is needed.

Verhaegen and Liu (2001) later developed the PPS method to simulate the Varian enhanced dynamic wedge (EDW) delivery. The jaw positions as a function of cumulative MUs in the STT are converted to a cumulative probability distribution function (see Figure 7.2). During initialization of each incident history, the jaw position is randomly sampled from this cpdf. The authors also performed simulations using the SCS method with and without a correction for the reduction in the number of delivered MUs due to backscattering into the monitor chamber. The incorporation of this correction was not found to make any difference in the phase space files, thus again confirming the validity of the assumption relating MUs to particle histories. Comparing the results of the PPS and SCS methods, no differences were found in the resulting phase space files created with the two methods.

The use of the particle weighting method to simulate a Varian EDW delivery has been reported by Ahmad et al. (2009) using the method of Ma et al. (2000). Compared to measurements of wedged profiles and output factors, they obtained agreement within 2% and 1 mm.

7.2.2.2 MLC-Based IMRT

With IMRT, nonuniform fluence patterns or intensity-modulated maps that are incident upon a patient from multiple

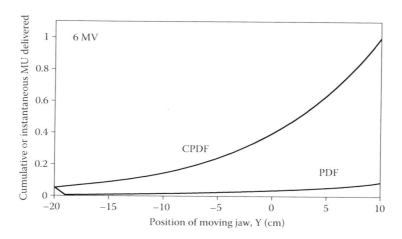

FIGURE 7.2 Calculated cumulative and noncumulative probability distribution functions for the position of the Y jaw for the delivery of a 60-wedged field derived from the STT. (From Verhaegen F. and H.H. Liu. Incorporating dynamic collimator motion in Monte Carlo simulations: An application in modeling a dynamic wedge. *Phys. Med. Biol.* 46, 287–296, 2001. With permission.)

directions are used to generate a uniform dose distribution that covers the target volume. These nonuniform fluence patterns are generated by using a continuous or discrete sequence of apertures formed using an MLC. In step-and-shoot delivery, each MLC-generated subfield is set prior to the beam turning on and there is no MLC motion during irradiation. In the case of the sliding window (or dynamic) delivery, the leaves move while the beam is on.

MC simulations of step-and-shoot and sliding window IMRT have been used by numerous authors to compare MC dose distributions with those produced by a commercial planning system. For example, Ma et al. (2000) used the particle weighting method to recalculate IMRT treatment plans from the CORVUS treatment planning system by using the leaf sequence files to calculate a two-dimensional (2D) intensity map.

Leal et al. (2003) and Seco et al. (2005) performed the full linac and patient simulation (i.e., particle transport through jaws, MLC, and patient), respectively, for the Siemens and Elekta linac using BEAMnrc, where step-and-shoot delivery was modeled using the SCS method. Each beam consisted of 5–15 segments, which were simulated individually with the number of histories proportional to the number of MUs delivered by the individual segment. Both authors performed their simulations on a personal computer (PC cluster). Scripts were developed to automate generation and distribution of separate files for the simulation of each segment phase space and the resulting dose distribution.

In Liu et al. (2001), the authors simulated both dynamic MLC and step-and-shoot deliveries of IMRT fields using the PPS method. First, the leaf positions as a function of cumulative MUs are used to create a probability function for each leaf (see Figure 7.3a). An MC simulation of an IMRT delivery is performed by randomly selecting on the Y-axis of Figure 7.3a an MU_{index} value between 0 and 1. The corresponding MLC subfield is then obtained by converting the MU_{index} (Y-axis) into a segment index (X-axis) using the curve of either

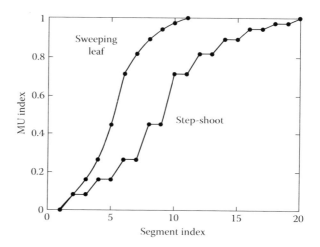

FIGURE 7.3 An example of the IMRT field generated by dynamic leaf sweeping or step and shoot, where in the *Y*-axis, MU index represents the cumulative probability function of the MLC motion. (From Liu H.H., Verhaegen F., and L. Dong. A method of simulating dynamic multileaf collimators using Monte Carlo techniques for intensity-modulated radiation therapy. *Phys. Med. Biol.* 46, 2283–2298, 2001. With permission.)

step-and-shoot or sweeping leaf curve. If the *X*-axis value falls between two segment index values, then the leaf positions at this fractional index value will be linearly interpolated from the neighboring segments. Note that the "step-like" appearance of the step-and-shoot curve results from the beam being turned off when the segment index and the leaf positions change. A similar approach was used by Heath and Seuntjens (2003) for simulations of IMRT delivery with the Varian Millennium 120-leaf MLC using BEAMnrc. They modified the standard VARMLC component module, which is available for modeling MLC geometry, to model specifically the new geometry of the

120-leaf MLC and to allow simulation of IMRT with both the SCS and PPS methods.

A fourth approach to simulating MLC-based IMRT is through a combination of the geometry sampling approaches and simplification of the particle transport in the MLC. The complexity of the leaf geometries typically leads to a high number of geometrical regions through which particles must be tracked. When particle transport in these detailed geometries is performed with low cut-off energies, these simulations can become very time consuming. Depending on the quantity being investigated, the contributions of electrons and multiple-scattered photons may not be important. Therefore, a significant improvement in computational efficiency can be realized by simplifying both the particle transport and the simulation geometry. An example of this is the fast MC code VCU-MLC (Siebers et al., 2002), which uses a single-Compton approximation of photon interactions within the MLC. No electron transport in the MLC is performed and the incident electrons have their weight reduced by the probability of striking an MLC leaf. The leaf geometry is approximated by segmenting each leaf into upper and lower halves, with a correct thickness to model the tongue and groove. Scattered photons are assumed to pass through the current leaf until they exit the MLC; thus, the adjacent leaves are ignored. The VCU-MLC model predicts correctly the field size dependence of MLC leakage, inter- and intraleaf leakage, and the tongue-and-groove dose effect. The model is applicable to both step-and-shoot and sweeping window deliveries. For each incident particle, the weight is modified by randomly sampling MLC positions 100 times from the leaf sequence file and averaging the transmission probabilities for each sample. Compared to the SCS and PPS methods that sample an MLC configuration for each incident particle, fewer source particles are required with the transmission probability weighting approach, which results in a gain in calculation efficiency. The VCU-MLC code is thus very useful to patient IMRT dose calculations because of its accuracy and speed of calculation. The MLC model has also been extended to simulate photon transport through both jaws and MLC simultaneously with the addition of multiple Compton interactions (Seco et al., 2008). Tyagi et al. (2007) have also implemented a similar approach in the DPM code to simplify particle transport in a model of the Varian Millennium MLC; however, they adopted a serial approach to simulating IMRT delivery by using the PPS method to determine the leaf positions for each incident particle.

An application of MC simulations of MLC-based IMRT delivery is to reconstruct the delivered dose based on recorded delivery log files (Fan et al., 2006; Tyagi et al., 2006). For example, on Varian linacs, the MLC leaf positions and beam state (on/off) are recorded every 0.050 s and written to a Dynalog file. The log files can be processed in a manner similar to leaf sequence files. This can be used as a quality assurance method to detect dosimetric errors resulting from errors in positioning of the MLC leaves. Owing to the large number of points within these files, calculation with an analytical algorithm would prove impractical.

7.2.2.3 Tomotherapy

The challenge of simulating tomotherapy, and other delivery techniques that involve dynamic gantry rotation, is that the gantry angle is continuously changing during the delivery. The geometry of a helical tomotherapy delivery is illustrated in Figure 7.4. Tomotherapy uses a rotating fan beam source whose intensity is modulated by a binary MLC. The helical beam delivery is specified in a sinogram file in which the fractional opening time for each leaf is specified at discrete gantry positions. Since the delivery is dynamic, leaf motion is occurring while the source rotates. Depending on the amount of leaf motion, subsampling of the sinogram may be needed to accurately simulate the dosimetric effects of the combined gantry and leaf motion. The amount by which the couch translates as a function of the gantry rotation is specified by the pitch. Static models of tomotherapy units without MLC modulation have been developed by Jeraj et al. (2004) and Thomas et al. (2005).

Sterpin et al. (2008) developed a PENELOPE user code called TomoPen for simulating helical tomotherapy delivery. First, a phase space file is created for each of the three jaw settings used during delivery. Then, for each entry in the sinogram file, a projection-specific phase space file is created by adjusting the weights of the particles in the phase space file associated with the appropriate jaw setting. To model the continuous delivery, each entry in the sinogram file is further subdivided into 11 subprojections for which the leaf openings are linearly interpolated. It should be noted that the treatment planning system used the sinogram entries to calculate the dose distribution, therefore

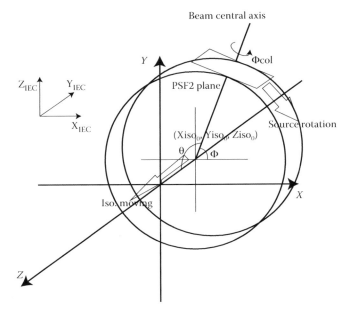

FIGURE 7.4 Geometry of the simulation of helical tomotherapy delivery. Phase space fields (PSF2) are calculated for each projection, which are rotated by the projection angle (θ) and shifted along the Z-axis according to the sinogram file. (From Zhong H. and J.V. Siebers. Monte Carlo dose mapping on deforming anatomy. *Phys. Med. Biol.* 54, 5815–5830, 2009. With permission.)

discretizing the delivery. For all the particles in the jaw-specific phase space, weighting factors are calculated, assuming linear attenuation, by a ray tracing through the MLC leaves. Specifically, the subprojection-specific phase space is calculated by dividing the jaw-specific phase space into 64 regions, one for each MLC leaf, and applying the weighting for leaf attenuation if the current leaf is closed or not adjusting the weight if the leaf is open. Only leaves that are open or adjacent to an open leaf are considered for this step, otherwise, the particles were discarded; thus, the leakage radiation in this situation is not modeled. Each of the resulting subprojection-specific phase space files is then rotated and translated according to the projection angle and couch translation before being applied to the patient geometry. TomoPen has been used to simulate the delivery of commissioning and clinical cases in homogeneous phantoms. A comparison with film and ionization chamber measurements (see Figure 7.5) showed a good agreement within 2% and 1 mm.

Zhao et al. (2008a) also simulated a helical tomotherapy delivery with EGSnrc/BEAMnrc by using the SCS method. They also subdivided each sinogram entry into static MLC subfields. An initial phase space is scored above the MLC. Then, the binary collimator is modeled using the VARMLC component module and a separate full simulation of particle transport in the MLC is performed for each subprojection resulting in a second set of phase spaces. The number of histories for each subprojection simulation was proportional to the calculated leaf opening time. The phase spaces are rotated in the *XY* plane and the isocenter is modified along the *Z* direction to simulate the helical delivery as shown in Figure 7.4. This model was used to compare with dose distributions calculated by their treatment planning system for clinical treatment plans (Zhao et al., 2008b).

7.2.2.4 Intensity-Modulated Arc Therapy

IMAT is a form of conventional MLC-based IMRT where the gantry rotates around the patient as the beam apertures are dynamically modified by the MLC. There are a number of different IMAT implementations, which are distinguished by the number of arcs that are delivered and whether the dose rate is variable or constant. The delivery is specified by an MLC leaf sequence file, which is indexed to the gantry rotation angle.

The approach for simulating IMAT delivery is similar to that for tomotherapy. An SCS-based approach to modeling IMAT delivery on an Elekta linac has been described by Li et al. (2001) using EGS4/BEAM. Emulating the dose calculation method used by the treatment planning system, the authors discretize each arc into 5–10° steps for which static simulations of the MLC apertures defined in the leaf sequence file are performed. A simplified geometry is used for the MLC, without modeling the tongue

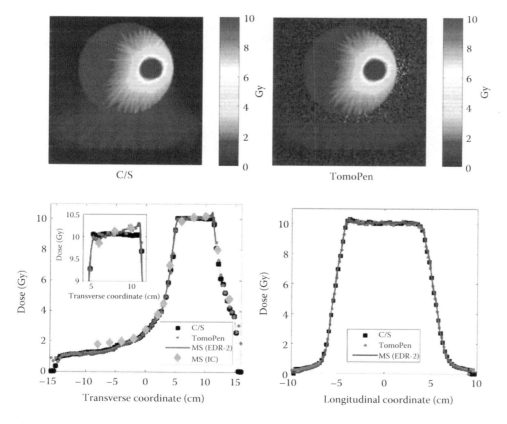

FIGURE 7.5 (**See color insert.**) Comparison of dose distributions calculated with a convolution–superposition algorithm (C/S) and Monte Carlo (TomoPen) and measured with film (MS EDR-2) and ionization chamber (MS IC) for a commissioning plan in a cylindrical phantom. (From Sterpin E. et al. Monte Carlo simulation of helical tomotherapy with PENELOPE. *Phys. Med. Biol.* 53, 2161–2180, 2008. With permission.)

and groove. For each phase space, a DOSXYZ calculation on the patient geometry is carried out with the appropriate source angle.

A model of a Varian RapidArc delivery has been developed by Bush et al. (2008), which used EGSnrc/BEAMnrc. They also modeled the gantry rotation as a series of discrete static simulations. However, owing to the significant leaf motion that occurs during RapidArc delivery, it is also necessary to model the leaf motion that occurs between the gantry control points. This is addressed by calculating, for each sequential pair of gantry angles specified in the sequence file, a mean gantry angle. The leaf openings of the adjacent gantry angles are then used to define a pair of control points for a sliding window dynamic multi-leaf collimator (DMLC) delivery at this mean angle, as illustrated in Figure 7.6. The variable dose rate is accounted for by weighting the subsequent dose calculation by the fractional MUs to be delivered by each segment. The authors used the VCU-MLC model (Siebers et al., 2002) to improve the computational efficiency of their simulations. The same group has used the recorded Dynalog files to reconstruct the delivered dose from a RapidArc treatment (Teke et al., 2009).

Simulation of tomotherapy and IMAT delivery using the SCS approach requires the calculation of phase space files for a significant number of geometry samples due to the additional dimension of gantry rotation. Even with automation of the input file preparation and patient dose calculation, the generation of hundreds to thousands of phase space files representing each subprojection is not optimal in terms of storage requirements.

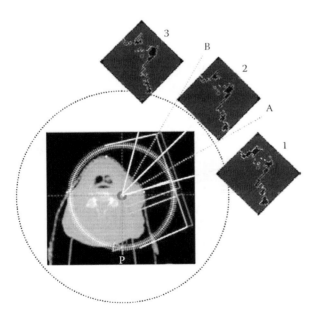

FIGURE 7.6 Illustration of the generation of DMLC segments for the simulation of leaf motion in volumetric arc therapy (VMAT) delivery. For each consecutive set of gantry positions (1 and 2), a mean gantry angle A is calculated and the leaf positions from 1 and 2 are used as control points for a dynamic leaf sequence to be simulated at position A. (From Bush K., Townson R., and S. Zavgorodni. Monte Carlo simulation of RapidArc radiotherapy delivery. *Phys. Med. Biol.* 53, N359–N370, 2008. With permission.)

Another drawback of using the SCS approach is that the dosimetric influence of the number of subsamples for simulating the continuous delivery must be carefully investigated.

Owing to the continuous motion of the gantry and collimators, the position-probability approach is better suited to the simulation of IMAT and tomotherapy. With the PPS approach, a randomly sampled monitor unit index at the beginning of each history can be used to determine the collimator settings as well as gantry angle and isocenter translation. These delivery settings can be obtained from the sinogram/leaf sequence file from the treatment planning system or from a Dynalog file recorded during a treatment delivery.

Two PPS-based approaches have been developed for simulating tomotherapy and IMAT using the BEAMnrc code. The first, by Lobo and Popescu (2010), makes use of the option in DOSXYZnrc to use a full BEAMnrc simulation as a particle source (ISOURCE 9). This DOSXYZnrc source has been modified so that the randomly sampled monitor unit index that is used to determine the geometrical configuration of any dynamic component module in the accelerator is passed to DOSXYZnrc, where it is used to sample the gantry and collimator angle as well as couch rotation and translation from lookup tables resembling the sequence files used to specify the MLC leaf motion. On the basis of this source configuration, the coordinates and trajectory of the particle exiting the accelerator are transformed to the patient coordinate system before being transported through the patient density matrix. These two new sources (ISOURCE 20 and ISOURCE 21) are now distributed with the BEAMnrc V4 package. Source 20 uses a phase space, scored below the secondary collimators (e.g., jaws) as an input. This is useful for simulations when the jaw configuration does not change during the delivery. Lobo and Popescu integrated the VCU-MLC code (Siebers et al., 2002) into Source 20 by compiling it as a shared library, which can be accessed by DOSXYZnrc to transport particles through the MLC. The randomly sampled monitor unit index is used to sample the MLC configuration for each incident particle. The other source (ISOURCE 21) consists of a full accelerator model simulation and is useful when there is more than one dynamic component. In source 21, the same monitor unit index that is randomly sampled when a particle is incident on the first dynamic component is used when that particle enters the second dynamic component. This ensures that the motion of the two collimators is synchronized.

For the second PPS-based approach that was developed by Belec et al. (2011), a time variable is stored in the phase space file. This time variable replaces the ZLAST variable, which records the z position of the last interaction of the particle. The time variable is actually the same monitor unit index that is used by all PPS approaches and will be randomly sampled when a particle passes through the first dynamic component module. Unlike the method of Lobo and Popescu, with this approach, the output of the accelerator simulation can be stored in an intermediate phase space file without losing the temporal information. When the phase space file is read by DOSXYZnrc, the time variable is used to sample from the lookup tables specifying the source configuration.

7.2.2.5 Protons

Before the introduction of scanning delivery systems for proton beams, temporal modulation in the form of spinning range modulator wheels was used for modulating the range of an incident proton beam to create a spread-out Bragg peak. In what is perhaps the first published report of a dynamic MC simulation, Palmans and Verhaegen (1998) reported implementing the PPS method in the PTRAN MC code to simulate a range modulator wheel. Paganetti (2004) described a detailed model of the Northwest Proton Therapy Center beam line using GEANT4. Dynamic elements such as the range modulator wheel and beam scanning magnets (see Figure 7.7) were modeled exploiting the feature of GEANT4 to update geometry parameter values during the simulation. The geometry updates were performed in a linear time fashion, with each time step assigned a number of histories. The modulator wheel rotation was simulated in 0.5° steps whereas the scanning magnet settings were updated in steps of 0.02 T. The authors noted that the calculation time for their dynamic MC simulations were virtually unaffected by the number of times that the geometry was updated since this involved only updating a pointer in the memory.

7.3 Dynamic Patient Simulations

7.3.1 Patient Motion in Radiotherapy

The need for 4D patient dose calculation methods was initiated by an interest in compensating for the effects of respiratory motion during treatment planning and delivery (Keall et al. 2004). 4D dose calculation methods are required to calculate the cumulative dose distribution, which is received by the varying anatomy. The effects of motion on the delivered dose distribution were discussed by Bortfeld et al. (2004). These include: (1) blurring of the dose distribution along the path of motion, (2) localized spatial deformations of the dose distribution at deforming organ boundaries or regions of density changes, and (3) interplay between tumor motion and dynamic beam delivery. As will be discussed in the following sections, the blurring effect can be simply modeled with analytical dose calculation algorithms by convolution methods; however, to quantify dose deformation effects, the influence of individual geometric states on the deposited dose distribution needs to be modeled. The methods discussed here are focused on respiratory motion; however, they can be extended to other examples of patient anatomical variations, such as interfractional geometry variations and applications in adaptive radiotherapy, with appropriate modifications.

7.3.2 Strategies for 4D Patient Simulations

7.3.2.1 Convolution-Based Methods

Similar to approaches used for simulating dynamic beam delivery, the cumulative dose can be calculated either by adjusting the particle fluence or from multiple calculations on different respiratory phases. If the dose distribution is assumed to be shift invariant, then the effect of motion on the dose delivered

for a large number of fractions can be estimated by convolving the dose distribution with a probability distribution describing the positional variations (Lujan et al., 1999). This models only the blurring effect of motion on the dose distribution, dose deformations, and differential motion that cannot be modeled by convolution. Furthermore, the assumption of spatial invariance of the dose distribution is not valid at tissue interfaces, which can lead to an underestimation of the dose at these locations (Craig et al. 2001; Chetty et al. 2004).

In the reference frame of the patient, patient motion is interpreted as motion of the beam; therefore, the approach of fluence convolution has also been proposed (Beckham et al., 2002; Chetty et al., 2004). This is analogous to the particle weighting approach; however, the (x, y, z) positions and direction cosines of the particles in the phase space file are modified by randomly sampling shifts from a probability distribution function (pdf) describing the respiratory motion. Fluence convolution attempts to overcome the limitation of the shift invariance assumption in the dose convolution method. However, it is also limited by the approximation of the patient as undergoing rigid body motion.

7.3.2.2 Dose-Mapping Methods: COM and Dose Interpolation

The limitation of the convolution approach is that it does not consider the differential organ motion, deformation, and density changes that cause the shape of the dose distribution to locally "deform" during the beam delivery. To accurately account for these effects, the dose distributions calculated on the different respiratory phases (e.g., exhale, 50% inhale, and 100% inhale) need to be calculated, similar to the SCS method of simulating an IMRT delivery, and weighted by the fraction of the total breathing cycle for which they occur. The number of samples of respiratory states needed to reconstruct the cumulative dose for a full respiratory cycle has been investigated by Rosu et al. (2007). They demonstrated no significant differences in dose volume histograms between cumulative doses calculated from 10 respiratory states and calculations based on only the inhale and exhale states. This conclusion is dependent on the treatment plan design; if dose perturbations due to motion are large, then nonlinearities of the motion between inhale and exhale can be expected to have a nonnegligible influence on the cumulative dose distribution.

Unlike dose deposition from a dynamic beam delivery where the reference frame of the dose deposition remains constant, for a dynamic patient simulation, the position and volume of tissue elements is changing (possibly, in concert with dynamic beam delivery changes). The dose deposited in each geometry needs to be mapped to a reference geometry to determine the cumulative dose deposition from irradiation of the multiple variable geometries. This requires knowledge of the geometrical transformation between tissue locations in the reference and current geometrical states. Image registration methods are commonly used to estimate these transformations. Briefly, the transformation is calculated by finding a transformation, which maximizes the similarity of the two images. This similarity may be based on image intensity or previously identified landmarks

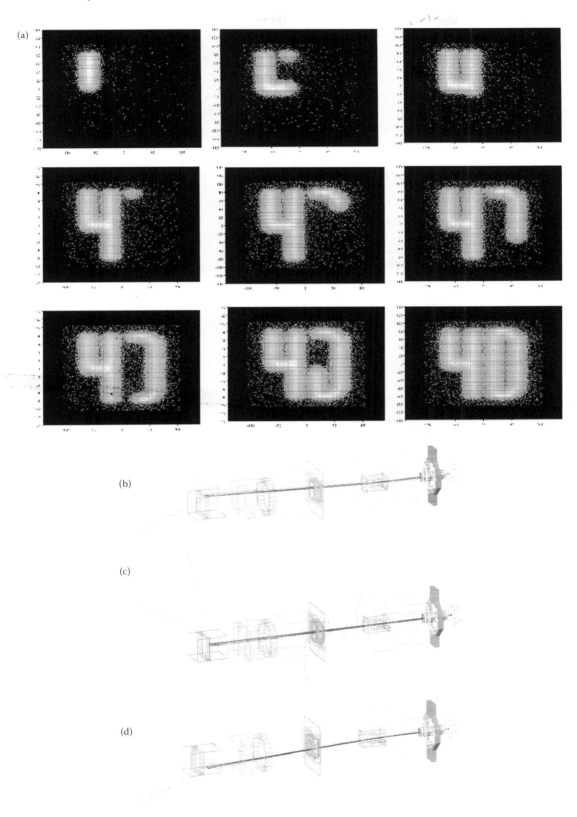

FIGURE 7.7 (a) Fluence distribution from simulation of scanned proton beam delivery, (b)–(d) simulation of proton trajectories in the treatment nozzle for different settings of the scanning magnets. (From Paganetti H. Four-dimensional Monte Carlo simulation of time-dependent geometries. *Phys. Med. Biol.* 49, N75–N81, 2004. With permission.)

such as points or contours. A wide range of image registration algorithms exist; mainly, they are classified as being rigid or nonrigid. For the types of anatomical changes observed in patient geometries, deformation of the tissue is observed and thus nonrigid or deformable image registration is best suited for this application. A review of deformable image registration algorithms used in medical imaging has been given by Hill et al. Holden (2008), and Xing et al. (2007). It should be obvious to the reader that the accuracy of the mapped dose calculation depends on the accuracy of the image registration. Some discussion of this topic will be made in Section 7.3.2.4.

There are a variety of *dose-mapping methods* that are used to map the dose from the geometry on which it is deposited to a reference geometry. These are applicable to any dose calculation method. Generally, dose mapping is discussed as applying to two single geometry samples, but it could easily be implemented as a "true 4D" simulation if the patient geometry is updated during the simulation and the mapping is applied to accumulate the dose depositions on the current scoring geometry to the reference dose grid.

The simplest approach to mapping the dose delivered on the target geometry to the reference geometry is to use the deformation map to determine the voxel in the target geometry that corresponds with the voxel of interest in the reference geometry. The reference voxel is assigned a dose value equal to the dose computed in the target voxel in which the transformed center of mass of the reference voxel lies (see Figure 7.8a). This "center-of-mass (COM)" remapping method ignores contributions in other voxels, which may be overlapped by the transformed reference voxel. This can easily occur if the transformed reference voxels do not exactly overlap the target voxels. Also, if there are any volume changes, then this exact voxel-to-voxel mapping is incorrect. In an application of the COM method, Paganetti et al. (2004) used the updateValue method in GEANT4 to update the patient geometry to one of the 10 different phases of the patient's breathing from a 4D computed tomography (CT) data set. To accumulate the dose, pointers in the dose matrix were modified using the COM method to ensure that dose depositions on the different geometries were accumulated in the corresponding voxels of the reference geometry.

To deal with possible deformations and the inexact overlap of transformed and target voxels, a trilinear interpolation of the dose from the local neighboring voxels at the transformed COM point in the target geometry can be used to assign the dose in the reference geometry (Figure 7.8b). Voxel volume changes can also be accounted for by subdividing each reference voxel, estimating the mapped dose for each subvolume, and then interpolating these values to assign the dose in the reference geometry (Figure 7.8c). Rosu et al. (2005) reported the application of these methods to dose distributions calculated using an MC algorithm, though the method is applicable to a dose distribution calculated with any algorithm. Dose interpolation methods are computationally efficient; however, like any interpolation method, they suffer from a reduced accuracy wherever dose gradients exist. Heath and Seuntjens (2007) showed that due to these interpolation errors, the dose interpolation method calculates the dose

incorrectly in the regions of large dose and deformation gradients. Siebers and Zhong (2008) also demonstrated that in the situation where voxels of the target geometry merge in the reference geometry (i.e., compression), dose interpolation methods do not conserve the integral dose and thus lead to errors in the regions of dose and density gradients.

7.3.2.3 Voxel Warping Method

Two MC approaches to dose remapping that ensure energy conservation have been developed by Heath et al. (2007, 2009), Heath and Kawrakow (2009), and Siebers and Zhong (2008). The first method, called voxel warping, entails deforming the nodes of the reference dose grid voxels using the deformation vectors obtained from registration of the reference to the target image, and was first implemented in a modified version of the DOSXYZnrc user code, called defDOSXYZnrc (Heath et al., 2007). Particle transport is performed on this deformed voxel geometry. As the voxel indices do not change between the reference and deformed state, the energy mapped between these states is conserved. The densities of these deformed voxels are adjusted according to the volume change to conserve mass. To describe the deformed voxels, the faces of the rectangular voxels are divided into two subplanes, forming dodecahedrons (Figure 7.9). This voxel face subdivision leads to a twofold increase in the number of distance-to-voxel boundary calculations that need to be performed by the boundary checking algorithm. Furthermore, because of ambiguities in the calculation of particle–plane intersection points, which assume that the plane has an infinite extent, these intersections need to be further tested to determine if they lie within the boundaries of each plane. These additional computations result in up to a 10-fold increase in computation times with defDOSXYZnrc compared to DOSXYZnrc.

A recent reimplementation of the voxel warping method in VMC++ (Heath and Kawrakow, 2009) achieved a 130-fold improvement in computational efficiency compared to defDOSXYZnrc by using a fast MC code and optimized geometry definitions based on tetrahedral elements. Each deformed voxel is divided into six tetrahedrons, which are used for particle tracking while dose deposition is scored in the voxels. The main advantage of using tetrahedral volume elements is the elimination of the plane–particle intersection ambiguity; furthermore, there are fewer planes to be checked in each tetrahedron compared to the dodecahedron geometry. Compared to VMC++ calculations in rectilinear geometry, the calculation time with the deformable VMC++ geometry is increased by a factor of 2 for the same level of statistical variance. The deformable geometry was implemented as a new geometry class for VMC++, which was compiled as a dynamic shared library to be loaded at run time. Two alternate definitions of the tetrahedral geometry were implemented, differing by the direction in which the voxel faces fold. Though it could be demonstrated that the geometry definition affects the dose deposition, no differences in patient dose distributions calculated with the two alternate geometries could be found (Heath and Kawrakow, 2009).

A possible limitation of the voxel warping method is that it requires a continuous deformation field, which is not always

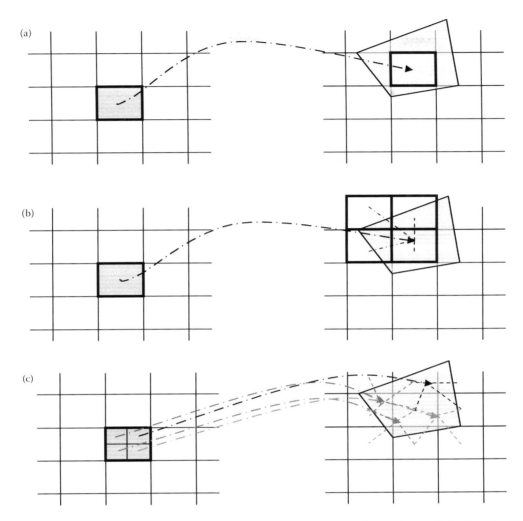

FIGURE 7.8 Illustration of different dose-mapping methods. The left grid represents the reference dose grid. On the right side is the dose grid from the irradiated geometry. The dose deposited on the source irradiated geometry is to be mapped to the reference geometry for accumulation. The shaded polygon represents the boundaries of the reference voxel after the reference-to-target geometry transformation is applied. (Parts of this figure adapted from Rosu M. et al. *Med. Phys.* 32, 2487–2495, 2005.) (a) COM mapping. (b) Dose mapping with trilinear interpolation. (c) Method (b) with subdivision of the reference voxels.

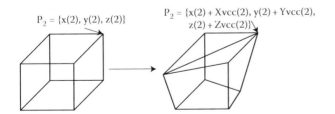

FIGURE 7.9 Deformed voxel geometry used in defDOSXYZnrc obtained by applying deformation vectors to reference voxels.

the physical situation. At the boundary between the lung and the chest wall tissue, sliding occurs, which, at the resolution of the dose calculation and image registration, appears as a discontinuous motion. Before using the displacement vector maps obtained from a deformable image registration, they must be checked for discontinuities in the region where the dose will be

calculated. This can be performed by calculating the determinant of the Jacobian matrix of the transformations in the neighborhood of each node of the reference voxel grid (Heath et al., 2007; Christensen et al., 1996). A negative value of the determinant (see Equation 7.1) at a particular node indicates a discontinuity in the transformation (u_x, u_y, u_z) at this node. These can be removed by smoothing the transformations either globally or in the vicinity of the discontinuous node.

$$
\det\left[J\left(N(\vec{x})\right)\right] = \begin{vmatrix} \dfrac{\partial u_x}{\partial x} + 1 & \dfrac{\partial u_x}{\partial y} & \dfrac{\partial u_x}{\partial z} \\[2mm] \dfrac{\partial u_y}{\partial x} & \dfrac{\partial u_y}{\partial y} + 1 & \dfrac{\partial u_y}{\partial z} \\[2mm] \dfrac{\partial u_z}{\partial x} & \dfrac{\partial u_z}{\partial y} & \dfrac{\partial u_z}{\partial z} + 1 \end{vmatrix} \quad (7.1)
$$

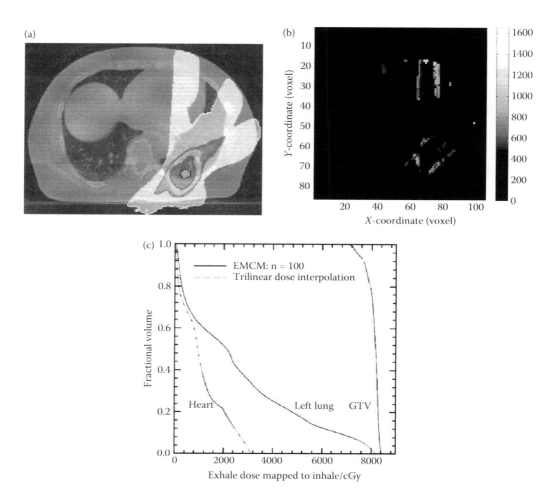

FIGURE 7.10 (a) Dose distribution mapped to Exhale geometry with ETM; (b) difference between dose interpolation and ETM for the same image slice; (c) dose volume histograms. (From Zhong H. and J.V. Siebers. Monte Carlo dose mapping on deforming anatomy. *Phys. Med. Biol.* 54, 5815–5830, 2009. With permission.)

7.3.2.4 Energy-Mapping Methods

Siebers and Zhong (2008) proposed an alternative energy-mapping-based approach in which particles could be transported in the rectilinear target geometry and the energy deposition points are mapped according to the deformation vectors to the target dose grid. To account for the possibility that particle steps in the target geometry may be split over two or more voxels if mapped to the reference geometry, the energy deposition location is randomly sampled along the step. This implementation, termed etmDOSXYZnrc, results in a 10–50% increase in computation time, depending on how the deformation vectors are interpolated, compared with the standard DOSXYZnrc computations. This approach also does not require a continuous deformation vector field.

However, if the transformation between the reference and target geometry is not exact, the energy mapped to the voxels of the reference phase will not be consistent with the mass assigned to those reference voxels. This means that in the lung, for example, energy could be deposited in a voxel with a density

of 1.0 g/cm³, but due to registration error, that energy deposition is mapped to a reference voxel with a density of 0.3 g/cm³. These inconsistent energy–mass mappings lead to a discontinuous dose distribution on the reference geometry. In deformable registration of patient images, determination of an exact transformation is not possible due to a lack of one-to-one correspondence between images arising from image artifacts, noise, partial volume effects, among other causes (Crum et al. 2003). In a further paper, Zhong and Siebers (2009) proposed methods to map the mass along with the energy to ensure that mass is also conserved; thus, the energy and mass distributions used to calculate the final mapped dose distribution are mapped in a consistent manner. This approach is termed the "energy and mass congruent mapping" (EMCM) method. The target voxels are subdivided into 100 subvoxels whose masses are mapped to the reference geometry using the same deformation vectors used for mapping the energy deposition points. The precision of the mass mapping was determined to be 99.95% on a simple deformable phantom geometry. Alternatively, a mass "pulling" approach based on the reference–target geometry transformation can be used. This requires a method to invert the deformation field. An energy-mapping method, which performs the energy and mass mapping in the reference–target direction is described below. Dose distributions mapped between the inhale and exhale phases of a lung cancer patient using the EMCM method and dose interpolation are shown in Figure 7.10. The mean difference between the EMCM and trilinear dose interpolation (TDI)-mapped dose distributions is 7% of the maximum dose.

Heath et al. (2009) proposed an alternative energy-mapping approach that can be performed, postcalculation, on a dose distribution. Instead of a point-by-point energy remapping, the energy mapped from the target geometry to each reference voxel is determined by the volume overlap of the deformed reference voxels on the target geometry. Each reference voxel is subdivided into tetrahedrons, which are deformed using the displacement vectors, and the volume intersected by each of these tetrahedral and the target voxels is calculated. The energy mapped to each reference voxel is simply the sum over all its tetrahedra of the energy deposited in each overlapped target voxel multiplied by the fractional volume overlap. This second method can be applied to a dose distribution calculated with any algorithm; however, information about the spatial distribution of energy depositions within a voxel is lost. This leads to a loss of accuracy when the energy deposited in one voxel is mapped to multiple voxels on the reference geometry.

7.4 Combining Dynamic Beam and Patient Simulations

Combining the simulation of dynamic beam deliveries and variable patient geometries is of interest to study the interplay effects, which arise from the relative motion between the MLC leaves and the moving target region and anatomy upstream from the target, for the case of dynamic MLC delivery. We designate this by "4D²-interplay" effects and they were initially studied by Yu et al. (1998). They demonstrated that when IMRT beams are delivered with dynamic collimation, the problem of intrafraction motion causes large errors in the locally delivered photon dose per fraction due to motion in the penumbra region of the beam. The magnitude of the photon dose variations was shown to be strongly dependent on the speed of the beam aperture relative to the speed of the target motion and on the width of the scanning beam relative to the amplitude of target motion. In the case of proton therapy beams, interplay effects may also occur for proton beam scanning where the dose distribution is affected by (i) patient breathing, (ii) proton energy change times, (iii) motion amplitude, and (iv) proton beam rescanning methodology used (Seco et al., 2009a).

Modeling of interplay effects with MC can only be done if every particle is tagged with information that indicates both the beam delivery (e.g., MLC leaf or beam spot positions) through which it travels and the patient geometry state (e.g., breathing phase) to which it will deliver the dose. Particle tagging is not straightforward to setup within an MC simulation because it requires the additional inputs of the (i) beam delivery sequence and (ii) patient breathing pattern. Interplay may be simulated using a motion convolved dose kernel; however, this does not account properly for interface effects or rapid changes in electron density within a specific tissue such as within the lung.

An MC method to study the dosimetric effects of interplay and delivery errors due to dropped segments, dose over/under shoot, faulty leaf motors, tongue-and-groove effect, rounded leaf ends, and communication delays in phantoms has been developed by Litzenberg et al. (2007). Real-time information about the tracking of the MLC leaf positions is obtained from the Dynalog MLC log. Real-time target volume position information is measured using wireless electromagnetic transponders (Calypso Medical Technologies, Inc), where three transponders were implanted in the target allowing position updates 10 times per second. The authors termed this method the "synchronized dynamic delivery" approach. In the MC simulation, each particle is transported through a selected MLC field segment and into a specific breathing phase of the dose cube. Therefore, all breathing phase dose cubes would have to reside in the central processing unit (CPU) memory to reduce input/output (I/O) exchanges. Although the study was performed in phantoms, it was a proof of principle that MC could be used to study interplay effects. However, in the case of patients, the method proposed would require large amounts of CPU memory to work efficiently because of the significantly larger patient CT data cubes.

MC modeling has been used for patients with lung cancer to study interplay effects (Seco et al., 2009b) with a more CPU-efficient method of tracking photons through MLC leaf segments and for a specific breathing phase. To reduce the memory requirements, each photon can be tagged with both MLC segment and the breathing phase to which it would deliver the dose. Photon tagging involves changing one or more bits of the particle information, which is written to the phase file. This can be done readily in codes such as BEAMnrc by altering the LATCH bits to include MLC and breathing phase information. Then, dose calculations can be performed separately on each CT cube using only the particles that were delivered on that phase. Preliminary MC results using this approach showed variations in the delivered dose of 5–8% per beam angle. However, the final dose plan for the lung patient made up of 30 or more fractions was shown to have small or negligible interplay dose effects due to the averaging that occurs over the multiple deliveries.

7.5 Summary and Outlook

Three main approaches exist to simulating dynamic beam and patient geometries: particle weighting/convolution, SCS, and PPS. Dynamic simulations of IMRT deliveries is a well-established technique compared to 4D MC calculations in deforming patient geometries. For the latter, dose interpolation methods have been generally adopted for dose accumulation in radiotherapy planning; however, the development of accurate 4D MC methods that conserve mass and energy have highlighted the inaccuracies of the interpolation approach.

While particle weighting approaches may appear to be more computationally efficient, comparable performances can be achieved with full "4D" MC simulations using the SCS or PPS method when fast MC codes, with appropriate variance reduction techniques and simplifications of particle transport, are used. The advantage of MC codes compared to analytical algorithms in this regard is that the computation time does not increase as a function of the number of geometries that are simulated since the number of simulated particles may be split between geometries; therefore, a higher temporal resolution can be achieved with MC simulations, which may be of importance for continuous delivery methods such as tomotherapy and IMAT.

The study of temporal interplay effects, which occur when both treatment beam and tumor motion occur is an ongoing issue particularly in proton radiotherapy. MC methods are well suited to examining these problems though an effort has to be made to reduce the CPU memory requirements. Some initial results have demonstrated that this is possible, which opens up further possibilities of using MC as a method to evaluate the magnitude of these interplay effects for different dynamic beam delivery techniques. Another important application is the ability to track deviations of the beam delivery and patient motion from their planned values through reconstruction of the dynamic dose delivery to the patient based on delivery and patient log files.

In summary, time-dependent treatment planning and delivery are being ever more exploited and scrutinized in radiotherapy. MC codes are well suited to the study of temporal effects and should play an important role in their accurate quantification.

References

Ahmad M., Deng J., Lund M.W., Chen Z., Kimmett J., Moran M.S., and R. Nath. Clinical implementation of enhanced dynamic wedges into the Pinnacle treatment planning system: Monte Carlo validation and patient-specific QA. *Phys. Med. Biol.* 54, 447–465, 2009.

Beckham W.A., Keall P.J., and J.V. Siebers. A fluence-convolution method to calculate radiation therapy dose distributions that incorporate random set-up errors. *Phys. Med. Biol.* 47, 3465–3473, 2002.

Belec J., Ploquin N., La Russa D., and B.G. Clark. Position-probability-sampled Monte Carlo calculations of VMAT, 3DCRT, step-shoot IMRT and helical tomotherapy distributions using BEAMnrc/DOSXYZnrc. *Med. Phys.* 32, 948–960, 2011.

Bortfeld T., Jiang S.B., and E. Rietzel. Effects of motion on the total dose distribution. *Semin. Radiat. Oncol.* 14, 41–51, 2004.

Bush K., Townson R., and S. Zavgorodni. Monte Carlo simulation of RapidArc radiotherapy delivery. *Phys. Med. Biol.* 53, N359–N370, 2008.

Chetty I.J., Rosu M., McShan D.L., Fraass B.A., Balter J.M., and R.K. Ten Haken. Accounting for center-of-mass target motion using convolution methods in Monte Carlo-based dose calculations of the lung. *Med. Phys.* 31, 925–932, 2004.

Christensen G.E., Rabbitt R.D., and M.I. Miller. Deformable templates using large deformation kinetics. *IEEE Trans. Image Process.* 5, 1435–1447, 1996.

Craig T., Battista J., and J. Van Dyk. Limitations of a convolution method for modeling geometric uncertainties in radiation therapy I. The effect of shift invariance. *Med. Phys.* 30, 2001–2011, 2001.

Crum W.R., Griffin L.D., Hill D.L.G., and D.J. Hawkes. Zen and the art of medical image registration: Correspondence, homology and quality. *NeuroImage* 20, 1425–1437, 2003.

Deng J., Pawlicki T., Chen Y., Li J., Jiang S.B., and C-M. Ma. The MLC tongue-and-groove effect on IMRT dose distributions. *Phys. Med. Biol.* 46, 1039–1060, 2001.

Fan J., Li J., Chen L., Stathakis S., Luo W., Du Plessis F., Xiong W., Yang J., and C-M. Ma. A practical Monte Carlo MU verification tool for IMRT quality assurance. *Phys. Med. Biol.* 51, 2503–2515, 2006.

Heath E., Collins D.L., Keall P.J., Dong L., and J. Seuntjens. Quantification of accuracy of the automated nonlinear image matching and anatomical labeling (ANIMAL) nonlinear registration algorithm for 4D CT images of lung. *Med. Phys.* 34, 4409–4421, 2007.

Heath E. and I. Kawrakow. Fast 4D Monte Carlo dose calculations in deforming anatomies. *Med. Phys.* 36, 2620, 2009.

Heath E. and J. Seuntjens. Development and validation of a BEAMnrc component module for accurate Monte Carlo modeling of the Varian dynamic Millennium multileaf collimator. *Phys. Med. Biol.* 48, 4045–63, 2003.

Heath E. and J. Seuntjens. A direct voxel tracking method for four-dimensional Monte Carlo dose calculations in deforming anatomy. *Med. Phys.* 33, 434–445, 2006.

Heath E., Tessier F., Siebers J., and I. Kawrakow. Investigation of voxel warping and energy mapping approaches for fast 4D Monte Carlo dose calculations in deformed geometry using VMC++. *Phys. Med. Biol.* 56, 5187–5202, 2011.

Holden M. A review of geometric transformations for nonrigid body registration. *IEEE Trans. Med. Imaging* 27, 111–128, 2008.

Jeraj R., Mackie T.R., Balog J., Olivera G., Pearson D., Kapatoes J., Ruchala K., and P. Reckwerdt. Radiation characteristics of helical tomotherapy. *Med. Phys.* 31, 396–404, 2004.

Keall P.J., Siebers J.V., Arnfield M., Kim J.O., and R. Mohan. Monte Carlo dose calculations for dynamic IMRT treatments. *Phys. Med. Biol.* 46, 929–941, 2001.

Keall P.J., Siebers J.V., Joshi S., and R. Mohan. Monte Carlo as a four-dimensional radiotherapy treatment-planning tool to account for respiratory motion. *Phys. Med. Biol.* 49, 3639–3648, 2004.

Leal A., Sanchez-Doblado F., Arran R., Rosello J., Pavon E., and J. Lagares. Routine IMRT verification by means of an automated Monte Carlo simulation system. *Int. J. Radiat. Oncol. Biol. Phys.* 56, 58–68, 2003.

Li X.A., Ma L., Naqvi S., Shih R., and C. Yu. Monte Carlo dose verification for intensity-modulated arc therapy. *Phys. Med. Biol.* 46, 2269–2282, 2001.

Litzenberg D.W, Hadley S.W., Tyagi N, Balter J.M., Ten Haken R., and I.J. Chetty. Synchronized dynamic dose reconstruction. *Med. Phys.* 34(1), 91–102, 2007.

Liu H.H., Mackie T.R., and E.C. McCullough. Modeling photon output caused by backscattered radiation into the monitor chamber from collimator jaws using a Monte Carlo technique. *Med. Phys.* 27(4), 737–744, 2000.

Liu H.H., Verhaegen F., and L. Dong. A method of simulating dynamic multileaf collimators using Monte Carlo techniques for intensity-modulated radiation therapy. *Phys. Med. Biol.* 46, 2283–2298, 2001.

Lobo J. and I. Popescu. Two new DOSXYZnrc sources for 4D Monte Carlo simulations of continuously variable beam configurations, with applications to RapidArc, VMAT, tomotherapy and CyberKnife. *Phys. Med. Biol.* 55, 4431–4443, 2010.

Lujan A.E., Larsen E.W., Balter J.M., and R. Ten Haken. A method for incorporating organ motion due to breathing into 3D dose calculations. *Med. Phys.* 26, 715–720, 1999.

Ma C-M., Pawlicki T., Jiang S.B., Li J.S., Deng J., Mok E., Kapur A., Xing L., Ma L., and A.L. Boyer. Monte Carlo verification of IMRT dose distributions from a commercial treatment planning optimization system. *Phys. Med. Biol.* 45, 2483–2495, 2000.

Paganetti H. Four-dimensional Monte Carlo simulation of time-dependent geometries. *Phys. Med. Biol.* 49, N75–N81, 2004.

Paganetti H., Jiang H., Adams J.A., Chen G.T., and E. Rietzel. Monte Carlo simulations with time-dependent geometries to investigate effects of organ motion with high temporal resolution. *Int. J. Radiat. Oncol. Biol. Phys.* 60, 942–950, 2004.

Palmans H. and F. Verhaegen. Monte Carlo study of fluence perturbation effects on cavity dose response in clinical proton beams. *Phys. Med. Biol.* 43, 65–89, 1998.

Rosu M., Balter J.M., Chetty I.J., Kessler M.L., McShan D.L., Balter P., and R. Ten Haken. How extensive of a 4D dataset is needed to estimate cumulative dose distribution plan evaluation metrics in conformal lung therapy? *Med. Phys.* 34, 233–245, 2007.

Rosu M., Chetty I.J., Balter J.M., Kessler M.L., McShan D.L., and R.K. Ten Haken. Dose reconstruction in deforming lung anatomy: Dose grid size effects and clinical implications. *Med. Phys.* 32, 2487–2495, 2005.

Seco J., Adams E., Bidmead M., Partridge M., and F. Verhaegen. Head & neck IMRT treatment assessed with a Monte Carlo dose calculation engine. *Phys. Med. Biol.* 50, 817–830, 2005.

Seco J., Sharp G.C., Wu Z., Gierga D., Buettner F., and H. Paganetti. Dosimetric impact of motion in free-breathing and gated lung radiotherapy: A 4D Monte Carlo study of intrafraction and interfraction effects. *Med. Phys.* 35, 356–356, 2008.

Seco J., Robertson D., Trofimov A., and H. Paganetti. Breathing interplay effects during proton beam scanning: Simulation and statistical analysis. *Phys. Med. Biol.* 54, N283–N294, 2009a.

Seco J., Sharp G.C., and H. Paganetti. Study of the variability of the dosimetric outcome produced by patient organ-movement and dynamic MLC with focus on intra-fraction effects. *Med. Phys.* 36, 2505, 2009b.

Shih R., Li X.A., and J.C.H. Chu. Dynamic wedge versus physical wedge: A Monte Carlo study. *Med. Phys.* 28, 612–619, 2001.

Siebers J.V, Keall P.J., Kim J.O., and R. Mohan. A method for photon beam Monte Carlo multileaf collimator particle transport. *Phys. Med. Biol.* 47, 3225–3249, 2002.

Siebers J.V. and H. Zhong. An energy transfer method for 4D Monte Carlo dose calculation. *Med. Phys.* 35, 4096–4105, 2008.

Sterpin E., Salvat F., Cravens R., Ruchala K., Olivera G.H., and S. Vynckier. Monte Carlo simulation of helical tomotherapy with PENELOPE. *Phys. Med. Biol.* 53, 2161–2180, 2008.

Teke T., Bergman A., Kwa W., Gill B., Duzenli C., and I.A. Popescu. Monte Carlo based RapidArc QA using LINAC log files. *Med. Phys.* 36, 4302, 2009.

Thomas S.D., Mackenzie M., Rogers D.W.O., and B.G. Fallone. A Monte Carlo derived TG-51, equivalent calibration for helical tomotherapy. *Med. Phys.* 32, 1346–1353, 2005.

Tyagi N., Litzenberg D., Moran J., Fraass B., and I. Chetty. Use of the Monte Carlo method as a comprehensive tool for SMLC and DMLC-based IMRT delivery and quality assurance (QA). *Med. Phys.* 33, 2148, 2006.

Tyagi N., Moran J.M., Litzenberg D.W., Bielajew A.F., Fraass B.A., and I.J. Chetty. Experimental verification of a Monte Carlo-based MLC simulation model for IMRT dose calculation. *Med. Phys.* 34, 651–663, 2007.

Verhaegen F. and I.J. Das. Monte Carlo modeling of a virtual wedge. *Phys. Med. Biol.* 44(12), N251–N259, 1999.

Verhaegen F. and H.H. Liu. Incorporating dynamic collimator motion in Monte Carlo simulations: An application in modeling a dynamic wedge. *Phys. Med. Biol.* 46, 287–296, 2001.

Xing L., Siebers J., and P. Keall. Computational challenges for image-guided radiation therapy: Framework and current research. *Semin. Radiat. Oncol.* 17, 245–257, 2007.

Yu C., Jaffray D.A., and J.W. Wong. The effects of intrafraction organ motion on the delivery of dynamic intensity modulation. *Phys. Med. Biol.* 43, 91–104, 1998.

Zhao Y-L., Mackenzie M., Kirkby C., and B.G. Fallone. Monte Carlo calculation of helical tomotherapy dose delivery. *Med. Phys.* 35, 3491–3500, 2008a.

Zhao Y-L., Mackenzie M., Kirkby C., and B.G. Fallone. Monte Carlo evaluation of a treatment planning system for helical tomotherapy in an anthropomorphic heterogeneous phantom and for clinical treatment plans. *Med. Phys.* 35, 5366–5374, 2008b.

Zhong H. and J.V. Siebers. Monte Carlo dose mapping on deforming anatomy. *Phys. Med. Biol.* 54, 5815–5830, 2009.

8

Patient Dose Calculation

Joao Seco
Harvard Medical School

Maggy Fragoso
Alfa-Comunicações

8.1 General Introduction

In the most general form, the Monte Carlo (MC) method is a statistical approach to solving mathematical and/or physics problems in the form of integration, derivation, and so on. In the case of radiation therapy of cancer, MC is ideally suited to solve the complex problem of particle transport and energy deposition within a heterogeneous medium, that is, the human body. An extensive review of the MC method in radiotherapy is given by Chetty et al. (2007) and references within. In this chapter, we will provide a review of the most widely used MC codes for patient dose calculation with emphasis on the statistical uncertainties, denoising and smoothing methods, Hounsfield unit (HU) to medium conversion, deformable registration, and inverse planning with MC.

There is a large variety of MC codes available now in radiation therapy of which the most widely used are EGSnrc (Kawrakow 2000), MCNP (Brown, 2003), GEANT4 (Agostinelli, 2003), and PENELOPE (Baro et al., 1995; Salvat et al., 2009). All these algorithms are based on the condensed history techniques first developed by Berger et al. (1963), where a large number of electron interactions are condensed into one because electrons lose very little energy in a single electromagnetic interaction. Condensed history implementation was divided into two main classes. In the class I condensed history approach, all collisions are subject to grouping, with the effect of secondary particle creation above a specified threshold energy taken into account after the fact (i.e., independently of the energy actually lost during the step). This is performed by transporting a number of

secondary particles. MCNP is an example of a class I type MC code, where the system adopted the electron transport algorithm from ETRAN (Berger and Seltzer, 1973; Seltzer, 1988). This algorithm was developed at NIST by Berger and Seltzer following the condensed history techniques proposed by Berger (1963). In the class II condensed history approach, interactions are divided into "hard" (also designated as "catastrophic") and "soft" collisions. Soft collisions are grouped in class I scheme, while for hard collisions, the particles are tracked individually as independent particles. EGSnrc, GEANT4, and PENELOPE are examples of class II MC simulations. An important aspect of condensed history is the simulation of a particle crossing a boundary between different regions. A lot of work has gone into MC research of boundary crossing and dependency of calculated results on step size. Step-size artifacts and boundary crossing issues are now well understood (Larsen, 1992; Kawrakow and Bielajew, 1998; Kawrakow, 2000) and have led to a significant improvement in the accuracy of MC code predictions. For further details of condensed history techniques, consult Chetty et al. (2007) and the literature within.

8.2 Statistical Uncertainties in Patient Dose Calculation

8.2.1 Introduction

The use of MC simulation techniques in radiation therapy planning introduces unavoidable statistical noise in the calculated dose because of the fluctuation of the dose around its mean value in a given volume element (usually termed voxel). This can have a profound effect on the information extracted from the dose distributions, namely, the isodose lines and dose–volume histograms (DVHs). These statistical fluctuations are characterized by the variance, σ^2, on the calculated quantity of interest, which is an estimate of the true variance, σ^2. In an ideal MC simulation, a zero variance is desirable but unachievable within a finite amount of time.

The performance of an MC calculation is characterized by its efficiency, e, which combines the estimated variance with the number of iterations, N, required to achieve that variance, within a given amount of time. In other words, the efficiency of an MC simulation (also called "figure of merit") is an indicator of how "fast" the MC code is achieving a specific σ^2 value. It is defined as

$$\varepsilon = \frac{1}{\sigma^2 T}, \tag{8.1}$$

where σ^2 is an estimate of the true variance on the quantity of interest and T is the central processing unit (CPU) time required to achieve this variance. The efficiency of an MC algorithm can be improved by either decreasing the σ^2 value for a given T, or by decreasing T for a given N, while not changing the variance. Variance reduction techniques are used to reduce the σ^2 value without changing the number of iterations used, N. The VRT methods use "clever" physics, mathematics, or numerical "tricks" to accelerate the overall MC algorithm calculation time, thus increasing its efficiency.

8.2.2 Dose-Scoring Geometries and Calculation of Uncertainties

Different types of scoring geometries are used for three-dimensional (3D) dose calculations, one cubic or parallelepiped voxels being the most common. Moreover, the dose is usually scored in the geometrical voxels that are based on the computed tomography (CT) information. One problem that arises is that decreasing the CT resolution enormously increases the memory usage. One alternative that is being currently adopted is to decouple the scoring grid from the geometrical grid. In the following paragraphs, some examples of scoring geometries are presented, followed by some methods of uncertainty calculation.

8.2.2.1 Dosels

In the PEREGRINE MC dose engine, the particle transport is performed in a patient transport mesh, which is a Cartesian map of material composition and density, obtained from the patient's CT images. The dose is scored in an array of overlapping spheres that are independent of the material transport mesh, termed dosels. During the dose calculation, the standard deviation in the dosel receiving the highest dose is tracked and when it reaches a level specified by the user, the simulation is terminated.

8.2.2.2 Kugels

The electron macro Monte Carlo (MMC) dose engine calculates the dose distribution from predetermined electron histories that are stored in look-up tables, which were obtained in spherical volume elements, termed kugels, of different materials. The concept was initially introduced by Neuenschwander et al. (1995) for simulating electron beam treatment planning. The MMC algorithm uses results derived from conventional MC simulations of electron transport through macroscopic spheres of various radii and consisting of a variety of media. On the basis of kugel spheres, the electrons are transported in macroscopic steps through the absorber. The energy loss is scored in the 3D dose matrix. The transport of secondary electrons and bremsstrahlung is also accounted for in the model.

8.2.2.3 Segmented Organs

MC dose scoring can be significantly improved if many of the voxels are combined into a large volume, that is, an organ. In this case, we lose the spatial resolution of the dose distribution, but we gain in the ability to predict organ dose very rapidly and accurately with MC. However, contouring CT volumes into organs can be very subjective and depends significantly on the clinician performing the contouring. Therefore, segment organ dose scoring can only accurately be applied in controlled simulation environments such as the NCAT phantom (Segars, 2001). The NCAT anthropomorphic computational phantom is based on the Visible Human dataset (Spitzer and Whitlock, 1998). The phantom uses nonuniform rational B-spline surfaces to represent 3D human anatomy and also allows the incorporation of four-dimensional (4D) motion of the cardiac and respiratory motions. The anatomical parameters of the phantom are set using an input file so that the anatomy parameters can be scaled to dimensions and values different to that of the default settings based on the Visible Human. Such details include the level of desired anatomical detail for the blood vessels and lung airway. The dynamic nature of the phantom is invoked by specifying whether respiratory and/or cardiac motion is to be included.

MC simulations using NCAT have been performed allowing both voxel scoring and organ scoring by Riboldi et al. (2008). A 4D MC framework was developed to study dose variations with motion, within a controlled environment of a phantom. Results show that the MC simulation framework can model tumor tracking in deformable anatomy with high accuracy, providing absolute doses for intensity-modulated radiotherapy (IMRT) and conformal radiation therapy. Badal et al. (2009) developed a realistic anatomical phantom based on NCAT on which MC simulation may be performed for imaging applications, from image-guided treatments to portal imaging. In this case, the organs' scoring volumes were segmented into triangular meshes.

8.2.2.4 Voxel Size Effects

The scoring voxel size has a large impact on the calculation time because of the amount of time spent in geometry boundary checks and scoring. Cygler et al. (2004) showed that the number of histories to be simulated, per unit area, is linearly proportional to the mass of the scoring voxels.

8.2.2.5 Concept of Latent Variance

There are generally two sources of statistical uncertainty in the MC calculations of patient dose: those resulting from the simulation of the accelerator treatment head and those arising from fluctuations in the phantom/patient dose calculation. Sempau et al. (2001) coined the term "latent variance" to describe the uncertainty due to the statistical fluctuation in the phase space generated from the simulation of the treatment head as opposed to the uncertainty due to the random nature of the dose calculation by MC techniques.

The statistical uncertainty in the dose calculated in a phantom by reusing the particles from the phase space file, and assuming that the particles are independent and ignoring the correlations between them, will approach the finite, latent variance associated with the phase space data, regardless of the number of times the phase space is used.

More importantly, in estimating the statistical uncertainty in the patient dose calculation, it is necessary to account for the latent variance from the phase space calculation as well as for the random uncertainty from the patient calculation. Should latent variance be a significant factor in the total uncertainty, more independent phase space particles need to be used in the patient simulations.

8.2.2.6 Batch Method

In the batch method, the estimate of the uncertainty, $s_{\bar{X}}$, of a scored quantity, X, is given by

$$s_{\bar{X}} = \sqrt{\frac{\sum_{i=1}^{N}\left(X_i - \bar{X}\right)^2}{N\left(N-1\right)}}, \tag{8.2}$$

where N is the number of batches, usually 10, X_i is the value of X in batch i, and $\bar{X}$ is the mean value of X evaluated over all batches. The sample size, N, is thus given by the number of batches.

As pointed out by Walters et al. (2002), three main problems can be identified with this approach for the uncertainty calculation:

1. A large number of batches must be used or else there will be significant fluctuations in the uncertainty itself because the sample size, N, is quite small.
2. Arbitrary grouping of histories into batches ignores any correlations between incident particles.
3. The batch approach adds an extra dimension to the scoring quantities of interest.

8.2.2.7 History-by-History Method

The history-by-history method, implemented accordingly with Salvat's approach (Salvat et al., 2009), allows the elimination of the above-mentioned problems when using the batch method to estimate the uncertainty in the MC dose calculation. The scored dose data are stored in a very efficient way, while the simulation is running. In this method, X_i now represents the scored quantity in history i rather than in batch i, and N is the number of independent events. Equation 8.2 can then be rewritten as follows:

$$s_{\bar{X}} = \sqrt{\frac{1}{N\left(N-1\right)}\left(\frac{\sum_{i=1}^{N}X_i^2}{N} - \left(\frac{\sum_{i=1}^{N}X_i}{N}\right)^2\right)}. \tag{8.3}$$

When using phase space sources to calculate the dose distribution, one history is defined to be all particle tracks associated with one initial particle. It should also be emphasized that the quantity X_i may be weighted quantities, if variance reduction techniques are used.

During the MC calculation, that is, on the fly, a record of the quantities $\sum_{i=1}^{N}X_i^2$ and $\sum_{i=1}^{N}X_i$ is kept. At the end of the calculation, the uncertainty of the calculation is determined according to Equation 8.3, without the need to store the scored quantity in batches. More details on the algorithm implementation can be found in Sempau et al. (2000) and Salvat et al. (2009).

8.3 Denoising and Smoothing Methods

8.3.1 Introduction

As previously mentioned, one possible way to reduce the statistical fluctuations is by performing the MC simulation for a very large number of histories, which is generally not practical without the application of powerful variance reduction techniques. Another possibility is the process of removing or reducing the statistical fluctuations from a noisy calculated data with smoothing or denoising algorithms that are routinely applied in a variety of disciplines that deal with noisy signals (e.g., imaging), thus speeding up the MC calculations.

Analogous to image restoration problems in the field of image processing, an improved estimate of the true image can be produced by smoothing or denoising the raw MC result, produced with fewer source particles than needed with the denoising process resulting in an accelerated image of the true dose distribution. Obviously, the goal of MC denoising algorithms is to be as aggressive as possible in locally smoothing the raw MC result while attempting to avoid the introduction of systematic errors—bias—especially near sharp features such as beam edges.

The work of Keall et al. (Keall, 1999, 2000), Jeraj and Keall (Jeraj et al., 1999), and Buffa et al. (Buffa, 1999) have been seminal in the study of the effect of statistical uncertainty on the evaluation of MC plans and the possibility of removing these effects on MC calculated DVHs.

Various methods related to image digital filtering, wavelet thresholding, adaptive anisotropic diffusion, and denoising based on a minimization problem have been proposed to solve this problem. These algorithms are an approximate efficiency-enhancing method since they can introduce a systematic bias into the calculation. Nonetheless, denoising techniques are useful as they can reduce the overall (systematic and random) uncertainty when the random component decreases more than the systematic component increases. However, it must be emphasized that the denoising techniques require proper validation under the full range of clinical circumstances before they can be used with MC dose algorithms.

Kawrakow (2002) introduced five accuracy criteria, which have become a standard, to evaluate the performance of smoothing algorithms, where smoothed and unsmoothed MC calculations must be compared to a "benchmark result" that has been obtained through the simulation of a large number of histories, namely

1. Visual inspection of the isodose lines, where the differences between the benchmark and the smoothed results should be minimal.
2. DVHs. The difference area between two DVHs, one corresponding to the smoothed data and another to the benchmark data, should be quantified. This difference should be small.
3. Maximum dose difference. The maximum dose difference to the benchmark for smoothed and not smoothed MC simulations should be obtained.
4. Root-mean-square difference (RMSD). This is a standard quantitative measure for the degree of agreement between two distributed quantities. The RMSD in an MC calculation without smoothing is solely due to the statistical uncertainties and, after smoothing, it will contain some systematic bias introduced by the denoising process. RMSD should be obtained between the benchmark and the smoothed and unsmoothed dose distribution.
5. $x\%/y$ mm test. The fraction of voxels with a smoothed dose value that differs more than $x\%$ from a benchmark dose value at the same point and there is no point in the benchmark dose distribution that is closer than y mm to the point that has the same dose value (Van Dyke, 1993). The current accuracy recommendation is 2–3%/2 mm (Fraass et al., 1998).

8.3.2 Denoising Integrated Dose Tallies

8.3.2.1 DVH Denoising Methods

Two groups have introduced techniques for denoising the DVH: Sempau and Bielajew (2000) and Jiang et al. (Jiang, 2000). Sempau and Bielajew (2000) used a deconvolution method, where the calculated DVH was considered as the "true" DVH, which would be obtained after an infinite number of histories, thus yielding zero variance, convolved by noise. The implications of this approach would be that decisions based upon DVHs could be made in a faster time and, more importantly, inverse treatment planning or even optimization methods could be employed using MC dose calculations at all stages of the iterative

process, since the long calculation times at the intermediate calculation steps could be eliminated.

In the work presented by Jiang et al. (Jiang, 2000), the MC calculated DVH is treated as blurred from the noiseless and "true" DVH. The technique described is similar to image restoration, where a deblurring function is employed to obtain the noiseless DVH. An estimate of the smoothed DVH is blurred and the difference between this image and the blurred and original image is minimized with a least-square minimization method iteratively.

8.3.3 Dose Distributions Denoising Methods

Denoising of DVHs cannot completely address the problem of statistical fluctuations as the dose distributions are frequently assessed and represented by other means (e.g., isodose lines, calculation of the tumor control, and/or normal tissue complication probabilities). This makes the removal of statistical uncertainties from the dose distribution itself more desirable and relevant.

8.3.3.1 Deasy Approach

The first to propose denoising of the 3D dose distribution was Deasy (2000), who suggested that MC results can be obtained from an infinite number of source particles and a noise source due to the statistics of particle counting. He proposed several digital filtering techniques to denoise electron beam MC dose distributions and demonstrated that these techniques improve the visual usability and the clinical reliability of MC dose distributions.

MC calculated dose distributions with high statistical uncertainty were subjected to various digital filters and a comparison was then performed between the resulting denoised distributions to calculations with much lower uncertainty using isoline representations. The concluding remarks were that smoothing with digital filters was a promising technique for dealing with the statistical noise of MC simulations. The main objection against the use of denoising for MC calculated dose distributions results from the concern that smoothing may systematically alter its distribution.

8.3.3.2 Wavelet Approach

Deasy et al. (2002) presented another approach, where wavelet threshold denoising was used to accelerate the radiation therapy MC simulations by factors of 2 or more. The dose distribution is separated into a smooth function and noise, represented by an array of dose values that are, subsequently, transformed into discrete wavelet coefficients. Below a positive threshold of the wavelet coefficient values, they were set equal to zero. They showed that with a suitable value of this threshold parameter, the statistical noise was suppressed with little introduction of bias.

8.3.3.3 Savitzky-Golay Method

Kawrakow (2002) presented a 3D generalization of a Savitzky–Golay filter with adaptive smoothing window size, that is, the number of surrounding voxels that is used, to reduce the probability for systematic bias. The size of the smoothing window is based on the statistical uncertainty in the voxel that is being smoothed. According to the author, this filter decreases the number of particle histories by factors of 2–20, concluding that smoothing is extremely valuable for the initial trial-and-error phase of the treatment planning process.

8.3.3.4 Diffusion Equation Method

Miao et al. (2003) investigated the denoising of MC dose distributions with a 3D adaptive anisotropic diffusion method. The conventional anisotropic diffusion method was extended by, adaptively, modifying the filtering parameters according to the local statistical noise. The work presented showed that this method can reduce statistical noise significantly, that is, 2–5 times, corresponding to a reduction in the simulation time by a factor of up to 20, while maintaining the characteristics of the gradient dose areas.

8.3.3.5 IRON Method

Fippel and Nusslin (2003) introduced the so-called iterative reduction of noise (IRON) algorithm, which iteratively reduces the statistical noise for MC dose calculations. By varying the dose in each voxel, this algorithm minimizes the second partial derivatives of the dose with respect to the three spatial coordinates. This algorithm requires as input MC dose distributions with or without known statistical uncertainties.

Smoothing of the MC dose distributions using IRON was proven to lead to an additional reduction of the MC calculation time by factors between 2 and 10. As with the application of the other smoothing techniques, this reduction is particularly useful if the MC dose calculation is part of an inverse treatment planning calculation.

8.3.3.6 Content Adaptive Median Hybrid Filter Method

In this method, linear filters were combined through a weighted sum, with the median operation to produce hybrid median filters (El Naqa et al., 2005). In regions with strong second derivative features, the median operator is used and in smoother regions, the mean operator is preferred. In general, median filters outperform the moving average and other linear filters in the process of removing impulsive noise (outliers), and in the preservation of edges, however, they fail to provide the same degree of smoothness in homogenous regions. An adaptive combination with mean value linear filters was then chosen in this study.

8.4 CT to Medium Conversion Methods

8.4.1 CT Stoichiometric Conversion Methods

The conversion of CT HUs into material composition and mass density may strongly influence the accuracy of patient dose calculations in MC treatment planning or in any other dose calculation algorithm. To establish an accurate relationship between CT HU and electron density of a tissue, the CT scale is usually divided into a number of subsets. The initial MC dose algorithms used six or fewer materials to define the conversion, for example, air, lung, fat, water, muscle, and bone (DeMarco et al., 1998; Ma et al., 1999). The influence of the differences in tissue composition on MC computation was investigated by du Plessis et al. (1998) for a set of 16 human tissues. They assessed the optimal number of tissue subsets to achieve 1% dose accuracy for millivolt photon beams, and concluded that seven basic tissue subsets were sufficient. By varying the physical mass density for lung and cortical bone, 57 different tissue types (excluding air) could be constructed, spanning a total CT range of 3000 HU. The division adopted by du Plessis et al. was 21 types of cortical bone ranging from 1100 to 3000 HU and 31 different lung tissues ranging from 20 to 950 HU. The remaining five types were chosen from the basic tissue subsets. The disadvantage of using the method proposed by du Plessis is that it does not use a stoichiometric conversion scheme and therefore is limited to the beam quality under consideration.

Stoichiometric conversion schemes were a large step forward in the accuracy of MC algorithms for patient dose calculations relative to the methods discussed previously. The stoichiometric conversion scheme from CT to electron (or proton stopping power) was initially proposed in Schneider et al. (1996) for proton MC dose algorithms. Initially, a set of materials of well-known atomic composition and density (usually plastic materials of varying composition and density to mimic human tissues) are CT scanned to measure the corresponding CT HU values. Next, the measured CT HU values are fitted by a theoretical equation interrelating the CT HU, mass density, atomic number Z, and atomic weight A of each material. Finally, the fitted parameters can be used to calculate the CT HU values for real tissues using tabulated composition data.

Schneider et al. (Schneider, 2000) proposed a CT conversion based on the stoichiometric conversion scheme, considering 71 human tissues and they adopted the total cross-section parameterization given by Rutherford et al. (1976). To reduce the effort of fitting the CT HU numbers to mass density and elemental weights, the authors created four sections on the CT scale within which the selected tissues were confined. Within each section, both the mass density and elemental weights of the tissues were interpolated.

8.4.1.1 Calculation of CT HU Number for Stoichiometric Conversion Schemes

In Rutherford et al. (1976), a parameterization of the cross section in the diagnostic x-ray energy range is presented:

$$\sigma_i(E) = Z_i\, K^{\mathrm{KN}}(E) + Z_i^{2.86}\, K^{\mathrm{sca}}(E) + Z_i^{4.62}\, K^{\mathrm{ph}}(E), \qquad (8.4)$$

where E is the energy, $\sigma_i(E)$ is the total cross section in units of barn/atom, K^{KN} denotes Klein–Nishina coefficient, K^{sca} is the Rayleigh or coherent scattering coefficient, and K^{ph} is the photoelectric coefficient. The linear attenuation coefficient, μ, is now obtained by replacing the total cross section:

$$\mu(E) = \rho N_A \sum_{i=1}^{n} \frac{w_i}{A_i} \sigma_i(E)$$

$$= \rho N_A \sum_{i=1}^{n} \frac{w_i}{A_i} \left[Z_i K^{KN}(E) + Z_i^{2.86} K^{sca}(E) + Z_i^{4.62} K^{ph}(E) \right], \quad (8.5)$$

where ρ (g/cm^3) is the mass density, N_A (mol^{-1}) is the Avogadro constant (6.022045×10^{23}), i is the element index, and w_i is the element weight. Recall that the μ values are converted in HU using

$$\mathrm{HU} = \left(\frac{\mu}{\mu_{H_2O}} - 1 \right) 1000, \quad (8.6)$$

where the CT HU is defined such that water has the value 0 and air has the value -1000. The equation then becomes

$$\frac{\mu}{\mu_{H_2O}} = \frac{\rho}{\rho_{H_2O}} \frac{\sum_{i=1}^{n} \frac{w_i}{A_i} (Z_i + Z_i^{2.86}k_1 + Z_i^{4.62}k_2)}{\frac{w_H}{A_H}(1 + k_1 + k_2) + \frac{w_O}{A_O}(8 + 8^{2.86}k_1 + 8^{4.62}k_2)}, \quad (8.7)$$

with $k_1 = K^{sca}/K^{KN}$ and $k_2 = K^{ph}/K^{KN}$, and the values of (k_1, k_2) are dependent on the CT kilovolt value used for imaging and are determined experimentally by a least-square fit of the measured HU values to the previous equation, that is, minimizing the following expression:

$$\sum_{i=1}^{n} \left[\left(\frac{\mu}{\mu_{H_2O}} \right)(k_1, k_2) - \left(\frac{H(meas.)}{1000} + 1 \right) \right]^2. \quad (8.8)$$

Additional predictions of the k_1 and k_2 values are also presented in Vanderstraten et al. (2007) for a large group of CT scanners of different models and manufacturers for several European radiation oncology centers.

8.4.1.2 CT Hounsfield Units Interpolation Method

In this section, we present a brief description of the Schneider et al. method of interpolating within a tissue range, for tissue samples composed of only two components, where the components are denoted as $(\rho_1, w_{1,i}, H_1)$ and $(\rho_2, w_{2,i}, H_2)$ and the composite by (ρ, w_i, H) and where H_1, H_2 represent CT HU values range and where $H_1 < H_2$. In addition, W_1 represents the proportion of the first component, while the second is obtained from $W_2 = 1 - W_1$. A brief note on the meaning of W_1 and W_2 and how they differ from $w_{1,i}$ and $w_{2,i}$: while $w_{1,i}$ and $w_{2,i}$ represent the weighting within a tissue of the chemical elements such as oxygen, hydrogen, and so on, W_1 and W_2 represent the weighting

of two reference tissues to form a new "interpolated" tissue such as combining osseous tissue and bone marrow to form different skeletal tissues. The new interpolated medium has the following weight per chemical element:

$$w_i = W_1 w_{1,i} + W_2 w_{2,i} = W_1(w_{1,i} - w_{2,i}) + w_{2,i} \quad (8.9)$$

and mass density:

$$\rho = \frac{m}{V} = \frac{m}{(m_1/\rho_1) + (m_2/\rho_2)}$$

$$= \frac{1}{(W_1/\rho_1) + (W_2/\rho_2)} = \frac{\rho_1 \rho_2}{W_1(\rho_2 - \rho_1) + \rho_1}. \quad (8.10)$$

The linear attenuation coefficient of the new media can now be generated by substituting Equations 8.9 and 8.10 in Equation 8.5:

$$\mu = \frac{\rho_1 \rho_2}{W_1(\rho_2 - \rho_1) + \rho_1} N_A \left[W_1 \sum_i \left(\frac{w_{1,i} - w_{2,i}}{A_i} \sigma_i \right) + \sum_i \left(\frac{w_{2,i}}{A_i} \sigma_i \right) \right] \quad (8.11)$$

$$= \frac{\rho_1 \rho_2}{W_1(\rho_2 - \rho_1) + \rho_1} N_A \left[W_1 \left(\frac{\mu_1}{\rho_1} - \frac{\mu_2}{\rho_2} \right) + \frac{\mu_2}{\rho_2} \right]. \quad (8.12)$$

The composite tissue with H, such that $H_1 < H < H_2$ is then given by

$$W_1 = \frac{\rho_1(H_2 - H)}{(\rho_1 H_2 - \rho_2 H_1) + (\rho_2 - \rho_1)H}. \quad (8.13)$$

The composite medium will then have the following mass density and elemental composition:

$$\rho = \frac{\rho_1 H_2 - \rho_2 H_1 + (\rho_1 - \rho_2)H}{H_2 - H_1} \quad (8.14a)$$

$$w_i = \frac{\rho_1(H_2 - H_1)}{(\rho_1 H_2 - \rho_2 H_1) + (\rho_2 - \rho_1)H}(w_{1,i} - w_{2,i}) + w_{2,i},$$

$$H_1 \leq H \leq H_2 \quad (8.14b)$$

8.4.1.3 Stoichiometric Conversion Method Based on Dose-Equivalent Tissue Subsets

In Vanderstraeten et al. (2007), the authors generalized the Schneider et al. (2000) method to generate subsets of tissue materials that were dosimetrically equivalent to within 1%. Their key assumption was that in the case of tissues that are situated close to the interface between two subsets, the dose differences

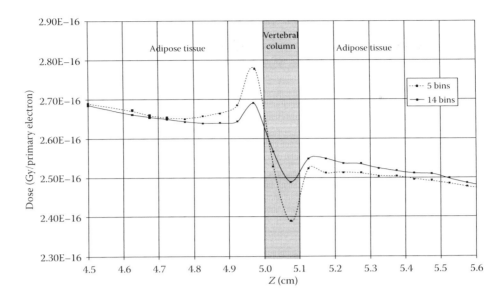

FIGURE 8.1 Depth dose curve with a slab phantom calculated with Monte Carlo for an interface between the adipose tissue and bone (thickness 1 mm). The results compare the five-bin CT with 14-bin CT proposed by Vanderstraeten (2007). (Courtesy of Vanderstraeten et al. 2007. Conversion of CT numbers into tissue parameters for Monte Carlo dose calculations: A multicenter study. *Phys. Med. Biol.* 52, 539–62.)

between them should not exceed 1%. In a previous study by Verhaegen and Devic (2005), the authors had showed the dosimetric importance of the CT calibration curve "tissue-binning" approach and the large dosimetric impact of tissue substitutes containing fluorine. Vanderstraeten et al. (2007) proposed the use of 14 dosimetrically equivalent tissue subsets (bins), where 10 were bone, to reduce significantly tissue-binning effects. The study was performed for a large number of different CT scanners obtained from nine different radiotherapy departments across Europe. These were then compared to the conventional five-bin scheme with only one bone bin. In Figure 8.1, the MC dose prediction for 5 and 14 bins of the CT is presented, where it is possible to see 5% differences at the interface.

The 10 bone tissue subsets used were generated to accurately model both the hydrogen (H) and calcium (Ca) content from 100 HU to beyond 1450 HU, consisting of cortical bone or denser bone. Both H and Ca contents strongly influence photon attenuation as pointed out in Seco and Evans (2006). In addition, the Ca content at the diagnostic energies will strongly increase the importance of the photoelectric and Rayleigh scattering terms because of their strong Z-dependence. The 14- and 5-bin approaches were then studied for a cohort of nine patients, where dosimetric differences of up to 5% were observed. However, DVH comparisons did not show any significant differences.

8.4.2 Dual-Energy X-Ray CT Imaging: Improved HU CT to Medium Conversions

Dual-energy x-ray CT imaging involves scanning an object with two significantly different tube voltages to obtain a better estimate of the effective atomic numbers, Z, and the relative electron

densities, ρ_e. In Torikoshi et al. (2003), the linear attenuation coefficient of a material at a set x-ray energy, E, is given by

$$\mu(E) = \rho_e'(Z^4 F(E,Z) + G(E,Z)), \qquad (8.15)$$

where ρ_e' is the electron density, $\rho_e' Z^4 F(E,Z)$ and $\rho_e' G(E,Z)$ are, respectively, the photoelectric and combined Rayleigh and Compton scattering terms of the linear attenuation coefficient. Both F and G are obtained by quadratic fits of the photoelectric and scattering terms of NIST attenuation coefficients (Berger et al., 2005).

A detailed method of performing dual-energy tissue segmentation is presented in Bazalova et al. (2008). Dual-energy CT material extraction was performed using the 100 and 140 kVp CT images with corresponding x-ray spectra. The authors found that the mean errors of extracting ρ_e' and Z are 1.8% and 2.8%, respectively. Dose calculations were then performed using various photon beam energies of 250 kVp, 6 MV, and 18 MV, and 18 MeV electron beams and a solid water phantom with tissue-equivalent inserts. The dose calculation errors were particularly high for the 250 kVp beam, leading to 17% error due to a misassigned soft bone tissue-equivalent cylinder. In the case of the 18 MeV electron beam and 18 MV photon beams, they found 6% and 3% dose calculation errors due to misassignment. In the case of the 6 MV electron beam, dose differences were below 1%.

8.5 Deformable Image Registration

(A brief overview is given of deformable image registration; for more detailed description, please consult the Heath and Seco Chapter 7).

The use of deformable image registration varies from 4D radiotherapy to contour propagation, treatment adaptation, dosimetric evaluation, and 4D optimization (Keall et al., 2004; Rietzel, 2005; Trofimov et al., 2005; Brock et al., 2006). A variety of different approaches exist to the nonrigid registration; Wang et al. (2005) used an accelerated Demons algorithm, Yang (2007) employed B-spline image registration with normalized cross-correlation metric, and Rietzel and Chen (2006) used a B-spline method that respects discontinuity at the pleural interface. A variety of other methods exist, which use optical flow, thin-plate spline, calculus of variations, and finite element methods with different motion algorithms (Meyer et al., 1997; Guerrero et al., 2004; Lu et al., 2004).

8.5.1 Developing Dose Warping Approaches for 4D Monte Carlo

In Keall et al. (2004), the authors performed one of the first "pseudo" 4D MC dose calculations. The method used MC to calculate dose on each of the N (= 8) individual 3D CT cubes representing an individual breathing phase. The dose distribution of each breathing phase was then mapped back to the end-inhale CT image set, using deformable image registration. The flowchart presented in Figure 8.2 represents the various steps of performing 4D MC dose calculation with deformable image registration. There are several stages to the process; from deforming contours of the anatomy from the reference phase to all CT datasets, to performing dose warping from the CT of the various breathing phases back to the reference dataset.

Deformable image registration has mostly been used in lung cancer dosimetric studies with MC (Flampouri et al., 2005; Rosu, 2005; Seco et al., 2008). A major issue with deformable registration algorithms is that no true one-to-one correspondence exists between the initial dataset (image or dose) and the registered dataset. This is a consequence of the initial voxels being rearranged (split or merged) to produce a new registered dataset. This occurs because registration algorithms either are statistical in nature or involve interpolation methods. Heath (2006) developed a method to circumvent errors introduced by deformable image registration mapping dose distributions. They developed the MC code defDOSXYZnrc, which uses a vector field to directly warp the dose grid on the fly while tracking individual particles. The major disadvantage of this method was the increased calculation time of the MC dose calculation by a factor of 2 or more.

In a more recent study by Siebers and Zhong (2008), a different approach was adopted to circumvent dose warping, while maintaining a fast MC dose calculation algorithm. The authors designated their method as the energy transfer method (ETM), where the separation between radiation transport and energy deposition is performed. The authors showed that some compensation mechanism must be provided that accounts for the rearrangement of the energy deposited per unit mass, particularly in the dose gradient regions. As an example of the ETM approach, consider two adjacent voxels $A1$ and $A2$ that merge into one voxel $X1$, following image registration. The dose interpolation method calculates the dose, $d(X1)$, from $d(A1)$ and $d(A2)$ by

$$d^{\mathrm{INTERP}}(X1) = \frac{d(A1) + d(A2)}{2}$$

$$= \frac{(E(A1)\,/\,m(A1)) + (E(A2)\,/\,m(A2))}{2}, \quad (8.16)$$

where $E(A1)$, $E(A2)$, $m(A1)$, and $m(A2)$ are the energy deposited and masses of voxel $A1$ and $A2$, respectively. However, the authors point out that the correct interpretation of the dose should be

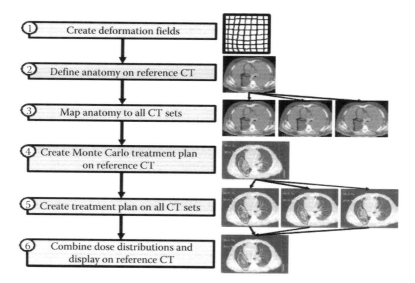

FIGURE 8.2 A flowchart of the 4D radiotherapy planning process using Monte Carlo. (Courtesy of Keall P. et al. 2004. Monte Carlo as a four-dimensional radiotherapy treatment-planning tool to account for respiratory motion. *Phys. Med. Biol.* 49, 3639–48.)

$$d^{\text{ETM}}(X1) = \frac{E(A1) + E(A2)}{m(A1) + m(A2)}, \qquad (8.17)$$

where the dose interpolation error is $\varepsilon = \left| d^{\text{INTERP}} - d^{\text{ETM}} \right|$. The ETM approach allows energy mapping from multiple different source anatomies to multiple different "reference" anatomies simultaneously. In addition, the authors indicate that the method could also be useful in 4D IMRT optimization for 4D lung cases. In this case, beam intensities are simultaneously optimized on multiple breathing phases; thus, accurate dose evaluation is desired not only at a reference phase but also at each phase of the breathing cycle.

8.5.2 Comparing Patient Dose Calculation between 3D and 4D

Flampouri et al. (2005), studied the effect of respiratory motion on the delivered dose for lung IMRT plans, where a group of six patients were studied. For each patient, a free breathing (FB) helical CT and a 10-phase 4D CT scan was acquired. A commercial planning system was then used to generate an IMRT plan. The authors found that conventional planning was sufficient for patients with tumor motion less than 12 mm, where FB helical CT had small or no artifacts relative to the 4D CT dataset. The authors also found that conventional planning was not adequate for patients with larger tumor motion or severe CT artifacts. Their study indicated that CT reconstruction artifacts have far bigger effects than tumor motion. Therefore, a major conclusion was that for accurate 4D CT MC dosimetry, CT datasets with less artifacts and distortions are required. In addition, the authors also evaluated the minimum number of breathing phases required to recreate the 10-phase composite dose. They found that a three-phase composite dose would allow a 3% error relative to a 10-phase dose prediction, while five phases would achieve 0.5% error relative to the 10-phase dose. These results are in agreement with predictions by Rosu et al. (2007), where two phases (inhale and exhale) where shown to be sufficient to predict dose distribution in the respiratory motion. In Seco et al. (2008), the authors compared the dose distributions for 3D CT FB with 4D CT FB and 4D CT-gated treatments, to assess if gated treatments provided improved delivered dose distributions. The question addressed was: do we require gating to mitigate motion effects in delivered dose distributions? The breathing pattern of each individual patient was accounted for while studying interfraction and intrafraction effects of breathing in the delivered dose. The respiratory motion was recorded by the RPM system of Varian (Varian Medical Systems, Palo Alto, CA). The authors showed that the largest dosimetric differences occurred between inhale and the other breathing phases, including the FB scan.

To analyze the spatial differences between two MC dose distributions for different breathing phases as a function of the density, the authors defined a parameter called omega index, $\Omega(\rho)$, or, $\Omega(\rho;f,\text{VOI})$, where the density, ρ, is the independent variable, and f is a dose or dose difference distribution and VOI is the volume of interest for which the omega index is calculated:

$$\Omega(\rho) = \frac{\displaystyle\int f(x) \star \delta(\rho(x) - \rho)\,\mathrm{d}x}{\displaystyle\int \delta(\rho(x) - \rho)\,\mathrm{d}x}, \qquad (8.18)$$

The parameter f can be either a 3D dose cube or a dose difference between two 3D cubes, that is, $f = D_{MC\varphi_1} - D_{MC\varphi_2}$, where φ_1 and φ_2 are any breathing phases. The omega index indicates dose differences as a function of the density. In Figure 8.3, a comparison is given of the dose differences between inhale and FB dose calculations.

It is possible to observe dose differences of the order 3–5 Gy between inhale and FB that occur in both the low- and high-density regions of lung and bone, respectively. This indicates that 4D MC can improve not only accurate dose predictions within the primary tumor but also for the surrounding tissue, that is, lung or bone, which is considered part of the planning target volume (PTV) because of the expanding margin to account for motion and setup errors. The expanding area can encompass a large volume of both lung and bone, which may affect the final dose estimate to the PTV due to movement.

In addition, the authors also addressed the question of whether 4D-gated treatments are better than 4D free-breathing treatments in providing a better coverage of the PTV, where gated treatments have a far longer total delivery time and subsequent patient time on the couch. No significant dosimetric differences were observed between 4D CT FB and 4D CT-gated dose distributions; however, only a small group of three patients were assessed; so no generalization can be made. Larger differences (of approximately 3%) were observed between 3D FB and 4D FB, which could be attributed to various factors such as (i) differences in image reconstruction of serial versus helical CT, (ii) importance of the inhale phase in the final 3D FB CT, and (iii) coughing or breath-hold by the patient during the serial imaging process.

8.5.3 Intercomparison of Dose Warping Techniques for 4D Dose Distributions

In Heath et al. (2008), an intercomparison is performed of the various dose warping methods for 4D MC dose calculations in the lung. The methods compared were center-of-mass tracking, trilinear interpolation, and the defDOSXYZ method (Heath, 2006). No clinically significant dose differences were observed by the three methods. However, for the extreme case where motion was not accounted for in the treatment plan, the authors noticed an underestimate of the target volume dose by up to 16% by the remapping techniques. The authors also pointed out that the accuracy of the dose calculation is significantly affected by the continuity of the deformation fields from nonlinear image registration.

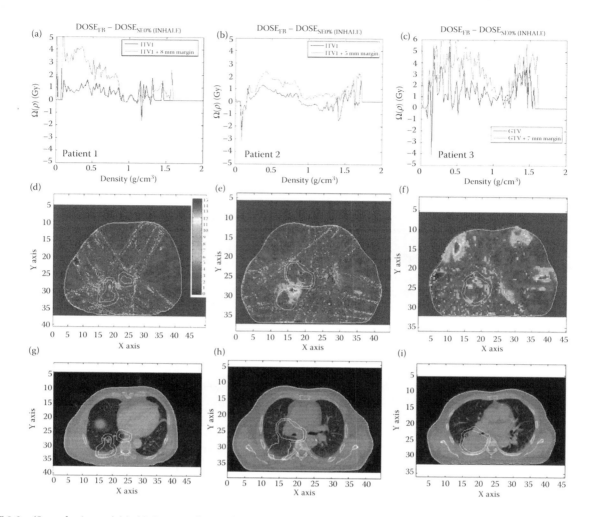

FIGURE 8.3 (**See color insert.**) (a)–(c) Omega index prediction, (d)–(f) gamma index predictions, and (g)–(i) an example of CT slice for patients 1–3 (gamma values >3 are represented as red, the maximum gamma value of the color scale). (Courtesy of Seco J. et al. 2008. Dosimetric impact of motion in free-breathing and gated lung radiotherapy: A 4D Monte Carlo study of intrafraction and interfraction effects. *Med. Phys.* 356–66.)

8.6 Inverse Planning with MC for Improved Patient Dose Calculations in Both 3D and 4D

Inverse planning for IMRT with any dose algorithm, including MC, involves the calculation of what is usually called the D_{ij} matrix, which allows the conversion of a fluence map into a dose distribution by the following relationship:

$$D_i(x) = \sum_{j \in Beamlet} D_{ij} x_j, \qquad (8.19)$$

where D_{ij} is the respective dose influence and x_j is a given fluence or beamlet intensity from a set of all beamlets. The D_{ij} can then be calculated using pencil beam (PB), superposition–convolution, or MC dose algorithms. A major issue with the formulation given in Equation 8.16 is that usually delivery constraints

and field size dependencies of output factors are not accounted for in the generation of D_{ij} values. These have to be added after the planning stage, which can lead to suboptimal delivered dose distributions.

One of the initial attempts in using MC to perform 3D inverse planning was done by Jeraj and Keall (Jeraj and Keall, 1999) using the MCNP code. The authors performed optimization on MC-generated PBs, where an initial back-projection estimate was obtained with MCNP to generate an initial IMRT fluence. EGS4 was then used to calculate dose within the patient and therefore generate the required PBs for optimization of all beams, using a narrow-beam approach to generate each PB. The major issues with the method proposed by Jeraj and Keall (1999) was that each MC-generated PB did exactly model the spectrum exiting the linac, MLC delivery constraints were not modeled, and head scatter variations with field sizes and motion effects were not taken into account. More recently, Siebers et al. (Siebers and Kawrakow, 2007) proposed a hybrid method of 3D MC inverse

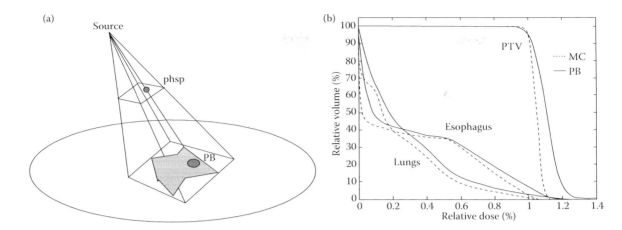

FIGURE 8.4 (a) Monte Carlo fluence-based pencil beam (PB) dose calculation method. (b) Dose–volume histogram of optimized IMRT plan based on Monte Carlo (full line) and standard PB-like dose algorithm (dashed line) normalized to 95% isodose line. (Courtesy of Nohadani O. et al. 2009. Dosimetry robustness with stochastic optimization. *Phys. Med. Biol.* 54, 3421–32.)

planning where an initial dose prediction was performed by PB algorithm. For subsequent iterations, the PB dose prediction was improved by using the MC algorithm such that the optimized plans were equivalent to an MC-based optimization. This produced a gain of 2.5 times over a full MC-based optimization.

Nohadani et al. (2009) proposed a 4D MC-based inverse planning approach using broad beam approach, in contrast to the narrow beam approach of Jeraj et al. (1999). For each incident beam, the jaw setting does not change during IMRT delivery for a Varian linac, making it possible to generate an open-field phase space file just after the *X*/*Y* jaws. Therefore, the MC phase space after the jaws can be split into PB-like pixels from which particles can be sampled to produce a 3D PB dose distribution within the patient. In Figure 8.4a, we present a diagram showing how sampling a pixel in the PHase SPace (phsp) produces an MC–PB distribution within the patient.

The advantages of the broad beam approach for generating the MC–PB are: (i) it will account for heterogeneity effects in the patient IMRT calculation; (ii) if all PBs are combined into a homogeneous open field, the output factor will be that of standard open field, no further output corrections are required; and (iii) when delivering the IMRT profile with, for example, a multileaf collimator, the changes in output, due to field size changes during delivery, will be small and therefore will not affect the planned optimal dose distribution. In Figure 8.4b, the DVH comparing IMRT obtained with MC–PB versus standard PB is given. It shows that an IMRT plan based on PB may appear to be a better plan in terms of PTV coverage, hot spots, and organs at risk sparing. However, it deviates significantly from an MC-based optimized plan that closely mimics the reality of the deposited dose. The figure also illustrates that an otherwise optimal IMRT plan based on PB dose calculation may not correspond to the actual dose distribution because of these dosimetric errors. Therefore, errors in dose calculation cannot be neglected.

References

Agostinelli S. 2003. GEANT4—A simulation toolkit. *Nucl. Instrum. Methods Phys. Res. A* 506, 250–303.

Badal A. et al. 2009. PenMesh–Monte Carlo radiation transport simulation in a triangle mesh geometry. *IEEE Trans. Med. Imaging* 28(12), 1894–901.

Baro J., Sempau J., Fernandez-Varea J.M. and Salvat F. 1995. PENELOPE—An algorithm for Monte Carlo simulation of the penetration and energy loss of electrons and positrons in matter. *Nucl. Instrum. Method Phys. Res. A* 100, 31–46.

Bazalova M. et al. 2008. Dual-energy CT-based material extraction for segmentation in Monte Carlo dose calculations. *Phys. Med. Biol.* 53, 2439–56.

Berger M. and Seltzer S. 1973. ETRAN Monte Carlo code system for electron and photon transport through extended media. Radiation Shielding Information Center (RSIC) Report CCC-107, Oak Ridge National Laboratory, Oak Ridge, TN.

Berger M.J. 1963. *Methods in Computational Physics*. B. Alder, S. Fernbach, and M. Rothenberg (eds.), Academic, New York, Vol. 1, p. 135.

Berger M.J. et al. 2005. XCOM: Photon Cross Section Database NBSIR 87-3597 (web version 1.3 (http://physics.nist.gov/PhysRefData/Xcom/Text/XCOM.html)).

Brock K.K. et al. 2006. Feasibility of a novel deformable image registration technique to facilitate classification, targeting and monitoring of tumor and normal tissue. *Int. J. Radiat. Oncol. Biol. Phys.* 64, 1245–54.

Brown F.B. 2003. MCNP—A general Monte Carlo particle transport code, Version 5. Report LA-UR-03 1987 Los Alamos National Laboratory, Los Alamos, NM.

Buffa F.M., Nahum A.E., and Mubata C., 1999. Influence of statistical fluctuations in Monte Carlo dose calculations on radiobiological modelling and dose volume histograms. *Med. Phys.* 26, 1120.

Chetty I.J. et al. 2007. Report of the AAPM Task Group No. 105: Issues associated with clinical implementation of Monte Carlo-based photon and electron external beam treatment planning. *Med. Phys.* 34(12), 4818–53.

Cygler J.E., Daskalov G.M., Chan G.H., and Ding G.X. 2004. Evaluation of the first commercial Monte Carlo dose calculation engine for electron beam treatment planning. *Med. Phys.* 31, 142–53.

Deasy J.O. 2000. Denoising of electron beam Monte Carlo dose distributions using digital filtering techniques. *Phys. Med. Biol.* 45, 1765–79.

Deasy J.O., Wickerhauser M.V., and Picard M. 2002. Accelerating Monte Carlo simulations of radiation therapy dose distributions using wavelet threshold denoising. *Med. Phys.* 29, 2366–73.

DeMarco J.J., Solberg T.D., and Smathers J.B., 1998. A CT-based Monte Carlo simulation tool for dosimetry planning and analysis. *Med. Phys.* 25, 1–11.

El Naqa I., Kawrakow I., Fippel M., Siebers J.V., Lindsay P.E., Wickerhauser M.V., Vicic M., Zakarian K., Kauffmann N., and Deasy J.O. 2005. A comparison of Monte Carlo dose calculation denoising techniques. *Phys. Med. Biol.* 50, 909–22.

Fippel M. and Nusslin F. 2003. Smoothing Monte Carlo calculated dose distributions by iterative reduction of noise. *Phys. Med. Biol.* 48, 1289–304.

Flampouri S. et al. 2005. Estimation of the delivered patient dose in lung IMRT treatment based on deformagistration of 4D-CT data and Monte Carlo simulations. *Phys. Med. Biol.* 51, 2763–79.

Fraass B., Doppke K., Hunt M., Kutcher G., Starkschall G., Stern R., and Van Dyke J. 1998. American association of physicists in medicine radiation therapy committee task group 53: Quality assurance for clinical radiotherapy treatment planning. *Med. Phys.* 25, 1773–829.

Guerrero T. et al. 2004. Intrathoracic tumor motion estimation from CT imaging using the 3D optical flow method. *Phys. Med. Biol.* 49, 4147–61.

Heath E. and Seuntjens J. 2006. A direct voxel tracking method for four-dimensional Monte Carlo dose calculations in deforming anatomy. *Med. Phys.* 33, 434–45.

Heath E., Seco J., Wu Z., Sharp G.C., Paganetti H., and Seuntjens J. 2008. A comparison of dose warping methods for 4D Monte Carlo dose calculations in lung. *J. Phys.: Conf Ser.* 102,102.

Jeraj R. and Keall P. 1999. Monte Carlo-based inverse treatment planning. *Phys. Med. Biol.* 44, 1885–96.

Jeraj R., Keall P., and Ostwald P.M. 1999. Comparisons between MCNP, EGS4 and experiment for clinical electron beams. *Phy. Med. Biol.* 44, 705–18.

Jiang S.B., Pawlicki T., and Ma C.-M. 2000. Removing the effect of statistical uncertainty on dosevolume histograms from Monte Carlo dose calculations. *Phys. Med. Biol.* 45, 2151–62.

Kawrakow I. 2000. Accurate condensed history Monte Carlo simulation of electron transport, I. EGSnrc, the new EGS4 version. *Med. Phys.* 27, 485–98.

Kawrakow I. 2002. On the denoising of Monte Carlo calculated dose distributions. *Phys. Med. Biol.* 47, 3087–103.

Kawrakow I. and Bielajew A.F. 1998. On the condensed history technique for electron transport. *Nucl. Instrum. Methods Phys. Rev. B* 142, 253–80.

Keall P., Siebers J.V., Libby B., Mohan R., and Jeraj R. 1999. The effect of Monte Carlo noise on radiotherapy treatment plan evaluation. *Med. Phys.* 26, 1149.

Keall P.J., Siebers J.V., Jeraj R., and Mohan R. 2000. The effect of dose calculation uncertainty on the evaluation of radiotherapy plans. *Med. Phys.* 27, 478–84.

Keall P. et al. 2004. Monte Carlo as a four-dimensional radiotherapy treatment-planning tool to account for respiratory motion. *Phys. Med. Biol.* 49, 3639–48.

Larsen E.W. 1992. A theoretical derivation of the condensed history algorithm. *Ann. Nucl. Energy* 19, 701–14.

Lu W. et al. 2004. Fast free-form deformable registration via calculus of variations. *Phys. Med. Biol.* 49, 3067–87.

Ma C.-M., Mok E., Kapur A., Pawlicki T., Findley D., Brain S., Forster K., and Boyer A. L. 1999. Clinical implementation of a Monte Carlo treatment planning system. *Med. Phys.* 26, 2133–43.

Meyer C.R. et al. 1997. Demonstration of accuracy and clinical versatility of mutual information for automatic multimodality image fusion using affine and thin-plate spline warped geometric deformation. *Med. Image Anal.* 1, 195–206.

Miao B., Jeraj R., Bao S., and Mackie T.R. 2003. Adaptive anisotropic diffusion filtering of Monte Carlo dose distributions. *Phys. Med. Biol.* 48, 2767–81.

Neuenschwander H. et al. 1995. MMC—A high performance Monte Carlo code for electron beam treatment planning. *Phys. Med. Biol.* 40, 543–74.

Nohadani O., Seco J., Martin B.C., and Bortfeld T. 2009. Dosimetry robustness with stochastic optimization. *Phys. Med. Biol.* 54, 3421–32.

du Plessis F.C.P. et al. 1998. The direct use of CT numbers to establish material properties needed for Monte Carlo calculation of dose distributions in patients. *Med. Phys.* 25, 1195–201.

Riboldi M. et al. 2008. Design and testing of a simulation framework for dosimetric motion studies integrating an anthropomorphic phantom into four-dimensional Monte Carlo. *Technol. Cancer Res. Treat.* 7(6), 449–56.

Rietzel E. and Chen G.T.Y. 2006. Deformable registration of 4D computed tomography data. *Med. Phys.* 33, 4423–30.

Rietzel E. et al. 2005. Four-dimensional image-based treatment planning: Target volume segmentation and dose calculation in the presence of respiratory motion. *Int. J. Radiat. Oncol. Biol. Phys.* 61, 1535–50.

Rosu M., Chetty I.J., Balter J.M., Kessler M.L., McShan D.L., and Ten Haken R.K. 2005. Dose reconstruction in deforming lung anatomy: Dose grid size effects and clinical implications. *Med. Phys.* 23, 2487–95.

Rosu M. et al. 2007. How extensive of a 4D dataset is needed to estimate cumulative dose distribution plan evaluation metrics in conformal lung therapy? *Med. Phys.* 34, 233–45.

Rutherford R.P., Pullan B.R., and Isherwood I. 1976. Measurement of effective atomic number and electron density using EMI scanner. *Neuroradiology* 11, 15–21.

Salvat F., Fernández-Varea J.M., and Sempau J. 2009. PENELOPE-2008: A code system for Monte Carlo simulation of electron and photon transport. *Proceedings of a Workshop/Training Course*, OECD.

Schneider W., Bortfeld T., and Schlegel W. 2000. Correlation between CT numbers and tissue parameters needed for Monte Carlo simulations of clinical dose distributions. *Phys. Med. Biol.* 45, 459–78.

Schneider U., Pedroni E., and Lomax A. 1996. The calibration of CT Hounsfield units for radiotherapy treatment planning. *Phys. Med. Biol.* 41, 111–24.

Seco J. and Evans P.M. 2006. Assessing the effect of electron density in photon dose calculations. *Med. Phys.* 33, 540–52.

Seco J. et al. 2008. Dosimetric impact of motion in free-breathing and gated lung radiotherapy: A 4D Monte Carlo study of intrafraction and interfraction effects. *Med. Phys.* 35, 356–66.

Segars W.P. 2001. Development of a new dynamic NURBS-based cardiac-torso (NCAT) phantom. PhD thesis. University of North Carolina, North Carolina.

Seltzer S.M. 1988. An overview of ETRAN Monte Carlo methods in T.M. Jenkins, W.R. Nelson, A. Rindi, A.E. Nahum, and D.W.O.Rogers (eds.) *Monte Carlo transport of Electrons and Photons Below 50MeV*, Plenum, New York, pp. 153–82.

Sempau J. and Bielajew A.F. 2000. Towards the elimination of Monte Carlo statistical fluctuation from dose volume histograms for radiotherapy planning. *Phys. Med. Biol.* 45, 131–158.

Sempau, J., Wilderman S.J., and Bielajew A.F. 2000. DPM, a fast accurate Monte Carlo code optimized for photon and electron radiotherapy treatment planning dose calculations, Phys. Med. Biol. 45, 2263–92.

Sempau J., Sanchez-Reyes A, Salvat F., ben Tahar H.O., Jiang S.B., and Fernandéz-Varea J.M. 2001. Monte Carlo simulation of electron beams from an accelerator head using PENELOPE. *Phys. Med. Biol.* 46, 1163–86.

Siebers J.V. and Kawrakow I. 2007. Performance of a hybrid MC dose algorithm for IMRT optimization dose evaluation. *Med. Phys.* 34, 2853–63.

Siebers J.V. and Zhong H. 2008. An energy transfer method for 4D Monte Carlo dose calculation. *Med. Phys.* 35(9), 4095–105.

Spitzer V. and Whitlock D. 1998. The Visible Human dataset: The anatomical platform for human simulation. *Anat. Rec.* 253, 49–57.

Torikoshi M. et al. 2003. Electron density measurements with dual-energy x-ray CT using synchroton radiation. *Phys. Med. Biol.* 48, 673–85.

Trofimov A. et al. 2005. Temporo-spatial IMRT optimization: Concepts implementation and initial results. *Phys. Med. Biol.* 50, 2779–98.

Van Dyk J., Barnett R.B., Cygler, J.E., and Shragge, P.C. 1993. Commissioning and quality assurance of treatment planning computers. *Int. J. Radiat. Oncol. Biol. Phys.* 26, 261–73.

Vanderstraeten B. et al. 2007. Conversion of CT numbers into tissue parameters for Monte Carlo dose calculations: A multicentre study. *Phys. Med. Biol.* 52, 539–62.

Verhaegen F. and Devic S. 2005. Sensitivity study for CT image use in Monte Carlo treatment planning. *Phys. Med. Biol.* 50, 937–46.

Walters B.R.W., Kawrakow I., and Rogers D.W.O. 2002. History by history statistical estimators in BEAM code system. *Med. Phys.* 34(12), 4818–53.

Wang H. et al. 2005. Validation of an accelerated "demons" algorithm for deformable image registration in radiation therapy. *Phys. Med. Biol.* 50, 2887–905.

Monte Carlo Methods and Applications for Brachytherapy Dosimetry and Treatment Planning

Guillaume Landry
Maastro Clinic

Mark J. Rivard
Tufts University School of Medicine

Jeffrey F. Williamson
VCU Massey Cancer Center

Frank Verhaegen
Maastro Clinic

9.1 Introduction

As in other branches of radiation therapy, Monte Carlo (MC) simulation has become an essential dosimetry tool in modern brachytherapy, playing key roles in both clinical practice and research. The most established application of MC methods in brachytherapy is the determination of dose rate distributions around individual radiation sources. Modern sources generally contain low energy radionuclides such as [103]Pd, ^{125}I, or [131]Cs (mean energies <0.05 MeV, referred to henceforth as low-energy sources) or higher energy radionuclides such as ^{192}Ir and [137]Cs (mean energies 0.355 and 0.662 MeV). There are also miniature x-ray sources generating at 50 kVp Bremsstrahlung spectrum. Both source geometries and clinical applications are quite variable. In low dose rate (LDR) brachytherapy, radioactive material and radio-opaque markers are encapsulated to form permanently implantable seeds. In high dose rate (HDR) brachytherapy using a remote afterloader, an iridium pellet is welded to the tip of a drive cable. Miniature x-ray sources with tungsten anodes also fall in the HDR category, even though they emit photons in the low-energy range. While inverse square law dependence is the dominating feature of brachytherapy dose distributions, photon attenuation and scatter build-up in the surrounding medium as well as radiation interactions within the source structure give rise to anisotropic dose distributions. The significant modulation of dose distributions must be properly modeled to attain clinically acceptable dosimetric accuracy, and these features are not readily derived from analytical methods such as the Sievert integral (Williamson, 1996). The complexities of experimentally measuring single-source dose distributions caused by the sharp dose gradients, low photon energies, and dose rates associated with brachytherapy make computational dosimetry techniques such as MC simulations an essential tool in brachytherapy (Williamson and Rivard, 2005). Furthermore, properly accounting for radiobiological effects in low-energy brachytherapy requires calculation of photon and electron spectra at various distances from sources, which is most accurately done by MC modeling.

The first computational efforts toward obtaining brachytherapy dose distributions are attributed to the 1960s work of Meisberger who derived 1D tissue-attenuation and scatter build-up factors for [198]Au, ^{192}Ir, ^{137}Cs, [226]Ra, and ^{60}Co point sources (Meisberger et al., 1968), while Dale was the first to apply similar techniques to modern ^{125}I in 1983 (Dale, 1983). Although MC modeling of a 3D brachytherapy source geometry was performed as early as 1971 by Krishnaswamy for [252]Cf needles (Krishnaswamy, 1971), it took another decade for the field to fully embrace 3D modeling. Williamson showed in 1983, using 3D MC simulations, that the Sievert integral deviated by 5–100% from MC results for monoenergetic photons of energies lower than 0.3 MeV emitted from an encapsulated line source, emphasizing the need for accurate computational dosimetry

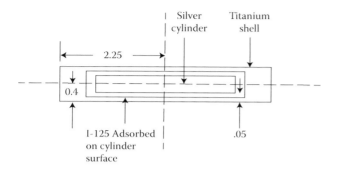

FIGURE 9.1 First 3D model of a 6711^{125}I source used in MC simulations by Burns and Raeside (1988). All dimensions in mm. The silver cylinder is 3 mm in length and 0.5 mm in diameter. (From Burns, G. S. and D. E. Raeside. 1988. *Med Phys* 15:56–60. With permission.)

(Williamson et al., 1983). Burns and Raeside were the first to fully model a commercial ^{125}I seed (model 6711, 3M, now GE HealthCare/Oncura), modeling the silver radiomarker, ^{125}I distribution and titanium encapsulation, as shown in Figure 9.1, to obtain a 2D dose rate distribution (Burns and Raeside, 1988). Since the range in water of secondary electrons generated by 30 keV photons is less than 20 μm, Burns and Raeside did not transport electrons in their simulations and scored collision kerma using a track-length estimator (Williamson, 1987). Approximating absorbed dose by collision kerma is commonly employed by most MC codes used for brachytherapy dosimetry.

While estimating relative 2D dose distributions in medium around ^{125}I sources from measurements and MC methods (Williamson and Quintero, 1988) was relatively common by the mid-1980s, the dose rate constant (dose at a reference point in medium per unit source strength) for low-energy seeds had not been definitively measured or calculated until the late 1980s. Williamson performed simulations in 1988 for models 6711, 6702, and 6701 3M sources as well as the National Institute of Standards and Technology (NIST) Ritz low-energy free-air

chamber (Figure 9.2), used as air-kerma strength standard, to obtain dose rate constants (Williamson, 1988). By including the effects of 4.5 keV Ti K x-rays produced in the source encapsulation, first experimentally observed by Kubo (1985), Williamson showed that ^{125}I absolute dose rates obtained from semiempirical methods were overestimated by 10–14%.

The 1990s saw several studies comparing MC simulations of brachytherapy sources and thermoluminescent dosimeter (TLD) measurements (Kirov et al., 1995; Valicenti et al., 1995). Williamson observed good agreement (1–5%) between simulations and measurements for ^{125}I when accounting for the measurement phantom medium in the simulations (Williamson, 1991). Excellent agreement was observed for ^{192}Ir (2–3%), when the shape and size of the measurement phantom were modeled, although the influence of its composition was found to be less important in this energy range (Williamson, 1991). In a significant series of articles, Williamson's group performed extensive benchmarking of MC photon-transport calculations against precision diode measurements in water showing that MC accurately (1–3%) reproduced both relative and absolute dose rates across the entire brachytherapy energy range in both homogeneous and heterogeneous phantom geometries (Williamson, 1991; Williamson et al., 1993; Perera et al., 1994; Das et al., 1996, 1997a,b). These results confirmed that the MC methodology applied to brachytherapy dosimetry was mature and sufficiently accurate and robust to support clinical dosimetry. This was reinforced by the American Association of Physicists in Medicine (AAPM) Task Group No. 43 (TG-43) requirement that at least one experimental and one MC determination of dosimetry parameters be published before using a source clinically, essentially making MC simulations an industry standard (Rivard et al., 2004). Investigators tackled evermore complex problems, such as the modeling of multicoil radioactive stents used in the treatment of arterial restenosis (Reynaert et al., 1999), illustrated in Figure 9.3. However, MC results should not be trusted blindly, as illustrated by the significant dose-estimation errors resulting from the use of obsolete low-energy cross-section data (Demarco

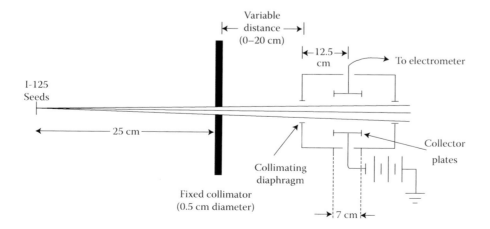

FIGURE 9.2 Williamson's model of the Ritz FAC. (From Williamson, J. F. 1988. *Med Phys* 15:686–694. With permission.)

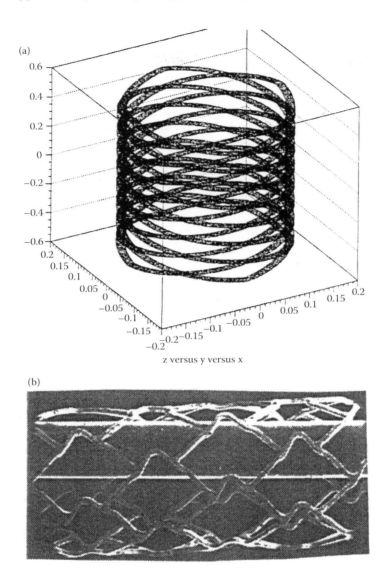

FIGURE 9.3   (a) EGS4 MC model of the ASC Multilink stent. (b) Scanning electron microscope image of the stent. (From Reynaert, N. et al. 1999. *Med Phys* 26:1484–1491. With permission.)

et al., 2002; Bohm et al., 2003; Reniers et al., 2004), a problem first pointed out by Williamson (1991). For brachytherapy dose distributions in the <50 keV energy range, where energy deposition is dominated by photoelectric absorption, even 1–2% errors in the photoelectric cross section can give rise to dose computation errors as large as 10–15% at 5 cm distance. This led to the adoption of modern cross-section libraries derived largely from theoretical quantum mechanical models (Cullen et al., 1997). The reader is referred to Williamson and Rivard for a more detailed discussion of this complex issue (Williamson and Rivard, 2009).

The rising popularity of prostate seed implantation in the United States, increasing from 5,000 in 1995 to about 50,000 in 2002, fuelled a rise in the number of commercially available brachytherapy seeds and designs (Grimm and Sylvester, 2004). While the initial 1995 TG-43 report presented consensus dosimetry parameters for one [103]Pd and two ^{125}I seeds (Nath et al., 1995), its 2004 low-energy seed update (TG-43U1) presented data for eight seed models (Rivard et al., 2004), while the 2007 supplement presented data from an additional eight seed models (Rivard et al., 2007). Yet another supplement is in preparation for the remaining commercially available low-energy photon-emitting sources. The recently published joint AAPM/European Society of Radiation Oncology (ESTRO) report (Perez-Calatayud et al., 2012) on high-energy brachytherapy dosimetry applied the AAPM prerequisite (Li et al., 2007) to 21 ^{192}Ir, ^{137}Cs, and ^{60}Co source models. This increasing proliferation of new brachytherapy sources and dosimetry datasets was associated with a rapidly growing number of MC-related brachytherapy publications in the peer-reviewed literature (Figure 9.4) and resulting in the 2001 "seed policy"

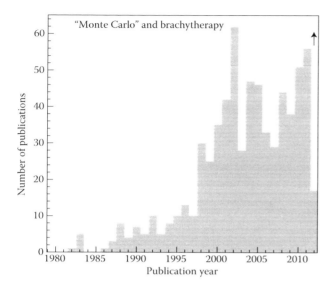

FIGURE 9.4 Results of a PubMed search (April 11, 2012) for the terms Monte Carlo and brachytherapy presented per publication year. The arrow indicates the extrapolated 2012 number of publications to a complete year (= 61).

of the *Medical Physics* journal, which limited TG-43 dosimetric parameter papers to technical notes and requested the authors to consider other publication venues for TG-43 seed parameter papers that did not exhibit sufficient scientific novelty to merit *Medical Physics* publication.

The use of MC methods in brachytherapy now goes beyond single-source dosimetry. An important area of current research is application of MC simulations to perform patient-specific dose calculations. This effort is necessary to overcome the limitations of TG-43-style table-based source-superposition algorithms, which include neglecting interseed attenuation (ISA) and tissue heterogeneities for low-energy seed implants (Burns and Raeside, 1989; Chibani et al., 2005; Carrier et al., 2007) and neglect of applicator shielding and partial scattering effects for higher-energy brachytherapy procedures (Valicenti et al., 1995; Gifford et al., 2005; Poon et al., 2006; Poon and Verhaegen, 2009). Several MC dose calculation platforms, generally based on computed tomography (CT) images, have been presented in the literature (Chibani and Williamson, 2005; Taylor et al., 2007; Afsharpour et al., 2012; Sampson et al., 2012). For low-energy sources, the most important challenge to accurate patient-specific dosimetry is accurate voxel-by-voxel assignment of photon cross-section tables.

In this chapter, we will cover reference quality and clinical dose calculation methods for single-source dosimetry, clinical patient-specific treatment planning, and thoughts on future technological advancements in brachytherapy.

9.2 Single-Source Dosimetry

The established application of MC simulations is determination of the single-source dose distribution for calculation of patient

dose distributions using table-lookup methods. In addition to describing the general process for source dosimetric characterization, this section will highlight issues unique to MC and other Boltzmann equation numerical solutions, including influence of seed design, phantom size and composition, and other effects not normally encountered in applying experimental and semiempirical dosimetry methods.

9.2.1 TG-43 Dosimetry Parameter Reference Datasets

Current brachytherapy treatment planning systems (TPS) utilize the dose calculation formalism of the AAPM TG-43 report (Nath et al., 1995; Rivard et al., 2004, 2007). The TG-43 report assumes that a brachytherapy source produces a cylindrically symmetric dose distribution and permits the treatment planner to determine patient dose distributions about the source via table lookup in a polar coordinate system using nonlinear interpolation.

When the source orientation is known, the 2D formalism of the TG-43 report can be used. In this formalism, contribution of each source to a point of interest (POI) is a function of polar angle θ and distance r from the source center to the POI (Figure 9.5). Sources in which the orientation is known include HDR ^{192}Ir or ^{60}Co sources (photon-emitting brachytherapy sources with average energy >50 keV, referred to henceforth as high-energy sources), that is, those delivered with a remote afterloading unit (RAU), LDR ^{137}Cs or ^{192}Ir sources, or low-energy LDR sources such as ^{125}I, ^{103}Pd, and ^{131}Cs when the imaging modality (typically CT or three-film projection techniques (Pokhrel et al., 2011) permits adequate resolution to discern the capsule orientation. When clinical imaging is inadequate to discern the capsule orientation, the 1D formalism of the TG-43 report is used, in which the polar angle dependence of single-seed contributions is ignored (Corbett et al., 2001; Lindsay et al., 2001).

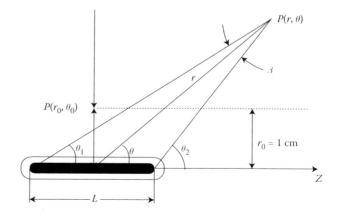

FIGURE 9.5 Coordinate system employed by the TG-43 formalism for a source with active length L. The POI is indicated by $P(r, \theta)$. (From Nath, R. et al. 1995. *Med Phys* 22:209–234. With permission.)

9.2.1.1 Dose Calculation Formalisms

The 2D dose calculation formalism of the AAPM TG-43 report is given as

$$\dot{D}(r,\theta) = S_K \cdot \Lambda \frac{G_L(r,\theta)}{G_L(r_0,\theta_0)} \cdot g_L(r) \cdot F(r,\theta) \qquad (9.1)$$

where the dose rate as a function of r and θ (see Figure 9.5) is equal to the product of the air-kerma strength S_K (a measure of the brachytherapy source strength (Rivard et al., 2004)); the dose rate constant Λ (defined as the ratio of the reference dose rate $\dot{D}(r_0,\theta_0)$, where $r_0 = 1$ cm and $\theta_0 = 90°$ and S_K as $\Lambda \equiv \dot{D}(r_0,\theta_0)/S_K$); the geometry function $G_L(r,\theta)$ at any point r and θ using the line-source (for a source of active length L) approximation divided by its value $G_L(r_0,\theta_0)$ at r_0 and θ_0; the radial dose function $g_L(r)$ based on the line-source approximation; and the 2D anisotropy function $F(r,\theta)$.

The 1D dose calculation formalism of the AAPM TG-43 report is given as

$$\dot{D}(r) = S_K \cdot \Lambda \frac{G_L(r,\theta_0)}{G_L(r_0,\theta_0)} \cdot g_L(r) \cdot \phi_{an}(r) \qquad (9.2)$$

where the dose rate as a function of r only is equal to the product of S_K, Λ, the geometry function $G_L(r,\theta_0)$ at any distance r using the line-source approximation divided by its value $G_L(r_0,\theta_0)$ at r_0 and θ_0, the radial dose function $g_L(r)$ using the line-source approximation, and the 1D anisotropy function $\phi_{an}(r)$.

These dosimetry parameters for the 2D and 1D dose calculation formalisms are described in great detail within the 2004 AAPM update to the TG-43 report (TG-43U1) (Rivard et al., 2004). Of note is that all parameters take a value of unity at the reference position (r_0, θ_0) except S_K and Λ such that $\dot{D}(r_0,\theta_0) = S_K \cdot \Lambda$ and $\dot{D}(r_0) = S_K \cdot \Lambda$.

Using equations and recommendations covered in TG-43U1, the MC dosimetry investigator extracts tables of TG-43 parameters from the measured or MC-computed dose distribution, which is typically performed using detectors arranged in a sparse polar coordinate grid. In turn, these parameters are then imported into a clinical brachytherapy TPS using the same formalism. Recovering the dose distribution in the original MC grid for each source modeled in the TPS is an essential institution-specific quality assurance test (Rivard et al., 2004).

9.2.1.2 MC Dose Calculation Method

The process for calculating dose rate distributions in the vicinity of a brachytherapy source has been described in the joint TG-138 report by the AAPM and the ESTRO on brachytherapy dosimetric uncertainties (DeWerd et al., 2011), the revised TG-43 report, and a recent review article (Williamson and Rivard, 2009). In short, the dosimetry parameters calculated in liquid water $\Delta D(r, \theta)$ (where ΔD means absorbed dose in the inherent units of the MC output, usually Gy/simulated primary photon or a multiple thereof) are determined separately from Λ. We describe a typical MC dose simulation process as a roughly chronological sequence.

9.2.1.2.1 Brachytherapy Source Design

Before initiating MC simulations of brachytherapy dose rate distributions, the investigator must have an accurate and complete geometric of model of the seed. This includes as many as possible of the following: the assembly process, the atomic composition, density, shape, dimensions, and location of each component along with tolerances in any of these features. In addition, the geometry and location of the radionuclide distribution, along with any radioactive contaminants or nonradioactive filler material, must be known. Often, this distribution is a very thin layer the thickness of which can only be known approximately. Typically, the MC investigator starts with manufacturer-provided construction drawings of an average source possibly along with proprietary descriptions of the chemical fabrication process, from which additional descriptive details can be inferred. TG-43U1 recommends that dosimetry consultants independently validate these specifications. As described elsewhere (Williamson and Rivard, 2009), validation techniques include direct mechanical measurement of nonradioactive seeds and preassembly internal components; pin hole autoradiography; transmission radiography; and low-magnification optical or electron microscopy of longitudinally sliced-open seeds or individual seed components. Because the MC simulation process tries to estimate the physical dose rate distribution, any differences between the simulated and real geometry may result in erroneous results. While some simulation errors are more forgiving than others (such as for thin-walled, high-energy photon-emitting brachytherapy sources), it is not always clear what is the quantitative sensitivity of a given design feature to the resultant dose rate distribution. TG-43U1 and TG-138 recommend that dosimetry investigators perform careful uncertainty analysis by performing MC simulations over an ensemble of source geometries sampled from the estimated tolerances of geometric parameters whose mean values cannot be accurately determined, as illustrated by a recent study of the Model 6711 ^{125}I seed (Dolan et al., 2006).

Some sources have internal components that move under the influence of the gravitational forces, resulting in an orientation-dependent geometry not evident upon external examination of the source (Rivard, 2001, 2007; Williamson, 2002; Rivard et al., 2004). This is most prevalent for low-energy LDR sources where the internal components are not rigidly attached to the capsule and the radiation emissions are highly attenuated by radio-opaque markers. This is illustrated in Figure 9.6 for [103]Pd sources. For these sources, the dose rate can vary by more than a factor of two within a few millimeters of the source. In this circumstance, MC simulation is a powerful tool for performing sensitivity analyses to understand how the dose rate distribution is modulated by repositioning of internal source components. Unlike LDR sources, HDR sources have their capsule attached to a drive cable for translating the source through catheter(s) in a patient. Because the source design may not be rigid, and since the HDR RAU drive cable is flexible, it may not be clear

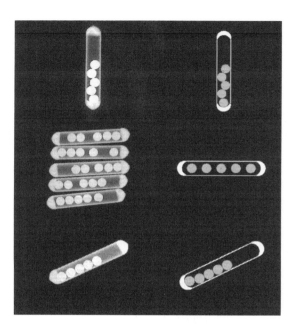

FIGURE 9.6 (Left) Radiographs of a brachytherapy source with ¹⁰³Pd adsorbed on Ag spheres in various orientations showing displacement of the Ag spheres. (Right) Source geometries simulated in MCNP. (From Rivard, M. J. et al. 2004. *Med Phys* 31:2466–2470. With permission.)

which source geometry best represents the most probable dose distribution. Simulations of ensemble source configurations are needed to address the impact of dynamic source geometries and to develop practical solutions (Mikell and Mourtada, 2010).

9.2.1.2.2 *Brachytherapy Source Radiation Emissions*

In general, brachytherapy sources utilize radionuclides as the radiation source; electronic brachytherapy sources are an exception. The nuclear disintegration process is well understood with photon energies, emission rates, and branching ratios often known to three or more significant digits for common radionuclides.

Brachytherapy sources are designed around a single radionuclide. However, radioactivity activation and fabrication processes often result in potentially significant concentrations of contaminant radionuclides even after efforts to improve radiopurity via downstream radiochemistry. Since radiopurification processes are not perfect, all brachytherapy sources contain radioimpurities. While the dosimetric influence of radioimpurities has not been sufficiently examined in the literature, the joint AAPM/ESTRO TG-167 report conservatively recommends that radioimpurities be minimized to such levels that their dosimetric contributions over the range of clinically relevant distances in the vicinity of the implant should be <5% of the dosimetric contributions of the primary radionuclide. In addition, the working life of the source must be considered for radioimpurities that are longer-lived than the intended radionuclide. The MC dosimetry investigator is advised to consider the impact of radioimpurities on the recommended dose distribution.

In addition to radioimpurities, all relevant radiation emissions for a given radionuclide should be included in the simulation, where "relevant" means affecting the dose distribution over the range of clinical interest by more than 2–3%. While neutrino emissions are dosimetrically irrelevant, many common radionuclides disintegrate via beta-decay with subsequent x-ray and gamma-ray emissions. Not all electrons (or positrons) generated within the capsule interior are fully attenuated by the source structure and may deposit dose (or bremsstrahlung x-rays) exterior to the capsule (Granero et al., 2011). Electronic brachytherapy sources generate characteristic x-rays in addition to bremsstrahlung photons in various internal components, which may also contribute to clinically significant energy deposition outside the source capsule (Liu et al., 2008). At a minimum, MC codes used in low-energy dosimetry should accurately model emission of K- and L-shell characteristic x-rays following ejection of inner-shell electrons via knock-on electron collisions, photoelectric effect, or inelastic photon scattering. The MC dosimetry investigator is advised to perform calculations to benchmark the code preceding simulation of a clinical brachytherapy source. This is discussed in greater detail in Section 9.2.1.2.5.

9.2.1.2.3 *Simulating Phantoms for Reference Quality or Experimental Dosimetry Measurements*

After source radiation is generated within the capsule, it may pass from the source into the surrounding phantom. A liquid water sphere is the standard phantom for presenting reference quality, single-source brachytherapy dose distributions. For low-energy sources, the sphere radius is 15 cm with the brachytherapy source being studied located at the center of this sphere (Rivard et al., 2004). For high-energy photon-emitting sources (>50 keV photons), the recommended reference phantom radius is 40 cm (Perez-Calatayud et al., 2012). Given the ranges of backscattered photons, the phantoms permit acquisition of dose rate data within 1% of the unbounded phantom dose distribution (Perez-Calatayud et al., 2004; Melhus and Rivard, 2006) for distances up to 10 and 20 cm, respectively, for low- and high-energy sources. In practice, simulating a larger phantom produces dose distributions at the phantom periphery quantitatively closer to that of an infinitely large phantom, with a penalty of longer simulation times for the same number of particle histories. Regardless, the MC dosimetry investigator should state the used phantom dimensions and employ actual-to-reference phantom corrections factors, so that the final recommended dose distribution adheres to the relevant AAPM recommendation.

The AAPM TG-43 dose calculation formalism ignores patient tissue composition and is based upon specification of absorbed dose and transport of radiation in liquid water. The AAPM TG-43U1 report (Rivard et al., 2004) specifies liquid water to consist of exactly two parts hydrogen and one part oxygen with a mass density of 0.998 g/cm³ at a temperature of 22°C. For low-energy brachytherapy dosimetry, cross sections for coherent scattering are very sensitive to intermolecular forces and molecular bonds. Hence, the atomic mixture rule, which generally supports accurate synthesis of cross sections for compounds and mixtures, is not applicable to coherent scattering. To accurately model coherent scattering, first, calculated σ_{coh} values should be recalculated using liquid or molecular water atomic form factors (Morin, 1982) and the water coherent scattering cross sections extracted from NIST XCOM or EPDL97 libraries discarded, as they are based upon the mixture rule. Additionally, dosimetry investigators utilizing experimental techniques need to be aware that the phantoms used for dose measurement often deviate significantly from liquid water (Williamson, 1991; Patel et al., 2001). As recommended by TG-43U1, brachytherapy experimentalists are expected to use MC simulation to calculate reference-to-experimental phantom corrections factors so that the final experimental results represent absorbed dose to liquid water in the spherical liquid water reference phantom. Finally, the reader must remember that the patient tissue compositions, densities, and dimensions deviate significantly from the reference geometry, which can result in large discrepancies between estimated and delivered dose, especially for low-energy sources. The AAPM TG-186 is the first concerted attempt to formulate guidelines for assigning cross sections and densities, other than unit-density water, to organs and tissues based upon patient-specific imaging (Beaulieu et al., 2012).

9.2.1.2.4 MC Radiation Transport Codes

Radiation transport following radioactive decay is a complex phenomenon, consisting of attenuation and scattering of photons, as well as transport of beta rays and secondary electrons via elastic collisions with nuclei; elastic and inelastic collisions with orbital electrons; and bremsstrahlung production. Because electrons experience on the order of 10^6 discrete collisions as they slow down, resulting in impractically long computing times, electron transport, if included at all, is modeled via the approximate "condensed history" approach (Berger, 1963; Kawrakow and Bielajew, 1998). In this approach, a much smaller number of randomly sampled condensed history steps is simulated, each of which represents the collective effects of many discrete collisions by means of multiple scattering distributions and the continuous slowing down approximation (CSDA), with either stochastic or analytical corrections to account for random fluctuations in energy loss about the CSDA energy loss. Despite these approximations, condensed history transport has been demonstrated to accurately characterize absorbed dose distributions over a wide range of circumstances with reasonable computing times.

Photon interactions may also be simplified. An MC code may simulate K-edge characteristic x-rays, but not consider L-edge x-rays for low-Z elements, or the atomic relaxation process may be simplified, including only the most probable K- and L-shell energy transitions for high-Z elements. Further, radiological interactions at energies less than a few keV will be subject to molecular binding effects, which are not accounted for in most current MC codes in use, with the possible exception of coherent scattering as described above. Even the impulse approximation, which models the influence of bound orbital electrons on Compton and Thomson scattering via the form factor approach, fails to model the influence of orbital electron

momentum distributions, which give rise to a spectrum of scattered photon energies centered about the Compton scattered photon energy, resulting in Compton Doppler broadening (Ribberfors and Carlsson, 1985). However, even in the low photon energy range, the dosimetric differences between the form factor and consistently implemented free-electron scattering models are small (Taylor and Rogers, 2008a; Williamson and Rivard, 2009).

Neglecting electron transport altogether, and simulating only photon transport, substantially reduces the computing burden in brachytherapy. This simplification is exploited by most dosimetry investigators and all specialized codes intended for treatment planning applications. It approximates absorbed dose by collision kerma, which in turn assumes equilibrium of secondary electrons. This approximation is valid everywhere for low-energy brachytherapy sources, where electron ranges are of the order of 10 μm. However, for higher-energy ^{192}Ir and ^{137}Cs sources, which have secondary electron and beta-ray ranges of 1–3 mm, charged-particle equilibrium approximation may introduce significant errors near metal–tissue interfaces and near sources. For example, dose errors exceeding 15% at distances less than 1 mm from an HDR ^{192}Ir source have been observed (Taylor and Rogers, 2008b; Ballester et al., 2009). In addition, hot spots due to beta-ray leakage near the end of LDR ^{192}Ir seeds have been observed experimentally (Chiu-Tsao et al., 2004). High-energy photon-emitting sources such as ^{192}Ir and ^{137}Cs may not be accurately simulated at distances within a few millimeters (DeWerd et al., 2011). While one must be aware that MC photon transport codes give rise to >5% errors at 1–2 mm distances from ^{192}Ir and ^{137}Cs sources (Perez-Calatayud et al., 2012), photon-only transport solutions produce acceptable results for conventional interstitial and intracavitary applications where the clinically relevant distance range is 3–50 mm. For applications where near-zone dosimetry is important, for example, ^{192}Ir-based intravascular brachytherapy or rigorous estimation of mucosal doses in contact with metal applicators, coupled photon–electron calculations and inclusion of the radionuclide beta-ray spectrum may be necessary.

9.2.1.2.5 Photon Cross-Section Libraries and Monte Carlo Scoring Functions

While both low- and high-energy source dose distributions are dominated by $1/r^2$ falloff, radiation interactions within the source capsule and in the surrounding medium significantly modulate the dose distribution, requiring MC simulation or other Boltzmann equation solutions to predict doses with clinically acceptable accuracies. Having finalized the source and phantom geometries and having chosen the collisional physics model, appropriate cross-section tables must be selected and assigned to each material in the simulation geometry. In addition, the detector geometries and estimator, that is, the mathematical function that estimates the absorbed dose contribution to each detector from each simulated history, must be selected.

Photon interaction cross sections are compiled into data libraries and are generally based on quantum mechanical models of each scattering and absorption process based upon approximate models of orbital electron wave functions, which have previously been validated by comparison to available experimental measurements (Hubbell, 1999). These libraries are maintained by organizations such as the National Nuclear Data Center of Brookhaven National Laboratory in New York, the Nuclear Data Section of the International Atomic Energy Agency in Vienna, and the Nuclear Energy Agency Data Bank of the Organization for Economic Cooperation and Development in Paris. However, the main clearinghouse for evaluated cross-section libraries and public domain transport codes is the Radiation Safety Information Computational Center (RSICC) (http://www-rsicc. ornl.gov), which is located at Oak Ridge National Laboratory. For modest fees, cross-section libraries and many codes for cross-section preprocessing and MC simulation can be obtained. Cross-section libraries typically consist of tables of partial cross sections for photoelectric effect, coherent scattering, incoherent scattering, and pair production for each element on a coarse logarithmic energy grid. In addition, atomic form factors and incoherent scattering factors for each element are available. The most modern libraries, for example, EPDL-97, have subshell-specific photoelectric cross sections and form factors and extensive tables of orbital electron transition and fluorescent yields so that atomic relaxation processes (the cascade of characteristic x-rays and Auger- and Coster-König electrons emitted following ejection of an inner shell electron by one of the photon or electron interaction processes) can be properly simulated. For brachytherapy, the most important issue is to select a library that contains up-to-date (post 1983) photoelectric effect and scattering cross sections. Fortunately, all of the modern libraries, for example, EPDL97 from Lawerence Livermore Laboratory, DLC-146 from RSICC, and XCOM from NIST, are all based on the same theoretical models, despite having many differences in format and extensiveness of compiled data. Specialized libraries of up-to-date cross sections in EGSnrc, Geant, and MCNP formats are also available. A more detailed review of both modern and obsolete cross-section libraries and formats is given elsewhere (Williamson and Rivard, 2009). These cross-section libraries can subtend several hundred megabytes and have complex numerical formats, for example, ENDF-B. These sparse tables therein are intended for interpolation onto finer grids using specific nonlinear interpolation schemes, most commonly log–log–linear interpolation for photon data. Most libraries come equipped with programs designed to access the database and preprocess the data into user-specified formats. When using the library data, the MC investigator must be careful to employ appropriate interpolation software so as to avoid large errors in the low-energy range (Demarco et al., 2002). Another related quantity that may be needed, depending on the estimator selected, is the mass–energy absorption coefficient. Tables of mass–energy absorption coefficients, consistent with the NIST XCOM or EPDL97 cross sections may be obtained from the NIST website (http://www.nist.gov/pml/data/xraycoef/index.cfm), in which elements $Z < 92$ are tabulated over a wide energy range. Tables for user-specified mixtures and compounds may also be obtained using the mixture rule. It is essential that the data used for scoring be derived from the same cross-section data used to transport the photons. For example, if the free-electron scattering model (no coherent scattering or electron-binding corrections to Klein–Nishina scattering) is used for transporting photons, then the NIST mass–energy absorption coefficients must be recalculated by replacing the term associated with incoherent scattering energy deposition with its Klein–Nishina counterpart.

Calculation of absorbed dose or collisional kerma is then a function of the radiation energy and medium (e.g., water). Uncertainties in the phantom composition or its radiological cross-section library will become more pronounced as distance from the source increases and systematic errors accumulate.

The final choices to be made are selection of a scoring grid and an estimator. More detailed treatments of this complex subject can be found elsewhere (Williamson, 1987; Williamson and Rivard, 2009). The simplest choice is the analog estimator, in which only those simulated collisions that occur within the detector volume contribute to dose. Invoking charged particle equilibrium (CPE), the dose contribution is computed as the difference between energy entering and leaving the detector divided by mass. While simple, the analog estimator is extremely inefficient. By using a variant of the track-length estimator (Williamson, 1987), efficiency can be improved 20- to 50-fold (Hedtjarn et al., 2002). The track-length estimator scores dose as $(\Delta l \cdot E/\Delta V) \cdot (\mu_{en}/\rho)$, where Δl is the distance within the detector of volume ΔV traversed by incoming photon with energy E. Efficiency is improved since every voxel intersected by a photon flight path produces a nonzero dose score, greatly increasing the information that can be extracted from a finite sample of histories.

Since the analog and track-length estimators converge to the same value, collision kerma integrated over the detector volume, the detector grid must be selected carefully to minimize volume-averaging artifacts and to maintain acceptable statistics out to distances of 5–7 cm for low-energy brachytherapy. For single-source dose estimation, typically a spherical grid is used, with very thin detector elements near the source and thicker ones far from the source so as to improve statistics. By ignoring azimuthal angle, the detectors become a set of spherical segment shells that can vastly improve efficiency by exploiting cylindrical symmetry of the source. Since the TG-43 protocol requires specification of dose at geometric points, the MC investigator must correct each MC estimate for volume averaging, typically by using simple inverse square law-based correction factors (Ballester et al., 1997). An alternative to volume detectors is to use more complex dose-at-a-point estimators (Williamson, 1987; Li et al., 1993), in which analytic formulas are used to estimate the contribution from each simulated collision to dose at a geometric point. When using any kind of estimator, the statistical precision of the dose estimates, quantified in terms of standard deviation about the mean (67% confidence interval) should be carefully tracked as a function of distance. For volume detector estimators, detector size is always a trade-off between statistical precision (larger is always better) and spatial resolution, that is, reduced volume-averaging artifact, for which smaller is always better.

MC estimation of brachytherapy dosimetry parameters such as $g(r)$, $F(r, \theta)$, and $\phi_{an}(r)$ is performed using the selected detector grid and estimator by obtaining estimates of $\Delta K(r_i, \theta_j) \approx \Delta D(r_i, \theta_j)$ (dose per simulated history) over the detector grid (r_i, θ_j) embedded in the reference liquid water sphere. By judicious choice of detector volumes and estimators, it should be possible to limit statistical uncertainties to <1% at distances less than 5 cm. Once volume-averaging corrections have been applied, the TG-43 relative parameters can be estimated using the equations given in the TG-43U1 report.

9.2.1.2.6 Dose Rate Constant

To estimate the value for Λ, the ratio of $\Delta D(r_0, \theta_0)$ to ΔS_K (the air-kerma strength per simulated history) is obtained. Because MC methods generally estimate absorbed dose or collisional kerma in terms of energy imparted per number of particle histories performed or radionuclide disintegrations simulated, the results from the liquid water simulations need to be combined with simulations of the air-kerma strength in a suitable free-air geometry (vacuum or an air sphere). This way, the inherent normalization used by the MC code is irrelevant since both $\Delta D(r_0, \theta_0)$ and ΔS_K are expressed in terms of the same clinically irrelevant normalization. However, the MC normalization constant must be known when absolute dose distribution results are needed in terms of units of air-kerma strength. This process is described by Melhus and Rivard (2006) and Williamson and Rivard (2009) for calculating dose rate distributions per unit source strength U, requiring knowledge of the number of photons emitted per disintegration. Thus, to compute dose around an actual brachytherapy source, the clinical medical physicist need only measure S_K (in a traceably calibrated reentrant well-type ionization chamber) and multiply this number by the MC calculated dose rate constant. Since the other TG-43 brachytherapy dosimetry parameters are ratios of doses computed in a single simulation run, it is crucial that the same source design and other starting conditions be used between the liquid water and vacuum (or air) simulations or else a scalar offset (percentage systematic dose error) will be present throughout the entire range of calculated dose rates.

Vacuum is easy to model. Air is more complicated, requiring a mass density (1.196 mg/cm^3 at a temperature of 22°C for dry air at 101.325 kPa) with mass compositions of 1.24%, 75.5268%, 23.1781%, and 1.2827% for C, N, O, and Ar, respectively (Rivard et al., 2004). As for the liquid water description in Section 9.2.1.2.4, this description for dry air is for standardization purposes since brachytherapy source strength (air-kerma strength) is calibrated for these conditions. There are several subtleties MC investigators must appreciate, which have been reviewed by Monroe and Williamson (2002). First, if air is employed, photon attenuation and scatter-build-up effects must be corrected in order to satisfy the AAPM's definition of S_K, which is specified in free space. Second, care must be taken to filter out or suppress any low-energy contaminant photons below 5 keV, as recommended by the TG-43U1. Especially if a vacuum-filled calibration range is simulated, failure to suppress Ti characteristic x-rays could inflate

ΔS_K by 20%. Finally, care should be taken to simulate any known departures of the NIST primary standard from the S_K. This was first demonstrated by Williamson in 1988 (Williamson, 1988), who found it necessary to simulate the Ritz free-air chamber and the NIST procedure (Loftus, 1984) for air-attenuation correction in order to correctly model the impact of Ti capsule characteristic x-ray generation on the comparison of theoretical and measured dose rate constants. More recently, it was found that the sharp edges of the palladium-plated carbon pellets of the model 200 ^{103}Pd seed induced significant and distance-dependent anisotropy of the polar fluence profile near the transverse axis (Monroe and Williamson, 2002). This necessitated a detailed simulation of the wide-angle free-air chamber (Seltzer et al., 2003), which integrated over this region of polar anisotropy, significantly deviating from the definition of air-kerma strength and altering the dose rate constant by as much as 12%.

9.2.2 Calculation of Spectral Effects and Relative Biological Effectiveness

9.2.2.1 Spectral Effects and Detector Response

MC simulations are also excellent sources of information on photon and electron spectra at various distances from brachytherapy sources. These data are needed to correct the readings of dosimetry detectors that may have an energy-dependent response (film, TLD, ionization chamber, etc.). This is essential for both accurate absolute and relative dosimetry near brachytherapy sources. Das et al. (1996) used the Monte Carlo photon transport (MCPT) (Williamson and Meigooni, 1995) code to investigate the energy response of diodes and two sizes of TLDs to a range of photon energies from superficial x-rays to ^{192}Ir and ^{60}Co sources. They found a constant ratio between measured detector signal and MC calculated dose to the detector-sensitive volume for all detectors used, independent of photon energy. This indicates that MC simulations are indeed able to model energy response of detectors in the brachytherapy energy range. The absolute response function of the detectors used was also found to be proportional to absorbed dose by the active volume of the detector. However, other recent investigators (Davis et al., 2003; Nunn et al., 2008) claim that the relative intrinsic TLD-100 response at ^{125}I energies (reading per unit dose to the active detector volume at ^{125}I relative to that at 4 MV x-rays or ^{60}Co) deviate from unity by 5–10% due to the impact of higher LET low-energy photons on the process of TLD signal formation. On the other hand, comparison of MC doses to TLD measurements for many published data sets suggests that a relative energy response factor closer to that computed by MC alone is more accurate (Williamson and Rivard, 2009). The appropriate choice of TLD-100 relative energy response correction remains unresolved at the moment.

A recent example of the utility of MC simulations to determine energy response with respect to ^{60}Co radiation of various detector materials embedded in various phantom materials employed in brachytherapy is the study by Selvam and Keshavkumar (2010). They used EGSnrc codes and a large cavity approach to calculate the energy response of ^{125}I and ^{169}Yb sources as a function of distance and detector material thickness, but they did not model detailed detector geometries. Their work (and references therein) clearly shows that energy correction factors need to be taken into account in brachytherapy and that MC simulation offers a convenient and accurate way to obtain them. Failure to account for these energy response effects may lead to severe over/underestimation of measured dose particularly in low-energy brachytherapy. In another recent application using MOSFET detectors, the mean photon energy from an ^{192}Ir source was found to decrease significantly with distance from the source by more than 100 keV over about 10 cm, leading to an increase in dose response by 60% (Reniers et al., 2012).

Another powerful application of MC simulations in brachytherapy is the determination of dosimetry detector perturbations in the—sometimes very close—vicinity of brachytherapy sources. Reynaert et al. (1998) modeled an NE2571 chamber in detail, which was used to perform dosimetry close to a pulsed dose rate (PDR) ^{192}Ir source in a water phantom. MC simulation is often the only method to reliably determine these perturbation factors, as discussed in other chapters in this volume. The use of MC-based correction factors for TLD dosimetry in brachytherapy has been reviewed in detail by Williamson and Rivard (2009).

9.2.2.2 Radiobiological Effectiveness

Knowledge of photon and electron spectra is also needed for applying biological damage models. It has been known for years from radiobiology that photons with energies below about 50 keV are more potent in inducing DNA damage than higher-energy photons (Hill, 2004). This finding, however, is rarely taken into account in clinical brachytherapy practice. As an example we discuss here an EGSnrc MC model of an electronic brachytherapy source, operated at 50 kVp. Reniers et al. (2008) simulated the initial yield of single- and double-strand DNA breaks (SSB and DSB), which allowed estimation of the relative biological effectiveness (RBE). The MC model was first used to generate x-ray spectra at various distances from the source and to calculate electron spectra from the photon spectra in several media (Figure 9.7). These data were then input into an empirical "Monte Carlo damage simulation" program derived from track structure considerations (Semenenko and Stewart, 2004) to estimate the initial yield of single- and double-strand breaks in DNA. From this, the RBE was derived.

The results indicate a substantially increased DSB yield for the electronic brachytherapy source compared to ^{60}Co or ^{192}Ir reference radiations, leading to an enhanced RBE for DSB by 40–50%. This RBE increase was mostly caused by Compton and Auger electrons with energies below 8 keV (Figure 9.7). The RBE estimate for the low-energy x-ray source was found to be very similar to the measured and calculated RBE for the low-energy gamma-ray brachytherapy radionuclide ^{125}I (Reniers et al., 2004). These findings should be taken into account if the HDR electronic brachytherapy source is intended to replace brachytherapy with the commonly used HDR ^{192}Ir radionuclide.

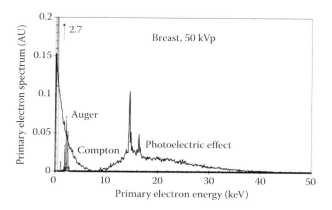

FIGURE 9.7 Primary electron spectrum from an electronic brachytherapy source (50 kV), scored at a distance of 0.5 cm from the source's x-ray target in breast tissue. Electrons arising from Compton, photoelectric, and Auger interactions are indicated. (From Williamson, J. F., R. L. Morin, and F. M. Khan. 1983. *Phys Med Biol* 28:1021–1032. With permission.)

9.3 Treatment Planning

9.3.1 Introduction

The TG-43 formalism, based on the work of the Interstitial Collaborative Working Group (ICWG) (Anderson et al., 1990), was a major step forward in brachytherapy dose calculation. It replaced the use of semiempirical calculation methods based on apparent activity, equivalent mass of radium, exposure rate constants, and tissue-attenuation coefficients, quantities that are not source model but only radionuclide dependent. By employing dosimetry parameters that are dependent on the radionuclide, its distribution in the source, and the source geometry itself, and by recommending measured or calculated consensus datasets, the formalism of the TG-43 protocol, and its subsequent updates, improved standardization and accuracy in brachytherapy dose calculations. Fifteen years later, the widely cited (over 900 citations [Web of Science October 24, 2012, Times cited = 1,051]) and widely adopted formalism is still heavily used, ensuring consistency, standardization, and comparability of brachytherapy dose calculations across institutions worldwide.

Since dose calculations based on the TG-43 formalism rely on the superposition of single-source dose distributions over the dwell or seed positions used for treatment, dose distributions can be obtained with minimal calculation time. This fast and practical method facilitates clinical practices such as transrectal ultrasound image-guided live planning and dose distribution optimization. However, the inherent simplicity of the TG-43 formalism can lead to inaccurate dose distributions when the calculation geometry deviates significantly from the reference water sphere used to derive the source's dosimetric parameters. Generally, TG-43 dose calculation limitations can be attributed to five phenomena: absorption, attenuation, shielding, scattering, and breakdown of the kerma approximation for absorbed dose. Depending on source energy and anatomic site, some or

TABLE 9.1   List of Anatomic Sites Where the Limitations of the TG-43 Dose Calculation Formalism Lead to Significant Dose Calculation Errors (Indicated by "Y")

Anatomic Site	Source Energy	Absorption	Attenuation	Shielding	Scattering	Breakdown Dose = Kerma Approximation
Prostate	High	N	N	N	N	N
	Low	Y	Y	Y	N	N
Breast	High	N	N	N	Y	N
	Low	Y	Y	Y	N	N
GYN	High	N	N	Y	N	N
	Low	Y	Y	N	N	N
Skin	High	N	N	Y	Y	N
	Low	Y	N	Y	Y	N
Lung	High	N	N	N	Y	Y
	Low	Y	Y	N	Y	N
Penis	High	N	N	N	Y	N
	Low	Y	N	N	Y	N
Eye	High	N	N	Y	Y	Y
	Low	Y	Y	Y	Y	N

Source: From Rivard, M. J. et al. 2009. *Med Phys* 36:2136–2153. With permission.

all of these phenomena may induce significant dose calculation errors. Table 9.1 lists, for both high- and low-energy brachytherapy sources, sites where the formalism leads to significant dose calculation errors (Rivard et al., 2009). The following sections briefly discuss these aspects.

9.3.1.1 Absorbed Dose Differences between Water and Human Tissues

At low photon energies, the mass–energy absorption coefficient μ_{en}/ρ varies significantly between tissues due to the importance of the approximately Z^3 dependence of the photoelectric cross section. For a given photon energy fluence under conditions of charged particle equilibrium, the absorbed dose-to-water and dose-to-tissue are related by $D_{tissue}/(\mu_{en}/\rho)_{tissue} = D_{water}/(\mu_{en}/\rho)_{water}$. As seen in Figure 9.8, the ratio $(\mu_{en}/\rho)_{tissue}/(\mu_{en}/\rho)_{water}$ differs from unity for most tissues in the energy range of low-energy sources. At high photon energies, the ratios converge to unity as the Compton cross section becomes more important. This gives rise to significant dependence of the fluence-to-kerma conversion factor on tissue elemental composition and mass density (Landry et al., 2011).

9.3.1.2 Attenuation Differences between Water and Human Tissues

As the linear attenuation coefficient μ is proportional to ρ, density differences between water and human tissues will result in different photon energy fluence and dose distributions, for both low- and high-energy sources. Mass attenuation coefficients μ/ρ also vary between tissues and water at low photon energies because of the aforementioned dependence of photoelectric cross section on atomic number. As μ/ρ is larger at low photon energies (disregarding the absorption edges), differences of elemental composition or density will result in larger dose differences for low-energy sources.

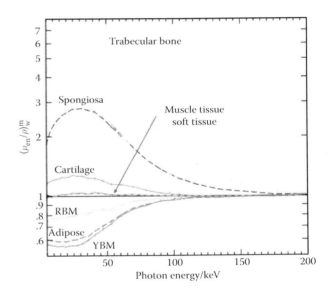

FIGURE 9.8   The EGSnrc (Kawrakow 2000) user code "g" was used to calculate μ_{en}/ρ for a series of human tissues taken from ICRU Report 46 (1992) and ICRP Report 89 (Valentin 2002). RBM and YBM stand for red and yellow bone marrow from "active and inactive marrow" in Valentin (2002). (From Beaulieu, L. et al. 2012. *Med Phys* 39:6208–6236. With permission.)

9.3.1.3 Shielding

In multisource implants, photons emitted from a given source can be absorbed by the radio-opaque markers (e.g., Au, Ag, Pb) or the radio-opaque components of adjacent sources, causing a lower dose than predicted by TG-43. This is generally referred to as ISA in the literature. For HDR high-energy sources, ISA is not an issue since a single source steps through the implanted applicators, treating one dwell position at a time. However, applicator materials such as stainless steel can cause deviations. Some applicators also contain high-atomic-number, high-density shielding material such as tungsten, used to protect organs at risk. The most common are shielded vaginal and intrarectal applicators.

9.3.1.4 Scattering Conditions

TG-43 parameters are calculated in water spheres with radii of 15 and 40 cm for low- and high-energy sources, respectively, ensuring that TG-43 calculations represent doses in unbounded water medium over the therapeutically relevant distance range. Situations where full scatter conditions are not met, sources close to the skin, for example, will result in deviation of delivered dose from TG-43. Given the longer pathlength and higher Compton scattering cross section, high-energy photons are more sensitive to geometries with tissue boundaries near the implanted sources. Example clinical geometries include the breast, skin, and lung (Poon and Verhaegen, 2009).

9.3.1.5 Breakdown of the Kerma Approximation for Absorbed Dose

The vast majority of MC studies deriving TG-43 parameters report collision kerma in water, relying on the equivalence of dose and kerma under conditions of charged particle equilibrium. For low-energy sources, electronic equilibrium is reached within 0.1 mm of sources and the kerma = dose approximation is accurate (Ballester et al., 2009). Since electron ranges are larger for high-energy sources, this assumption can lead to dose differences greater than 1% at distances of 7, 3.5, and 2 mm for ^{60}Co, and ^{137}Cs, and ^{192}Ir, respectively (Ballester et al., 2009). Studies performing measurements rarely report data so close to the source. In addition to lack of charged particle equilibrium, beta particles emitted from sources can also cause a breakdown of the kerma = dose approximation. Granero et al. investigated the contributions of both betas emitted from ^{192}Ir and lack of charged particle equilibrium for distances <2.5 mm from the source's center (Granero et al., 2011). They found that charged particle equilibrium is reached at 2 mm from the source, and that beta contribution to dose is negligible beyond 2.5 mm. Additionally, CPE breakdown may occur at tissue/high-Z interfaces.

9.3.2 Magnitude of Clinical Impact

Several studies have compared MC dose calculations to the results of the TG-43 formalism. Meigooni et al. were the first to investigate ISA in 1992 using TLD measurements and estimated that dose reductions of 6% could be expected at the edge of a ^{125}I prostate implant (Meigooni et al., 1992). Chibani et al. and Carrier et al. published results of MC simulations performed in real ^{125}I and ^{103}Pd prostate implant geometries in 2005 and 2006, respectively, finding D_{90} reductions of 2–5% due to ISA (Chibani et al., 2005; Carrier et al., 2006). Chibani et al. also investigated the dosimetric impact of the presence of calcifications in the

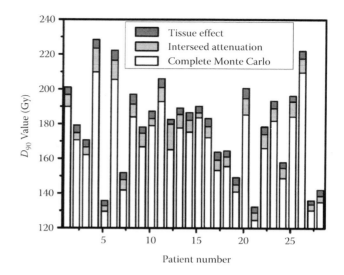

FIGURE 9.9 Impact of ISA and tissue nonwater equivalence on the clinical parameter D_{90} from postimplant dosimetry of 28 prostate cancer patients treated with ^{125}I implants. *Tissue effect* and *Interseed attenuation* represent the D_{90} reductions from the TG-43 estimate from each effect. *Complete Monte Carlo* represents the D_{90} parameter when accounting for both effects. (From Carrier, J. F. 2007. *Int J Radiat Oncol Biol Phys* 68:1190–1198. With permission.)

prostate and found D_{90} reductions of up to 37%. Carrier et al. subsequently performed a retrospective study of 28 prostate cancer patients implanted with ^{125}I using postimplant CT data (Figure 9.9) based upon international commission on radiological protection (ICRP) recommended tissue compositions (ICRP, 1975). They found an average D_{90} decrease of 7% due to ISA and tissue composition (Carrier et al., 2007). Afsharpour et al. performed a similar study for breast brachytherapy with ^{103}Pd implants (Afsharpour et al., 2010). They found D_{90} reductions ranging from 4% for an all-gland breast to 35% for an all-adipose breast. This increased sensitivity of breast brachytherapy compared to prostate implants is caused in part by the lower energy of ^{103}Pd, but mostly by the larger deviation of adipose tissue's effective atomic number to that of water compared to the relatively small difference between ICRP prostate and water. All studies listed above reported dose to tissue.

Melhus and Rivard showed that tissue effects were negligible for ^{192}Ir over clinically relevant distances in soft tissue, muscle, and breast using point sources in uniform spherical phantoms (Melhus and Rivard, 2006). This is due to the higher photon energies of ^{192}Ir. On the other hand, applicator shielding, designed to spare organs at risk, can cause dose reductions of 5–11% for gynecology (GYN) applicators (Markman et al., 2001) and up to 24% for endorectal applicators when compared to the TG-43 predicted dose (Poon et al., 2008). Poon et al. showed that TG-43 dose distributions for ^{192}Ir breast implants overestimated dose to the skin by 5%, due to differences from full scatter conditions with MC simulations (Poon and Verhaegen, 2009), confirming earlier findings from Pantelis et al. (2005).

9.3.3 Available MC Dose Calculation Tools

9.3.3.1 Monte Carlo Dose Calculation for Prostate Implant

MCPI (Monte Carlo dose calculation for prostate implant) was developed by Chibani and coworkers in 2005 to perform patient-specific prostate implant dose calculations with ^{125}I and ^{103}Pd (Chibani and Williamson, 2005). The code is based on the general-purpose MC software GEPTS (Chibani, 1995), using only photon transport for low-energy (<1 MeV) sources. Several variance reduction techniques, in addition to the track-length collision kerma estimator, were developed for MCPI. Photons are transported only once in the source geometry; a phase space file is subsequently used to emit photons from each source surface, while still modeling the source geometry to include ISA. The code handles both voxel (rectangular mesh) and seed (nested cylindrical objects, facilitating analytical ray-tracing) geometries simultaneously using a hybrid geometry model, where voxels intersecting seeds are flagged, thus avoiding the need to query every seed in the geometry when transporting photons, a technique the authors call "voxel indexing." Finally, MCPI uses ray-tracing instead of analog transport of photons; primary and secondary photon trajectories are projected through the entire mesh of voxels, regardless of whether they have an interaction or not, thereby enhancing the frequency of energy deposition in voxels at larger distances, a technique based upon Williamson's expected-value track-length estimator (Williamson, 1987). Calculation times are reported to be about 60 s for a $2 \times 2 \times 2$ mm^3 voxel mesh (Chibani and Williamson, 2005).

9.3.3.2 PTRAN_CT

The correlated sampling technique (Hedtjarn et al., 2002) has recently been implemented in Williamson's PTRAN_CT (Dolan et al., 2006) by Sampson and coworkers (Sampson et al., 2012). The PTRAN_CT implementation also included the variance reduction techniques of MCPI described above and expected-value track-length scoring. Dose calculation times of about 3 s for a prostate implant have been reported to achieve 2% statistical uncertainty in a $2 \times 2 \times 2$ mm^3 voxel mesh. This code can input digital imaging and communications in medicine (DICOM) CT images and currently uses the EGSnrc CTcreate software to assign cross-section files based upon single-energy CT. An alternative could be the user interface BrachyGUI, developed at McGill University (Poon et al., 2008).

9.3.3.3 BrachyDose

BRACHYDOSE is an EGSnrc (Kawrakow, 2000) user code utilizing Yegin's multigeometry package (Yegin, 2003) and was developed to perform prostate implant dose calculations (Yegin and Rogers, 2004). The geometry package permits detailed modeling of brachytherapy source geometries, and the code is capable of transporting both photons and electrons from 0.01 to 10 MeV, although the higher energies are not used in brachytherapy. Collision kerma is scored with a pathlength estimator, but dose can also be scored at the cost of longer calculation times. The code has been extensively benchmarked by generating single-source TG-43 parameters for several commercially available ^{125}I, ^{103}Pd, and ^{192}Ir sources (Taylor et al., 2007). BRACHYDOSE is also used for treatment planning of eye plaques with low-energy brachytherapy sources (Thomson et al., 2008). Calculation times of 30 s are reported to produce statistical uncertainties less than 2% for a $2 \times 2 \times 2$ mm^3 voxel mesh (Thomson et al., 2010).

9.3.3.4 ALGEBRA

ALGEBRA (algorithm for heterogeneous dosimetry based on Geant4 for brachytherapy) is based on the general-purpose MC code Geant4 and the DICOM radiotherapy (RT) standard. Initially developed by Carrier and coworkers (Carrier et al., 2007), and referred to as Geant4/DICOM-RT in the recent review by Rivard et al. (2010), the code was later modified by Afsharpour and given its current name (Afsharpour et al., 2012). The code can import planning data from a TPS (seed position, air-kerma strength, etc.) and structural data (contours). CT images are imported and a semiautomatic segmentation method uses the CT calibration curve to assign densities and organ contours to assign elemental compositions to each voxel. Seed geometries and voxels do not overlap in ALGEBRA; voxels intersecting seeds are discarded, and the missing volume is replaced by water. The new layered mass geometry functionality of Geant4, where the overlap of seeds and voxels is permitted, should eliminate the need to perform this step (Enger et al., 2012a). Dose scoring is performed using the "parallel world" functionality of Geant4 by creating a scoring mesh independent of the transport geometry. The resolution of the CT-derived transport geometry can be modified, as well as that of the scoring mesh, and they need not be equivalent. Secondary electrons are immediately discarded and collision kerma is scored with a pathlength estimator. Both kerma in water or medium by transporting photons in medium, $K_{w,m}$ and $K_{m,m}$, can be scored. The use of source phase space files is also possible. Calculation times in ALGEBRA for $2 \times 2 \times 2$ mm^3 voxels are about 6 and 12 min for 2% statistical uncertainty for a breast ^{103}Pd implant and a prostate ^{125}I implant, respectively.

9.3.4 Issues Involved in Clinical Implementation

9.3.4.1 $D_{w,m}$ or $D_{m,m}$

An excellent overview of this issue can be found in the report of AAPM Task Group 186 (Beaulieu et al., 2012). MC simulations provide the ability to calculate the energy deposition in the medium present in each voxel of a treatment geometry, either from photons ($K_{m,m}$) or from electrons set in motion by photons ($D_{m,m}$). An alternative approach to reporting dose to medium would be to calculate the energy deposition in a unique reference medium. For historical/practical reasons, this reference medium has consistently been water. While in external beam radiotherapy (EBRT) $D_{w,m}/D_{m,m}$ differences between soft tissues and water are of the order of 2%, at the energies of brachytherapy sources these differences can be much larger. In EBRT, conversion is performed by using ratios of mass collision stopping power under the assumptions of Bragg–Gray cavity theory (electron ranges larger than size of cavity). For brachytherapy, the method of conversion depends on the cavity size. Millimeter-sized voxels are several times larger than the range of secondary electrons produced by photons emitted from low- or high-energy sources; thus, ratios of mass energy absorption coefficients μ_{en}/ρ (large cavity theory) may be used to convert $K_{m,m}$ to $K_{w,m}$. The μ_{en}/ρ differences between tissue and water are generally important for low-energy sources (up to 80% for certain soft tissues), while the situation is similar to EBRT for high-energy brachytherapy sources (3–5%).

Complications arise when we consider that the radiation response of tissues and tumors correlates with the energy deposited in the nuclei of cells. The dimensions of mammalian cells (~10 μm) mean that they act as Bragg–Gray or small cavities at the energies of high-energy brachytherapy sources and as intermediate cavities bracketed by large and small cavity theories at the energies of low-energy brachytherapy sources. This makes the conversion of $D_{m,m}$ to $D_{w,m}$ using μ_{en} ratios questionable for low-energy sources and means that different conversion methods are required across the brachytherapy energy range. There is also no clear indication that water is the best material to represent the nuclei of various cells (Enger et al., 2012b). For these reasons, it has been recommended to report both $D_{m,m}$ and $D_{w,m}$ when performing MC dose calculations for brachytherapy (Beaulieu et al., 2012).

9.3.4.2 Tissue Composition Selection

MC dose calculations generally require voxel-by-voxel assignment of tissue density and elemental composition (mass fraction of each element composing the tissue). While tissue composition essentially plays no role for brachytherapy MC dose calculations with high-energy sources, it becomes very important for low-energy sources due to the Z dependence of μ/ρ and μ_{en}/ρ (photoelectric effect). While tissue density also affects dose distributions, it is readily obtained from CT images of the treated anatomy using a Hounsfield unit (HU) to density calibration (similar to the HU to electron density calibration used in external beam).

Soft tissue composition, on the other hand, is difficult to correlate with HU, as several human tissues possessing different compositions have HU falling in the [−100, 100] range around water (Schneider et al., 2000). In most MC brachytherapy studies, tissue composition is assigned on the basis of anatomical organ contours, that is, prostate tissue is uniformly assigned to the prostate contour and breast tissue is uniformly assigned to the breast contour. With this practice, the question of what constitutes prostate or breast tissue arises. Literature reports (ICRU, 1992; ICRP, 1975) provide compositions for a wide range of human tissues; however, these compositions are based on old (1930–1970) measurements on a small number of subjects and may be of questionable accuracy in characterizing average tissue composition in the population served by brachytherapy. Some reports provide a range of composition for a given tissue, an indication of intersample variability (Woodard and White, 1986). Landry et al. investigated the sensitivity of dose distributions from low-energy sources for tissues of interest in prostate and breast brachytherapy

using clinical ^{125}I and [103]Pd implants, respectively, and postimplant CT images (Landry et al., 2010). For prostate, two compositions were found in the literature resulting in $K_{w,m} D_{90}$ variations of 3.5%. For breast, which is composed of adipose and glandular tissue, Landry et al. generated a uniform tissue. The variation of the proportion of each tissue from 70/30 to 30/70 yielded $K_{w,m} D_{90}$ variations of 10% (or about 6% when considering $K_{m,m}$). For a given adipose/gland proportion, the variability of each tissue's composition resulted in dose variations of 10%. Recent studies on breast brachytherapy with low-energy sources suggest segmenting adipose and mammary gland on CT images, as the uniform tissue approximation may fail to reproduce the dose distributions obtained from segmented geometries (Sutherland et al., 2011; Afsharpour et al., 2012). Only one modern study of elemental composition of tumors appears to exist, that of Maughan et al. (1997), which demonstrated a large variation in carbon content (8–32% by weight) and mineral ash (0.9–3% by weight), which gives to variation in $(\mu_{en}/\rho)^{tumor}_{water}$ of 20% at 30 keV.

9.3.5 Image Usage

9.3.5.1 CT

CT images are generally the standard input to MC calculation, as CT image intensity approximately tracks the relative linear attenuation coefficient, evaluated at the effective scanning energy, of the underlying tissue. Converting HU to electron or mass density is a relatively straightforward step, as explained above. For low-energy sources, the issues of the previous section demand special attention, but for high-energy sources, where tissue composition effects are essentially negligible, CT images provide all the necessary information to accurately calculate the dose. Additionally, the location of implanted seeds, catheters, or applicators is easily derived from CT images.

CT artifacts, especially those caused by the high-density components of implanted LDR seeds (streaking artifacts) and other foreign metal bodies implanted in the body, can seriously degrade the information required for accurate dose calculations. Metal artifact correction algorithms exist and can alleviate this problem (Xu et al., 2011). However, metal artifact reduction cannot be considered to be a solved problem, especially at the level of quantitative accuracy needed to specify tissue composition for low-energy brachytherapy. Even relatively subtle streaks can degrade voxel-by-voxel single-energy CT tissue identity analyses (Schneider et al., 1996).

9.3.5.2 MRI

With its excellent soft tissue contrast, magnetic resonance imaging (MRI) permits the identification of lesions and organs at risk not visible on CT, and currently plays an important role in GYN brachytherapy (Potter et al., 2008). By differentiating tissues, it permits a better distribution of radiation dose at the time of planning. Prostate delineation is also acknowledged as being superior on magnetic resonance (MR) compared to CT imaging (Viswanathan et al., 2007). However, the lack of density information currently limits the use of MR images for MC dose calculations. Work has been performed to employ MR images in EBRT treatment planning and for PET attenuation correction, and two types of approaches have been developed (Johansson et al., 2011). Methods based on anatomy attempt to deform patient MR images to a reference MR image, and apply the resulting deformation on a corresponding reference CT image (Kops and Herzog, 2007). Voxel-based approaches attempt to classify voxels of MR images into tissue types (Keereman et al., 2010). Following classification, bulk electron density assignment or direct conversion to HU or electron density is performed. These techniques have not been evaluated in the context of brachytherapy dose calculations so far. For prostate EBRT, Lee et al. reported dose calculation errors of less than 2% when segmenting images into bone and soft tissue compared to using the full CT (Lee et al., 2003). Given that high-energy photons employed in brachytherapy are relatively insensitive to tissue type assignment (see Table 9.1), we can expect similar results. For low-energy photons, the added complexity of assigning a correct tissue composition would need to be considered. Geometric distortion of MR images due to magnetic field inhomogeneities, gradient nonlinearity, susceptibility effects, and chemical shifts have also been considered in EBRT (Lee et al., 2003). Brachytherapy is possibly less sensitive to distortions than EBRT, as the distortions are generally acknowledged to be low in the center of the field of view (FOV), where MR compatible applicators or seeds would be positioned.

9.3.5.3 Ultrasound

Transrectal ultrasound images are used to guide needle implantation for LDR seed implants and catheter implantation for HDR brachytherapy of the prostate. Ultrasound images provide information on the location and dimensions of certain organs of interest and permit TG-43 dose calculations (Chew et al., 2009). While ultrasound-based tissue typing is an active area of research (Zhou et al., 2009), it does not yet provide patient-specific density and tissue composition information necessary to perform MC simulations. Additionally, the presence of the TRUS probe deforms the organs of interest. External transducers, permitting noninvasive image acquisition, can avoid this issue but are currently not used in brachytherapy applications. One approach could be to assign generic densities and compositions to organ contours, as explained above for MR images. Using the TRUS-derived seed positions, ISA could also be modeled in prostate implants.

9.4 Future Use of Monte Carlo Methods for Brachytherapy

Like the previous two sections, opportunities for future applications of MC methods for brachytherapy may be divided into two categories: (a) single-source dosimetry for preclinical or research purposes and (b) patient-oriented applications such as for treatment planning.

9.4.1 Single-Source Dosimetry

Over the past decade, there have been significant advances in characterizing radiation dose distributions for individual brachytherapy sources. These studies have examined the sensitivity of phantom composition on absorbed dose, the influence of motion of dynamic internal components within the source, effects of manufacturing variations on resultant dose distributions, and previously ignored radiations such as beta penetration and characteristic x-ray influence. Forthcoming advances upon using MC methods for brachytherapy may include assessment of dosimetric contributions from radioimpurities, characterization of dissolvable brachytherapy sources, design of multiradionuclide sources, integration of brachytherapy dosimetry and patient imaging, and of course evaluation of new radionuclides. Since the focus can range from methodical improvements on existing paradigms to highly creative blue sky ideas, the possibilities seem endless.

9.4.2 Treatment Planning

The first decade of the twenty-first century has seen the exciting birth of image-based MC treatment planning for brachytherapy. Early advances have examined MC code variance reduction techniques while maintaining calculation accuracy, patient tissue assignment and sensitivity of results to such assignments, consideration for dose specification choice such as dose to medium in medium or dose to water in medium, and retrospective studies trying to correlate clinical outcomes with the more accurate dose distributions using MC methods.

Future directions for MC-based brachytherapy treatment planning will continue the organized improvements on existing methods, integration of MC computations during the dose optimization phase, use of MC techniques in inverse optimization, resolution of imaging artifacts such as from seeds or high-Z applicator, correcting for limited image datasets as needed to provide full radiation scatter conditions to simulate the clinical environment, and accounting for changes in brachytherapy applicator positioning relative to the relevant tissues in the patient. Perhaps MC methods can even play a role in accounting for organ motion during treatment or between treatments, as has been briefly explored for external beam irradiation. Multicenter brachytherapy dosimetry comparison trials versus MC as the gold standard are now just on the horizon. Radiobiological modeling on the submicron scale for microdosimetry and improved understanding of RBE effects, as well as source design for optimal exploitation of RBE effects, are also of interest.

Given the increased interest in brachytherapy dosimetry and the improved coordination of international research teams, the future is bright for using MC methods in the field of brachytherapy.

References

Afsharpour, H., G. Landry, M. D'Amours, S. Enger, B. Reniers, E. Poon, J. F. Carrier, F. Verhaegen, and L. Beaulieu. 2012. ALGEBRA: ALgorithm for the heterogeneous dosimetry based on GEANT4 for BRAchytherapy. *Phys Med Biol* 57:3273–3280.

Afsharpour, H., J. P. Pignol, B. Keller, J. F. Carrier, B. Reniers, F. Verhaegen, and L. Beaulieu. 2010. Influence of breast composition and interseed attenuation in dose calculations for post-implant assessment of permanent breast 103Pd seed implant. *Phys Med Biol* 55:4547–4561.

Afsharpour, H., B. Reniers, G. Landry, J. P. Pignol, B. M. Keller, F. Verhaegen, and L. Beaulieu. 2012. Consequences of dose heterogeneity on the biological efficiency of ^{103}Pd permanent breast seed implants. *Phys Med Biol* 57:809–823.

Anderson, L. L., R. Nath, and K. A. Weaver. 1990. Interstitial Collaborative Working Group (ICWG). Interstitial Brachytherapy: Physical, Biological, and Clinical Considerations. Raven, New York.

Ballester, F., D. Granero, J. Perez-Calatayud, C. S. Melhus, and M. J. Rivard. 2009. Evaluation of high-energy brachytherapy source electronic disequilibrium and dose from emitted electrons. *Med Phys* 36:4250–4256.

Ballester, F., C. Hernandez, J. Perez-Calatayud, and F. Lliso. 1997. Monte Carlo calculation of dose rate distributions around ^{192}Ir wires. *Med Phys* 24:1221–1228.

Beaulieu, L., A. C. Tedgren, J. F. Carrier, S. D. Davis, F. Mourtada, M. J. Rivard, R. M. Thomson, F. Verhaegen, T. A. Wareing, and J. F. Williamson. 2012. Report of the Task Group 186 on model-based dose calculation methods in brachytherapy beyond the TG-43 formalism: Current status and recommendations for clinical implementation. *Med Phys* 39:6208–6236.

Berger, M. J. 1963. Monte Carlo calculation of the penetration and diffusion of fast charged particles. In *Methods in Computational Physics, Volume 1*. B. Adler, S. Fernbach, and M. Rotenburg, editors. Academic Press, New York. 135–215.

Bohm, T. D., P. M. DeLuca, Jr., and L. A. DeWerd. 2003. Brachytherapy dosimetry of ^{125}I and ^{103}Pd sources using an updated cross section library for the MCNP Monte Carlo transport code. *Med Phys* 30:701–711.

Burns, G. S., and D. E. Raeside. 1988. Two-dimensional dose distribution around a commercial ^{125}I seed. *Med Phys* 15:56–60.

Burns, G. S., and D. E. Raeside. 1989. The accuracy of single-seed dose superposition for I-125 implants. *Med Phys* 16:627–631.

Carrier, J. F., L. Beaulieu, F. Therriault-Proulx, and R. Roy. 2006. Impact of interseed attenuation and tissue composition for permanent prostate implants. *Med Phys* 33:595–604.

Carrier, J. F., M. D'Amours, F. Verhaegen, B. Reniers, A. G. Martin, E. Vigneault, and L. Beaulieu. 2007. Postimplant dosimetry using a Monte Carlo dose calculation engine: A new clinical standard. *Int J Radiat Oncol Biol Phys* 68:1190–1198.

Chew, M. S., J. Xue, C. Houser, V. Misic, J. Cao, T. Cornwell, J. Handler, Y. Yu, and E. Gressen. 2009. Impact of transrectal ultrasound- and computed tomography-based seed localization on postimplant dosimetry in prostate brachytherapy. *Brachytherapy* 8:255–264.

Chibani, O. 1995. Electron depth-dose distributions in water, iron and lead—The GEPTS system. *Nucl Instrum Meth B* 101:357–378.rPoon, E., Y. Le, J. F. Williamson, and F. Verhaegen. 2008. BrachyGUI: An adjunct to an accelerated

Monte Carlo photon transport code for patient-specific brachytherapy dose calculations and analysis. *J Phys Conf Ser* 102:012018.

Chibani, O. and J. F. Williamson. 2005. MCPI: A sub-minute Monte Carlo dose calculation engine for prostate implants. *Med Phys* 32:3688–3698.

Chibani, O., J. F. Williamson, and D. Todor. 2005. Dosimetric effects of seed anisotropy and interseed attenuation for ^{103}Pd and ^{125}I prostate implants. *Med Phys* 32:2557–2566.

Chiu-Tsao, S. T., T. L. Duckworth, N. S. Patel, J. Pisch, and L. B. Harrison. 2004. Verification of Ir-192 near source dosimetry using GAFCHROMIC film. *Med Phys* 31:201–207.

Corbett, J. F., J. J. Jezioranski, J. Crook, T. Tran, and I. W. Yeung. 2001. The effect of seed orientation deviations on the quality of ^{125}I prostate implants. *Phys Med Biol* 46:2785–2800.

Cullen, D., J. H. Hubbell, and L. Kissel. 1997. EPDL97: The Evaluated Photon Data Library, 97 version. UCRL-50400 Vol. 6, Rev 5.

Dale, R. G. 1983. Some theoretical derivations relating to the tissue dosimetry of brachytherapy nuclides, with particular reference to iodine-125. *Med Phys* 10:176–183.

Das, R. K., D. Keleti, Y. Zhu, A. S. Kirov, A. S. Meigooni, and J. F. Williamson. 1997b. Validation of Monte Carlo dose calculations near ^{125}I sources in the presence of bounded heterogeneities. *Int J Radiat Oncol Biol Phys* 38:843–853.

Das, R. K., Z. Li, H. Perera, and J. F. Williamson. 1996. Accuracy of Monte Carlo photon transport simulation in characterizing brachytherapy dosimeter energy-response artefacts. *Phys Med Biol* 41:995–1006.

Das, R., A. S. Meigooni, V. Mishra, M. A. Langton, and J. F. Williamson. 1997a. Dosimetric characteristics of the type 8 Ytterbium-169 interstitial brachytherapy source. *J Brachyther Int* 13:219–234.

Davis, S. D., C. K. Ross, P. N. Mobit, L. Van der Zwan, W. J. Chase, and K. R. Shortt. 2003. The response of LiF thermoluminescence dosemeters to photon beams in the energy range from 30 kV x rays to ^{60}Co gamma rays. *Radiat Prot Dosim* 106:33–43.

Demarco, J. J., R. E. Wallace, and K. Boedeker. 2002. An analysis of MCNP cross-sections and tally methods for low-energy photon emitters. *Phys Med Biol* 47:1321–1332.

DeWerd, L. A., G. S. Ibbott, A. S. Meigooni, M. G. Mitch, M. J. Rivard, K. E. Stump, B. R. Thomadsen, and J. L. Venselaar. 2011. A dosimetric uncertainty analysis for photon-emitting brachytherapy sources: Report of AAPM Task Group No. 138 and GEC-ESTRO. *Med Phys* 38:782–801.

Dolan, J., Z. Lia, and J. F. Williamson. 2006. Monte Carlo and experimental dosimetry of an ^{125}I brachytherapy seed. *Med Phys* 33:4675–4684.

Enger, S. A., A. Ahnesjö, F. Verhaegen, and L. Beaulieu. 2012b. Dose to tissue medium or water cavities as surrogate for the dose to cell nuclei at brachytherapy photon energies. *Phys Med Biol* 57:4489.

Enger, S. A., G. Landry, M. D'Amours, F. Verhaegen, L. Beaulieu, M. Asai, and J. Perl. 2012. Layered mass geometry: A novel technique to overlay seeds and applicators onto patient geometry in Geant4 brachytherapy simulations. *Phys Med Biol* 57:6269–6277.

Gifford, K. A., J. L. Horton, Jr., C. E. Pelloski, A. Jhingran, L. E. Court, F. Mourtada, and P. J. Eifel. 2005. A three-dimensional computed tomography-assisted Monte Carlo evaluation of ovoid shielding on the dose to the bladder and rectum in intracavitary radiotherapy for cervical cancer. *Int J Radiat Oncol Biol Phys* 63:615–621.

Granero, D., J. Vijande, F. Ballester, and M. J. Rivard. 2011. Dosimetry revisited for the HDR ^{192}Ir brachytherapy source model mHDR-v2. *Med Phys* 38:487–494.

Grimm, P., and J. Sylvester. 2004. Advances in brachytherapy. *Rev Urol* 6 Suppl 4:S37–S48.

Hedtjarn, H., G. A. Carlsson, and J. F. Williamson. 2002. Accelerated Monte Carlo based dose calculations for brachytherapy planning using correlated sampling. *Phys Med Biol* 47:351–376.

Hill, M. A. 2004. The variation in biological effectiveness of x-rays and gamma rays with energy. *Radiat Prot Dosim* 112:471–481.

Hubbell, J. H. 1999. Review of photon interaction cross section data in the medical and biological context. *Phys Med Biol* 44:R1–R22.

ICRP. 1975. Report of the task group on reference man, ICRP Report 23, Washington D.C.

ICRU. 1992. Report 46: Photon, electron, proton and neutron interaction data for body tissues, Bethesda, MD.

Johansson, A., M. Karlsson, and T. Nyholm. 2011. CT substitute derived from MRI sequences with ultrashort echo time. *Med Phys* 38:2708–2714.

Kawrakow, I. 2000. Accurate condensed history Monte Carlo simulation of electron transport. I. EGSnrc, the new EGS4 version. *Med Phys* 27:485–498.

Kawrakow, I. and A. F. Bielajew. 1998. On the condensed history technique for electron transport. *Nucl Instrum Meth Phys Res B, Beam Interact Mater At* 142:253–280.

Keereman, V., Y. Fierens, T. Broux, Y. De Deene, M. Lonneux, and S. Vandenberghe. 2010. MRI-based attenuation correction for PET/MRI using ultrashort echo time sequences. *J Nucl Med* 51:812–818.

Kirov, A., J. F. Williamson, A. S. Meigooni, and Y. Zhu. 1995. TLD, diode and Monte Carlo dosimetry of an ^{192}Ir source for high dose-rate brachytherapy. *Phys Med Biol* 40:2015–2036.

Kops, E. R., and H. Herzog. 2007. Alternative methods for attenuation correction for PET images in MR-PET scanners. In Nuclear Science Symposium Conference Record, 2007. NSS '07. IEEE. 4327–4330.

Krishnaswamy, V. 1971. Calculation of the dose distribution about californium-252 needles in tissue. *Radiology* 98:155–160.

Kubo, H. 1985. Exposure contribution from Ti K x rays produced in the titanium capsule of the clinical I-125 seed. *Med Phys* 12:215–220.

Landry, G., B. Reniers, L. Murrer, L. Lutgens, E. B. Gurp, J. P. Pignol, B. Keller, L. Beaulieu, and F. Verhaegen. 2010. Sensitivity of low energy brachytherapy Monte Carlo dose calculations to uncertainties in human tissue composition. *Med Phys* 37:5188–5198.

Landry, G., B. Reniers, J. P. Pignol, L. Beaulieu, and F. Verhaegen. 2011. The difference of scoring dose to water or tissues in Monte Carlo dose calculations for low energy brachytherapy photon sources. *Med Phys* 38:1526–1533.

Lee, Y. K., M. Bollet, G. Charles-Edwards, M. A. Flower, M. O. Leach, H. McNair, E. Moore, C. Rowbottom, and S. Webb. 2003. Radiotherapy treatment planning of prostate cancer using magnetic resonance imaging alone. *Radiother Oncol* 66:203–216.

Li, Z., R. K. Das, L. A. DeWerd, G. S. Ibbott, A. S. Meigooni, J. Perez-Calatayud, M. J. Rivard, R. S. Sloboda, and J. F. Williamson. 2007. Dosimetric prerequisites for routine clinical use of photon emitting brachytherapy sources with average energy higher than 50 keV. *Med Phys* 34:37–40.

Li, Z., J. F. Williamson, and H. Perera. 1993. Monte Carlo calculation of kerma to a point in the vicinity of media interfaces. *Phys Med Biol* 38:1825–1840.

Lindsay, P., J. Battista, and J. Van Dyk. 2001. The effect of seed anisotrophy on brachytherapy dose distributions using ^{125}I and ^{103}Pd. *Med Phys* 28:336–345.

Liu, D., E. Poon, M. Bazalova, B. Reniers, M. Evans, T. Rusch, and F. Verhaegen. 2008. Spectroscopic characterization of a novel electronic brachytherapy system. *Phys Med Biol* 53:61–75.

Loftus, T. P. 1984. Exposure standardization of Iodine-125 seeds used for brachytherapy. *J Res Natl Bur Stand* 89:295–303.

Markman, J., J. F. Williamson, J. F. Dempsey, and D. A. Low. 2001. On the validity of the superposition principle in dose calculations for intracavitary implants with shielded vaginal colpostats. *Med Phys* 28:147–155.

Maughan, R. L., P. J. Chuba, A. T. Porter, E. Ben-Josef, and D. R. Lucas. 1997. The elemental composition of tumors: Kerma data for neutrons. *Med Phys* 24:1241–1244.

Meigooni, A. S., J. A. Meli, and R. Nath. 1992. Interseed effects on dose for ^{125}I brachytherapy implants. *Med Phys* 19:385–390.

Meisberger, L. L., R. J. Keller, and R. J. Shalek. 1968. The effective attenuation in water of the gamma rays of gold 198, iridium 192, cesium 137, radium 226, and cobalt 60. *Radiology* 90:953–957.

Melhus, C. S., and M. J. Rivard. 2006. Approaches to calculating AAPM TG-43 brachytherapy dosimetry parameters for ^{137}Cs, ^{125}I, ^{192}Ir, ^{103}Pd, and ^{169}Yb sources. *Med Phys* 33:1729–1737.

Mikell, J. K., and F. Mourtada. 2010. Dosimetric impact of an ^{192}Ir brachytherapy source cable length modeled using a grid-based Boltzmann transport equation solver. *Med Phys* 37:4733–4743.

Monroe, J. I., and J. F. Williamson. 2002. Monte Carlo-aided dosimetry of the theragenics TheraSeed model 200 ^{103}Pd interstitial brachytherapy seed. *Med Phys* 29:609–621.

Morin, L. R. M. 1982. Molecular form factors and photon coherent scattering cross sections of water. *J Phys Chem Ref Data* 11:1091–1098.

Nath, R., L. L. Anderson, G. Luxton, K. A. Weaver, J. F. Williamson, and A. S. Meigooni. 1995. Dosimetry of interstitial brachytherapy sources: Recommendations of the AAPM Radiation Therapy Committee Task Group No. 43. American Association of Physicists in Medicine. *Med Phys* 22:209–234.

Nunn, A. A., S. D. Davis, J. A. Micka, and L. A. DeWerd. 2008. LiF:Mg,Ti TLD response as a function of photon energy for moderately filtered x-ray spectra in the range of 20–250 kVp relative to ^{60}Co. *Med Phys* 35:1859–1869.

Pantelis, E., P. Papagiannis, P. Karaiskos, A. Angelopoulos, G. Anagnostopoulos, D. Baltas, N. Zamboglou, and L. Sakelliou. 2005. The effect of finite patient dimensions and tissue inhomogeneities on dosimetry planning of ^{192}Ir HDR breast brachytherapy: A Monte Carlo dose verification study. *Int J Radiat Oncol Biol Phys* 61:1596–1602.

Patel, N. S., S. T. Chiu-Tsao, J. F. Williamson, P. Fan, T. Duckworth, D. Shasha, and L. B. Harrison. 2001. Thermoluminescent dosimetry of the Symmetra ^{125}I model I25.S06 interstitial brachytherapy seed. *Med Phys* 28:1761–1769.

Perera, H., J. F. Williamson, Z. Li, V. Mishra, and A. S. Meigooni. 1994. Dosimetric characteristics, air-kerma strength calibration and verification of Monte Carlo simulation for a new Ytterbium-169 brachytherapy source. *Int J Radiat Oncol Biol Phys* 28:953–970.

Perez-Calatayud, J., F. Ballester, R. K. Das, L. A. Dewerd, G. S. Ibbott, A. S. Meigooni, Z. Ouhib, M. J. Rivard, R. S. Sloboda, and J. F. Williamson. 2012. Dose calculation for photon-emitting brachytherapy sources with average energy higher than 50 keV: Report of the AAPM and ESTRO. *Med Phys* 39:2904–2929.

Perez-Calatayud, J., D. Granero, and F. Ballester. 2004. Phantom size in brachytherapy source dosimetric studies. *Med Phys* 31:2075–2081.

Pokhrel, D., M. J. Murphy, D. A. Todor, E. Weiss, and J. F. Williamson. 2011. Reconstruction of brachytherapy seed positions and orientations from cone-beam CT x-ray projections via a novel iterative forward projection matching method. *Med Phys* 38:474–486.

Poon, E., and F. Verhaegen. 2009. Development of a scatter correction technique and its application to HDR ^{192}Ir multicatheter breast brachytherapy. *Med Phys* 36:3703–3713.

Poon, E., B. Reniers, S. Devic, T. Vuong, and F. Verhaegen. 2006. Dosimetric characterization of a novel intracavitary mold applicator for 192Ir high dose rate endorectal brachytherapy treatment. *Med Phys* 33:4515–4526.

Poon, E., Y. Le, J. F. Williamson, and F. Verhaegen. 2008a. BrachyGUI: An adjunct to an accelerated Monte Carlo photon transport code for patient-specific brachytherapy dose calculations and analysis. *J Phys Conf Ser* 102:012018.

Poon, E., J. F. Williamson, T. Vuong, and F. Verhaegen. 2008b. Patient-specific Monte Carlo dose calculations for high-dose-rate endorectal brachytherapy with shielded intracavitary applicator. *Int J Radiat Oncol Biol Phys* 72:1259–1266.

Potter, R., E. Fidarova, C. Kirisits, and J. Dimopoulos. 2008. Image-guided adaptive brachytherapy for cervix carcinoma. *Clin Oncol (R Coll Radiol)* 20:426–432.

Reniers, B., D. Liu, T. Rusch, and F. Verhaegen. 2008. Calculation of relative biological effectiveness of a low-energy electronic brachytherapy source. *Phys Med Biol* 53:7125–7135.

Reniers, B., F. Verhaegen, and S. Vynckier. 2004. The radial dose function of low-energy brachytherapy seeds in different solid phantoms: Comparison between calculations with the EGSnrc and MCNP4C Monte Carlo codes and measurements. *Phys Med Biol* 49:1569–1582.

Reniers, B., G. Landry, R. Eichner, A. Hallil, and F. Verhaegen. 2012. *In vivo* dosimetry for gynaecological brachytherapy using a novel position sensitive radiation detector: Feasibility study. *Med Phys* 39:1925–1935.

Reniers, B., S. Vynckier, and F. Verhaegen. 2004. Theoretical analysis of microdosimetric spectra and cluster formation for 103Pd and 125I photon emitters. *Phys Med Biol* 49:3781–3795.

Reynaert, N., F. Verhaegen, Y. Taeymans, M. Van Eijkeren, and H. Thierens. 1999. Monte Carlo calculations of dose distributions around 32P and 198Au stents for intravascular brachytherapy. *Med Phys* 26:1484–1491.

Reynaert, N., F. Verhaegen, and H. Thierens. 1998. In-water calibration of PDR 192Ir brachytherapy sources with an NE2571 ionization chamber. *Phys Med Biol* 43:2095–2107.

Ribberfors, R., and G. A. Carlsson. 1985. Compton component of the mass-energy absorption coefficient: Corrections due to the energy broadening of compton-scattered photons. *Radiat Res* 101:47–59.

Rivard, M. J. 2001. Monte Carlo calculations of AAPM Task Group Report No. 43 dosimetry parameters for the MED3631-A/M125I source. *Med Phys* 28:629–637.

Rivard, M. J. 2007. Brachytherapy dosimetry parameters calculated for a ^{131}Cs source. *Med Phys* 34:754–762.

Rivard, M. J., B. M. Coursey, L. A. DeWerd, W. F. Hanson, M. S. Huq, G. S. Ibbott, M. G. Mitch, R. Nath, and J. F. Williamson. 2004. Update of AAPM Task Group No. 43 Report: A revised AAPM protocol for brachytherapy dose calculations. *Med Phys* 31:633–674.

Rivard, M. J., L. Beaulieu, and F. Mourtada. 2010. Enhancements to commissioning techniques and quality assurance of brachytherapy treatment planning systems that use model-based dose calculation algorithms. *Med Phys* 37:2645–2658.

Rivard, M. J., W. M. Butler, L. A. DeWerd, M. S. Huq, G. S. Ibbott, A. S. Meigooni, C. S. Melhus, M. G. Mitch, R. Nath, and J. F. Williamson. 2007. Supplement to the 2004 update of the AAPM Task Group No. 43 Report. *Med Phys* 34:2187–2205.

Rivard, M. J., C. S. Melhus, and B. L. Kirk. 2004. Brachytherapy dosimetry parameters calculated for a new ^{103}Pd source. *Med Phys* 31:2466–2470.

Rivard, M. J., J. L. Venselaar, and L. Beaulieu. 2009. The evolution of brachytherapy treatment planning. *Med Phys* 36:2136–2153.

Sampson, A., Y. Le, and J. F. Williamson. 2012. Fast patient-specific Monte Carlo brachytherapy dose calculations via the correlated sampling variance reduction technique. *Med Phys* 39:1058–1068.

Schneider, U., E. Pedroni, and A. Lomax. 1996. The calibration of CT Hounsfield units for radiotherapy treatment planning. *Phys Med Biol* 41:111–124.

Schneider, W., T. Bortfeld, and W. Schlegel. 2000. Correlation between CT numbers and tissue parameters needed for Monte Carlo simulations of clinical dose distributions. *Phys Med Biol* 45:459–478.

Seltzer, S. M., P. J. Lamperti, R. Loevinger, M. G. Mitch, J. T. Weaver, and B. M. Coursey. 2003. New national air-kerma-strength standards for ^{125}I and ^{103}Pd brachytherapy seeds. *J Res Natl Inst Stand Technol* 108:337–358.

Selvam, T. P. and B. Keshavkumar. 2010. Monte Carlo investigation of energy response of various detector materials in ^{125}I and ^{169}Yb brachytherapy dosimetry. *J Appl Clin Med Phys* 11:3282.

Semenenko, V. A. and R. D. Stewart. 2004. A fast Monte Carlo algorithm to simulate the spectrum of DNA damages formed by ionizing radiation. *Radiat Res* 161:451–457.

Sutherland, J. G., R. M. Thomson, and D. W. Rogers. 2011. Changes in dose with segmentation of breast tissues in Monte Carlo calculations for low-energy brachytherapy. *Med Phys* 38:4858–4865.

Taylor, R. E. and D. W. Rogers. 2008a. An EGSnrc Monte Carlo-calculated database of TG-43 parameters. *Med Phys* 35:4228–4241.

Taylor, R. E. and D. W. Rogers. 2008b. EGSnrc Monte Carlo calculated dosimetry parameters for ^{192}Ir and ^{169}Yb brachytherapy sources. *Med Phys* 35:4933–4944.

Taylor, R. E., G. Yegin, and D. W. Rogers. 2007. Benchmarking brachydose: Voxel based EGSnrc Monte Carlo calculations of TG-43 dosimetry parameters. *Med Phys* 34:445–457.

Thomson, R. M., R. E. Taylor, and D. W. Rogers. 2008. Monte Carlo dosimetry for ^{125}I and ^{103}Pd eye plaque brachytherapy. *Med Phys* 35:5530–5543.

Thomson, R., G. Yegin, R. Taylor, J. Sutherland, and D. Rogers. 2010. Fast Monte Carlo dose calculations for brachytherapy with BrachyDose. *Med Phys* 37:3910(abs).

Valentin, J. 2002. Basic anatomical and physiological data for use in radiological protection: Reference values: ICRP Publication 89. *Ann ICRP* 32:1–277.

Valicenti, R. K., A. S. Kirov, A. S. Meigooni, V. Mishra, R. K. Das, and J. F. Williamson. 1995. Experimental validation of Monte Carlo dose calculations about a high-intensity Ir-192 source for pulsed dose-rate brachytherapy. *Med Phys* 22:821–829.

Viswanathan, A. N., J. Dimopoulos, C. Kirisits, D. Berger, and R. Potter. 2007. Computed tomography versus magnetic resonance imaging-based contouring in cervical cancer brachytherapy: Results of a prospective trial and preliminary guidelines for standardized contours. *Int J Radiat Oncol Biol Phys* 68:491–498.

Williamson, J. F. 1987. Monte Carlo evaluation of kerma at a point for photon transport problems. *Med Phys* 14:567–576.

Williamson, J. F. 1988. Monte Carlo evaluation of specific dose constants in water for ^{125}I seeds. *Med Phys* 15:686–694.

Williamson, J. F. 1991. Comparison of measured and calculated dose rates in water near I-125 and Ir-192 seeds. *Med Phys* 18:776–786.

Williamson, J. F. 1996. The Sievert integral revisited: Evaluation and extension to ^{125}I, ^{169}Yb, and ^{192}Ir brachytherapy sources. *Int J Radiat Oncol Biol Phys* 36:1239–1250.

Williamson, J. F. 2002. Dosimetric characteristics of the DRAX-IMAGE model LS-1 I-125 interstitial brachytherapy source design: A Monte Carlo investigation. *Med Phys* 29:509–521.

Williamson, J. F. and A. S. Meigooni. 1995. Quantitative dosimetry methods for brachytherapy. In *Brachytherapy Physics*. J. F. Williamson, B. T. Thomadsen, and R. Nath, editors. Medical Physics Publishing, Madison, WI. 87–134.

Williamson, J. F., R. L. Morin, and F. M. Khan. 1983. Monte Carlo evaluation of the Sievert integral for brachytherapy dosimetry. *Phys Med Biol* 28:1021–1032.

Williamson, J. F., H. Perera, Z. Li, and W. R. Lutz. 1993. Comparison of calculated and measured heterogeneity correction factors for ^{125}I, ^{137}Cs, and ^{192}Ir brachytherapy sources near localized heterogeneities. *Med Phys* 20:209–222.

Williamson, J. F. and F. J. Quintero. 1988. Theoretical evaluation of dose distributions in water about models 6711 and 6702 ^{125}I seeds. *Med Phys* 15:891–897.

Williamson, J. F. and M. J. Rivard. 2005. Quantitative dosimetry methods for brachytherapy. In *Brachytherapy Physics: Second Edition*. B. R. Thomadsen, M. J. Rivard, and W. M. Butler, editors. Medical Physics Publishing, Madison, WI. 233–294.

Williamson, J. F. and M. J. Rivard. 2009. Thermoluminescent detector and Monte Carlo techniques for reference-quality brachytherapy dosimetry. In *Clinical Dosimetry Measurements in Radiotherapy (AAPM 2009 Summer School)*. D. W. O. Rogers, and J. Cygler, editors. Medical Physics Publishing, Madison, WI. 437–499.

Woodard, H. Q. and D. R. White. 1986. The composition of body tissues. *Br J Radiol* 59:1209–1218.

Xu, C., F. Verhaegen, D. Laurendeau, S. A. Enger, and L. Beaulieu. 2011. An algorithm for efficient metal artifact reductions in permanent seed implants. *Med Phys* 38:47–56.

Yegin, G. 2003. A new approach to geometry modeling for Monte Carlo particle transport: An application to the EGS code system. *Nucl Instrum Meth Phys Res B, Beam Interact Mater At* 211:331–338.

Yegin, G. and D. W. O. Rogers. 2004. A fast Monte Carlo code for multi-seed brachytherapy treatments including interseed effects. *Med Phys* 31:1771(abs).

Zhou, J., P. Zhang, K. S. Osterman, S. A. Woodhouse, P. B. Schiff, E. J. Yoshida, Z. F. Lu, E. fR. Pile-Spellman, G. J. Kutcher, and T. Liu. 2009. Implementation and validation of an ultrasonic tissue characterization technique for quantitative assessment of normal-tissue toxicity in radiation therapy. *Med Phys* 36:1643–1650.

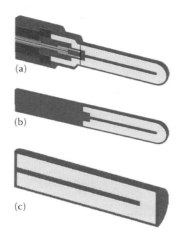

FIGURE 4.6

(a)   (b)

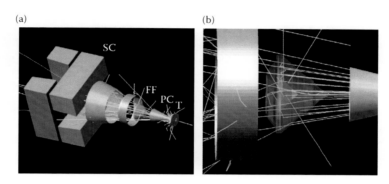

FIGURE 5.2

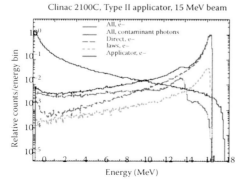

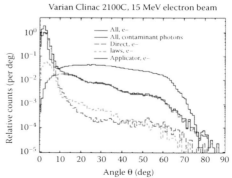

FIGURE 6.5

(a)

(b)

(c)

(d)

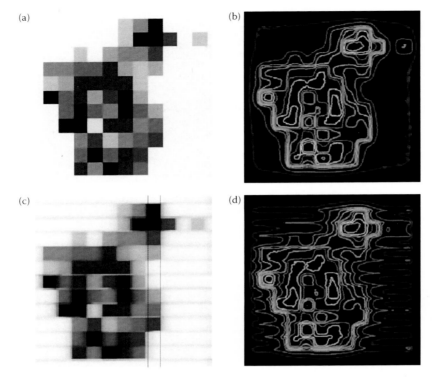

FIGURE 7.1

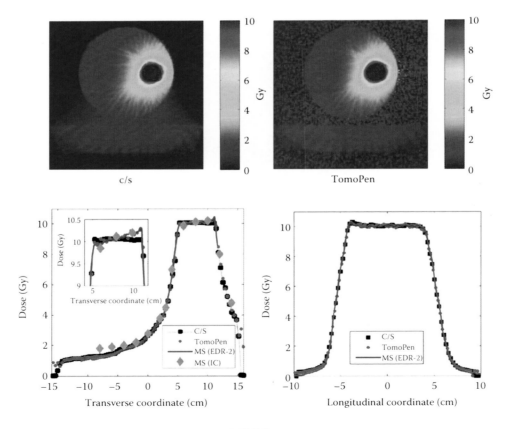

FIGURE 7.5

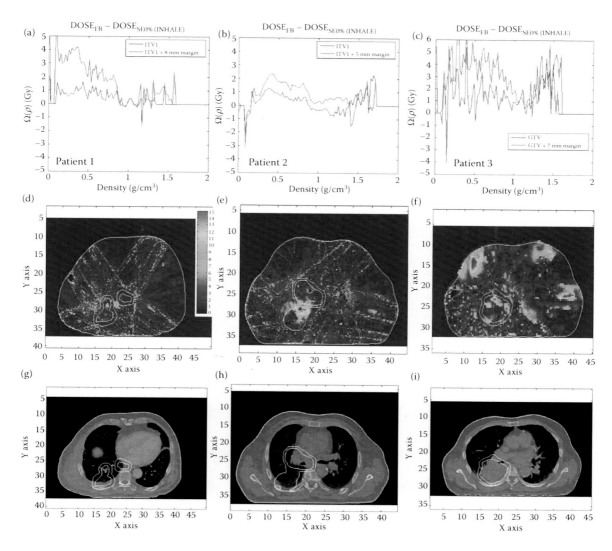

FIGURE 8.3

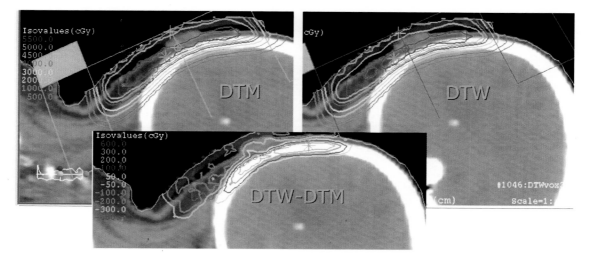

FIGURE 11.5

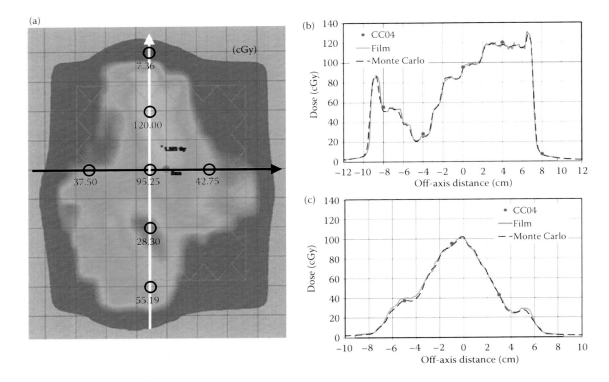

FIGURE 12.5

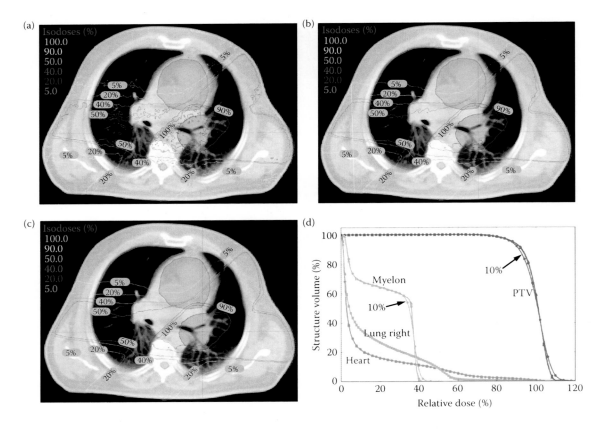

FIGURE 12.7

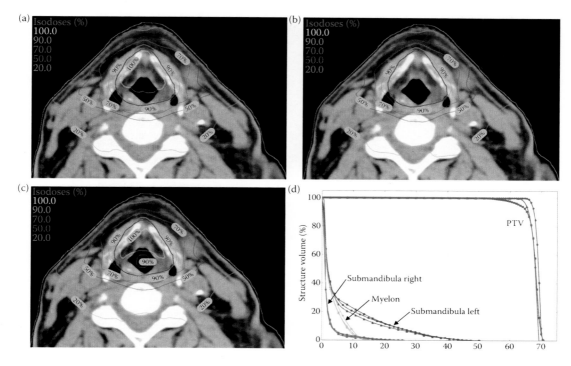

FIGURE 12.9

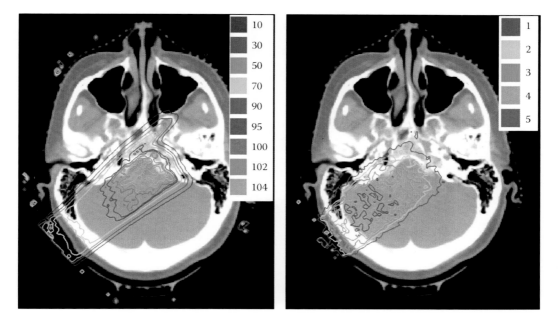

FIGURE 14.8

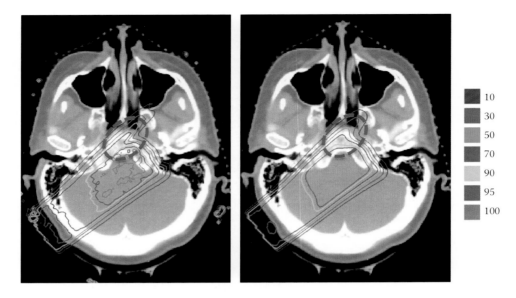

FIGURE 14.9

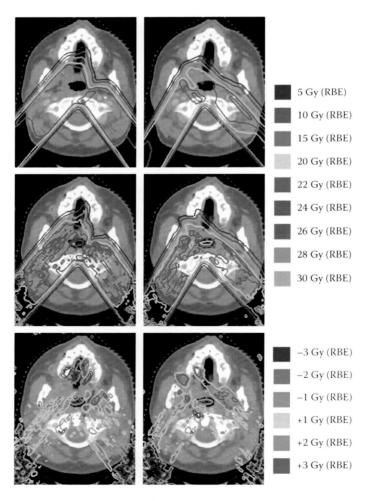

FIGURE 14.10

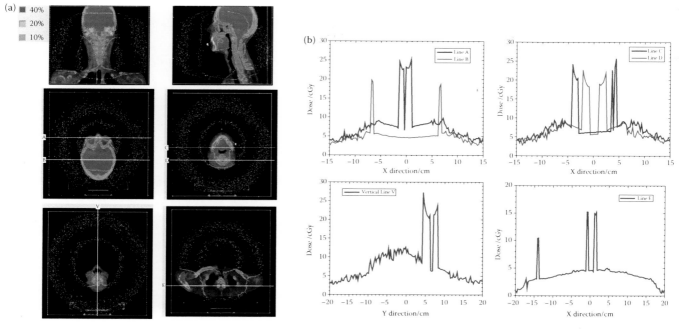

FIGURE 16.5

(a) MatriXX measurement (b) TPS calculation

(c) EPID measurement (d) Monte Carlo simulation

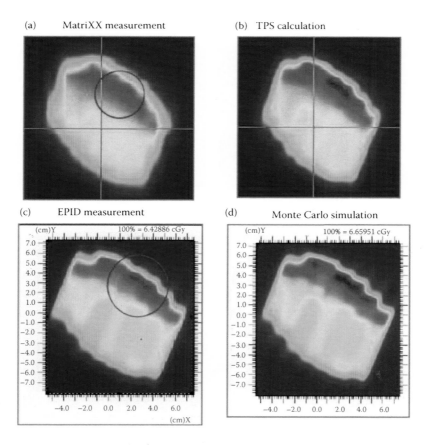

FIGURE 16.10

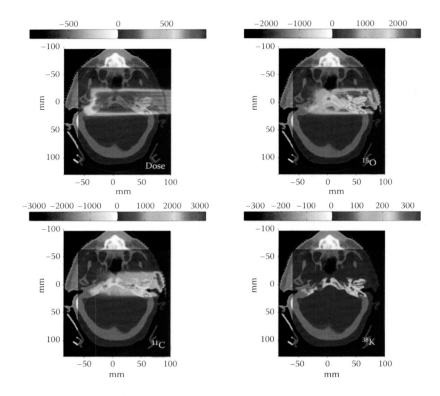

FIGURE 17.4

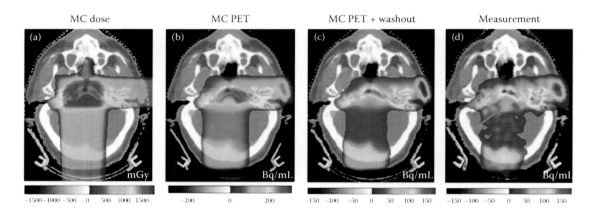

MC dose   MC PET   MC PET + washout   Measurement

FIGURE 17.7

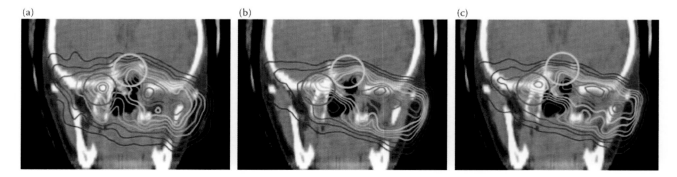

FIGURE 17.8

10

Monte Carlo as a QA Tool for Advanced Radiation Therapy

JinSheng Li
Fox Chase Cancer Center

C.-M. Charlie Ma
Fox Chase Cancer Center

10.1 Introduction

Advanced radiation therapy (RT), such as intensity-modulated radiation therapy (IMRT) (Purdy, 1996; Burman et al., 1997; Webb, 1998; Boyer and Yu, 1999) and intensity-modulated rotational therapy, including volumetric modulated arc therapy (VMAT) and RapidArc (Lagerwaard et al., 2008), has been widely accepted as the standard treatment strategy for many treatment sites in the field of radiation oncology because of the ability to provide quality conformal dose distributions. Advanced RT treatment techniques offer better sparing for the surrounding normal tissues than conventional treatment methods and, therefore, lead to less normal tissue complications (Salama et al., 2006; Studer et al., 2008; Zelefsky et al., 2008) and the possibility for dose escalation to the treatment target (Al-Mamgani et al., 2009). To provide a conformal dose distribution, advanced RT treatment techniques require a more complex treatment planning process utilizing computers in the plan optimization and a more complex beam delivery system, which may utilize a multileaf collimator (MLC) with complex leaf motion sequences. For intensity modulation purposes, the radiation field is created with a number of small beamlets with a typical size of 10×10 or 5×5 mm^2 delivered through many small, irregular, and asymmetric MLC fields, which obscure the relationship between the accelerator monitor unit (MU) setting and the radiation dose received by the patient. When on-line images or other localization and tracking systems are used for treatment guidance and intervention, the beam delivery may become even more complex. Furthermore, patient anatomy heterogeneity, organ motion, and deformation may add additional uncertainties to the actual dose distribution received by the patient. Overall, potential errors associated with advanced RT include dose calculation inaccuracies, plan transfer errors, beam delivery errors, and target localization uncertainties due to patient setup errors and organ motion during the treatment. Considering the serious consequences of these errors, a comprehensive quality assurance (QA) should be performed before and/or during the patient treatment.

The ultimate goal of RT QA is to ensure that the patient will receive the planned dose distribution. However, in clinical routine practice, only a part of the radiotherapy process is usually verified. Techniques, methods, and protocols for RT QA vary from facility to facility in current practices. These methods can be categorized as measurement based, calculation based, and simulation based. The measurement-based method includes measuring and verifying the fluence map for all fields using a film or a two-dimensional (2D) detector array (Grein et al., 2002; Greer and Popescu, 2003; Iori et al., 2007; Pallotta et al., 2007) to measure and compare point doses or a 2D dose distribution using an ion chamber or a 2D detector array in a dummy phantom (Dong et al., 2003; Ma et al., 2003; Dobler et al., 2010). The measurement-based method can verify the treatment plan, dose calculation in the phantom, and plan transfer and delivery. However, it is unable to verify the treatment dose to the patient, especially for a treatment site with heterogeneity in which the absolute dose distribution can be significantly different from that in a homogeneous phantom. In addition, patient setup error, organ motion and deformation, and their effects on the patient's dose distribution cannot be adequately evaluated since no real-time imaging is available to record the variation of the patient geometry during treatment. Other concerns include the requirement of treatment machine time and the off-hour workload to perform these QA measurements. The calculation-based method can be performed in the patient geometry as specified by the patient's computed tomography (CT) scan or other image data set. This technique can be performed at any time using computers, making it less labor intensive. Most calculation-based QA methods currently used clinically employ simple dose/MU calculation algorithms. The effects of patient heterogeneity, setup error, and organ motion on the dose distribution are not included in the calculation. They are frequently used as a secondary MU check tool. In addition, plan transfer and beam delivery cannot be verified by this method. Computerized Monte Carlo simulation can be used as a comprehensive method for RT treatment QA because it can simulate the measurement by performing numerical experiments in the patient geometry and the treatment machine head based on fundamental physics principles. The simulation-based method can perform many tasks that cannot be done with measurement-based methods, such as determining the dose in a patient, and it can provide more information in an accurate manner. Together with measured or recorded beam delivery information, it can verify the plan transfer and treatment delivery as well.

The Monte Carlo method is a well-known computer simulation method utilizing random sampling techniques for particle transport and interaction with material. Its accuracy for radiation dose calculation in a complex and heterogeneous geometry has been extensively benchmarked (Rogers and Bielajew, 1990; Rogers, 2006; Andreo, 1991). Compared with those correction-based and even model-based methods used in conventional algorithms, the Monte Carlo method can model the details of the beam delivery system more precisely because each individual particle is simulated separately in the accelerator head and patient geometry (Rogers et al., 1995). The computation speed of the Monte Carlo method is always a major concern. However, the recent developments in computer techniques and efficiency-enhancing techniques have made it practical for clinical use. Therefore, the Monte Carlo method can be employed as an accurate and comprehensive QA tool for advanced IR applications.

The following sections will discuss Monte Carlo dose calculation, techniques and implementations, and the clinical applications of Monte Carlo-based QA for advanced radiotherapy treatments.

10.2 Techniques and Implementation Methods

10.2.1 Overview of Monte Carlo-Based QA

Monte Carlo dose calculation is performed by transporting the primary radiation particles and their descendants through the patient phantom geometry and recording their energy deposition along the path with a random sampling-based computer simulation process. One can perform Monte Carlo-based patient dose calculation for IR starting with the electrons exiting from the accelerator vacuum window and following them and their descendants through the rest of the accelerator head including the beam

collimation and monitoring devices and then the patient geometry. This method is very inefficient, since most of the particles are blocked by the collimators. Realistically, the radiation beam can be represented by a phase space data set (Rogers et al., 1995) or a source model (Ma et al., 1997) at the plane below the fixed components of the accelerator head. The phase space data is generated by direct Monte Carlo simulation of the patient-independent components in the accelerator head and the source model parameters are derived from the phase space data (Ma et al., 1997; Ma and Jiang, 1999; Chetty et al., 2000; Deng et al., 2000; Fix et al., 2001c; Chetty, 2007) or measured beam data (Jiang et al., 2001). The phase space data and the source model can be used for all patients to be treated on this accelerator. Properties of the particles, including type, energy, direction, and location can be retrieved from the phase space data file or reconstructed from the source model. Monte Carlo-based RT QA can start from the phase space plane or the source model and follow all the particles and their descendants through the patient-dependent accelerator components (e.g., jaws, compensators, and MLC) and the patient geometry. This is referred to as direct Monte Carlo simulation of the beam delivery. Because many particles will be blocked by the beam collimation system, direct Monte Carlo simulation for advanced radiotherapy QA is not efficient. An alterative method is to use the beam intensity map. Instead of simulating a particle through the collimation system, the particle is assigned a weighting factor based on the probability for it to go through the collimation system. The distribution of the weighting factor is defined by the transmission properties of the compensator or the MLC leaf geometry and the sequence of the leaf movements. The particle's weighting factor will determine the particle's contribution to the dose. The intensity map can be obtained from the 2D fluence measurement or reconstruction from the treatment delivery information as discussed in the following sections.

10.2.2 Treatment Delivery Information

As illustrated in Figure 10.1, in addition to the phase space file/source model and the patient geometry as specified by a CT or other imaging data set, the treatment delivery information is required by the Monte Carlo-based QA process. First, the radiation beam and its direction relative to the patient geometry need to be established. The phase space data or source model will be selected based on the actual radiation beam used for the treatment. The beam direction is usually described using the gantry, collimator, and couch angles for a conventional medical linear accelerator. It is sometimes described differently for other treatment machines, such as the nodes for the CyberKnife or the rotation angles for a tomotherapy machine. This information can be found in the treatment plan, record and verify (R&V) system, and the treatment log file, as illustrated in Figure 10.1. The jaw setting or MLC leaf positions, which define the shape of a radiation field or an MLC field segment, and the MUs used for the radiation field are required by the direct Monte Carlo simulation and the intensity map reconstruction program for the entire treatment. The MU information can be found in the

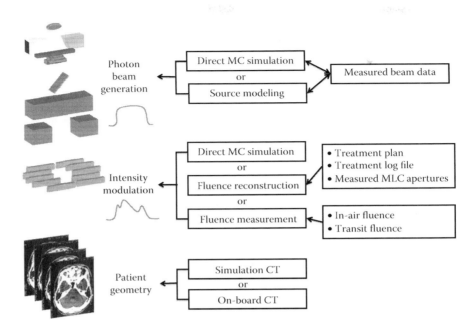

FIGURE 10.1 Treatment beam information used for Monte Carlo-based dose verification.

treatment plan, the R&V record, and/or the machine treatment log file. The jaw setting and MLC leaf position information can be found in the treatment plan or the machine treatment log file, or measured directly with a film, an electronic portal imaging device (EPID), or a 2D detector array.

Simulation-based and calculation-based QA methods can verify the various steps of a radiation treatment procedure. If the treatment field information is taken from the treatment plan, the dosimetry accuracy of the treatment plan can be verified but the plan transfer among different systems involved in the treatment delivery, for example, the treatment planning system, the R&V system, and the accelerator control system, are not ensured. If the treatment field information is taken from the R&V database and/ or the accelerator log file, both the treatment plan and the plan transfer accuracy are verified but the treatment delivery is still not ensured because the actual dose received by the patient also depends on the treatment machine conditions (beam output variation and collimator positional errors) and the patient geometry (setup uncertainty and inter- and intrafractional organ motion). If the treatment field information is derived from the measured portal images of the treatment beam, the treatment delivery can also be verified. If the patient's three-dimensional (3D) image data is available in real time, the actual dose received by the patient can be determined using Monte Carlo simulation and compared with that of the original treatment plan. This will verify the entire radiotherapy process for a particular treatment, which will be the most complete patient-specific treatment QA procedure.

10.2.3 Direct Monte Carlo Simulation

Direct Monte Carlo simulation means to simulate the particle transport and interaction in the patient-dependent field-defining components in the treatment delivery system together with patient dose calculation using the Monte Carlo method. The radiation particles and their descendents are transported in the beam collimation system and the patient geometry until they exhaust their energy or escape to the free space. Geometry parameters of the physical components of the treatment delivery system are determined based on the treatment delivery sequence (jaw setting, MLC leaf positions, and MUs for individual segments) and the geometry of machine components. During the simulation, the beam collimation geometry can be considered to be stationary at any particular moment (treatment segment). This works for "stationary" beam delivery, for example, conventional radiotherapy and step-and-shoot IMRT, and for "dynamic" beam delivery, such as dynamic IMRT and intensity-modulated arc therapy (IMAT)/VMAT/RapidArc. The particles are weighted by the MUs of each treatment segment and, thus, produce "absolute" rather than "relative" dose distributions. Various beam modifiers, such as the block, wedge, and MLC, have been incorporated in the patient simulation geometry for IMRT dose verification studies by several research groups (Li et al., 2000; Fix et al., 2001a,b; Keall et al., 2001; Kim et al., 2001; Liu et al., 2001; Spezi et al., 2001; Verhaegen and Liu, 2001; Aaronson et al., 2002; Siebers et al., 2002; Heath and Seuntjens, 2003; Leal et al., 2003; Fragoso et al., 2009). With direct Monte Carlo simulation, the details of the beam collimation device, such as the tongue-and-groove structure and the leaf end shape of the MLC, can be considered in the patient dose calculation accurately if the treatment delivery system is modeled precisely. However, direct Monte Carlo simulation of the particle transport through the collimator system is time consuming because most particles are blocked before they can reach the patient geometry. To improve the simulation efficiency, a variety of strategies have been applied. One method is to terminate the particle transport in the beam collimation system using high-energy cutoffs for photon and electron particles. Another method is to disregard the secondary Compton scattering photons (Siebers et al., 2002) or the secondary electrons (Tyagi et al., 2007) because they deposit most of their energy locally. Some researchers have also investigated methods to only model the radiation transport at the field edge to account for the effects of partial transmission and photon scattering (Chetty et al., 2000, Fix et al., 2000, 2001c). Significant time saving has been observed using these methods.

10.2.4 Beam Intensity Map for Monte Carlo-Based QA

As mentioned above, instead of performing direct Monte Carlo simulations, the beam intensity maps obtained by direct measurements or analytical reconstruction methods can be used in Monte Carlo patient dose calculation.

10.2.4.1 Beam Intensity Map Measurement

The beam intensity map for each beam angle can be measured using a film, EPID, or 2D detector arrays, such as Map CHECK (Sun Nuclear, Melbourne, FL) and MatriXX (IBA, Schwarzenbruck, Germany). Measurements can be performed at various locations, for example, the beam exiting window, the isocenter plane or any extended distances, pretreatment without a patient in the beam, or during the treatment with a patient in the beam. All the measured intensity maps without a patient in the beam can be used for dose calculation after some corrections are made (e.g., the detector response and back scattering from surrounding materials). The map measured on-line behind the patient is called the transmission (transit) intensity map, which also includes the effect of beam attenuation and scattering by the patient. This effect must be corrected or removed from the transit intensity maps before they can be used for dose calculation. Ray-tracing algorithms are frequently used to derive the

entrance intensity map in front of the patient from the transit images. Overall, a careful calibration of the 2D detector array and proper consideration of the patient scattering and attenuation will improve the accuracy of the patient dose distribution calculated this way. Some investigators have reported the use of the EPID for routine machine QA, pretreatment dose verification, and *in vivo* dosimetry (Boellaard et al., 1998; Kroonwijk et al., 1998; Pasma et al., 1999a,b; van Zijtveld et al., 2006; van Elmpt et al., 2008).

10.2.4.2 Intensity Map Reconstruction

In addition to the use of direct measurements, the beam intensity map can also be reconstructed based on the treatment delivery information. Reconstructing a 2D intensity map based on the recorded MLC leaf sequence and MU information for advanced radiotherapy dose calculation can save a great deal of simulation time compared with direct Monte Carlo simulations and can be more accurate and versatile compared with on-line intensity map measurements because the corrections are patient specific. The beam delivery information, including MU and MLC leaf sequences, can be derived from various sources as described in Section 10.2.2.

The 2D beam intensity map can be built at the midplane of the MLC and can be expressed at the isocenter plane for convenience extending over the maximum treatment field dimensions. The pixel size of the beam intensity map may affect the simulation results of the patient dose distribution. It was shown (Li et al., 2010) that the pixel size should be as small as 0.2 mm. Figure 10.2 demonstrates that the MUs for each MLC segment are added to the beam intensity values of all pixels in the open area of an MLC segment (particle B, D, and H). Variable fractions of the MUs are added to the beam intensity values of all affected pixels, based on the attenuation and scattering effect of the MLC leaves, for example, using the effective transmission

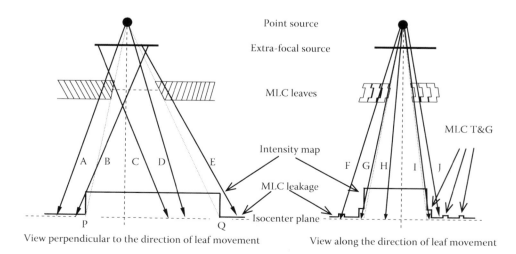

FIGURE 10.2 Diagram showing how the intensity map is built based on the leaf sequence information to include the effects of photon partial transmission through the MLC leaf end and the tongue-and-groove geometry. A through J are photons emitted from the point source and the extra-focal source.

factors of the MLC leaf (particle A, C, E, and J) and the MLC leaf tongue-and-groove structure (particle F, G, and I). The dimensions of the leaf opening are set according to the treatment beam information plus an offset to consider the partial transmission effect of the leaf end or leaf side. The leaf offset can be calculated based on a picket-fence film measurement as described in the literature (Li et al., 2010). The transmission factor for the areas under the photon jaws can be assumed to be zero. The dimensions of the photon jaw opening are set according to the actual treatment setups for each MLC segment. The leaf sequences are read sequentially and the beam intensity map is built for every beam direction. It was shown (Li et al., 2010) that it is essential to carefully consider the effects of the details of the beam delivery system, including the source distribution, MLC thickness, MLC transmission and scattering, MLC tongue-and-groove structure, and the leaf end shape. These factors can affect the doses at various field locations differently due to the complexity of the MLC leaf sequence and, thus, they can result in not only different mean doses but also different dose uncertainties at individual field locations. Their effects are case dependent because the MLC leaf sequence is specific to the treatment plan of a particular patient.

10.2.5 Monte Carlo Dose Calculation

During a Monte Carlo simulation, the spatial source distribution and its effects can be considered by sampling particles from multiple subsources, which include a primary point source at the target location, an extra-focal source to account for photon scattering in the primary collimator and flattering filter (Jiang et al., 2001) and an electron contamination source at a location above the jaws to account for the electron component in the beam (Yang et al., 2004). The MLC collimation can be considered either by a direct Monte Carlo simulation of particle transport and interaction in the MLC leaf geometry or by ray tracing through the MLC leaf geometry to determine whether the particles can go through the MLC opening (Li et al., 2010).

When the beam intensity map was initially implemented for Monte Carlo-based dose calculation, the weight of a phase space particle was altered based on the value of the pixel in the intensity map through which the particle was traveling (Ma et al., 1999, 2000). The gantry/couch/collimator angles were changed automatically after the simulation of each treatment field. A 3D voxelized rectilinear phantom converted from the patient/phantom CT scan (Ma et al., 1999; Chetty et al., 2007) has been used for the dose calculation. The density and material for each voxel are converted based on the CT number. The conversion function for any individual CT scanner is unique and is obtained based on the CT calibration with a standard phantom. To avoid the spatial averaging effects (Ai-Dong et al., 2005), a proper voxel size needs to be used when heterogeneity presents. Since the Monte Carlo method is based on random sampling, the calculated doses always contain statistical uncertainties. The doses calculated using analytical methods do not have statistical uncertainties; however, they usually contain larger systematic

errors in heterogeneous phantoms compared to the Monte Carlo method. By simulating a sufficient number of particles, the statistical uncertainty of a Monte Carlo result can be reduced to such a degree (e.g., <2%) that will be considered clinically insignificant. Such a result will be useful and meaningful clinically. Otherwise, large statistical uncertainties may lead to confusing results as shown by some researchers (Jeraj and Keall, 2000; Jiang et al., 2000).

Another issue that is often encountered in the patient dose calculation is whether the dose should be reported as dose-to-water or dose to the local medium. By default, the initial calculated Monte Carlo dose is dose to the local medium. Compared with the results calculated by conventional methods, significant differences are usually seen since conventional methods are reporting dose-to-water by default. If one converts the dose value calculated by Monte Carlo simulation of millivolt photon beams in a bone voxel to dose-to-water using the mass electron stopping power ratio between bone and water, more significant differences may result in the dose value (Siebers et al., 2000, Ma and Li, 2011). It is still debatable as to whether to use dose-to-water or dose-to-medium for advanced IR using megavoltage photon and electron beams since it depends on the actual situation (e.g., the particle type and treatment depth).

10.3 Clinical Applications

The radiation treatment procedure for a patient includes treatment planning using a treatment planning system (TPS), treatment plan transfer, and verification using an R&V system, and treatment delivery on a clinical treatment machine. Accordingly, as shown in Figure 10.3, the Monte Carlo method can be applied in the TPS commissioning and verification, patient-specific plan QA, and on-line/off-line patient treatment QA.

10.3.1 MC for TPS QA

Since computers were employed in IR treatment planning, many dose calculation algorithms have been developed, which become more accurate but also more sophisticated and computation intensive. Those early correction-based algorithms can compute the patient dose distributions based on the dose data measured in water for a clinical treatment machine with simple corrections for the differences in beam attenuation and scattering between a uniform water phantom irradiated by radiation beams with regular field sizes and the heterogeneous patient geometry irradiated by the actual treatment beams with irregular field shapes. More recent model-based algorithms predict patient dose distributions from primary particle fluence and a dose kernel, which can be derived from direct measurements, analytical calculations, or Monte Carlo simulations. The accuracy of these algorithms as implemented in different TPS depends critically on the details of the beam fluence modeling for individual treatment delivery systems and the kernel variation to account for the effect of particle attenuation and scattering in the patient's heterogeneous geometry. The same applies to the Monte Carlo

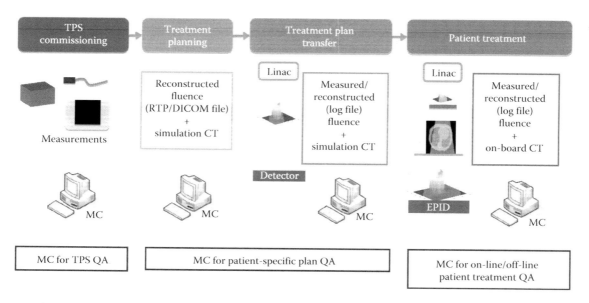

FIGURE 10.3 Clinical applications of Monte Carlo-based QA tool.

dose calculation method, in which the dose calculation accuracy can be affected by the implementation details such as the particle phase space reconstruction and patient geometry setup. To ensure the accuracy of the dose calculation algorithms of a TPS, a series of tests must be performed during the commissioning process before its clinical use, which has been summarized by Anders Ahnesjö and Maria Mania Aspradakis in their review paper on external photon beam dose calculation (Ahnesjo and Aspradakis, 1999).

Source modeling and beam commissioning are important elements of TPS commissioning, especially for model-based algorithms and Monte Carlo techniques. The source can be modeled as a single point source or multiple subsources in which the source size and spatial source distribution can be considered in the dose calculation. One application of Monte Carlo-based QA is to evaluate the accuracy of the source modeling and commissioning by comparing the dose distributions calculated by the TPS and the Monte Carlo code for some representative fields defined by the jaws and/or MLC. Percentage depth dose (PDD) distributions for different square fields, lateral dose profiles at different depths, and beam output factors for different field sizes are usually used for the evaluation.

The dose calculation algorithms implemented in a TPS can be verified by standard phantom measurements. The QA measurements for a typical patient's treatment plan are usually performed at a reference point with an ionization chamber and a plane with a film or a 2D detector array in a phantom. Assorted treatment plans, including 3D conformal RT, IMRT, and dynamic rotational therapy for conventional accelerators or other kinds of treatment plans for various specialty treatment machines such as CyberKnife and tomotherapy, must be tested thoroughly. Though measurement is a good standard for dose verification, it has some major limitations. First, it is still a challenge to measure the 3D dose distribution in a phantom,

although some 3D detector arrays and image-based measurement techniques (e.g., gel dosimetry) have been developed. It is also difficult to mimic the real patient geometry using a heterogeneous phantom. *In vivo* dosimetry is typically performed with detectors on the patient's surface and cavities or inserted invasively in the patient's body. Corrections have to be made to the detector readings to account for the variation of detector response and other factors due to noncalibration conditions. In comparison, the Monte Carlo method can be used to simulate the 3D dose distribution in the patient geometry represented by the patient's CT or in a heterogeneous phantom that can represent the patient geometry accurately. The dose volume histograms (DVH) for the treatment targets and organs at risk (OAR) can be compared together with the isodose distributions between the TPS and the Monte Carlo method. This comparison can also be used to verify the MUs for the patient's treatment plan as calculated by the TPS, which serves as a part of the independent physics check in the treatment planning QA process. Several research groups have published their results of this application for different TPS (Ma et al., 2000; Li et al., 2001; Weber and Nilsson, 2002; Francescon et al., 2003; Wang et al., 2006; Bush et al., 2008; Sarkar et al., 2008).

The details of the beam delivery system and their effects on the patient dose distributions may not be precisely accounted for in the TPS dose calculation due to simplification in the implementation and uncertainties resulting from the commissioning procedure. The TPS usually employs a global correction factor (a fudge factor) based on the mean ratio of measured point doses for a number of test plans to those predicted by the TPS calculation. This correction factor can only reduce but not eliminate the mean dose difference because this point dose ratio is case dependent and position dependent as a result of the leaf sequence for a specific treatment plan. The Monte Carlo method is more advantageous in the verification of the absolute dose and for determination of the correction factor since the mean dose ratio can be evaluated based on the dose in a region rather than at one or more points.

10.3.2 MC for Patient-Specific Plan QA

Dose measurement in a phantom is frequently used to verify the dosimetry accuracy of the phantom plan, plan transfer, and machine delivery for a specific patient. However, this plan QA process only ensures the dosimetry accuracy of the phantom plan, which is calculated based on the patient's treatment plan and the phantom geometry. This process does not really verify the dose to be received by the patient. Monte Carlo dose calculation based on the patient geometry and the treatment delivery information can verify the dose distribution to be received by the patient and, thus, it can be utilized as a patient-specific QA tool. The treatment delivery information obtained at the various steps of the radiotherapy process can be used for the Monte Carlo QA calculation. A good agreement between the patient dose distribution predicted by the TPS and that by Monte Carlo simulation has been observed for various treatment sites and different treatment planning systems. However, significant differences in the doses to the target and the critical structures have been observed for some treatment cases, especially in the region where heterogeneity exists. The dose difference is mainly caused by the systematic error of the simple dose calculation algorithm in the heterogeneous region and inadequate consideration of the effects of the beam delivery system details such as the leaf thickness, the MLC leaf end, and tongue-and-groove structure. The faults in the implementation and programming of the treatment optimization, dose calculation, and leaf sequencing algorithms can, of course, produce a significant dose error. Many researchers have proposed and demonstrated their methods for Monte Carlo-based patient dose verification. For example, Ma et al. (2000), Pawlicki and Ma (2001), Wang et al. (2002), Li et al. (2004), Yang et al. (2005), and Jiang et al. (2006) reconstructed the intensity map using the leaf sequence file or radiotherapy treatment plan (RTP) file from the TPS. Luo et al. (2006) used the treatment log file and R&V record and Lin (Lin et al., 2009) used the measured EPID images for intensity map reconstruction. Their results show that the Monte Carlo method is a useful and practical tool for patient dose verification. Together with measurement or recorded beam delivery information, the treatment delivery and plan transfer can also be verified.

On the basis of the patient's four-dimensional (4D) CT data, dose calculation can be performed for each phase of the breathing cycle. After performing image deformation and registration for all the breathing phases, the dose can be summed up for the entire treatment with a proper weighting factor based on the patient's average breathing cycle and, thus, a 4D dose distribution can be obtained (Keall et al., 2004). The time association between the beam-on time and the breathing phase can be associated with the real-time positioning management (RPM) system or other respiratory tracking signals. Although dosimetric uncertainties still exist due to the uncertainties of the 4D CT images, the correlation between the external surrogate

and internal organ/target motion, and the deformation registration of CT images at various phases, 4D dose calculation using the Monte Carlo method is still valuable for the investigation of the dosimetric effects of the patient's intrafractional motion and organ deformation.

10.3.3 MC for On-Line and Off-Line Treatment QA

Utilizing a film, 2D detector arrays, EPID, and 3D patient imaging for pretreatment or on-line treatment measurement, the Monte Carlo method can be used for pretreatment, and on-line and off-line patient treatment QA for advanced RT. As described in the previous sections, a film, 2D detector arrays, and EPID can be used to measure the entrance fluence map in the absence of the patient or to reconstruct the entrance fluence map based on the transit intensity map measured with the patient in place during the treatment. The entrance fluence map as directly measured by a film, 2D detector arrays, and EPID must be corrected for the detector response variation and other effects due to attenuation and scattering by the measurement device itself. The entrance fluence map can be derived from the transit intensity maps by ray tracing after correcting the beam attenuation and scattering from the patient, or reconstructed based on the MLC leaf positions and MUs recorded by the R&V system or the treatment log file. Subsequently, on-line or off-line dose calculation using the Monte Carlo method can be performed based on this information. Patient CT images acquired with the in-room CT (cone-beam or CT-on-rails) or other 3D imaging systems (ultrasound and MRI) before and during the treatment can be used for the on-line dose calculation and the calculated dose distribution can be used to guide the treatment. Adaptive radiotherapy is possible if the dose calculation and image processing is sufficiently fast. The advances of computer techniques and variance reduction techniques have accelerated the Monte Carlo dose calculation and image processing dramatically in the recent years. The overall computation time for this procedure can be within several minutes. One example is the COMPASS system developed by Boggula et al. (2011), which reconstructed the 3D dose distribution for treatment verification using the fluence maps measured off-line or on-line with a 2D detector array system. Other systems similar to this can be employed for off-line or on-line patient dose verification and for adaptive radiotherapy.

10.4 Conclusions

Advanced RT techniques have received widespread clinical applications for cancer treatments. A comprehensive and practical QA program is essential to ensure the dosimetry accuracy of treatment planning, plan transfer, and treatment delivery for advanced RT. The Monte Carlo method can serve as a useful QA tool in the various steps of the advanced RT process. Techniques and methods to perform Monte Carlo-based QA have been discussed in this chapter. With these methods, a patient's treatment can be verified specifically with confidence. Off-line and on-line dose verification can be performed with on-line imaging and real-time measurement techniques, which, in turn, can provide useful information for the oncologist to make on-line or off-line treatment adjustments. The Monte Carlo calculated dose distributions incorporated into the patient's setup and organ motion/deformation information during the entire treatment course can be used for treatment assessment and outcome analysis, which will be essential to the design and execution of any retrospective and prospective clinical trials utilizing advanced RT treatment techniques.

References

Aaronson, R. F., Demarco, J. J., Chetty, I. J., and Solberg, T. D. 2002. A Monte Carlo based phase space model for quality assurance of intensity modulated radiotherapy incorporating leaf specific characteristics. *Med Phys*, 29, 2952–8.

Ahnesjo, A. and Aspradakis, M. M. 1999. Dose calculations for external photon beams in radiotherapy. *Phys Med Biol*, 44, R99–155.

Ai-Dong, W., Yi-Can, W., Sheng-Xiang, T., and Jiang-Hui, Z. 2005. Effect of CT image-based voxel size on Monte Carlo dose calculation. *Conf Proc IEEE Eng Med Biol Soc*, 6, 6449–51.

Al-Mamgani, A., Heemsbergen, W. D., Peeters, S. T. H., and Lebesque, J. V. 2009. Role of intensity-modulated radiotherapy in reducing toxicity in dose escalation for localized prostate cancer. *Int J Radiat Oncol Biol Phys*, 73, 685–91.

Andreo, P. 1991. Monte Carlo techniques in medical radiation physics. *Phys Med Biol*, 36, 861–920.

Boellaard, R., Essers, M., Van Herk, M., and Mijnheer, B. J. 1998. New method to obtain the midplane dose using portal *in vivo* dosimetry. *Int J Radiat Oncol Biol Phys*, 41, 465–74.

Boggula, R., Jahnke, L., Wertz, H., Lohr, F., and Wenz, F. 2011. Patient-specific 3D pretreatment and potential 3D online dose verification of Monte Carlo-calculated IMRT prostate treatment plans. *Int J Radiat Oncol Biol Phys*, 81, 1168–75.

Boyer, A. L. and Yu, C. X. 1999. Intensity-modulated radiation therapy with dynamic multileaf collimators. *Semin Radiat Oncol*, 9, 48–59.

Burman, C., Chui, C. S., Kutcher, G., Leibel, S., Zelefsky, M., Losasso, T., Spirou, S. et al. 1997. Planning, delivery, and quality assurance of intensity-modulated radiotherapy using dynamic multileaf collimator: A strategy for large-scale implementation for the treatment of carcinoma of the prostate. *Int J Radiat Oncol Biol Phys*, 39, 863–73.

Bush, K., Townson, R., and Zavgorodni, S. 2008. Monte Carlo simulation of RapidArc radiotherapy delivery. *Phys Med Biol*, 53, N359–70.

Chetty, I. 2007. Virtual source modelling in Monte Carlo-based clinical dose calculations: Methods and issues associated with their development and use. *Radiother Oncol*, 84, S38–9.

Chetty, I., Demarco, J. J., and Solberg, T. D. 2000. A virtual source model for Monte Carlo modeling of arbitrary intensity distributions. *Med Phys*, 27, 166–72.

Chetty, I. J., Curran, B., Cygler, J. E., Demarco, J. J., Ezzell, G., Faddegon, B. A., Kawrakow, I. et al. 2007. Report of the AAPM Task Group No. 105: Issues associated with clinical implementation of Monte Carlo-based photon and electron external beam treatment planning. *Med Phys*, 34, 4818–53.

Deng, J., Jiang, S. B., Kapur, A., Li, J., Pawlicki, T., and Ma, C. M. 2000. Photon beam characterization and modelling for Monte Carlo treatment planning. *Phys Med Biol*, 45, 411–27.

Dobler, B., Streck, N., Klein, E., Loeschel, R., Haertl, P., and Koelbl, O. 2010. Hybrid plan verification for intensity-modulated radiation therapy (IMRT) using the 2D ionization chamber array I'mRT MatriXX—A feasibility study. *Phys Med Biol*, 55, N39–55.

Dong, L., Antolak, J., Salehpour, M., Forster, K., O'Neill, L., Kendall, R., and Rosen, I. 2003. Patient-specific point dose measurement for IMRT monitor unit verification. *Int J Radiat Oncol Biol Phys*, 56, 867–77.

Fix, M. K., Keller, H., Ruegsegger, P., and Born, E. J. 2000. Simple beam models for Monte Carlo photon beam dose calculations in radiotherapy. *Med Phys*, 27, 2739–47.

Fix, M. K., Manser, P., Born, E. J., Mini, R., and Ruegsegger, P. 2001a. Monte Carlo simulation of a dynamic MLC based on a multiple source model. *Phys Med Biol*, 46, 3241–57.

Fix, M. K., Manser, P., Born, E. J., Vetterli, D., Mini, R., and Ruegsegger, P. 2001b. Monte Carlo simulation of a dynamic MLC: Implementation and applications. *Z Med Phys*, 11, 163–70.

Fix, M. K., Stampanoni, M., Manser, P., Born, E. J., Mini, R., and Ruegsegger, P. 2001c. A multiple source model for 6 MV photon beam dose calculations using Monte Carlo. *Phys Med Biol*, 46, 1407–27.

Fragoso, M., Kawrakow, I., Faddegon, B. A., Solberg, T. D., and Chetty, I. J. 2009. Fast, accurate photon beam accelerator modeling using BEAMnrc: A systematic investigation of efficiency enhancing methods and cross-section data. *Med Phys*, 36, 5451–66.

Francescon, P., Cora, S., and Chiovati, P. 2003. Dose verification of an IMRT treatment planning system with the BEAM EGS4-based Monte Carlo code. *Med Phys*, 30, 144–57.

Greer, P. B. and Popescu, C. C. 2003. Dosimetric properties of an amorphous silicon electronic portal imaging device for verification of dynamic intensity modulated radiation therapy. *Med Phys*, 30, 1618–27.

Grein, E. E., Lee, R., and Luchka, K. 2002. An investigation of a new amorphous silicon electronic portal imaging device for transit dosimetry. *Med Phys*, 29, 2262–8.

Heath, E. and Seuntjens, J. 2003. Development and validation of a BEAMnrc component module for accurate Monte Carlo modelling of the Varian dynamic millennium multileaf collimator. *Phys Med Biol*, 48, 4045–63.

Iori, M., Cagni, E., Nahum, A. E., and Borasi, G. 2007. IMAT–SIM: A new method for the clinical dosimetry of intensity-modulated arc therapy (IMAT). *Med Phys*, 34, 2759–73.

Jang, S. Y., Liu, H. H., Wang, X., Vassiliev, O. N., Siebers, J. V., Dong, L., and Mohan, R. 2006. Dosimetric verification for intensity-modulated radiotherapy of thoracic cancers using experimental and Monte Carlo approaches. *Int J Radiat Oncol Biol Phys,* 66, 939–48.

Jeraj, R. and Keall, P. 2000. The effect of statistical uncertainty on inverse treatment planning based on Monte Carlo dose calculation. *Phys Med Biol,* 45, 3601–13.

Jiang, S. B., Boyer, A. L., and Ma, C. M. 2001. Modeling the extra-focal radiation and monitor chamber backscatter for photon beam dose calculation. *Med Phys,* 28, 55–66.

Jiang, S. B., Pawlicki, T., and Ma, C. M. 2000. Removing the effect of statistical uncertainty on dose–volume histograms from Monte Carlo dose calculations. *Phys Med Biol,* 45, 2151–61.

Keall, P. J., Siebers, J. V., Arnfield, M., Kim, J. O., and Mohan, R. 2001. Monte Carlo dose calculations for dynamic IMRT treatments. *Phys Med Biol,* 46, 929–41.

Keall, P. J., Siebers, J. V., Joshi, S., and Mohan, R. 2004. Monte Carlo as a four-dimensional radiotherapy treatment-planning tool to account for respiratory motion. *Phys Med Biol,* 49, 3639–48.

Kim, J. O., Siebers, J. V., Keall, P. J., Arnfield, M. R., and Mohan, R. 2001. A Monte Carlo study of radiation transport through multileaf collimators. *Med Phys,* 28, 2497–506.

Kroonwijk, M., Pasma, K. L., Quint, S., Koper, P. C., Visser, A. G., and Heijmen, B. J. 1998. *In vivo* dosimetry for prostate cancer patients using an electronic portal imaging device (EPID); demonstration of internal organ motion. *Radiother Oncol,* 49, 125–32.

Lagerwaard, F. J., Verbakel, W. F. A. R., Van der Hoom, E., Slotman, B. J., and Senan, S. 2008. Volumetric modulated arc therapy (RapidArc) for rapid, non-invasive stereotactic radiosurgery of multiple brain metastases. *Int J Radiat Oncol Biol Phys,* 72, S519–30.

Leal, A., Sanchez-Doblado, F., Arrans, R., Capote, R., Carrasco, E., Lagares, J. I., Rosello, J., Perucha, M., Molina, E., and Terron, J. A. 2003. Influence of the MLC leaf width on the dose distribution: A Monte Carlo study. *Radiother Oncol,* 68, S97.

Li, J. S., Lin, T., Chen, L., Price, R. A., Jr., and Ma, C. M. 2010. Uncertainties in IMRT dosimetry. *Med Phys,* 37, 2491–500.

Li, J. S., Pawlicki, T., Deng, J., Jiang, S. B., Mok, E., and Ma, C. M. 2000. Validation of a Monte Carlo dose calculation tool for radiotherapy treatment planning. *Phys Med Biol,* 45, 2969–85.

Li, J. S., Wang, L., Chen, L., Yang, J., and Ma, C. M. 2004. Monte Carlo dose verification for IMRT plan delivered using micro-multileaf collimators. *Med Phys,* 31, 1844.

Li, X. A., Ma, L., Naqvi, S., Shih, R., and Yu, C. 2001. Monte Carlo dose verification for intensity-modulated arc therapy. *Phys Med Biol,* 46, 2269–82.

Lin, M. H., Chao, T. C., Lee, C. C., Tung, C. J., Yeh, C. Y., and Hong, J. H. 2009. Measurement-based Monte Carlo dose calculation system for IMRT pretreatment and on-line transit dose verifications. *Med Phys,* 36, 1167–75.

Liu, H. H., Verhaegen, F., and Dong, L. 2001. A method of simulating dynamic multileaf collimators using Monte Carlo

techniques for intensity-modulated radiation therapy. *Phys Med Biol,* 46, 2283–98.

Luo, W., Li, J., Price, R. A. Jr., Chen, L., Yang, J., Fan, J., Chen, Z., Mcneeley, S., Xu, X., and Ma, C. M. 2006. Monte Carlo based IMRT dose verification using MLC log files and R/V outputs. *Med Phys,* 33, 2557–64.

Ma, C. M., Faddegon, B. A., Rogers, D. W. O., and Mackie, T. R. 1997. Accurate characterization of Monte Carlo-calculated electron beams for radiotherapy. *Med Phys,* 24, 401–16.

Ma, C. M. and Jiang, S. B. 1999. Monte Carlo modelling of electron beams from medical accelerators. *Phys Med Biol,* 44, R157–89.

Ma, C. M., Jiang, S. B., Pawlicki, T., Chen, Y., Li, J. S., Deng, J., and Boyer, A. L. 2003. A quality assurance phantom for IMRT dose verification. *Phys Med Biol,* 48, 561–72.

Ma, C. M. and Li, J. 2011. Dose specification for radiation therapy: Dose to water or dose to medium? *Phys Med Biol,* 56, 3073–89.

Ma, C. M., Mok, E., Kapur, A., Pawlicki, T., Findley, D., Brain, S., Forster, K., and Boyer, A. L. 1999. Clinical implementation of a Monte Carlo treatment planning system. *Med Phys,* 26, 2133–43.

Ma, C. M., Pawlicki, T., Jiang, S. B., Li, J. S., Deng, J., Mok, E., Kapur, A., Xing, L., Ma, L., and Boyer, A. L. 2000. Monte Carlo verification of IMRT dose distributions from a commercial treatment planning optimization system. *Phys Med Biol,* 45, 2483–95.

Pallotta, S., Marrazzo, L., and Bucciolini, M. 2007. Design and implementation of a water phantom for IMRT, arc therapy, and tomotherapy dose distribution measurements. *Med Phys,* 34, 3724–31.

Pasma, K. L., Dirkx, M. L., Kroonwijk, M., Visser, A. G., and Heijmen, B. J. 1999a. Dosimetric verification of intensity modulated beams produced with dynamic multileaf collimation using an electronic portal imaging device. *Med Phys,* 26, 2373–8.

Pasma, K. L., Kroonwijk, M., Quint, S., Visser, A. G., and Heijmen, B. J. 1999b. Transit dosimetry with an electronic portal imaging device (EPID) for 115 prostate cancer patients. *Int J Radiat Oncol Biol Phys,* 45, 1297–303.

Pawlicki, T. and Ma, C. M. 2001. Monte Carlo simulation for MLC-based intensity-modulated radiotherapy. *Med Dosim,* 26, 157–68.

Purdy, J. A. 1996. Intensity-modulated radiation therapy. *Int J Radiat Oncol Biol Phys,* 35, 845–6.

Rogers, D. W. 2006. Fifty years of Monte Carlo simulations for medical physics. *Phys Med Biol,* 51, R287–301.

Rogers, D. W., Faddegon, B. A., Ding, G. X., Ma, C. M., We, J., and Mackie, T. R. 1995. BEAM: A Monte Carlo code to simulate radiotherapy treatment units. *Med Phys,* 22, 503–24.

Rogers, D.W.O. and Bielajew, A. F. 1990. Monte Carlo techniques of electron and photon transport for radiation dosimetry. In: Kase, K., Bjarngard, B., and Attix, F. H. (eds.) *Dosimetry of Ionizing Radiation.* Academic, New York.

Salama, J. K., Mundt, A. J., Roeske, J., and Mehta, N. 2006. Preliminary outcome and toxicity report of extended-field, intensity-modulated radiation therapy for gynecologic malignancies. *Int J Radiat Oncol Biol Phys*, 65, 1170–6.

Sarkar, V., Stathakis, S., and Papanikolaou, N. 2008. A Monte Carlo model for independent dose verification in serial tomotherapy. *Technol Cancer Res Treat*, 7, 385–92.

Siebers, J. V., Keall, P. J., Kim, J. O., and Mohan, R. 2002. A method for photon beam Monte Carlo multileaf collimator particle transport. *Phys Med Biol*, 47, 3225–49.

Siebers, J. V., Keall, P. J., Nahum, A. E., and Mohan, R. 2000. Converting absorbed dose to medium to absorbed dose to water for Monte Carlo based photon beam dose calculations. *Phys Med Biol*, 45, 983–95.

Spezi, E., Lewis, D. G., and Smith, C. W. 2001. Monte Carlo simulation and dosimetric verification of radiotherapy beam modifiers. *Phys Med Biol*, 46, 3007–29.

Studer, G., Graetz, K. W., and Glanzmann, C. 2008. Outcome in recurrent head–neck cancer treated with salvage-IMRT. *Radiat Oncol*, 3, 43.

Tyagi, N., Moran, J. M., Litzenberg, D. W., Bielajew, A. F., Fraass, B. A., and Chetty, I. J. 2007. Experimental verification of a Monte Carlo-based MLC simulation model for IMRT dose calculation. *Med Phys*, 34, 651–63.

van Elmpt, W., Mcdermott, L., Nijsten, S., Wendling, M., Lambin, P., and Mijnheer, B. 2008. A literature review of electronic portal imaging for radiotherapy dosimetry. *Radiother Oncol*, 88, 289–309.

van Zijtveld, M., Dirkx, M. L., de Boer, H. C., and Heijmen, B. J. 2006. Dosimetric pre-treatment verification of IMRT using an EPID; clinical experience. *Radiother Oncol*, 81, 168–75.

Verhaegen, F. and Liu, H. H. 2001. Incorporating dynamic collimator motion in Monte Carlo simulations: An application in modelling a dynamic wedge. *Phys Med Biol*, 46, 287–96.

Wang, L., Li, J., Paskalev, K., Hoban, P., Luo, W., Chen, L., Mcneeley, S., Price, R., and Ma, C. 2006. Commissioning and quality assurance of a commercial stereotactic treatment-planning system for extracranial IMRT. *J Appl Clin Med Phys*, 7, 21–34.

Wang, L., Yorke, E., and Chui, C. S. 2002. Monte Carlo evaluation of 6 MV intensity modulated radiotherapy plans for head and neck and lung treatments. *Med Phys*, 29, 2705–17.

Webb, S. 1998. Intensity-modulated radiation therapy: Dynamic MLC (DMLC) therapy, multisegment therapy and tomotherapy. An example of QA in DMLC therapy. *Strahlenther Onkol*, 174(Suppl 2), 8–12.

Weber, L. and Nilsson, P. 2002. Verification of dose calculations with a clinical treatment planning system based on a point kernel dose engine. *J Appl Clin Med Phys*, 3, 73–87.

Yang, J., Li, J., Chen, L., Price, R., Mcneeley, S., Qin, L., Wang, L., Xiong, W., and Ma, C. M. 2005. Dosimetric verification of IMRT treatment planning using Monte Carlo simulations for prostate cancer. *Phys Med Biol*, 50, 869–78.

Yang, J., Li, J. S., Qin, L., Xiong, W., and Ma, C. M. 2004. Modelling of electron contamination in clinical photon beams for Monte Carlo dose calculation. *Phys Med Biol*, 49, 2657–73.

Zelefsky, M. J., Yamada, Y., Kollmeier, M. A., Shippy, A. M., and Nedelka, M. A. 2008. Long-term outcome following three-dimensional conformal/intensity-modulated external-beam radiotherapy for clinical stage T3 prostate cancer. *Eur Urol*, 53, 1172–9.

11

Electrons: Clinical Considerations and Applications

Joanna E. Cygler
The Ottawa Hospital Cancer Centre

George X. Ding
Vanderbilt University School of Medicine

11.1 Introduction: Rationale for Monte Carlo-Based Treatment Planning Systems for Electron Beams

This chapter covers the practical aspects of Monte Carlo (MC) implementation, the challenges encountered, the main application areas, comparison with conventional algorithms, and so on. Ideally, the discussion will include implementation of commercial packages and research MC packages.

The goal and at the same time the greatest challenge of radiation therapy is to kill the tumor and to spare healthy tissues that do not require treatment. Lots of effort have been devoted to technological developments that allow achieving that goal and improve treatment outcomes. As a result, modern radiation therapy is a complex process that consists of multiple steps, with each step containing inherent uncertainties and assumptions. It has been stated that the uncertainty of dose delivered to the patient should be less than 5% (ICRU-29 1978; Brahme 1984; Papanikolaou et al. 2004). This in turn requires that the uncertainty of dose calculation be less than 2% (Papanikolaou et al. 2004). Most of the commercial treatment planning systems (TPS) cannot provide this level of accuracy for all clinically encountered situations. This has been especially true for electron

beams for which traditional dose calculation algorithms fail in many cases (Hogstrom et al. 1981, 1984, 1989; Hogstrom and Almond 1983; Cygler et al. 1987; Ding et al. 2005). In principle, MC gives an accurate answer to within the statistical uncertainty, as there are no significant approximations (except in beam models) involved in the calculations, no approximate scaling of dose kernels is needed and the electron transport is fully modeled. Both treatment machines and patient geometries can be modeled accurately. All types of inhomogeneities are properly handled in the calculations. It has been well documented that MC calculations can be very accurate (Ma et al. 1997; Zhang et al. 1999; Reynaert et al. 2007). In spite of that, for many years, MC calculations were available only for research purposes and used in more advanced departments for special projects or as a quality assurance tool for commercial non-MC systems. This has been mostly due to the long calculation times for MC algorithms. That problem is now overcome with the arrival of modern fast computers and efficient calculation algorithms. Finally, at the dawn of the twenty-first century, state-of-the-art MC dose calculations for electron beams were implemented in commercial TPS. Such systems are making their way into the cancer clinics and are becoming a routine tool used for radiotherapy patients. This is truly a major breakthrough in radiotherapy, since no other existing algorithm can calculate the dose from electron beams accurately for all treatment situations. There have been several papers describing commissioning and clinical implementation of MC-based TPS for electron beams (Cygler et al. 2004; Ding et al. 2005, 2006; Popple et al. 2006; Fragoso et al. 2008; Edimo et al. 2009; Ali et al. 2011). Recognizing that MC-based TPS are becoming a reality, the American Association of the Physicists in Medicine, AAPM, formed a special task group, TG-105, to address issues related to clinical implementation of such systems. The TG-105 report provides a useful framework for commissioning of MC-based TPS (Chetty et al. 2007).

11.1.1 Advantages of MC versus Pencil Beam Algorithm

Until recently, the commercial TPS for electron beams were based on pencil beam (PB) algorithms. Although at the time of its implementation, the PB approach was a big step forward, it has been known to have very serious limitations. PB-based planning system cannot really handle monitor unit (MU) calculations for arbitrary source-to-surface distance (SSD) values when only one machine of a single SSD is configured in the system. This is due to the fact, that unlike in the case of high-energy photon beams, the electron beam source is not a point, but rather an extended source. The virtual position of this source in the accelerator head depends on beam energy and field size and so the simple inverse square law, ISL, applicable in case of point source does not work for electron beams (Khan et al. 1991; Cygler et al. 1997). To handle monitor unit calculations for extended SSDs in PB systems, one has to configure several virtual machines, one for each SSD. This is not the case in MC treatment planning systems. In such systems, users need to install just a single virtual machine of a

standard SSD for each beam energy to achieve monitor units and dose distributions within clinically acceptable accuracy not only at a standard, but also at any extended SSD (>110 cm) (Cygler et al., 2004; Ding et al., 2006).

In addition, the dose distributions in heterogeneous media have large errors for the complex geometries, frequently encountered in human anatomy. This has been well documented in the literature (Cygler et al. 1987; Hogstrom et al. 1989; Starkschall et al. 1991; Shiu and Hogstrom 1991; Keall and Hoban 1996; Ding et al. 2005; Fragoso et al. 2008). Figure 11.1 shows one of the most striking examples of the failure of PB algorithm (Ding et al. 2005). It is the "trachea and spine" phantom used in electron beam commissioning by Cygler et al. (1987) and Ding et al. (2005, 2006). The phantom consists of overlying low- and high-density heterogeneities, as frequently encountered in the human body. In a solid water slab of overall thickness of 6.2 cm, there is an air pipe of 2.5 cm diameter and 10 cm length and about half centimeter below it, there are four 1 cm thick bone cylinders of 2.5 cm diameter each. Figure 11.1 presents the dose profiles measured and calculated at various depths behind the "trachea and spine" phantom for a 9 MeV beam. As one can see, the PB cannot even remotely resolve the shape of the hot and cold spots for this complex geometry, while MC can do it very well. The MC-measured disagreement that we see here in the peaks is of about 4% and can be attributed to the fact that the measurement resolution was 1 mm and the MC calculation voxel was 3.9 mm. The measurement and calculation resolutions could not be the same, because in this software (Nucletron), users have no control over the calculation voxel size. For 2 mm calculation voxel size, one sees excellent agreement with the measured data (Ding et al. 2005, 2006). In addition to their limited accuracy, PB algorithms require relatively long calculation times for multislice

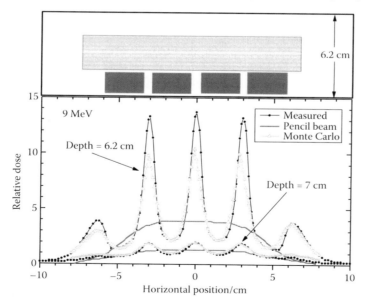

FIGURE 11.1 Comparison between measured and calculated cross-beam dose profiles at various depths for a 9 MeV beam incident on the trachea and spine phantom with a 10 × 10 cone and SSD = 110 cm. (From *Int J Radiat Oncol Biol Phys*, 63 (2), Ding, G. X. et al. A comparison of electron beam dose calculation accuracy between treatment planning systems using either a pencil beam or a Monte Carlo algorithm, 622–33, Copyright 2005, with permission from Elsevier.)

anatomies. For example, on a standard Eclipse TPS work station (with 8 Xeon processors 2.8 MHz), the MC calculations (with 2% statistical uncertainty) took 4 min whereas the PB calculations took 21 min for the same single beam of 9 MeV electron beam and a 10×10 cm^2 field size. The calculation speed gain for MC calculations is partially due to more efficient algorithms and to employing multiple processors of the work station (Eclipse). The PB algorithm calculations for a single beam on Eclipse utilize a single processor.

11.2 Research and Commercial Monte Carlo Treatment Planning Systems

11.2.1 Meeting the Challenges: The OMEGA Project

The MC method is regarded as the "gold standard" in dose calculation, since it is capable of accurately predicting the dose distributions under almost all circumstances. It has played a significant role in electron treatment planning because it provides accurate dose predictions even in complex three-dimensional heterogeneous geometries, where the model-based dose calculations algorithms, such as PB, have shown significant limitations. Recognizing the requirement of accurate patient treatment planning, in 1990 the NIH funded the OMEGA research project. It was a collaborative project between National Research Council of Canada and the University of Wisconsin to develop a fully 3D electron beam TPS based on MC simulation techniques to calculate the dose to patients. The result of this project was a general purpose MC code, BEAM (Rogers et al. 1995), to simulate radiation beams from radiotherapy machines, including high-energy photon and electron beams, Co-60 beams and kilovoltage units. The BEAM code is currently used worldwide for research. It is also employed to generate and verify beam models used in commercial TPS.

11.2.2 Methods to Speed Up the MC Calculations

Computing time used to be considered a major drawback of the MC technique. In order to use MC in routine clinical treatment planning, special methods were developed to speed up MC calculations for these applications.

One of such methods, called "a voxel based electron beam Monte Carlo algorithm," introduced some simplifications and approximations into the transport algorithm (Fippel et al. 1997; Kawrakow 1997, 2001). These methods have been implemented into commercial TPS for electron dose calculations (Nucletron and Elekta Software) (Cygler et al. 2004; Ali et al. 2011).

The other technique developed to speed up the electron beam calculations is called the macro Monte Carlo method (MMC) (Neuenschwander and Born 1992; Neuenschwander et al. 1995). The key features of the MMC transport model are: (a) Conventional Monte Carlo simulations of electron transport are performed in well-defined local geometries, namely, spheres ("kugels"). The result of these calculations is a library of probability distribution

functions (PDFs) of particles emerging from the "kugels." The PDFs used in a commercial TPS (Eclipse) were generated in extensive precalculations by employing the EGSnrc (Kawrakow and Rogers, 2002) software to simulate the transport of vertically incident electrons of variable energies through macroscopic spheres of various sizes and five materials (air, lung, water, Lucite, and solid bone). They were calculated only once for a variety of clinically relevant energies and the above-mentioned five materials; (b) patient-specific MC calculations are performed in a global geometry. Electrons are transported through the patient in macroscopic steps based on the PDFs generated in the above described local calculations. As already mentioned, this method has been implemented in a commercial TPS by Varian (Ding et al. 2006; Popple et al. 2006).

Another method to speed up MC dose calculations is to use source (beam) models instead of direct particle transport through the machine head.

11.2.3 EGS-Based MC Research Package

MC particle transport techniques were first introduced in the 1940s and have since evolved into many different areas of applications including radiotherapy physics. In the last several decades, the MC technique has become ubiquitous in medical physics (Rogers 2006). There are many MC codes available for the applications in radiotherapy (Nelson et al. 1985; Seltzer 1991; Sempau et al. 1997; Briesmeister 2000; Kawrakow 2000; GEANT 2006). BEAM/EGS4, a general purpose MC user code for simulation of radiotherapy beams from treatment units was made available in 1995 (Rogers et al. 1995). This code makes it easier to simulate radiotherapy units and has been widely used for the study of photon and electron beams from accelerators. Owing to the easy interface for using the BEAM/EGS4 research package, the BEAM code has been used extensively to characterize therapy beams, including megavoltage electron and photon beams from linear accelerators and kilovoltage photon beams from x-ray units (Ding 1995; Ding and Rogers 1995; Zhang et al. 1998; Zhang et al. 1999; Francescon et al. 2000, 2009; Sheikh-Bagheri and Rogers 2002a,b; Keall et al. 2003; Ding et al. 2006; Jarry et al. 2006; Bazalova and Verhaegen 2007; Pimpinella et al. 2007; Francescon et al. 2008; Bazalova et al. 2009; Iaccarino et al. 2011; Pasciuti et al. 2011). Not only is the BEAM/EGS4 (now BEAMnrc/EGSnrc) research package capable of simulating the treatment beams, but also uses the simulated realistic beams to calculate dose to a patient using the information from CT images. MC research packages have also been used in validating the accuracy of dose calculations of commercial TPS (Ma et al. 2000; Wieslander and Knoos 2007; Ali et al. 2011).

11.2.4 Commercial MC TPS

MC calculations of dose distribution in a patient can be divided into three steps (Chetty et al. 2007):

1. Particle transport through the top of machine head (patient-independent step).

2. Particle transport through the patient-specific part of the machine.

3. Dose calculations in the patient.

In research, for particle transport through the machine head, BEAM or BEAMnrc are frequently used. BEAM generates a particle phase-space file, which is used as the input file to steps 2 and 3, dose calculations in a patient. In step 2, codes such as VMC, XVMC, VMC++ are used.

Direct simulation of particles in the machine head is not used in commercial TPS. The main reason is that simulations of the direct particle transport through the machine head strongly depend on knowing accurately the machine head details (materials, dimensions, etc.), which may not be easily available to users (Schreiber and Faddegon 2005).

So, typically commercial MC-based TPS use source models, which describe the beams with several fitting parameters derived from the fully simulated phase-space file and/or measured dose profiles.

Presently, there are three commercial MC-based TPS available: Nucletron Oncentra MasterPlan (VMC++), Varian Eclipse (eMC), and Elekta Software XiO® (XVMC).

11.2.4.1 Nucletron VMC++

In 2002, Nucletron* was the first in the world to release a commercial MC-based TPS for electron beams. The MC option is available in the Dose Calculation Module, DCM, of Theraplan Plus and Oncentra MasterPlan. The beam transport is divided into two components: treatment independent and treatment dependent (*Oncentra MasterPlan—Physics Reference Manual*). The treatment-independent component, called phase-space engine, handles the particle transport through the linac head down to the lowest patient-independent collimation level. The patient-dependent component, called the dose engine, is based on the VMC++ algorithm (Kawrakow 2000). The accelerator beam model in Nucletron TPS was developed by Traneus and colleagues (Traneus et al. 2001). It consists of the coupled multisource beam model. It is based on five parameters and requires a limited set of clinic-measured data to configure the beam model. This beam model is flexible enough to support current collimation types such as applicators with optional inserts, variable trimmer machines, and potential multileaf collimator (MLC), applications in the future. The modeling of a particular beam starts upstream of the uppermost changeable/movable collimating element with a source phase space (SPS). The SPS is propagated through the treatment head to an exit phase space (EPS) plane located at the lowest collimating element thereby defining the interface between the phase space and the dose engines, see Figure 11.2.

For a treatment unit with fixed applicators with optional patient-specific cutouts, the block collimators and possible multileaf collimator have fixed positions per energy/applicator combination. The SPS for these applicators is prepropagated as part of the beam data-handling process to an EPS plane located just in front of the insert position. The in-patient dose calculation is then driven

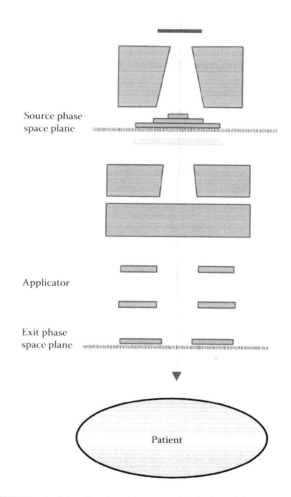

FIGURE 11.2 Linac head and beam model. (Adapted from *Oncentra MasterPlan—Physics Reference Manual 92.724ENG-00.*)

by sampling from a parameterized description of this EPS. In this way, the transport through the treatment head does not have to be performed for every dose calculation. For more general treatments that are not using fixed collimation setups (e.g., MLC fields) the dose calculation can be driven by sampling directly from the SPS.

To by-pass time-consuming in-collimator transport during dose planning, scattering effects in the applicator are accounted for by using precalculated collimator scatter kernels, describing the coupled angular and energy probability distributions of indirect electrons coming from applicator elements.

Therefore, for the Nucletron planning system, the accelerator model provides the EPS file, which serves as the input file for the in-phantom/patient calculations performed with VMC++ (Kawrakow et al. 1996; Kawrakow 1997, 2001).

The code can handle fixed applicators with optional, arbitrary inserts of any shape. It also can handle the variable size fields defined by an applicator like DEVA (Siemens). The code always calculates the absolute dose per MU, cGy/MU.

Theraplan Plus and Oncentra planning systems can handle all major accelerators available on the market. The beam modeling is performed by Nucletron.

---

* It actually was MDS Nordion who later sold the TPS to Nucletron.

The user interface for the Nucletron planning systems does not allow for selection of any calculation parameters, except the number of particle histories/cm². The larger this number, the lower the statistical uncertainty in the dose distribution. The user has no control over the calculation's voxel size, which is automatically assigned by the system. The voxel size assignment is based on the CT anatomy volume and the total number of allowed voxels being 800,000. So for smaller CT volumes, one achieves smaller calculation voxel sizes. The larger overall volumes result in being divided into larger calculation voxels. A database of 21 materials is used in segmentation of patient CT anatomy.

In Theraplan Plus, the user can select to calculate and report dose-to-medium or dose-to-water. Unfortunately, the user does not have this choice in Oncentra, which reports dose to water.

11.2.4.2 Varian eMC

The electron Monte Carlo (eMC) algorithm employed by Eclipse is an implementation of the MMC method (Neuenschwander et al. 1995) for calculation of dose from high-energy electron beams. The algorithm consists of: (a) an electron beam phase-space model describing the electrons that emerge from the treatment head of the linear accelerator. This model is based on precalculated data for a machine type and configured using measured beam data; (b) electron transport/dose deposition in the patient is calculated using the MMC method.

The patient's anatomy is based on CT images. The CT image volume is first converted into a mass density volume with a user-defined resolution (0.1–0.5 cm), applying appropriate CT-to-mass density conversion factors. The resulting density volume is then scanned for heterogeneities. To each voxel of the density-volume, a sphere index is assigned. A voxel of the density volume is considered to be part of a heterogeneous volume if the density ratio of the voxel and its neighbors exceed a limit (typically 1.5). If the densities in both voxels are below a threshold (typically 0.05 g/cm³), the ratio is not evaluated. The material assigned to each sphere depends on the average mass density within the sphere. If the average mass density of a sphere is exactly equal to the mass density of one of the preset materials, the preset material is selected for the sphere. If the mass density of a sphere is between two preset materials, the material is randomly selected for these two materials each time a particle enters the sphere. The probability for a material to be selected is proportional to the closeness of a sphere's average mass density to the mass density of the material. There are five preset different materials (air, lung, water, Lucite, and bone) in the MMC calculation database (*The Reference Guide for Eclipse Algorithms Version 6.5* (P/N B401653R01M). For example, if the average density within a sphere is 1.12, there is a 12/19 chance that the material is selected to be Lucite and a 7/19 chance that is selected to be water. Unlike conventional TPS, the reported doses are the doses calculated to the material (Ding et al. 2006) and not the dose-to-water of different densities. However, the user can define other calculation parameters, such as number of particle histories or required statistical uncertainty of calculation, voxel size, or whether to apply isodose smoothing.

11.2.4.3 Elekta Software XiO eMC

The XiO eMC module is based on XVMC (Kawrakow et al. 1996; Fippel 1999; Fippel et al. 1999). The beam model has been developed by Elekta Software and it consists of a weighted combination of particle sources (direct and indirect electrons and photons).

Beam modeling based on user measurements is performed by Elekta Software. As mentioned above, the dose calculation in the phantom/patient is based on XVMC. The user interface in XiO allows for full control over the dose calculation parameters. The user can define calculation voxel dimensions and can control the statistical uncertainty in the dose distributions by setting either the maximum number of particle histories or the goal MRSU (the mean relative statistical uncertainty) defined as

$$\text{MRSU} = \sqrt{\frac{1}{N} \sum_{\substack{D(\vec{r_i}) \geq aD_{\max}}}^{N} \left(\frac{\sigma(\vec{r_i})}{D_i} \right)^2},$$

where D_i is the dose in ith voxel, $\sigma(\vec{r_i})$ is the statistical uncertainty of dose for the ith voxel. The sum is computed over all N voxels in which the dose is greater than or equal to some fraction, α, of the maximum dose $D_{\max}$. It is commonly accepted that for tumor dose calculations $\alpha = 50\%$. However, for accurate calculation of dose to organs at risk, α should be set to a much lower value, which will of course increase the calculation time. XiO eMC allows the user to calculate and report either dose-to-water or dose-to medium, which makes it compliant with TG-105 recommendations (Chetty et al. 2007). XiO can handle calculations for linacs from all major vendors.

11.3 Commissioning of MC-Based TPS

Commissioning of MC-based TPS can be divided into three logical steps:

1. Collection of data to allow beam modeling, as outlined in the user manual.
2. Verification of the beam model.
3. Verification of dose calculations in heterogeneous media and under typical clinical conditions.

11.3.1 Measurements Required for Beam Characterization

To create a beam model, the user has to provide the following information about the treatment unit to the treatment planning vendor:

- Position and thickness of jaw collimators and MLC.
- For each applicator scraper layer: thickness, position, shape (perimeter and edge), and composition.
- For inserts: thickness, shape, and composition.

The open-field measurements (without the applicator, with collimator jaws wide open) consist of:

- Depth–dose curves in water at SSD = 100 cm.

- Absolute doses (reference dose), expressed in cGy/MU, at a specified point on the depth–dose curve.
- Dose profiles in air at source-to-detector distance = 95 cm (applicator level).

The applicator measurements for each energy/applicator combination are:

- Depth–dose curves in water at SSD = 100 cm.
- Absolute doses (reference dose), expressed in cGy/MU, at a specified point on the depth–dose curve.

The reference dose (in cGy/MU) for the beam defined by an applicator is normally greater than the reference dose of the open beam (no applicator) due to the applicator scatter.

The electron applicator field size is entered in the beam configuration task. The photon jaw settings are set corresponding to beam energy and electron applicator size based on the accelerator manufacturer's specifications.

11.3.2 Additional Beam Data Measurements for Commissioning MC-Based TPS

To commission a MC-based TPS, additional input data related to the beam characteristics may be needed. For example, to configure the eMC algorithm in the Varian Eclipse system, measurements are required for all open-field and applicator combinations for each beam energy. In addition, the dose profiles in air at a specified source-to-detector distance equal to 95 cm are also required (Ding et al. 2006). In the MC-based TPS, the user typically needs to measure for each beam energy dose profiles in air and water, both with and without the applicators (Cygler et al. 2004; Ding et al. 2006). The open-field (beam without electron applicator) beam data are necessary in order for the TPS system to accurately model the incident electron beams for dose calculations. In addition, the in-air profiles provide incident electron beam information (without the effect of an applicator) needed to configure the beam (Cygler et al. 2004; Ding et al. 2006).

The beam modeling procedure in XiO is based on in-water measurements only. It requires for each energy and applicator depth–dose curves and profiles at 7 depths for both an SSD of 100 cm and for an extended SSD commonly used in the clinic. The same set of scans is also required for a 15×15 cm^2 applicator with a 5×5 cm^2 cutout in place. In addition, depth–dose curves and profiles at 7 depths are also required for SSD = 100 cm with no applicator present and the jaws set to their maximum. The user has to also provide point dose values at a reference depth (typically d_{max}) for each field size and SSD listed above.

11.3.3 Beam Measurements Required for In-Phantom/Patient Dose Calculation Verification

Good clinical practice requires careful commissioning of a TPS. During the commissioning, the user should verify the accuracy

of dose calculations of a TPS before such a system is put into clinical use. This can be done by performing calculations and comparing the results with the data measured in homogeneous and some heterogeneous phantoms.

11.3.3.1 Homogeneous Phantom: Dose Profiles, MU Calculations at Various SSDs

Water is normally the choice of homogeneous phantom because the dose distributions can be obtained through depth–dose curve or dose profile scanning measurements by using commercially available scanning systems with an ionization chamber, diode or other type of detector. The measurements of depth–dose curves and dose profiles at several different SSD that are clinically relevant (i.e., SSD = 100 cm, 115 cm) are necessary not only for beam modeling but also for verification of dose calculations at standard and extended SSDs. The validation should also include the accuracy of MU calculations at standard and extended SSDs for open applicators and for a variety of cutout-defined field sizes for each specified electron applicator. One of the important strengths of the MC-based electron TPS is that it is capable of accurately calculating both dose distributions and MUs not only at a standard SSD but also at any extended SSD (Cygler et al. 2004; Ding et al. 2006). This is a significant advantage because conventional electron dose calculation algorithms are not capable of accurately predicting either the dose distributions or MUs at an extended SSD when the beam is commissioned only at the standard SSD (Ding et al. 1999, 2005).

11.3.3.2 Heterogeneous Phantoms

In addition to the validation of dose calculation in homogeneous phantoms, some tests should be carried out for heterogeneous phantoms before clinical implementation of the TPS. Such tests should include heterogeneous phantoms of various complexities such as 1D-type or slab, 2D-type, and more complex 3D-type geometries. The density of the heterogeneous materials must include high-(bone) and low-(lung) density materials that are clinically relevant in the patient treatment planning calculations. The validation involves the comparison of doses calculated by the TPS and the measured ones. Studies by Cygler et al. (2004) and Ding et al. (2006) showed some possible methods of evaluating the accuracy of a commercial electron TPS in a complex phantom containing 3D-type heterogeneities.

11.4 Issues Arising from the Clinical Implementation of MC Dose Calculations

11.4.1 Calculation Normalization, Dose Prescription, and Isodose Lines

As part of the commissioning of an electron TPS for clinical use, we also validated the accuracy of this system to calculate output factors for various cutout sizes for arbitrary SSDs for all available beam energies. A clinically acceptable agreement between measured and calculated output factors should be obtained. This includes both

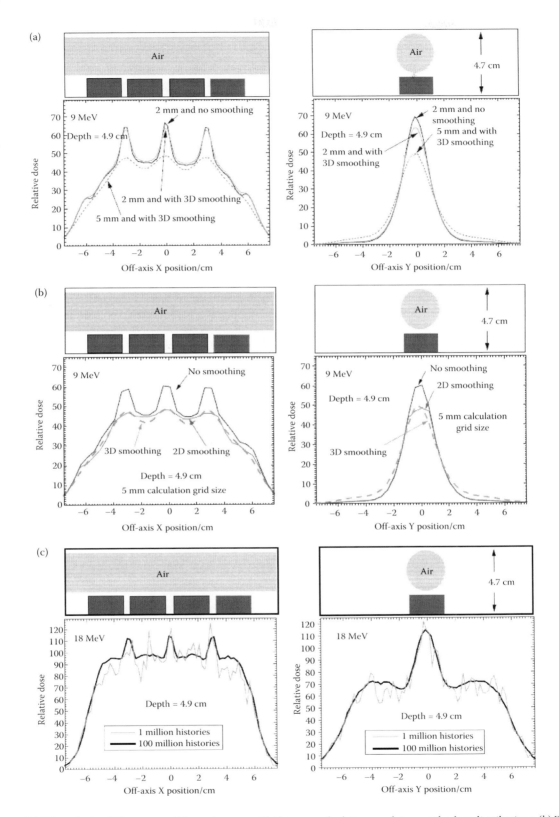

FIGURE 11.3 (a) Effect of using different smoothing techniques with the same calculation voxel sizes on the dose distributions. (b) Effect of using different calculation voxel sizes and smoothing techniques on the depth–dose distributions. (c) Effect of statistical uncertainties on dose distributions in the MC-based TPS. (Reproduced from Ding et al. 2006. *Phys Med Biol* 51(11):2781–99. With permission.)

open applicator and cutout-defined fields. When a calculated dose distribution is not normalized appropriately in a TPS, user applied renormalization may be necessary in order to obtain accurate MUs. It has been shown that MC methods available in commercial TPS are capable of calculating output factors accurately even at extended SSDs. This is a significant improvement over the PB algorithm implemented in CADPLAN (Ding et al. 2005).

11.4.2 Statistical Uncertainty, Smoothing and Calculation Voxel Size

It is noteworthy that any MC calculation includes a statistical uncertainty. It is not easy to have a rigorous method to obtain an accurate statistical uncertainty of the calculated dose distribution. Therefore, if a target prescription dose is based on a point dose calculation, the statistical uncertainty may have a significant effect on the dose prescription. The tools provided for smoothing dose distributions in some commercial TPS do not discriminate between real dose gradients and statistical variations. When applying the smoothing tools or using large calculation voxels to reduce the computation time, caution must be exercised or the resultant details of the calculated dose distributions may not be accurate, especially in regions of high-dose gradients (Ding et al. 2006a,b). On the other hand, the statistical uncertainty can be misinterpreted as dose variation when the real statistical uncertainty of the calculation is not well known. Figure 11.3 illustrates the effect of using different calculation voxel sizes, smoothing techniques, and statistical uncertainties on dose distributions in the MC-based TPS.

11.4.3 Dose-to-Medium versus Dose-to-Water Calculations

One of the issues arising when using MC-based TPS is "dose-to-water vs. dose-to-medium" calculations (Liu 2002; Keall 2002; Ding et al. 2006; Gardner et al. 2007). The two are conceptually different and which one is used has a significant impact on the reported doses to different organs, such as bone and lung. In conventional TPS, CT numbers are converted to electron densities

of water-like material before calculations are carried out. Unlike conventional TPS, MC inherently calculates the dose to medium. There can be deviations from unity (>10%) in reported doses when dose to water as opposed dose to medium is calculated. This is due to the fact that the stopping-power ratios of water/medium may not be unity for electron beams. There could be significant differences. The study by Ding et al. (2006) provided some examples of such situations. Their investigation found significant differences for lung and bone, respectively, when "dose-to-medium" is reported instead of "dose-to-water" as done in the conventional planning systems. In addition, the magnitude of this difference is not constant but depends on the beam energy, calculation depth, and medium. This variation is due in part to the fact that water/medium stopping-power ratios are strongly dependent on beam energy compared to photon beams. Figures 11.4 and 11.5 illustrate the magnitude of this difference for hard bone. This large variation due only to the method of dose reporting may have significant clinical implications. As MC-based TPS begin to enter routine clinical practice, it is imperative that a consistent approach to dose reporting is used. Only then can the prescription dose and treatment outcomes can be compared in a meaningful way between different centers.

11.4.4 Examples of CT-Based Dose Calculations

Figure 11.6 shows a calculated dose distribution for a tumor near the left ear. It can be seen that the MC calculated dose distributions show a significant hot and cold regions due to the presence of low- and high-density media whereas the PB dose calculation algorithm is not able to accurately predict the dose changes caused by the inhomogeneity. The ability to accurately predict hot and cold regions in patient treatment planning is essential to provide the best possible choice for treatment options.

11.4.5 Typical Calculation Times

MC dose calculation times depend on the computer hardware, dose calculation algorithm, beam energy, field size, and

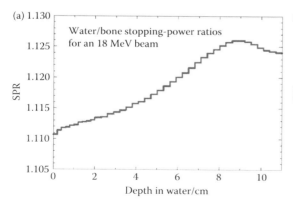
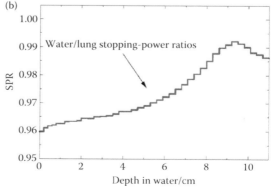

FIGURE 11.4 Illustration of water-to-medium stopping power radios as a function of depth for an 18 MeV electron beam in water. It is seen that the magnitude of differences between (a) dose-to-bone or (b) dose-to-lung and dose-to-water can be up to 12% or 4%, respectively. (Reproduced from Ding et al. 2006. *Phys Med Biol* 51(11):2781–99. With permission.)

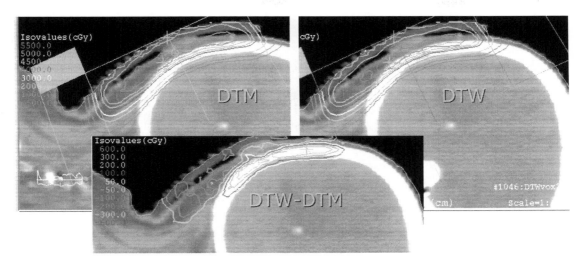

FIGURE 11.5 **(See color insert.)** Clinical example of differences between dose-to-medium and dose-to-water calculations for 6 MeV beam (XiO eMC).

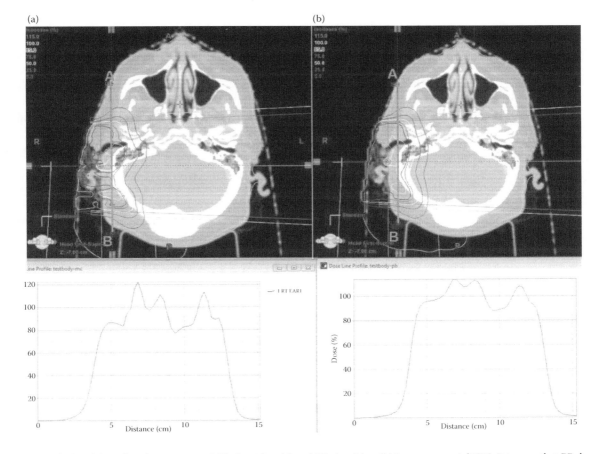

FIGURE 11.6 Calculated dose distributions using MC algorithm (a) and PB algorithm (b) in a commercial TPS. It is seen that PB dose calculation algorithm is not able to accurately predict dose changes caused by the inhomogeneity. The dose profiles start at point A and end at point B as shown in the axial CT images.

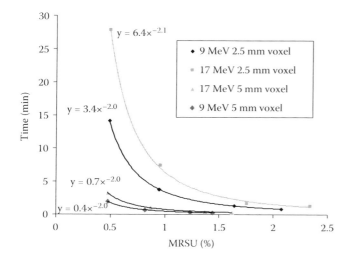

FIGURE 11.7 Timing results for a clinical XiO workstation (Linux OS, 8 CPU, 3 GHz each, 8.29 GB of RAM).

calculation parameters. The parameters that affect the speed of dose calculation are the number of particle histories or required statistical uncertainty and the voxel size. Users of a commercial TPS do not always have control over all parameters that affect the speed of dose calculations. Elekta Software is the vendor that allows users to set the largest range of calculation parameters. The proper choice of parameters depends on an anatomical site and required statistical uncertainty. Figure 11.7 shows typical calculation times for the trachea and spine phantom using a clinical XiO (version 4.51) workstation (Linux OS, 8 CPU 3 GHz each, 8.29 GB of RAM) for two beam energies and two voxel sizes as a function of MRSU. Typical calculation times for other planning systems can be found in Cygler et al. (2004), Ding et al. (2005), Chetty et al. (2007), and Fragoso et al. (2008).

11.5 Summary

Commercial MC-based TPS are becoming more common in the clinics. They provide significant improvement in dose calculation accuracy over the PB algorithms. It has been shown that in homogeneous phantoms both PB and MC systems are able to predict dose distribution accurately for standard SSDs. However, PB calculations are not accurate in a homogeneous water phantom for extended SSDs. For such cases, MC provides significant improvements. Only one virtual electron beam for each energy is used for accurate dose and MU calculations at any arbitrary SSD as demonstrated in the study by Cygler et al. (2004) and Ding et al. (2005). In heterogeneous phantoms, MC dose calculation method offers superior accuracy as compared to PB algorithm, especially where small 3D heterogeneities or overlying low-(air) and high-(bone) density materials are present. MC results have generally much better agreement with measurements, especially in high-dose gradient regions caused by the perturbation of adjacent 3D inhomogeneities. The overall results reported in the literature clearly underscore the potential

of MC to become the method of choice in the electron beam treatment planning compared to PB algorithms. However, it is important for the user to understand the MC implementation in a particular commercial TPS in order to select appropriate calculation settings for a given clinical situation. It is also worth noting that there is no absolute guarantee that the result from the MC calculation is always accurate.

References

Ali, O. A., C. A. Willemse, W. Shaw, F. H. O'Reilly, and F. C. du Plessis. 2011. Monte Carlo electron source model validation for an Elekta Precise linac. *Med Phys* 38(5):2366–73.

Bazalova, M. and F. Verhaegen. 2007. Monte Carlo simulation of a computed tomography x-ray tube. *Phys Med Biol* 52(19):5945–55.

Bazalova, M., H. Zhou, P. J. Keall, and E. E. Graves. 2009. Kilovoltage beam Monte Carlo dose calculations in submillimeter voxels for small animal radiotherapy. *Med Phys* 36(11):4991–9.

Brahme, A. 1984. Dosimetric precision requirements in radiation therapy. *Acta Radiol Oncol* 23(5):379–91.

Briesmeister, J. F. 2000. MCNP—A general Monte Carlo *n*-particle transport code, Version 4C. Technical Report No LA-13709-M. Los Alamos National Laboratory.

Chetty, I. J., B. Curran, J. E. Cygler, J. J. DeMarco, G. Ezzell, B. A. Faddegon, I. Kawrakow et al. 2007. Report of the AAPM Task Group No. 105: Issues associated with clinical implementation of Monte Carlo-based photon and electron external beam treatment planning. *Med Phys* 34(12):4818–53.

Cygler, J., J. J. Battista, J. W. Scrimger, E. Mah, and J. Antolak. 1987. Electron dose distributions in experimental phantoms: A comparison with 2D pencil beam calculations. *Phys Med Biol* 32(9):1073–86.

Cygler, J. E., G. M. Daskalov, G. H. Chan, and and G. X. Ding. 2004. Evaluation of the first commercial Monte Carlo dose calculation engine for electron beam treatment planning. *Med Phys* 31(1):142–53.

Cygler, J., X. A. Li, G. X. Ding, and E. Lawrence. 1997. Practical approach to electron beam dosimetry at extended SSD. *Phys Med Biol* 42(8):1505–14.

Ding, G. X. 1995. An investigation of radiotherapy electron beams using Monte Carlo techniques, PhD thesis, Department of Physics, Carleton University, Ottawa Carleton Institute for Physics. PhD, Ottawa, Canada.

Ding, G. X. and D. W. O. Rogers. 1995. Energy spectra, angular spread, & dose distributions of electron beams from various accelerators used in radiotherapy. National Research Council of Canada, Report No. PIRS-0439, Ottawa; see also http://www.irs.inms.nrc.ca/inms/irs/papers/PIRS439/pirs439.html.

Ding, G. X., J. E. Cygler, C. W. Yu, N. I. Kalach, and G. Daskalov. 2005. A comparison of electron beam dose calculation accuracy between treatment planning systems using either a pencil beam or a Monte Carlo algorithm. *Int J Radiat Oncol Biol Phys* 63(2):622–33.

Ding, G. X., J. E. Cygler, G. G. Zhang, and M. K. Yu. 1999. Evaluation of a commercial three-dimensional electron beam treatment planning system. *Med Phys* 26(12):2571–80.

Ding, G. X., D. M. Duggan, C. W. Coffey, P. Shokrani, and J. E. Cygler. 2006. First macro Monte Carlo based commercial dose calculation module for electron beam treatment planning: New issues for clinical consideration. *Phys Med Biol* 51(11):2781–99.

Edimo, P., C. Clermont, M. G. Kwato, and S. Vynckier. 2009. Evaluation of a commercial VMC++ Monte Carlo based treatment planning system for electron beams using EGSnrc/BEAMnrc simulations and measurements. *Phys Med* 25(3):111–21.

Fippel, M. 1999. Fast Monte Carlo dose calculation for photon beams based on the VMC electron algorithm. *Med Phys* 26(8):1466–75.

Fippel, M., I. Kawrakow, and K. Friedrich. 1997. Electron beam dose calculations with the VMC algorithm and the verification data of the NCI working group. *Phys Med Biol* 42(3):501–20.

Fippel, M., W. Laub, B. Huber, and F. Nusslin. 1999. Experimental investigation of a fast Monte Carlo photon beam dose calculation algorithm. *Phys Med Biol* 44(12):3039–54.

Fragoso, M., S. Pillai, T. D. Solberg, and I. J. Chetty. 2008. Experimental verification and clinical implementation of a commercial Monte Carlo electron beam dose calculation algorithm. *Med Phys* 35(3):1028–38.

Francescon, P., C. Cavedon, S. Reccanello, and S. Cora. 2000. Photon dose calculation of a three-dimensional treatment planning system compared to the Monte Carlo code BEAM. *Med Phys* 27(7):1579–87.

Francescon, P., S. Cora, and C. Cavedon. 2008. Total scatter factors of small beams: A multidetector and Monte Carlo study. *Med Phys* 35(2):504–13.

Francescon, P., S. Cora, C. Cavedon, and P. Scalchi. 2009. Application of a Monte Carlo-based method for total scatter factors of small beams to new solid state micro-detectors. *J Appl Clin Med Phys* 10(1):2939.

Gardner, J. K., J. V. Siebers, and I. Kawrakow. 2007. Comparison of two methods to compute the absorbed dose to water for photon beams. *Phys Med Biol* 52(19):N439–47.

GEANT. 2006. Geant4 developments and applications. *IEEE Trans Nucl Sci* 53 (1):270–78.

Hogstrom, K. R. and P. R. Almond. 1983. Comparison of experimental and calculated dose distributions. Electron beam dose planning at the M.D. Anderson Hospital. *Acta Radiol Suppl* 364:89–99.

Hogstrom, K. R., R. G. Kurup, A. S. Shiu, and G. Starkschall. 1989. A two-dimensional pencil-beam algorithm for calculation of arc electron dose distributions. *Phys Med Biol* 34(3):315–41.

Hogstrom, K. R., M. D. Mills, and P. R. Almond. 1981. Electron beam dose calculations. *Phys Med Biol* 26(3):445–59.

Hogstrom, K. R., M. D. Mills, J. A. Meyer, J. R. Palta, D. E. Mellenberg, R. T. Meoz, and R. S. Fields. 1984. Dosimetric evaluation of a pencil-beam algorithm for electrons employing a two-dimensional heterogeneity correction. *Int J Radiat Oncol Biol Phys* 10(4):561–9.

Iaccarino, G., L. Strigari, M. D'Andrea, L. Bellesi, G. Felici, A. Ciccotelli, M. Benassi, and A. Soriani. 2011. Monte Carlo simulation of electron beams generated by a 12 MeV dedicated mobile IORT accelerator. *Phys Med Biol* 56(14):4579–96.

ICRU-29. 1978. Dose specifications for reporting external beam therapy with photons and electrons. In *ICRU: Report 29*. Bethesda, MD: International Committee on Radiation Units and Measurements.

Jarry, G., S. A. Graham, D. J. Moseley, D. J. Jaffray, J. H. Siewerdsen, and F. Verhaegen. 2006. Characterization of scattered radiation in kV CBCT images using Monte Carlo simulations. *Med Phys* 33(11):4320–9.

Kawrakow, I. 1997. Improved modeling of multiple scattering in the voxel Monte Carlo model. *Med Phys* 24(4):505–17.

Kawrakow, I. 2000. Accurate condensed history Monte Carlo simulation of electron transport. I. EGSnrc, the new EGS4 version. *Med Phys* 27(3):485–98.

Kawrakow, I. 2001. VMC++ electron and photon Monte Carlo calculations optimized for radiation treatment planning. Paper read at Kling Aea, editor. *Proceedings of the Monte Carlo 2000 Meeting*, Lisbon, Springer, Berlin.

Kawrakow, I., M. Fippel, and K. Friedrich. 1996. 3D electron dose calculation using a voxel based Monte Carlo algorithm (VMC). *Med Phys* 23(4):445–57.

Kawrakow, I. and D. W. O. Rogers. 2002. The EGSnrc Code System: Monte Carlo Simulation of Electron and Photon Transport. Ionizing Radiation Standards, National Research Council of Canada, NRCC Report PIRS-701.

Keall, P. 2002. Dm rather than Dw should be used in Monte Carlo treatment planning. Against the proposition. *Med Phys* 29(5):923–4.

Keall, P. J. and P. W. Hoban. 1996. Super-Monte Carlo: A 3-D electron beam dose calculation algorithm. *Med Phys* 23(12):2023–34.

Keall, P. J., J. V. Siebers, B. Libby, and R. Mohan. 2003. Determining the incident electron fluence for Monte Carlo-based photon treatment planning using a standard measured data set. *Med Phys* 30(4):574–82.

Khan, F. M., K. P. Doppke, K. R. Hogstrom, G. J. Kutcher, R. Nath, S. C. Prasad, J. A. Purdy, M. Rozenfeld, and B. L. Werner. 1991. Clinical electron-beam dosimetry: Report of AAPM Radiation Therapy Committee Task Group No. 25. *Med Phys* 18(1):73–109.

Liu, H. H. 2002. Dm rather than Dw should be used in Monte Carlo treatment planning. For the proposition. *Med Phys* 29(5):922–3.

Ma, C. M., B. A. Faddegon, D. W. Rogers, and T. R. Mackie. 1997. Accurate characterization of Monte Carlo calculated electron beams for radiotherapy. *Med Phys* 24(3):401–16.

Ma, C. M., T. Pawlicki, S. B. Jiang, J. S. Li, J. Deng, E. Mok, A. Kapur, L. Xing, L. Ma, and A. L. Boyer. 2000. Monte Carlo verification of IMRT dose distributions from a commercial treatment planning optimization system [in process citation]. *Phys Med Biol* 45(9):2483–95.

Nelson, W. R., H. Hirayama, and D. W. O. Rogers. 1985. The EGS4 Code System. Report SLAC-265, Stanford Linear Accelerator Center, Stanford, California.

Neuenschwander, H. and E. J. Born. 1992. A macro Monte Carlo method for electron beam dose calculations. *Phys Med Biol* 37(1):107–125.

Neuenschwander, H., T. R. Mackie, and P. J. Reckwerdt. 1995. MMC—A high-performance Monte Carlo code for electron beam treatment planning. *Phys Med Biol* 40(4):543–74.

Papanikolaou, N., J. Battista, A. Boyer, C. Kappas, E. Klein, T. R. Mackie, M. Sharpe, and J. Van Dyk. 2004. *Tissue Inhomogeneity Corrections for Megavoltage Photon Beams*. Madison, WI: Task Group No. 65 of the Radiation Therapy Committee of the American Association of Physicists in Medicine.

Pasciuti, K., G. Iaccarino, L. Strigari, T. Malatesta, M. Benassi, A. M. Di Nallo, A. Mirri, V. Pinzi, and V. Landoni. 2011. Tissue heterogeneity in IMRT dose calculation for lung cancer. *Med Dosim* 36(2):219–27.

Pimpinella, M., D. Mihailescu, A. S. Guerra, and R. F. Laitano. 2007. Dosimetric characteristics of electron beams produced by a mobile accelerator for IORT. *Phys Med Biol* 52(20):6197–214.

Popple, R. A., R. Weinber, J. A. Antolak, S. J. Ye, P. N. Pareek, J. Duan, S. Shen, and I. A. Brezovich. 2006. Comprehensive evaluation of a commercial macro Monte Carlo electron dose calculation implementation using a standard verification data set. *Med Phys* 33(6):1540–51.

Reynaert, N., S. C. van der Marck, D. R. Schaart, W. W. Van der Zee, C. Van, Vliet-Vroegindeweij, M. Tomsej et al. 2007. Monte Carlo treatment planning for photon and electron beams. *Radiat Phys and Chem* 76:643–86.

Rogers, D. W. O. 2006. Fifty years of Monte Carlo simulations for medical physics. *Phys Med Biol* 51(13):R287–301.

Rogers, D. W. O., B. A. Faddegon, G. X. Ding, C. M. Ma, J. We, and T. R. Mackie. 1995. BEAM: A Monte Carlo code to simulate radiotherapy treatment units. *Med Phys* 22(5):503–24.

Schreiber, E. C. and B. A. Faddegon. 2005. Sensitivity of large-field electron beams to variations in a Monte Carlo accelerator model. *Phys Med Biol* 50(5):769–78.

Seltzer, S. M. 1991. Electron–photon Monte Carlo calculations: The ETRAN code. *Appl. Radiat. Isot.* 42:917–41.

Sempau, J, E. Acosta, J. Baro, J. M. Fernandez-Varea, and F Salvat. 1997. An algorithm for Monte Carlo simulation of coupled electron–photon transport. *Nucl Instrum Methods B* 132 377–90.

Sheikh-Bagheri, D. and D. W. Rogers. 2002a. Monte Carlo calculation of nine megavoltage photon beam spectra using the BEAM code. *Med Phys* 29(3):391–402.

Sheikh-Bagheri, D. and D. W. Rogers. 2002b. Sensitivity of megavoltage photon beam Monte Carlo simulations to electron beam and other parameters. *Med Phys* 29(3):379–90.

Shiu, A. S. and K. R. Hogstrom. 1991. Dose in bone and tissue near bone–tissue interface from electron beam. *Int J Radiat Oncol Biol Phys* 21(3):695–702.

Starkschall, G., A. S. Shiu, S. W. Bujnowski, L. L. Wang, D. A. Low, and K. R. Hogstrom. 1991. Effect of dimensionality of heterogeneity corrections on the implementation of a three-dimensional electron pencil-beam algorithm. *Phys Med and Biol* 36(2):207.

Traneus E, A. Ahnesjö, M. Fippel, I. Kawrakow, F. Nusslin, G. Zeng et al. 2001. Application and verification of a coupled multi-source electron beam model for Monte Carlo based treatment planning. *Radiother Oncol* 61:S102.

Wieslander, E. and T. Knoos. 2007. A virtual-accelerator-based verification of a Monte Carlo dose calculation algorithm for electron beam treatment planning in clinical situations. *Radiother Oncol* 82(2):208–17.

Zhang, G. G., D. W. Rogers, J. E. Cygler, and T. R. Mackie. 1998. Effects of changes in stopping-power ratios with field size on electron beam relative output factors. *Med Phys* 25(9):1711–6.

Zhang, G. G., D. W. Rogers, J. E. Cygler, and T. R. Mackie. 1999. Monte Carlo investigation of electron beam output factors versus size of square cutout. *Med Phys* 26(5):743–50.

12

Photons: Clinical Considerations and Applications

Michael K. Fix
Inselspital and University of Bern

12.1 Introduction

The principal aim of radiotherapy is to give a tumoricidal dose to the cancer-bearing tissue and to minimize the dose to normal healthy tissue. Given the dose response function of the tumor and the healthy tissue, the requirement stated in the ICRU report 50 (International Commission on Radiation Units and Measurements Report 50, 1993) is to apply the dose to the tumor within −5% and 7% of the prescribed dose. A detailed analysis of uncertainties associated with radiation treatment shows that 3% accuracy is required in the dose calculation to yield ±5% accuracy in the dose delivered to the patient (International Commission on Radiation Units and Measurements Report 24, 1976; Brahme, 1984; Dutreix, 1984; Van Dyk et al., 1993). Some studies have even concluded that for certain types of tumors the uncertainty in dose delivery should be smaller than 3.5% (Brahme, 1984; Mijnheer et al., 1987, 1989), which in turn means that the clinically implemented dose calculation algorithm should be accurate within ±2%. Thus, accurate dose calculation algorithms for treatment planning systems are of critical importance in radiation therapy.

Within the last two decades, huge developments have been seen in the field of dose calculation algorithms. The early clinical treatment planning systems used correction-based algorithms (Mackie et al., 1996), whereas nowadays mostly model-based algorithms such as convolution or convolution/superposition algorithms are used. Algorithms using the Monte Carlo (MC) method can also be considered as model based. Although over

the last few years many applications of MC techniques have been published in the field of medical physics, the clinical impact of MC calculated patient dose distributions for photon beams still remains unclear.

In contrast to the other common techniques, the MC method starts from first principles and tracks individual particle histories; thus, it takes into account the transport of secondary particles. For a long time, a major drawback of using MC methods for photon beam treatment planning in clinical routine was the long calculation time needed to achieve a dose distribution with a reasonable statistical uncertainty. However, recent advances in MC patient dose calculation algorithms coupled with increasing computer processing speed has made MC patient dose calculation speed acceptable for the radiotherapy clinic. In principle, the MC method produces accurate results in regions of tissue heterogeneities such as lung and surface irregularities, thus providing the most accurate method for the simulation of patient treatment dose distributions, especially for complex techniques such as intensity modulated radiotherapy (IMRT) or intensity-modulated arc therapy (Andreo, 1991; DeMarco et al., 1998; Wang et al., 1998; Arnfield et al., 2000; Bush et al., 2008; Oliver et al., 2008; Sarkar et al., 2008; Teke et al., 2010; Fix et al., 2011). The National Cancer Institute recognizes the need for research and development on MC techniques in radiation therapy and it is anticipated that MC will become a necessary dose calculation tool (Fraass et al., 2003). However, ultimately the accuracy of the treatment planning dose calculation algorithm depends strongly on the implementation and the accuracy of the input data. For instance, the accuracy depends on the quality of the anatomical information of the patient, since this affects the irradiating geometry as well as the tissue cross sections. Another prerequisite is the accurate information about the radiation beam incident on the patient. In fact, given the availability of fast radiotherapy-specific MC codes, the major limitation to the widespread implementation of MC dose calculation algorithms is the lack of a general, accurate, and user-specific scalable source model of the accelerator radiation source. More precisely, a user with an arbitrary linear accelerator should be able to commission the source model so that the MC dose calculation algorithm meets predefined accuracy requirements compared to measurements prior to using the algorithm for patient dose calculation, for example, 2% or 2 mm.

12.2 Requirements for Clinical MC Treatment Planning

A clinically usable MC treatment planning system for photon beams is more than just an MC dose calculation algorithm together with a beam model to describe the radiation beam. Additionally, it is important to also have the capability, that is, a tool or a platform, for beam setup, dose display, and dose evaluation. Whereas these platforms are available in commercial treatment planning systems, they are in general missing in the preclinical or research MC dose calculation packages. If MC dose calculation is used only in some special rare cases, this might be acceptable. However, if MC treatment planning should be used on a large scale, automation of the MC dose calculation is needed. Consequently, some research systems have also been extended to provide capabilities for beam setup, dose display, and dose evaluation. The following sections will cover the requirements for clinical MC treatment planning for photon beams.

12.2.1 Beam Setup Capability

For the commercial treatment planning systems, this is the same as for any other dose calculation algorithm. But research systems, if intended for clinical use, also provide such a platform. Most commonly, this has been realized by interfacing the external MC dose calculation with an already existing commercial treatment planning system automatically (Fix et al., 2007; Siebers et al., 2000), by another interfacing program (Ma et al., 2002), or by a DICOM-RT interface (Alexander et al., 2007).

12.2.2 Beam Model

Certainly, an accurate characterization of the radiation beam exiting the treatment head is a prerequisite for accurate dose calculations in the patient. For MC photon treatment planning, several beam models have been developed and investigated. They are either measurement or MC based (Verhaegen and Seuntjens, 2003). The measurement-based beam models use analytical models for which the parameters are determined from the

measurements (Ahnesjo et al., 1992, 1995; Ahnesjo, 1994; Jiang et al., 2001; Fippel et al., 2003). The MC-based beam models can be further divided into those using full MC (Hasenbalg et al., 2008) or phase space files (Rogers et al., 1995; Sheikh-Bagheri and Rogers, 2002a,b) as input and those using histogram-based beam models (Ma, 1998; Schach von Wittenau et al., 1999; Deng et al., 2000; Fix et al., 2004). These MC-based beam models usually model the primary beam, that is, the part of the accelerator head that covers the patient-independent part.

Besides the characterization of the primary beam, the beam model also has to accurately model the clinically used patient-specific beam modifiers. For photon beams, these are blocks, hard wedges, dynamic wedges, multileaf collimators (MLCs), and so on (Magaddino et al., 2011). For the radiation transport through these beam modifiers, a number of transport parameter settings and dedicated variance reduction methods can be applied depending on the treatment case (Schmidhalter et al., 2010). In general, the research systems provide a higher degree of flexibility compared to commercial treatment planning systems with respect to transport options and also on how to use the beam modifiers. For example, the reduction in the MLC transmission for moving jaws in IMRT has been investigated (Schmidhalter et al., 2007). Those settings need to be considered very carefully before using them for clinical MC dose calculations.

Since errors in the beam model propagate through all the following processes for the dose calculation, it is very important to extensively verify the performance of the beam model during the commissioning and validation process.

12.2.3 Patient Model

Another important issue affecting the accuracy of the MC calculated dose distribution in the patient is the anatomical patient representation, which is the basis for the geometrical and interaction data specification used for the dose calculation. Patient computed tomography (CT) scans are used to extract the geometrical information as well as the patient-specific tissue characterization leading to the interaction data for the dose calculation within the patient. Therefore, image artifacts lead to inaccurate patient representation and consequently to inaccurate dose distributions. For non-MC dose calculation algorithms, CT conversion curves from Hounsfield values to electron density or physical density are used. However, for MC algorithms, the interaction data also have to be known. Different conversion methods are used starting from a simple relation between mass density and interaction coefficients (Fippel, 1999) to explicit segmentation of the tissue followed by anatomic composition assignment to determine the interaction data (Chetty et al., 2007). The impact of material misassignment on dose distributions is not fully investigated yet and there are only a few studies in the literature (Verhaegen and Devic, 2005; Vanderstraeten et al., 2007; Ottosson and Behrens, 2011). Detailed information about this issue can be found in Chapter 8 of this book. Another issue with CT images is the grid resampling, since typically a different voxel size is used for the dose calculation than provided by the CT images. The applied interpolation method can introduce errors, which in turn lead to misassignments in the materials and finally in the interaction data. A recent study by Volken et al. (2008) investigating different interpolation algorithms shows that an integral conservative Hermitian curve interpolation improves the interpolation accuracy compared with typically used linear or cubic interpolation functions. Finally, it is worth mentioning that changes occurring between the image acquisition and the dose delivery need to be taken into account. This could be changes to the patient, for example, weight loss, accessories only used during CT image acquisition, or different couches used for the CT scan and radiation delivery. Consequently, it might be necessary to remove that information from the CT images or overwrite the information with more accurate data before using them for dose calculations.

12.2.4 Dose Calculation

The MC dose calculation algorithm needs to be interfaced with the beam model. In principle, there exist several methods. One approach is to store phase space files as output of the beam model and use these phase space files afterwards as input for the dose calculation, that is, the phase space file is used as beam model in this case. Another approach is to pass the particle from the beam model directly, that is, in memory, to the dose calculation algorithm. The latter has the advantage of being faster and that no large phase space files have to be stored on hard discs. The fact that MC dose calculation is dealing with random walks offers the possibility to reuse the same particle coming from the beam model. However, by this method, the statistical uncertainty of the calculated dose distribution is affected, since those particles are no longer independent. Thus, whereas this method saves some computing time and possibly disc space, the statistical uncertainty of the representation of the radiation beam by the beam model reveals some lower limit. One example is the latent variance of a phase space file (Sempau et al., 2001). Alternatively, modifications of the particles from the phase space utilizing existing symmetries (Rogers et al., 1995; Fix et al., 2004; Bush et al., 2007) or small perturbations (Tyagi et al., 2006) have been investigated.

One of the main advantages of using MC dose calculation is that such an algorithm is suitable not only to calculate the dose for static patient situations but also for dynamic situations, like patient motion. Since MC is able to include a time component—in the beam model as well as in the patient model—MC can be used straightforward for dynamic patient dose calculations (Paganetti et al., 2004, 2005; Seco et al., 2008). Recently, new dynamic arc therapy techniques such as RapidArc, IMAT, or VMAT have been introduced in clinical routine. It is worth mentioning that for these applications the dose calculations with MC methods can be performed continuously instead of using a discrete number of situations approximating the dynamic delivery, for example, multiple static beams every 5° to approximate the gantry rotation. It is also worth mentioning that the calculation time for MC dose calculations does not scale with the number of beams if the statistical uncertainty in the target is considered as compared to other conventional algorithms.

However, the calculation time might increase to some extent if a certain statistical uncertainty has to be achieved in organs at risk (OARs).

12.2.5 Dose Evaluation Capability

Similar to the beam setup, the dose evaluation can also be performed as for any other dose calculation algorithm; thus, the commercial treatment planning systems are able to use their already existing tools for dose display and dose evaluation. This is true also for those research systems that are automatically interfaced with commercial treatment planning systems or those using DICOM files as interfaces. However, additional important information is needed, which is generally not included in commercial dose display and evaluation tools. Since every MC simulation is associated with a statistical uncertainty, the documentation of this information for the dose evaluation is essential when dealing with MC. Bearing in mind that in the dose evaluation different structures are quantitatively analyzed, it is important to take into account the statistical uncertainty of the dose values within the structure, for example, by displaying the uncertainty distribution. Additionally, it should be stated whether dose to media or dose to water has been calculated independent of which dose calculation algorithm is used. Another possibility to display and evaluate the resulting dose distribution is external viewing and analysis tools, for example, the CERR software development (Deasy et al., 2003).

12.3 Commissioning and Validation

Usually, vendors of commercial treatment planning systems provide a recommendation of the commissioning and validation procedure for their planning systems. Thereby, the commissioning consists of the configuration of the treatment planning system. The commissioning and validation are performed by comparing measured with the corresponding calculated dose distributions for different setups. However, especially for MC implementations, it is not fully explored what the least sufficient set of comparisons is for the commissioning and validation process. Of course this depends strongly on the clinical use foreseen. For example, which type of treatment technique should be covered: 3D conformal, IMRT, stereotactic radiotherapy, and so on. It is further important to determine what kind of beam modifiers have to be provided. Finally, it also depends on the accuracy requested, detector types used for the measurements, and so on. The following sections provide an overview for the commissioning and validation of MC treatment planning systems.

12.3.1 Tolerances and Acceptance Criteria

During the commissioning and acceptance procedure, it is always a difficult task to determine the tolerance and acceptance criteria for the dose comparisons. Several quantities have been used for comparing two dose distributions, for example, dose difference, distance to agreement, gamma index (Low et al., 1998), or slightly modified gamma index values (Bakai et al., 2003; Jiang et al., 2006; Blanpain and Mercier, 2009). For these quantities, a certain tolerance level or acceptance criteria have to be defined. Often, 2% or 2 mm is chosen. However, if such criteria are required for patient dose calculation, the estimation for the error due to beam modeling may not be larger than 1% or 1 mm (Keall et al., 2003). It is also questionable to apply the same criteria throughout the comparison, that is, for all field sizes, at every location, and so on. Given the statistical nature of MC, there is a certain probability of dose values with large random errors that might not fulfill the criteria. But there might also be areas in which the accuracy is more important than in others, for example, build-up region or outside the direct radiation field. Thus, different criteria at different locations and setups might be justified. In summary, it is of great importance to carefully choose the tolerance and acceptance criteria ahead of the commissioning and validation procedure and in relation of the clinical usage of the machine.

12.3.2 CT Conversion

As already mentioned in previous sections, it is of great importance to have accurate anatomical and material information. Depending on the MC algorithm, different conversion methods are needed to finally receive the interaction data needed for the MC dose calculation. The according input information for the conversion method used has to be provided to the system by the user. Most MC treatment planning systems, including the research systems, use a mapping of Hounsfield values to mass density and material composition. In the beginning of MC photon treatment planning, often a default CT to material conversion has been used. This conversion is not appropriate if highly accurate dose calculation is required. Incorrect assignment of media and/or density can lead to significant dosimetric errors (see Section 12.2.3). A more accurate CT to material conversion requires more segmentation bins compared with the default scheme (Vanderstraeten et al., 2007) and depends on the CT scanner used. Hence, CT calibration phantoms applied for the CT scanner in use lead to more accurate patient representations and material specifications. However, the phantom has to be carefully evaluated to determine if it is suitable for such a calibration procedure, for example, Teflon is not an appropriate representation for cortical bone (Verhaegen and Devic, 2005).

12.3.3 Beam Model and Dose Calculation

The specific beam model commissioning depends strongly on the implemented beam model. Most research-based systems have a slightly different implementation and thus an individual beam model commissioning procedure. For beam models based on phase space data, usually the energy and the width of the Gaussian intensity distribution of the initial electron beam are used as parameter. Note that perfect knowledge of the treatment head geometry and material is assumed. Beam models derived from these phase space data usually have settings available for a wide range of different parameterized initial electron beams.

During the commissioning, the beam model created is determined by interpolation. Analytical or measurement-based beam models have a specified set of measurements needed for the commissioning. The scope of these measurements is provided either by the vendors or by the research group.

Beam modifiers typically represent the patient-specific part of the linear accelerator within the beam model. For the radiation transport, mainly either MC transport or transmission filters in multiple layers are used. The parameters that are adjusted during the commissioning phase of the beam model depend on the beam modifier considered. Typically, the density is tuned for the secondary collimators and blocks, whereas additionally, the thickness is adjusted for the hard wedges. For the tuning of the MLC, the gap between the leaves is used as a third parameter. As for the open beam, and for the beam modifiers too, a set of measurement is used to determine the settings of the tuning parameters. Since these parameters are not depending on the photon beam energy, the commissioning could in principle be done for only one beam energy. Alternatively and depending on the clinical demands, the commissioning could determine tuning parameters that provide the best fit for all beam energies on average. The validation, of course, has to be done for all beam energies used clinically.

The validation of the beam model is done by comparing calculated with measured dose distributions. The comparisons include the measurements that have been used during the commissioning process to verify that the treatment planning system is able to reproduce the input dose distributions, that is, a consistency test is performed. For the validation, it is important to cover the dosimetric range for which treatment plans are used clinically. Typically, the following measurements are compared:

- Relative depth dose curves and lateral dose profiles for different field sizes and shapes, including off-axis fields. These measurements are needed for open fields as well as for all beam modifiers considered such as wedged, blocked, dynamic wedged, and static and dynamic MLC-shaped fields. These measurements are performed in water or water equivalent phantoms. Since the primary aim is to validate the beam model and the dose calculation, disturbing side effects due to CT conversion or volume averaging and so on should be avoided or minimized. This could be realized by using digital phantoms.
- Since relative depth dose curves and lateral dose profiles are used above, it is necessary to check output factors for the same situations as well as an absolute dose calibration. Alternatively, all dose curves could be compared in absolute units, for example, cGy or Gy per monitor unit (MU).
- Relative measurements using simple inhomogeneous phantoms.
- Clinical treatment plan comparisons between dose distributions calculated with MC and with other available conventional dose calculation algorithms. This should involve simple cases as well as complex cases. Thereby, the patient is set to water with density 1 g/cm^3 or with the density

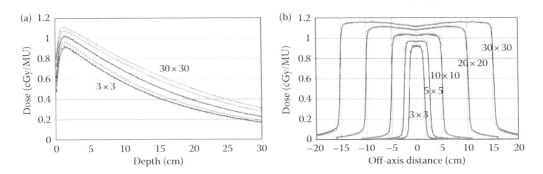

FIGURE 12.1 Comparison of MC calculated (symbols) and measured (lines) dose distributions in units of cGy/MU for different field sizes (in cm²) in water of a 6 MV photon beam: depth dose curves (a) and lateral dose profiles (b).

resulting from the CT conversion scheme. This provides some consistency checks with conventional dose calculation algorithms.

Some remarks with respect to the commissioning and validation:

- The measurement procedure should be consistent with those for conventional treatment planning algorithms, for example, Fraass et al. (1998). Appropriate detectors should be used for the different situations, for example, measurements at shallow depths. Additionally, the detector type has to be carefully selected to receive a consistent measured data set. Possible dependencies in energy response, effective point of measurement, dose rate, and so on need to be taken into account. Furthermore, the voxel size for the calculated dose distributions should be approximately the same as the sensitive volume of the detector used for the measurements.
- Since MC calculations are always associated with statistical uncertainties, single-point dose comparisons are critical. For example, absolute dose calibration using a single point is inappropriate. A method using multiple points is much more robust (Siebers et al., 1999; Fix et al., 2007). This is also true for dose prescription in clinical treatment plans.
- Generally, it is advisable to perform dose comparisons at shallow depths, since these measurements are highly sensitive with respect to the beam model settings. Thus, they are useful to verify the accuracy of the beam model. Additionally, in-air profile measurements can be used for investigating the performance of the beam model since the impact of scatter is reduced.
- Although in the commissioning and validation ultimately the calculated dose distribution has to match measurements, it is important to bear in mind that the measurements themselves are also associated with errors.

Overall, the result of the commissioning and validation process strongly depends on the quality of the measurements. Figure 12.1 shows a typical comparison between MC calculated and measured absolute depth dose curves and lateral dose profiles. Using an incompatible voxel size for the calculation compared

with the detector used for the measurement lead to dosimetric inaccuracies in the penumbra region. A study by Sahoo et al. (2008) shows that the penumbra width differs by a factor of 2 when different detectors are used. Figure 12.2 illustrates an example of the comparison between calculated and measured absolute dose profiles for 45° hard wedged beams. The comparison at different depths provides some indication whether or not the energy spectrum, including the beam hardening, has been characterized appropriately in the beam model. The validation of the MLC has to be done with great care since this complex beam modifier is often used in both static and dynamic treatment applications. Transmission, leakage, and various MLC-shaped fields should be included in the validation. Figure 12.3 demonstrates calculated and measured transmission and leakage profiles. If the interleave leakage is not taken into account for the MLC, the average transmission might be correct; however, the shape of the transmission profile cannot be reproduced. On the other hand, it is not clear if such a detailed modeling of the MLC is clinically needed. That dosimetric comparisons between dose calculation algorithms are important is demonstrated by a study by Reynaert et al. (2005). They encountered a 10% difference in dose volume histograms (DVHs) for the optical chiasm

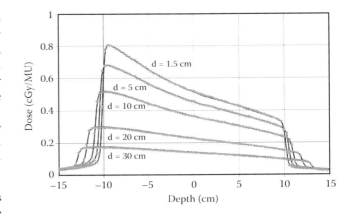

FIGURE 12.2 Comparison of MC calculated (symbols) and measured (lines) dose distributions in units of cGy/MU for a beam with a 45° hard wedge at several depths in water.

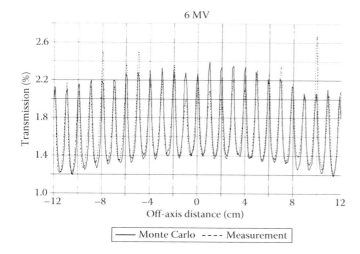

FIGURE 12.3 Comparison of MC calculated (lines) and measured (broken lines) transmission curves for an 80 leaf Varian MLC. (Adapted from Fix M. K. et al. 2007. *Phys Med Biol* 52: N425–37, with permission from IOP Publishing.)

measured dose values for one IMRT field from a verification plan using dynamic MLC, that is, one IMRT field is applied with gantry angle zero on a water phantom. For each measurement point using an ionization chamber (CC04), the whole IMRT field has to be delivered. Additionally, a corresponding film measurement has been performed in a homogeneous solid water phantom.

Apart from dose comparisons in homogeneous phantoms, it is also important to validate the treatment planning system in inhomogeneous phantoms. This allows the validation of the transport in nonwater materials (density and composition) as well as the characterization of the material through CT data sets. The first issue can be verified if digital phantoms are used for the calculation. Figure 12.6 shows an example of an inhomogeneous phantom comparing relative depth dose curves for a slab phantom with water–bone–water interfaces (Carrasco et al., 2007). A 6 and 18 MV beam with a 10×10 cm² field was used for the irradiation. Measurements are performed with TLDs, MOSFET, and several ionization chambers. These measurements are compared with calculations using different dose calculation algorithms, including MC, pencil beam, and convolution/superposition algorithms.

when dose distributions are compared using Peregrine and their in-house MC system and assigned the obtained dose differences found to an inaccurate Elekta MLC modeling. Additional studies demonstrating the importance of accurate MLC modeling are described in the literature (Schwarz et al., 2003; Webster et al., 2007).

Besides static beam modifiers, there are also dynamic beam modifiers such as dynamic wedges or the MLC. For the EDW (dynamic wedge for Varian linear accelerators), one jaw of the secondary collimators is traveling from one side to the other while the beam is on, thus generating a wedged beam. Figure 12.4 shows some examples from the validation of EDWs for a 6 MV Varian beam. Measurements with an ionization chamber are taken at some points along the depth dose curves and dose profiles. Since the MLC can be used for both static and dynamic treatment applications, it is not enough to validate the static characteristics through transmission measurements as mentioned above. Furthermore, dynamic applications have to be validated. Figure 12.5 illustrates the comparison between MC calculated and

12.4 Research and Commercial MCTP Systems

12.4.1 Research MCTP Systems

Over the years, many research institutions have implemented MC dose calculation algorithms. Those implementations have been coupled with different kinds of source models to allow photon beam MC treatment planning. These research planning systems are described in the literature and have been used to quantify the dosimetric difference between MC calculated dose distributions and those calculated with traditional treatment planning systems. Research planning systems developed at research institutions include the following:

- University of California, Los Angeles, RTMCNP (DeMarco et al., 1998): MC treatment planning based on MCNP4A using a speed-optimized photon particle transport and dose scoring within the standard lattice geometry. The RTMCNP preprocessor provides a user-friendly interface between the user and the MCNP4A command structure.

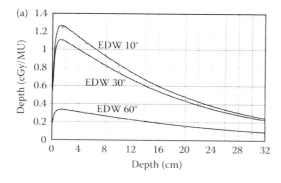
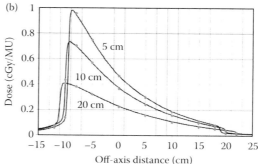

FIGURE 12.4 Comparison of MC calculated (lines) and measured (symbols) dose distributions for a 6 MV beam in units of cGy/MU for different enhanced dynamic wedges (EDWs) in water: depth dose curves (a) and lateral dose profiles for the EDW 60° at several depths (b).

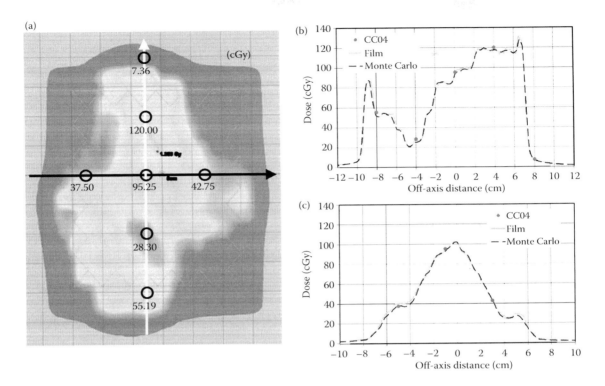

FIGURE 12.5 (**See color insert.**) Comparison of MC calculated and measured dose distributions for one head and neck IMRT field using a 6 MV beam in units of cGy in water: (a) calculated dose distribution and measurement points (with dose values in cGy), (b) calculated dose profile and measurements along the white arrow, and (c) calculated dose profile and measurements along the black arrow.

- Memorial Sloan-Kettering Cancer Center (Wang et al., 1998, 1999): An EGS4-based MC treatment planning environment using a dual-source beam model. The sources describe the primary and the scatter radiation of the treatment head, and are based on full MC simulation of the accelerator.

- Stanford University and Fox Chase Cancer Center, MCDOSE (Ma et al., 2002): This MC treatment planning system for electron and photon beams is coupled with the FOCUS system (Computerized Medical Systems (CMS), Inc., recently purchased by Elekta AB). MCDOSE is based

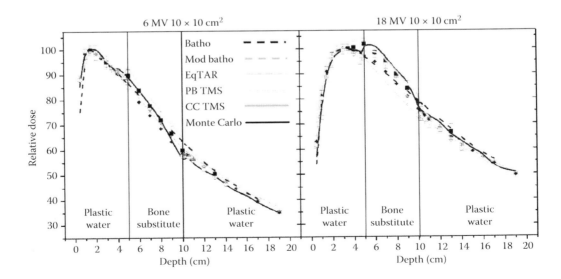

FIGURE 12.6 Comparison of measured and calculated relative depth dose curves in a water–bone–water phantom: 6 MV beam (left) and 18 MV beam (right). Measurements are: TLD, white circles; MOSFET, black squares; NACP02, black rhombus; and NE2571, white triangles. (Adapted from Carrasco P. et al. 2007. *Med Phys* 34: 3323–33, with permission from American Association of Physicists in Medicine.)

on EGS4 and as beam model either phase space files or multiple source models can be used.

- Virginia Commonwealth University (Siebers et al., 2000, 2002): An MC treatment planning system using EGSnrc as transport code. For the radiation transport through the MLC, a dedicated transport method has been developed (Siebers et al., 2002).
- University of Michigan, RT_DPM (Chetty et al., 2003): BEAMnrc phase space files coupled with the DPM MC code (Sempau et al., 2000) is used as dose calculation engine in this MC treatment planning system.
- University of Tübingen (Alber et al., 2003): An MC treatment system using MC also within the optimization. As beam model, the virtual fluence model (Fippel et al., 2003) is utilized and the XVMC code (Fippel, 1999) is used as a dose calculation algorithm. Recently, a dedicated source model for the electron contamination has been included in the beam model (Sikora and Alber, 2009).
- McGill University, MMCTP (Alexander et al., 2007): The McGill Monte Carlo treatment planning is a flexible software package with import options, for example, DICOM-RT, tools for contouring, beam editing, visualization, and analysis. MC dose calculations for electron and photon beams are supported using BEAMnrc for the simulation of the accelerator head and XVMC for the dose calculation.
- Inselspital and University of Bern, SMCP (Fix et al., 2007): The Swiss Monte Carlo Plan is an MC treatment planning system interfaced with Eclipse (Varian Medical Systems, Inc.). Transport methods of different complexity levels for the treatment head combined with MC dose calculation algorithms of EGSnrc or VMC++ allow automatic and flexible MC treatment planning.

12.4.2 Commercial MCTP Systems

In the recent decade, there have been only a few commercial treatment planning systems available offering MC dose calculation for photon beams:

- Peregrine is an MC dose calculation system developed at the Lawrence Livermore National Laboratory. Peregrine as implemented within the commercial treatment planning system Corvus (NOMOS Corporation) uses a four source beam model based on a phase space simulation of the accelerator head (Schach von Wittenau et al., 1999; Hartmann Siantar et al., 2001). Three photon subsources representing the target, the primary collimator, and the flattening filter are combined with the fourth source representing the electron contamination of the beam. A series of correlated histograms is used to reproduce the phase space information directly above the secondary collimator jaws. These sets of histograms are stored in a device file describing the accelerator according to the beam model and the beam modifiers (material, geometry, location, etc.). During the MC dose calculation, the

particle is generated by sampling from the histograms and then transported through the beam modifier and the patient taking into account the backscatter into the monitor chamber. For the commissioning process, a whole class of possible beam models is provided by the Corvus system. Based on the linear accelerator specification provided by the user, a certain beam model object is created. The tuning of the beam model is performed by interpolation of the parameter data generated for different initial electron beam energies such that the calculated depth dose curve for a 10×10 cm^2 field matches the corresponding measurement. Additionally, the 40×40 cm^2 field is used for the commissioning. Validation of the Peregrine system for clinical use is shown by several groups (Hartmann Siantar et al., 2001; Heath et al., 2004; Lehmann et al., 2006; Rassiah-Szegedi et al., 2007).

- Monaco is the MC treatment planning option offered by CMS (now Elekta AB). The beam model in Monaco is based on the virtual fluence model for photons developed by Fippel et al. (2003) and for the electron contamination the approach developed by Sikora and Alber (2009) is used. The inputs for this beam model are depth dose curves, output factors, and lateral dose profiles at several depths. Commissioning uses small-field depth dose curves to determine the primary energy spectrum, output factors for a reference field as well as for the smallest and largest field size to determine the primary source diameter. Finally, cross profiles for the largest field size at several depths are needed to extract the off-axis energy fluence variations. This leads to a beam model that is characterized by 11 parameters. The radiation transport through the MLC is performed by using transmission filters in multiple layers (Sikora et al., 2007). This method results in a speed increase of about a factor of 100 compared with full MC transport and allows simulating the geometry of the MLC but not the scatter radiation produced in the MLC. For the patient dose calculation, the XVMC code is used (Fippel, 1999). Clinical validations have been reported in the literature (Sikora et al., 2007, 2009; Fotina et al., 2009; Sikora and Alber, 2009).
- In December 2008, Brainlab AG released their MC treatment planning system within iPlan. The beam model is based on a virtual beam model developed by Fippel et al. (2003) and supports linear accelerators, including their MLCs of all major vendors. For the MLC beam modifier, a full MC simulation is used as radiation transport. However, the user is able to select between a speed-optimized (simplified MLC without leakage) and an accuracy-optimized MLC model. Further parameters in iPlan are the voxel size for the dose calculation, the requested mean variance of the MC dose distribution, and whether dose to medium or dose to water is calculated. Currently, the MC dose calculation algorithm requires the commissioning of the pencil beam algorithm in iPlan. The commissioning and validation of the iPlan system is done by Brainlab itself using a total of 93 in-air measurements and 97 measurements in water. The in-air measurements include depth profiles, cross profiles, and output factors whereas the

measurements in water include absolute calibration measurements, depth dose curves, and cross profiles all at a source to surface distance (SSD) of 90 and 100 cm as well as output factor measurements at SSD 90. Most of these measurements in water are used for the validation of the beam model. The results of the commissioned beam model are finally provided to the customer. Recently, validations of the MC implementation in iPlan have been published in the literature (Fotina et al., 2009; Kunzler et al., 2009).

- In early 2011, the treatment planning system ISOgray (DOSIsoft) received FDA clearance. ISOgray offers MC dose calculations based on the general-purpose MC code PENELOPE (Salvat et al., 2006). The beam model consists of two parts: The first part includes the treatment head components of the linear accelerator that are patient independent beginning with the target and the second part is the patient-specific part, including the secondary collimator jaws and beam modifiers such as blocks, wedges, and MLC. For the patient-independent part, a phase space file is generated using PENELOPE. This computation is performed by DOSIsoft during the commission process in which the incident energy and the spot size of the initial electron beam impinging on the target are tuned to fulfill the acceptance criteria according to the measurements from the user. The phase space file is then used as input for the patient-dependent part of the treatment head where an efficient selective particle tracking method is used (Brualla et al., 2009). Thereby, the geometry of the beam modifier is divided into areas depending on whether or not secondary particles or scatter radiation can emerge from the beam modifier leading to skin and nonskin areas, respectively. Whereas in skin areas, the radiation transport is modeled accurately, in nonskin areas, the transport is simplified. In this manner, linear accelerator models for Elekta, Siemens, and Varian are available based on the according manufacturer information. For the dose calculation within the patient, PENFAST is used, which is based on PENELOPE but utilized particle radiation transport techniques that are optimized with respect to calculation speed (Habib et al., 2010). The work from Habib et al. (2010) also serves as a preclinical validation of this treatment planning system.

Furthermore, Nucletron (now Elekta AB) has announced that MC photon treatment planning will be included in Oncentra in their upcoming releases.

12.5 Clinical Examples and Applications

12.5.1 MC as Treatment Planning Tool

There are many subjects that are worthwhile to be discussed in this section and most of them are described in detail in other chapters of this book. However, there are some issues that have to be carefully considered from a clinical perspective when using MC as a treatment planning tool. The two main issues involving the physicians' point of view are discussed in the following sections.

12.5.1.1 Noise in Dose Distributions

While conventional dose calculation algorithms determine the 3D dose distributions deterministically, the MC method leads to a result associated with a statistical uncertainty. Thus, viewing and evaluating MC calculated dose distributions for the first time within a clinical environment is challenging and needs some time to get used to. The statistical uncertainty influences many quantities with respect to dose distributions such as isodose lines, DVHs, dose indices, convergence of cost functions, and so on. The statistical uncertainty is coupled with the number of independent particle tracks used for the dose calculation and is usually determined by using the history by history method (Walters et al., 2002). Roughly, the square number of histories is needed to halve the statistical uncertainty of the dose distribution. Figure 12.7 demonstrates the impact of the statistical uncertainty, that is, the number of independent histories, on the dose distribution for a lung case using three wedged beams. Thereby, the MC uncertainty is the average of the statistical uncertainties (added in quadrature) of all dose values in the dose distribution with more than 50% of its maximum dose. Although the difference in quality of the isodose line is significant, the differences are less pronounced in the DVHs as shown in the lower right plot for the three cases. Some differences for the planning target volume (PTV) and myelon are visible when the statistical uncertainty is reduced from 10% to 2%. However, in the DVH, there is virtually no difference if the statistical uncertainty is further reduced to 0.5%. From this example, it is clear that for MC dose calculations, all point values, such as D_{max}, D_{min}, and so on, are highly critical, since those values might be associated with a high statistical uncertainty in the considered calculation. The uncertainty might be different if the calculation is done a second time using different initial settings for the random number generator. This impact is also important for the dose prescription to a specific point. Thus, quantities relating to a volume or a larger number of dose values, such as D_{median} or D_{mean}, are more reliable quantities when dealing with MC dose distributions. In general, a statistical uncertainty of 2% per beam results in a reasonable precision within the target volume, since usually three or more beams are used within a treatment plan. However, this might be no longer the case if the number of beams is below three or if OARs are considered. Since generally the OARs receive a lower particle fluence than the PTV, the statistical uncertainty in OARs can be much higher than in the target volume. Consequently, the evaluation for the OARs, for example, the DVHs or NTCP values, has to be done very carefully. It might be necessary to increase the number of histories to receive an acceptable statistical uncertainty in the normal tissue.

Other methods influencing the statistical noise are denoising techniques. Several different denoising algorithms have been developed. El Naqa et al. (2005) and the references therein provide a broad overview of existing denoising techniques as well

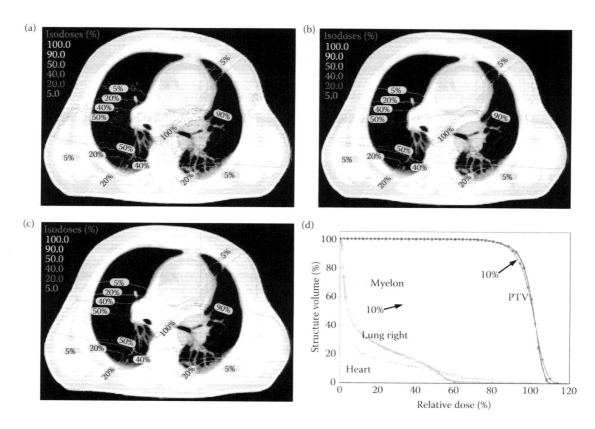

FIGURE 12.7 **(See color insert.)** Comparison of MC calculated dose distribution with different statistical uncertainties: 10% (a), 2% (b), 0.5% (c), and the impact on the DVHs (d), 10% squares, 2% dots, 0.5% triangles. For the DVH, there is virtually no difference between the 2% and 0.5% calculations.

as a detailed comparison between the different methods. The main difficulty is to keep the true dose gradients and remove only statistical, that is, random, noise. Additionally, it has to be mentioned that the denoising algorithm introduces some systematic errors. Thus, for clinical applications, denoising can be used during the initial treatment planning process and should not be used for the final dose calculation.

12.5.1.2 Calculation Time

The statistical uncertainty and the number of histories needed for the simulation are directly related to the CPU time needed for the calculation of the dose distribution. However, there are also some additional factors affecting the dose calculation time, for example, the dose voxel size, the dose scoring volume, the used beam modifiers, and the CPU. For clinical applications, it is important to have fast dose calculation algorithms. A summary of timing results for clinical treatment plans are given in the Task Group Report 105 of the AAPM (Chetty et al., 2007). Bearing in mind that this timing comparison has been performed several years ago and that the calculation speed of the computers is still rapidly increasing, nowadays, the calculation time for photon MC treatment planning is acceptable for clinical routine. Apart from increased computer speed, the MC dose calculation codes themselves have been optimized using variance reduction methods (Chetty et al., 2007) or due to simplification of the transport

code itself. For instance, the radiation transport within the beam modifier provides some potential in optimizing the calculation time, since there is also a trade-off between accuracy and CPU time. For example, the radiation transport through the secondary collimator jaws could be very simple (i.e., fast) if additionally the MLC is used below the jaws (Schmidhalter et al., 2010). Recently, MC algorithms have been implemented on graphical processing units demonstrating that the efficiency of the MC method could be further increased significantly (Badal and Badano, 2009; Jia et al., 2011). More information about this issue can be found in Chapter 19 of this book. Despite all these optimizations, the inverse planning might be the only exception where—given a large number of iterations—the overall calculation time might become unacceptable.

12.5.2 Comparisons of Dose Calculation Engines

12.5.2.1 Lung

Owing to the difficulties of the modeling of electron transport in low densities in the conventional dose calculation algorithms, lung is potentially the site showing the largest differences between dose distributions calculated using MC and conventional algorithms. There are many studies published in the literature investigating the differences between several dose calculation algorithms. These include investigations using inhomogeneous phantoms (Carrasco et al., 2004; Fogliata et al., 2007, 2011; Panettieri et al., 2007; Aarup et al., 2009; Fotina et al., 2009; Bush et al., 2011) as well as patient treatment planning studies (Wang et al., 2002a; Dobler et al., 2006; Vanderstraeten et al., 2006; Hasenbalg et al., 2007; Madani et al., 2007). As an example for a phantom study, Fogliata et al. (2007) demonstrate that even for simple lung inhomogeneities, algorithms based on pencil beam convolution can lead to dose differences of up to 30% compared with MC. For more advanced algorithms, this difference is reduced to about 8%. This difference could even be further reduced for a grid-based Boltzmann equation solver (Bush et al., 2011; Fogliata et al., 2011). In summary, the studies with inhomogeneous phantoms show that the higher the energy and the smaller the field sizes, the larger the differences between MC and conventional dose calculation algorithms become. In terms of lung patient studies, Wang et al. (2002a) have shown differences of more than 10% when MC is compared with an algorithm using the equivalent path length (EPL) inhomogeneity correction method. Additionally, they report that due to the reduced lateral electron range for lower energies, using 6 MV photons instead of 15 MV is advantageous (Wang et al., 2002b). This has been confirmed in a more recent study by Madani et al. (2007). If highly accurate algorithms are available, the energy selection depends on priority ranking of endpoints.

Vanderstraeten et al. (2006) compared MC calculated dose distributions for lung cancer patients with those calculated with the two different convolution/superposition algorithms Pinnacle (Philips Electronics N.V.) and Oncentra MasterPlan (Nucletron). Differences above 5% are reported. However, if the patient tissue is replaced by water, the differences above 5% disappear, thus demonstrating some limitations of dose calculation algorithms in conventional treatment planning algorithms. Hasenbalg et al. (2007) concluded that using a pencil beam algorithm may be inappropriate for lung cancer patients due to inaccurate calculated dose distributions. The more advanced dose algorithms AAA in Eclipse (Varian Medical Systems, Inc.) and collapsed cone convolution in MasterPlan (Nucletron) show dose differences of 5%. For stereotactic lung lesions, Dobler et al. (2006) show that the pencil beam algorithm overestimates measurements by up to 15% whereas the collapsed cone and MC underestimates measurements by 8% and 3%, respectively. In summary, the discrepancies concern differences in the build-up region, and the underestimation of the target coverage and of the isodose lines within the low-density tissue. The latter could also lead to differences in the dose distribution within the OARs themselves, which are located tangential to the incident fields. As an example with respect to the target coverage and isodose lines, Figure 12.8 shows the comparison between an MC and PB calculated dose distribution for a lung case within

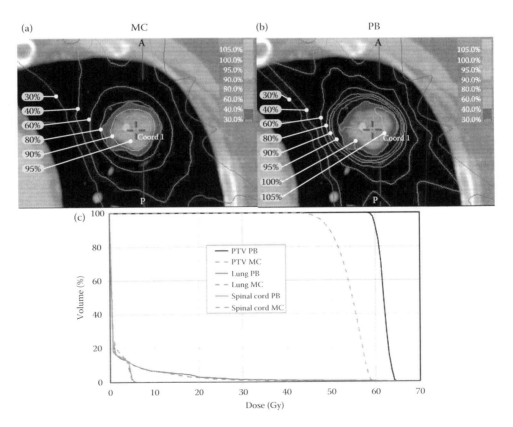

FIGURE 12.8 Comparison of an MC and pencil beam (PB) calculated dose distribution for a lung case within iPlan: (a) MC isodoses, (b) PB isodoses, and (c) DVH comparison.

iPlan. The MC calculated dose distribution demonstrates that the dose coverage of the target is not achieved when using the PB algorithm for the dose calculation. This is also expressed in the DVHs shown in Figure 12.8 if for both dose calculation algorithms the same number of MUs is used. Whereas the dose to the PTV is overestimated when using the PB algorithm, the differences for the OARs are small.

12.5.2.2 Head and Neck

Although most comparison studies between MC and conventional dose calculation algorithms have been performed for lung, there is also a motivation of investigating this comparison for head and neck cancer patients. Since critical structures such as spinal cord are often located close to the target volumes, dose distributions with high dose gradients are applied. Additionally, air cavities are in these regions that lead to difficulties for many conventional dose calculation algorithms due to inaccurate dose calculations around air cavities, especially when Batho or ETAR correction methods are used (du Plessis et al., 2001). Examples of studies including investigations for head and neck cancer are the following: Wang et al. (2002a), Francescon et al. (2003), Seco et al. (2005), Boudreau et al. (2005), Knoos et al. (2006), Partridge et al. (2006), Sakthi et al. (2006), Mihaylov et al. (2007), and Zhao et al. (2008). The differences are smaller for head and neck compared with those for lung cancer and it has not been proven that these differences are clinically relevant. However, in cases with air cavities, differences may have to be carefully considered when choosing the dose calculation algorithm (Paelinck et al., 2006). As an example, Figure 12.9 shows a dynamic MLC IMRT

larynx case. Comparisons between MC, AAA, and PB (Batho correction method applied) are shown together with the corresponding DVHs using the same number of MUs. The PB calculation results in a higher dose to the PTV compared with the AAA and MC algorithm. The MC method predicts a less homogeneous dose distribution within the PTV compared with the dose prediction when using the AAA algorithm. This observation corresponds with findings from other research groups (Seco et al., 2005; Sakthi et al., 2006). The dose to the OAR is predicted higher when MC is used. Overall, the differences are larger for the more simple pencil beam than for the more accurate AAA algorithm. Those differences need to be carefully considered in clinical routine.

12.5.3 Inclusion of Time Dependencies

One of the major advantages when using MC methods concerns the time dependencies of an irradiation in a 4D dose calculation. Whereas conventional algorithms usually discretize this problem, leading to increased calculation time, MC is able to sample the time with almost no increase in calculation time when considering the statistical uncertainty of the dose distribution in the target volume. Moreover, several approaches have been investigated to incorporate additionally the temporal patient movement into MC. One approach is to map the time-dependent dose distributions to a reference time point by using some kind of deformable registration between the time steps (Keall et al., 2004; Flampouri et al., 2006). Rosu et al. (2005) applied this method in conjunction with reducing the dose grid

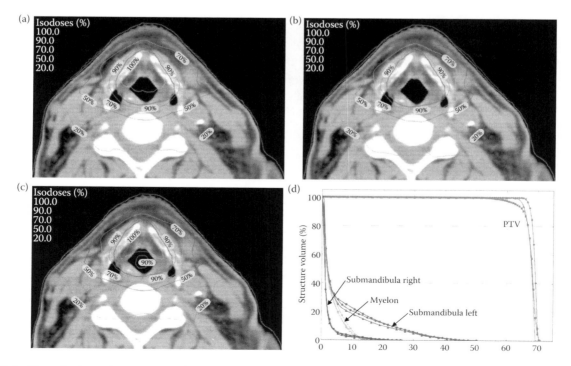

FIGURE 12.9 **(See color insert.)** Comparison of an MC, AAA, and PB calculated dose distribution for a head and neck case: PB isodoses (a), AAA isodoses (b), MC isodoses (c), DVH comparison with PB (dots), AAA (triangles), and MC (squares) (d). The red lines show the PTV.

size to increase the accuracy of the method. Alternatively, the deformation could be taken into account by adjusting the voxel geometry, that is, using deformed voxel instead of a rectilinear dose grid (Heath et al., 2007; Peterhans et al., 2011). A method that separates the particle transport from the energy deposition scoring is the energy transfer method. Thereby, the particle transport is done in a rectilinear grid. The scoring, that is, the energy deposition, is done in a reference geometry at a certain time point using deformable image registration (Siebers and Zhong, 2008). A further development of this method is called the energy and mass congruent mapping (Zhong and Siebers, 2009). By this method, not only the energy deposition but also the mass is mapped to a reference geometry by deformable image registration. Apart from using a time to resolve the patient movements, there are other temporal dependencies that can be handled within the same simulation, for example, dynamic MLCs of an IMRT application and gantry rotation (Paganetti et al., 2004; Manser et al., 2010). This behavior opens the door not only to accurate dose calculation but also to a highly valuable tool for investigations such as interplay effects (Paganetti et al., 2005). Another example is reported in a study by Schmidhalter et al. (2007). They investigated the dosimetric effect in the case where the secondary collimator jaws follow the open window of the MLC during dynamic MLC IMRT to reduce transmission radiation through the MLC.

12.5.4 Reevaluation of Studies

In order to investigate the impact of more accurately calculated dose distributions on outcome, MC might be the most appropriate method for reevaluation of clinical studies. There are only a few studies reported in the literature dealing with lung cancer patients (Lindsay et al., 2007). For lung patients, the outcome analysis is difficult, since the treatment plans for many lung studies are lacking inhomogeneity corrections. However, as already mentioned, the impact of these corrections is large for low-density tissue; thus, it is very important to use accurate dose distributions for investigating the mapping to clinical outcome. These correlations between the dose distribution and effects for tumor and normal tissue gained from retrospective studies may also help to define protocols for prospective studies. Additionally, such retrospective studies provide the opportunity to learn how to use MC for routine treatment planning, since switching from one dose calculation algorithm to another is always a much debated issue in the clinical environment. This is mainly because the treatment protocols used are based on the clinical experience. Most likely the change of the dose calculation algorithm will also change the dose prescription, that is, a simple scaling, due to changes of the dose distribution in the tumor. However, the situation is more difficult for the normal tissue. Generally, the OARs are often located at the field edges and the impact of the dose calculation algorithm on their dose distribution due to different beam models, implemented lateral electron transport, and so on is not obvious, especially in low-density tissues. Consequently, accurate dose calculations in OARs are of critical importance, particularly if protocols for dose escalation are considered. Thus, many more studies are needed.

12.5.5 MC as QA Tool

Another application of MC dose calculations is its use as a QA tool. This could be done for different purposes. As a first example, MC dose calculations can be used as an independent MU calculation (Seco et al., 2005; Fix et al., 2007; Yamamoto et al., 2007; Pisaturo et al., 2009). Thereby, the treatment plan of the conventional algorithm is recalculated with MC and the resulting MUs are compared. Instead of using the MC as an independent MU check, it could be just used to recalculate some complex or unclear treatment plans, especially if it is assumed that for such a plan the conventional algorithm is likely to fail. Bearing in mind that these are some individual cases, not the calculation time but rather the accuracy is highly important. Electronic portal imaging systems (EPIDs) are also often used for IMRT QA procedures. Thereby, measurements performed with the EPID are compared with a reference. One approach is to use MC to determine the reference image for this comparison (Siebers et al., 2004; Frauchiger et al., 2007; Jarry and Verhaegen, 2007; Beck et al., 2009; Wang et al., 2009). Owing to the increased distance of the EPID to the photon source in the linear accelerator head and the high resolution of the EPID detector, the calculation time is very long and thus optimized transport methods have to be used, for example, MC calculated kernels (Wang et al., 2009). Another purpose for the use of MC dose calculation algorithms is benchmarking, since it is assumed that MC is the most accurate method to predict dose distributions (Vanderstraeten et al., 2006; Fix et al., 2007). Since, for benchmarking, the accuracy is most important and there are generally only a few situations to be simulated, the calculation time is usually not very critical.

12.5.6 MC in Optimization

In the previous sections, the use of MC dose calculation for forward treatment planning has been described. However, in principle, the optimization also benefits from the accuracy achieved when using MC dose calculations in the inverse treatment planning process for IMRT or VMAT applications. In general, IMRT delivery is performed by a sequence of small fields or a dynamically moving and reshaping aperture. This leads to a great demand on the accuracy of the dose calculation algorithm. Furthermore, conventional dose calculation algorithms may give rise to dose errors, since a certain fraction of the dose is deposited by leakage and scatter radiation of the MLC. Most IMRT systems use simple but fast dose calculation algorithms, for example, pencil beam, within the optimization process. In many systems, a more sophisticated algorithm is used for the final dose calculation. The limited accuracy of these algorithms leads to dose prediction and convergence errors in the IMRT optimization (Mihaylov and Siebers, 2008).

Although the computing time of the MC dose calculation for a single field is very long compared with conventional dose calculation algorithms using a pencil beam or convolution/superposition, the MC calculation time is almost independent of the numbers of beams. This is not the case for the conventional dose calculation algorithms. However, the main disadvantage of using MC in the optimization remains to be the long calculation time, bearing in mind that many dose calculations during the optimization have to be done. As a consequence, one approach is to use MC dose calculation not in each iteration step but always after a certain number of iterations. In all other iteration steps, a fast dose calculation algorithm is used. This approach is called a hybrid scheme for IMRT optimization (Siebers et al., 2007; Siebers, 2008). Another approach is the use of MC calculated beamlets that are then used in the IMRT optimization process (Bergman et al., 2006; Sikora et al., 2009).

A few studies demonstrate that the use of MC in the optimization leads to improved dose distributions (Bergman et al., 2006; Dogan et al., 2006; Mihaylov and Siebers, 2008; Siebers, 2008; Sikora et al., 2009). Nevertheless, more detailed studies are needed to determine whether these improvements are clinically relevant.

12.6 Conclusions

MC dose calculation offers a great potential in photon treatment planning. Recently, the first few commercial treatment planning systems have MC methods for photon treatment planning available and more will come in the near future. Thus, MC dose calculation will be of increased importance in the future. Along with the shift to new dose calculation algorithms, there will be adjustments in clinical routine. It has been demonstrated that when using an MC dose calculation algorithm, not only the algorithm itself is changed, but almost the whole radiotherapy process is affected, for example, outlining, dose prescription, beam model, and CT conversion. There are still many open questions and studies required to carefully investigate all the impacts of this new technology and its impacts on outcome. This has to be done in close collaboration between clinical and research physicists.

References

Aarup L. R., Nahum A. E., Zacharatou C. et al. 2009. The effect of different lung densities on the accuracy of various radiotherapy dose calculation methods: Implications for tumour coverage. *Radiother Oncol* 91: 405–14.

Ahnesjo A. 1994. Analytic modeling of photon scatter from flattening filters in photon therapy beams. *Med Phys* 21: 1227–35.

Ahnesjo A., Knoos T., and Montelius A. 1992. Application of the convolution method for calculation of output factors for therapy photon beams. *Med Phys* 19: 295–301.

Ahnesjo A., Weber L., and Nilsson P. 1995. Modeling transmission and scatter for photon beam attenuators. *Med Phys* 22: 1711–20.

Alber M., Birkner M., Bakai A. et al. 2003. Routine use of Monte Carlo dose computation for head and neck IMRT optimization. *Int J Radiat Oncol Biol Phys* 57: S208.

Alexander A., Deblois F., Stroian G. et al. 2007. MMCTP: A radiotherapy research environment for Monte Carlo and patient-specific treatment planning. *Phys Med Biol* 52: N297–N308.

Andreo P. 1991. Monte Carlo techniques in medical radiation physics. *Phys Med Biol* 36: 861–920.

Arnfield M. R., Siantar C. H., Siebers J. et al. 2000. The impact of electron transport on the accuracy of computed dose. *Med Phys* 27: 1266–74.

Badal A. and Badano A. 2009. Accelerating Monte Carlo simulations of photon transport in a voxelized geometry using a massively parallel graphics processing unit. *Med Phys* 36: 4878–80.

Bakai A., Alber M., and Nusslin F. 2003. A revision of the gamma-evaluation concept for the comparison of dose distributions. *Phys Med Biol* 48: 3543–53.

Beck J. A., Budgell G. J., Roberts D. A., and Evans P. M. 2009. Electron beam quality control using an amorphous silicon EPID. *Med Phys* 36: 1859–66.

Bergman A. M., Bush K., Milette M. P. et al. 2006. Direct aperture optimization for IMRT using Monte Carlo generated beamlets. *Med Phys* 33: 3666–79.

Blanpain B. and Mercier D. 2009. The delta envelope: A technique for dose distribution comparison. *Med Phys* 36: 797–808.

Boudreau C., Heath E., Seuntjens J., Ballivy O., and Parker W. 2005. IMRT head and neck treatment planning with a commercially available Monte Carlo based planning system. *Phys Med Biol* 50: 879–90.

Brahme A. 1984. Dosimetric precision requirements in radiation therapy. *Acta Radiol Oncol* 23: 379–91.

Brualla L., Salvat F., and Palanco-Zamora R. 2009. Efficient Monte Carlo simulation of multileaf collimators using geometry-related variance-reduction techniques. *Phys Med Biol* 54: 4131–49.

Bush K., Gagne I. M., Zavgorodni S., Ansbacher W., and Beckham W. 2011. Dosimetric validation of Acuros XB with Monte Carlo methods for photon dose calculations. *Med Phys* 38: 2208–21.

Bush K., Townson R., and Zavgorodni S. 2008. Monte Carlo simulation of RapidArc radiotherapy delivery. *Phys Med Biol* 53: N359–70.

Bush K., Zavgorodni S. F., and Beckham W. A. 2007. Azimuthal particle redistribution for the reduction of latent phase-space variance in Monte Carlo simulations. *Phys Med Biol* 52: 4345–60.

Carrasco P., Jornet N., Duch M. A. et al. 2007. Comparison of dose calculation algorithms in slab phantoms with cortical bone equivalent heterogeneities. *Med Phys* 34: 3323–33.

Carrasco P., Jornet N., Duch M. A. et al. 2004. Comparison of dose calculation algorithms in phantoms with lung equivalent heterogeneities under conditions of lateral electronic disequilibrium. *Med Phys* 31: 2899–911.

Chetty I. J., Charland P. M., Tyagi N. et al. 2003. Photon beam relative dose validation of the DPM Monte Carlo code in lung-equivalent media. *Med Phys* 30: 563–73.

Chetty I. J., Curran B., Cygler J. E. et al. 2007. Report of the AAPM Task Group No. 105: Issues associated with clinical implementation of Monte Carlo-based photon and electron external beam treatment planning. *Med Phys* 34: 4818–53.

Deasy J. O., Blanco A. I., and Clark V. H. 2003. CERR: A computational environment for radiotherapy research. *Med Phys* 30: 979–85.

DeMarco J. J., Solberg T. D., and Smathers J. B. 1998. A CT-based Monte Carlo simulation tool for dosimetry planning and analysis. *Med Phys* 25: 1–11.

Deng J., Jiang S. B., Kapur A. et al. 2000. Photon beam characterization and modelling for Monte Carlo treatment planning. *Phys Med Biol* 45: 411–27.

Dobler B., Walter C., Knopf A. et al. 2006. Optimization of extracranial stereotactic radiation therapy of small lung lesions using accurate dose calculation algorithms. *Radiat Oncol* 1: 45.

Dogan N., Siebers J. V., Keall P. J. et al. 2006. Improving IMRT dose accuracy via deliverable Monte Carlo optimization for the treatment of head and neck cancer patients. *Med Phys* 33: 4033–43.

du Plessis F. C., Willemse C. A., Lotter M. G., and Goedhals L. 2001. Comparison of the Batho, ETAR and Monte Carlo dose calculation methods in CT based patient models. *Med Phys* 28: 582–9.

Dutreix A. 1984. When and how can we improve precision in radiotherapy? *Radiother Oncol* 2: 275–92.

El Naqa I., Kawrakow I., Fippel M. et al. 2005. A comparison of Monte Carlo dose calculation denoising techniques. *Phys Med Biol* 50: 909–22.

Fippel M. 1999. Fast Monte Carlo dose calculation for photon beams based on the VMC electron algorithm. *Med Phys* 26: 1466–75.

Fippel M., Haryanto F., Dohm O., Nusslin F., and Kriesen S. 2003. A virtual photon energy fluence model for Monte Carlo dose calculation. *Med Phys* 30: 301–11.

Fix M. K., Keall P. J., Dawson K., and Siebers J. V. 2004. Monte Carlo source model for photon beam radiotherapy: Photon source characteristics. *Med Phys* 31: 3106–21.

Fix M. K., Manser P., Frei D. et al. 2007. An efficient framework for photon Monte Carlo treatment planning. *Phys Med Biol* 52: N425–37.

Fix M. K., Volken W., Frei D. et al. 2011. Monte Carlo implementation, validation, and characterization of a 120 leaf MLC. *Med Phys* 38: 5311.

Flampouri S., Jiang S. B., Sharp G. C. et al. 2006. Estimation of the delivered patient dose in lung IMRT treatment based on deformable registration of 4D-CT data and Monte Carlo simulations. *Phys Med Biol* 51: 2763–79.

Fogliata A., Nicolini G., Clivio A., Vanetti E., and Cozzi L. 2011. Dosimetric evaluation of Acuros XB Advanced Dose Calculation algorithm in heterogeneous media. *Radiat Oncol* 6: 82.

Fogliata A., Vanetti E., Albers D. et al. 2007. On the dosimetric behaviour of photon dose calculation algorithms in the presence of simple geometric heterogeneities: Comparison with Monte Carlo calculations. *Phys Med Biol* 52: 1363–85.

Fotina I., Winkler P., Kunzler T. et al. 2009. Advanced kernel methods vs. Monte Carlo-based dose calculation for high energy photon beams. *Radiother Oncol* 93: 645–53.

Fraass B., Doppke K., Hunt M. et al. 1998. American Association of Physicists in Medicine Radiation Therapy Committee Task Group 53: Quality assurance for clinical radiotherapy treatment planning. *Med Phys* 25: 1773–829.

Fraass B. A., Smathers J., and Deye J. 2003. Summary and recommendations of a National Cancer Institute workshop on issues limiting the clinical use of Monte Carlo dose calculation algorithms for megavoltage external beam radiation therapy. *Med Phys* 30: 3206–16.

Francescon P., Cora S., and Chiovati P. 2003. Dose verification of an IMRT treatment planning system with the BEAM EGS4-based Monte Carlo code. *Med Phys* 30: 144–57.

Frauchiger D., Fix M. K., Frei D. et al. 2007. Optimizing portal dose calculation for an amorphous silicon detector using Swiss Monte Carlo Plan. *J Phys Conf Ser* 74: 021005.

Habib B., Poumarede B., Tola F., and Barthe J. 2010. Evaluation of PENFAST—A fast Monte Carlo code for dose calculations in photon and electron radiotherapy treatment planning. *Phys Med* 26: 17–25.

Hartmann Siantar C. L., Walling R. S., Daly T. P. et al. 2001. Description and dosimetric verification of the PEREGRINE Monte Carlo dose calculation system for photon beams incident on a water phantom. *Med Phys* 28: 1322–37.

Hasenbalg F., Fix M. K., Born E. J., Mini R., and Kawrakow I. 2008. VMC++ versus BEAMnrc: A comparison of simulated linear accelerator heads for photon beams. *Med Phys* 35: 1521–31.

Hasenbalg F., Neuenschwander H., Mini R., and Born E. J. 2007. Collapsed cone convolution and analytical anisotropic algorithm dose calculations compared to VMC++ Monte Carlo simulations in clinical cases. *Phys Med Biol* 52: 3679–91.

Heath E., Collins D. L., Keall P. J., Dong L., and Seuntjens J. 2007. Quantification of accuracy of the automated nonlinear image matching and anatomical labeling (ANIMAL) nonlinear registration algorithm for 4D CT images of lung. *Med Phys* 34: 4409–21.

Heath E., Seuntjens J., and Sheikh-Bagheri D. 2004. Dosimetric evaluation of the clinical implementation of the first commercial IMRT Monte Carlo treatment planning system at 6 MV. *Med Phys* 31: 2771–9.

International Commission on Radiation Units and Measurements Report 24. 1976. Measurement of Absorbed Dose in a Phantom Irradiation by a Single Beam of X or Gamma Rays. Bethesda, MD: ICRU.

International Commission on Radiation Units and Measurements Report 50. 1993. Prescribing, Recording and Report in Photon Beam Therapy. Bethesda, MD: ICRU.

Jarry G. and Verhaegen F. 2007. Patient-specific dosimetry of conventional and intensity modulated radiation therapy using a novel full Monte Carlo phase space reconstruction method from electronic portal images. *Phys Med Biol* 52: 2277–99.

Jia X., Gu X., Graves Y. J., Folkerts M., and Jiang S. B. 2011. GPU-based fast Monte Carlo simulation for radiotherapy dose calculation. *Phys Med Biol* 56: 7017–31.

Jiang S. B., Boyer A. L., and Ma C. M. 2001. Modeling the extrafocal radiation and monitor chamber backscatter for photon beam dose calculation. *Med Phys* 28: 55–66.

Jiang S. B., Sharp G. C., Neicu T. et al. 2006. On dose distribution comparison. *Phys Med Biol* 51: 759–76.

Keall P. J., Siebers J. V., Joshi S., and Mohan R. 2004. Monte Carlo as a four-dimensional radiotherapy treatment-planning tool to account for respiratory motion. *Phys Med Biol* 49: 3639–48.

Keall P. J., Siebers J. V., Libby B., and Mohan R. 2003. Determining the incident electron fluence for Monte Carlo-based photon treatment planning using a standard measured data set. *Med Phys* 30: 574–82.

Knoos T., Wieslander E., Cozzi L. et al. 2006. Comparison of dose calculation algorithms for treatment planning in external photon beam therapy for clinical situations. *Phys Med Biol* 51: 5785–807.

Kunzler T., Fotina I., Stock M., and Georg D. 2009. Experimental verification of a commercial Monte Carlo-based dose calculation module for high-energy photon beams. *Phys Med Biol* 54: 7363–77.

Lehmann J., Stern R. L., Daly T. P. et al. 2006. Dosimetry for quantitative analysis of the effects of low-dose ionizing radiation in radiation therapy patients. *Radiat Res* 165: 240–7.

Lindsay P. E., El Naqa I., Hope A. J. et al. 2007. Retrospective Monte Carlo dose calculations with limited beam weight information. *Med Phys* 34: 334–46.

Low D. A., Harms W. B., Mutic S., and Purdy J. A. 1998. A technique for the quantitative evaluation of dose distributions. *Med Phys* 25: 656–61.

Ma C. M. 1998. Characterization of computer simulated radiotherapy beams for Monte-Carlo treatment planning. *Radiat Phys Chem* 53: 329–44.

Ma C. M., Li J. S., Pawlicki T. et al. 2002. A Monte Carlo dose calculation tool for radiotherapy treatment planning. *Phys Med Biol* 47: 1671–89.

Mackie T. R., Reckwerdt P., and McNutt T. 1996. *Photon Beam Dose Computation*. Madison, WI: Advanced Medical Publishing.

Madani I., Vanderstraeten B., Bral S. et al. 2007. Comparison of 6 MV and 18 MV photons for IMRT treatment of lung cancer. *Radiother Oncol* 82: 63–9.

Magaddino V., Manser P., Frei D. et al. 2011. Validation of the Swiss Monte Carlo Plan for a static and dynamic 6 MV photon beam. *Z Med Phys* 21: 124–34.

Manser P., Schmidhalter D., Born E. J. et al. 2010. In-silico verification of RapidArc using Swiss Monte Carlo Plan. *Med Phys* 37: 3240.

Mihaylov I. B., Lerma F. A., Fatyga M., and Siebers J. V. 2007. Quantification of the impact of MLC modeling and tissue heterogeneities on dynamic IMRT dose calculations. *Med Phys* 34: 1244–52.

Mihaylov I. B. and Siebers J. V. 2008. Evaluation of dose prediction errors and optimization convergence errors of deliverable-based head-and-neck IMRT plans computed with a super-position/convolution dose algorithm. *Med Phys* 35: 3722–7.

Mijnheer B. J., Battermann J. J., and Wambersie A. 1987. What degree of accuracy is required and can be achieved in photon and neutron therapy? *Radiother Oncol* 8: 237–52.

Mijnheer B. J., Battermann J. J., and Wambersie A. 1989. Reply to Precision and accuracy in radiotherapy. *Radiother Oncol* 14: 163–7.

Oliver M., Gladwish A., Staruch R. et al. 2008. Experimental measurements and Monte Carlo simulations for dosimetric evaluations of intrafraction motion for gated and ungated intensity modulated arc therapy deliveries. *Phys Med Biol* 53: 6419–36.

Ottosson R. O. and Behrens C. F. 2011. CTC-ask: A new algorithm for conversion of CT numbers to tissue parameters for Monte Carlo dose calculations applying DICOM RS knowledge. *Phys Med Biol* 56: N263–74.

Paelinck L., Smedt B. D., Reynaert N. et al. 2006. Comparison of dose-volume histograms of IMRT treatment plans for ethmoid sinus cancer computed by advanced treatment planning systems including Monte Carlo. *Radiother Oncol* 81: 250–6.

Paganetti H., Jiang H., Adams J. A., Chen G. T., and Rietzel E. 2004. Monte Carlo simulations with time-dependent geometries to investigate effects of organ motion with high temporal resolution. *Int J Radiat Oncol Biol Phys* 60: 942–50.

Paganetti H., Jiang H., and Trofimov A. 2005. 4D Monte Carlo simulation of proton beam scanning: Modelling of variations in time and space to study the interplay between scanning pattern and time-dependent patient geometry. *Phys Med Biol* 50: 983–90.

Panettieri V., Wennberg B., Gagliardi G. et al. 2007. SBRT of lung tumours: Monte Carlo simulation with PENELOPE of dose distributions including respiratory motion and comparison with different treatment planning systems. *Phys Med Biol* 52: 4265–81.

Partridge M., Trapp J. V., Adams E. J. et al. 2006. An investigation of dose calculation accuracy in intensity-modulated radiotherapy of sites in the head & neck. *Phys Med* 22: 97–104.

Peterhans M., Frei D., Manser P., Aguirre M. R., and Fix M. K. 2011. Monte Carlo dose calculation on deforming anatomy. *Z Med Phys* 21: 113–23.

Pisaturo O., Moeckli R., Mirimanoff R. O., and Bochud F. O. 2009. A Monte Carlo-based procedure for independent monitor unit calculation in IMRT treatment plans. *Phys Med Biol* 54: 4299–310.

Rassiah-Szegedi P., Fuss M., Sheikh-Bagheri D. et al. 2007. Dosimetric evaluation of a Monte Carlo IMRT treatment planning system incorporating the MIMiC. *Phys Med Biol* 52: 6931–41.

Reynaert N., Coghe M., De Smedt B. et al. 2005. The importance of accurate linear accelerator head modelling for IMRT Monte Carlo calculations. *Phys Med Biol* 50: 831–46.

Rogers D. W., Faddegon B. A., Ding G. X. et al. 1995. BEAM: A Monte Carlo code to simulate radiotherapy treatment units. *Med Phys* 22: 503–24.

Rosu M., Chetty I. J., Balter J. M. et al. 2005. Dose reconstruction in deforming lung anatomy: Dose grid size effects and clinical implications. *Med Phys* 32: 2487–95.

Sahoo N., Kazi A. M., and Hoffman M. 2008. Semi-empirical procedures for correcting detector size effect on clinical MV x-ray beam profiles. *Med Phys* 35: 5124–33.

Sakthi N., Keall P., Mihaylov I. et al. 2006. Monte Carlo-based dosimetry of head-and-neck patients treated with SIB-IMRT. *Int J Radiat Oncol Biol Phys* 64: 968–77.

Salvat F., Fernandez-Varea J. M., and Sempau J. 2006. PENELOPE-2006: A Code System for Monte Carlo Simulation of Electron and Photon Transport. OECD Nuclear Energy Agency, Issy-les-Moulineaux, France.

Sarkar V., Stathakis S., and Papanikolaou N. 2008. A Monte Carlo model for independent dose verification in serial tomotherapy. *Technol Cancer Res Treat* 7: 385–92.

Schach von Wittenau A. E., Cox L. J., Bergstrom P. M., Jr. et al. 1999. Correlated histogram representation of Monte Carlo derived medical accelerator photon-output phase space. *Med Phys* 26: 1196–211.

Schmidhalter D., Fix M. K., Niederer P., Mini R., and Manser P. 2007. Leaf transmission reduction using moving jaws for dynamic MLC IMRT. *Med Phys* 34: 3674–87.

Schmidhalter D., Manser P., Frei D., Volken W., and Fix M. K. 2010. Comparison of Monte Carlo collimator transport methods for photon treatment planning in radiotherapy. *Med Phys* 37: 492–504.

Schwarz M., Bos L. J., Mijnheer B. J., Lebesque J. V., and Damen E. M. 2003. Importance of accurate dose calculations outside segment edges in intensity modulated radiotherapy treatment planning. *Radiother Oncol* 69: 305–14.

Seco J., Adams E., Bidmead M., Partridge M., and Verhaegen F. 2005. Head-and-neck IMRT treatments assessed with a Monte Carlo dose calculation engine. *Phys Med Biol* 50: 817–30.

Seco J., Sharp G. C., Wu Z. et al. 2008. Dosimetric impact of motion in free-breathing and gated lung radiotherapy: A 4D Monte Carlo study of intrafraction and interfraction effects. *Med Phys* 35: 356–66.

Sempau J., Sanchez-Reyes A., Salvat F. et al. 2001. Monte Carlo simulation of electron beams from an accelerator head using PENELOPE. *Phys Med Biol* 46: 1163–86.

Sempau J., Wilderman S. J., and Bielajew A. F. 2000. DPM, a fast, accurate Monte Carlo code optimized for photon and electron radiotherapy treatment planning dose calculations. *Phys Med Biol* 45: 2263–91.

Sheikh-Bagheri D. and Rogers D. W. 2002a. Monte Carlo calculation of nine megavoltage photon beam spectra using the BEAM code. *Med Phys* 29: 391–402.

Sheikh-Bagheri D. and Rogers D. W. 2002b. Sensitivity of megavoltage photon beam Monte Carlo simulations to electron beam and other parameters. *Med Phys* 29: 379–90.

Siebers J. V. 2008. The effect of statistical noise on IMRT plan quality and convergence for MC-based and MC-correction-based optimized treatment plans. *J Phys* 102: 12020.

Siebers J. V., Kawrakow I., and Ramakrishnan V. 2007. Performance of a hybrid MC dose algorithm for IMRT optimization dose evaluation. *Med Phys* 34: 2853–63.

Siebers J., Keall P. J., Kim J. O., and Mohan R. 2000. Performance Benchmarks of the MCV Monte Carlo System. In *XIII International Conference on the Use of Computers in Radiation Therapy*, ed W. Schlegel and T. Bortfeld, 129–31. Heidelberg: Springer-Verlag.

Siebers J. V., Keall P. J., Kim J. O., and Mohan R. 2002. A method for photon beam Monte Carlo multileaf collimator particle transport. *Phys Med Biol* 47: 3225–49.

Siebers J. V., Keall P. J., Libby B., and Mohan R. 1999. Comparison of EGS4 and MCNP4b Monte Carlo codes for generation of photon phase space distributions for a Varian 2100C. *Phys Med Biol* 44: 3009–26.

Siebers J. V., Kim J. O., Ko L., Keall P. J., and Mohan R. 2004. Monte Carlo computation of dosimetric amorphous silicon electronic portal images. *Med Phys* 31: 2135–46.

Siebers J. V. and Zhong H. 2008. An energy transfer method for 4D Monte Carlo dose calculation. *Med Phys* 35: 4096–105.

Sikora M. and Alber M. 2009. A virtual source model of electron contamination of a therapeutic photon beam. *Phys Med Biol* 54: 7329–44.

Sikora M., Dohm O., and Alber M. 2007. A virtual photon source model of an Elekta linear accelerator with integrated mini MLC for Monte Carlo based IMRT dose calculation. *Phys Med Biol* 52: 4449–63.

Sikora M., Muzik J., Sohn M., Weinmann M., and Alber M. 2009. Monte Carlo vs. pencil beam based optimization of stereotactic lung IMRT. *Radiat Oncol* 4: 64.

Teke T., Bergman A. M., Kwa W. et al. 2010. Monte Carlo based, patient-specific RapidArc QA using Linac log files. *Med Phys* 37: 116–23.

Tyagi N., Martin W. R., Du J., Bielajew A. F., and Chetty I. J. 2006. A proposed alternative to phase-space recycling using the adaptive kernel density estimator method. *Med Phys* 33: 553–60.

Van Dyk J., Barnett R. B., Cygler J. E., and Shragge P. C. 1993. Commissioning and quality assurance of treatment planning computers. *Int J Radiat Oncol Biol Phys* 26: 261–73.

Vanderstraeten B., Chin P. W., Fix M. et al. 2007. Conversion of CT numbers into tissue parameters for Monte Carlo dose calculations: A multi-centre study. *Phys Med Biol* 52: 539–62.

Vanderstraeten B., Reynaert N., Paelinck L. et al. 2006. Accuracy of patient dose calculation for lung IMRT: A comparison of Monte Carlo, convolution/superposition, and pencil beam computations. *Med Phys* 33: 3149–58.

Verhaegen F. and Devic S. 2005. Sensitivity study for CT image use in Monte Carlo treatment planning. *Phys Med Biol* 50: 937–46.

Verhaegen F. and Seuntjens J. 2003. Monte Carlo modelling of external radiotherapy photon beams. *Phys Med Biol* 48: R107–64.

Volken W., Frei D., Manser P. et al. 2008. An integral conservative gridding—Algorithm using Hermitian curve interpolation. *Phys Med Biol* 53: 6245–63.

Walters B. R., Kawrakow I., and Rogers D. W. 2002. History by history statistical estimators in the BEAM code system. *Med Phys* 29: 2745–52.

Wang L., Chui C. S., and Lovelock M. 1998. A patient-specific Monte Carlo dose-calculation method for photon beams. *Med Phys* 25: 867–78.

Wang S., Gardner J. K., Gordon J. J. et al. 2009. Monte Carlo-based adaptive EPID dose kernel accounting for different field size responses of imagers. *Med Phys* 36: 3582–95.

Wang L., Lovelock M., and Chui C. S. 1999. Experimental verification of a CT-based Monte Carlo dose-calculation method in heterogeneous phantoms. *Med Phys* 26: 2626–34.

Wang L., Yorke E., and Chui C. S. 2002a. Monte Carlo evaluation of 6 MV intensity modulated radiotherapy plans for head and neck and lung treatments. *Med Phys* 29: 2705–17.

Wang L., Yorke E., Desobry G., and Chui C. S. 2002b. Dosimetric advantage of using 6 MV over 15 MV photons in conformal therapy of lung cancer: Monte Carlo studies in patient geometries. *J Appl Clin Med Phys* 3: 51–9.

Webster G. J., Rowbottom C. G., and Mackay R. I. 2007. Development of an optimum photon beam model for head and-neck intensity-modulated radiotherapy. *J Appl Clin Med Phys* 8: 2711.

Yamamoto T., Mizowaki T., Miyabe Y. et al. 2007. An integrated Monte Carlo dosimetric verification system for radiotherapy treatment planning. *Phys Med Biol* 52: 1991–2008.

Zhao Y. L., Mackenzie M., Kirkby C., and Fallone B. G. 2008. Monte Carlo evaluation of a treatment planning system for helical tomotherapy in an anthropomorphic heterogeneous phantom and for clinical treatment plans. *Med Phys* 35: 5366–74.

Zhong H. and Siebers J. V. 2009. Monte Carlo dose mapping on deforming anatomy. *Phys Med Biol* 54: 5815–30.

Monte Carlo Calculations for Proton and Ion Beam Dosimetry

Hugo Palmans
National Physical Laboratory

13.1 Introduction

Proton and ion beam therapies have gained importance recently with, at the time of this writing, 38 centers in operation and 22 under construction (http://ptcog.web.psi.ch/). Given the nature of curative radiotherapy (the aim of achieving as high a cure rate as possible with as low as possible severe side effects), uncertainty requirements for proton and ion beam dosimetry are as stringent as for any other form of radiotherapy. Proton and ion Monte Carlo simulations play a crucial role at different levels of the dosimetry chain: in primary standards and reference instruments for the calculation of perturbations and in relative dosimetry for the calculation of both perturbations and the energy dependence of many detectors.

With respect to electromagnetic interactions, the transport simulation of protons and ions is similar as for electrons with the simplifying condition that both scattering angles and energy straggling are much reduced. This makes multiple scattering and straggling theories developed for electron transport work well for protons and ions, and major Monte Carlo codes for the transport of protons and ions such as Geant4, MCNPX, and SHIELD-HIT use similar models. In general, artifacts induced by simplifications in the implementation of these theories, for example, in boundary crossing algorithms, are less severe as for electrons. Of course, if the transport of the delta electrons themselves plays an important role in a dosimetry problem, as for example in transport simulations through small cavities, accurate electron transport is still crucial and the same considerations as brought up in Chapter 5 hold. A major difference with electron transport is that nuclear interactions play an important role. These interactions can lead to break up of the target nuclei

and in the case of ions also to break up of the projectile. The cross sections of these nuclear interactions are not well known (e.g., for protons, ICRU Report 63 (2000) quotes an uncertainty of 10% on the total interaction cross sections and 40% uncertainty on the production cross sections of secondary particles). Given that for a 200 MeV proton about 5% of the dose is deposited by charged secondary particles resulting from those nuclear interactions, the quoted uncertainties are potentially substantial for dosimetry problems.

This chapter reviews the application of proton and ion Monte Carlo simulations in the process of establishing primary standards, in calculating perturbations for ionization chambers in proton and carbon ion beam reference dosimetry, and in calculating absorbed dose response factors for relative dosimeters like alanine and radiochromic film using illustrations from the literature or new examples.

13.2 Primary Standards

Primary standards for proton and ion beam dosimetry can be divided in direct calorimetric methods and fluence-based methods (Palmans et al., 2009; Karger et al., 2010; Palmans, 2011).

In calorimetry, the temperature rise in the medium, ΔT_{med}, as a result of the energy deposited by ionizing radiation is measured with high precision. Absorbed dose to the medium, D_{med}, is then obtained by multiplying with the specific heat capacity of the medium, c_{med}, and the necessary correction factors:

$$D_{\mathrm{med}} = c_{\mathrm{med}} \cdot \Delta T_{\mathrm{med}} \cdot \frac{1}{1-h} \cdot \Pi k_i \qquad (13.1)$$

where *h* is the heat defect that represents the (fractional) energy that has been deposited by ionizing radiation but is taken away from the medium by any changes of its physical or chemical state and Πk_i is the product of correction factors for heat transported away or toward the measurement point, field nonuniformity, changed scatter, and attenuation due to the presence of nonmedium equivalent materials. The specific heat capacity is normally obtained from measuring the temperature rise in the medium due to electrical energy dissipation, either in a separate experiment or within the calorimeter setup. If the medium is water, the quantity of interest in reference dosimetry for radiotherapy, absorbed dose to water, is directly measured. If another medium is used, a conversion procedure is required.

Fluence-based methods rely on a measurement of the particle fluence at the surface, Φ, and knowledge of the stopping power in water at a shallow depth, *z*:

$$D_w(z) = \varphi \cdot \frac{S_w(z)}{\rho_w} \cdot \Pi k_i \qquad (13.2)$$

where $S_w(z)$ is the stopping power of the proton spectrum at depth *z* and Πk_i is the product of correction factors for beam divergence, scatter, field nonuniformity, beam contamination, and secondary particle buildup. It is obvious that this method relies on accurate values of the stopping power in water for which the uncertainty is estimated to be 1–2% for protons according to ICRU Report 49 (Palmans et al., 2009). The main instrument in use to measure the incident particle fluence is the Faraday cup, which enables an accurate measurement of the number of charged particles provided it is well designed to capture the entire beam and to repel charges generated in the entrance window and other components as well as to repel charges ejected from the cup back into it. For broad beams, an additional major uncertainty is due to the determination of the field area. For pencil beams, this uncertainty vanishes when the derived quantity is a laterally integrated dose. A major concern is the influence of electrons and protons generated in the entrance window that reach the collecting electrode (and thus contribute to the signal) as well as electrons liberated in and escaping from the collecting electrode (which enhance the signal). Both sources of perturbation to the measurement are usually suppressed by a guard electrode with a negative potential with respect to the electrode and casing sometimes with addition of a magnetic field.

13.2.1 Water Calorimetry

Water calorimeters require high-purity water to control the chemical heat defect (Ross and Klassen, 1996), which is usually contained in a small sealed glass or PMMA container. The perturbation by these container vessels can be measured or calculated by Monte Carlo simulations. The latter provides more information on the source of the perturbations. An example is shown here, where for three vessels used in proton dosimetry experiments (Palmans et al., 1996; Medin et al., 2006; Sarfehnia et al., 2010), these perturbations are calculated using MCNPX

2.4.0 (Pelowitz, 2005) and McPTRAN.CAVITY (Palmans, 2005), an in-house adaptation of PTRAN (Berger, 1993) that can simulate nonwater materials and inhomogeneous geometries. The vessels are a hollow PMMA cylinder of 40 mm diameter and 0.5 mm wall thickness (Palmans et al., 1996), a hollow glass cylinder of 67 mm diameter and 1 mm wall thickness (Medin et al., 2006), and a pancake cell of 24.9 mm thickness and 1.12 mm wall thickness (Sarfehnia et al., 2010). The resulting perturbation correction factors, obtained as the ratio of dose to water in a small volume around the measurement point in the absence and in the presence of the vessel, are shown in Figure 13.1 as a function of the residual range in water. For the PMMA vessels, the perturbations are negligible for all situations (note that PMMA is not preferred for other reasons (Seuntjens et al., 1993b)). For the glass vessels, the perturbation depends only on the thickness of the vessel, not on the shape, and a sigmoid fit to the data for the glass vessels is given by

$$k_p = 0.555 + \frac{0.4455}{1 + e^{-3\{\log(R_{res})+0.75\}}}, \qquad (13.3)$$

where R_{res} is the residual range.

The results are dominated by the shift of the effect depth of the measurement points due to the different water equivalent thickness of the glass vessel expressed as ratio of differential ranges in the continuous slowing down approximation (csda).

Sarfehnia et al. (2010) assumed this to be the only contribution and calculated this effect analytically. A small contribution, however, is also due to the different attenuation by nuclear interactions in the glass wall as compared to the equivalent water layer. This contribution was not more than 0.1% and the data in Figure 13.1 were corrected for it. Including this contribution, the simulated results are in good agreement with the measurement performed by Medin et al. (2006).

13.2.2 Graphite Calorimetry

13.2.2.1 Volume and Gap Corrections

In graphite calorimetry, the diffusivity of the medium (graphite) is such that a sample (the core) of it has to be thermally isolated from the environment to be able to measure its temperature rise. This is usually achieved with vacuum gaps introducing two perturbations that are difficult to study experimentally: the gap effect and the volume averaging effect. Both were studied for a small-body graphite calorimeter in a 60-MeV low-energy proton beam (Palmans et al., 2004), where the effects are expected to be larger than in a high-energy proton beam, using variants of PTRAN that later led to the earlier-mentioned McPTRAN.CAVITY code. The correction factor for the gap effect can be defined in two ways. The most obvious is the substituted gap correction factor derived as the ratio of dose in the core with the gaps filled up with graphite and the dose in the core with the gaps present. A more favorable, but also more subtle, definition is the compensated gap correction factor, which is defined as the

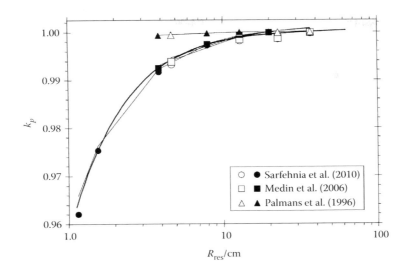

FIGURE 13.1 Perturbation correction factors for glass and PMMA vessels used in past experiments calculated for monoenergetic proton beams with energies 50, 60, 100, 150, 200, and 250 MeV at 3 cm depth (hollow data points) and at depths halfway between the phantom surface and the Bragg peak (solid data points), calculated using MCNPX with type-A uncertainties of less then 0.2%. The thin curves represent data calculated for the same conditions using McPTRAN.CAVITY, which are almost indistinguishable from the MCNPX results. The thick smooth curve represents a sigmoid fit to all the data using glass vessels with a thickness of about 1 mm, regardless the shape.

ratio of the average dose over the core volume when the core is moved toward the inner surface of the gap facing toward the beam source and the (new) gap is filled up with graphite and the average dose when the gap is filled up with air. The method with the compensated gap assures that, apart from the difference due to the air gap, the attenuation the particles that enter the core have undergone is the same for both dose determinations in the ratio. The volume averaging correction factor, k_{vol}, accounting for the finite dimensions of the core, has been calculated as the ratio of the dose in a small sphere at the center of the core and the average dose over the core volume. The simulated geometries for the gap and volume averaging are shown in Figure 13.2.

The calculated correction factors are shown in Figure 13.3 for modulated and nonmodulated beams. For the modulated beams, both substituted and compensated gap corrections are smaller than 0.1% except when part of the core is in the distal fall-off region of the spread-out Bragg peak (SOBP). For the

nonmodulated beam, the compensated gap correction is also small, but the substituted gap correction is substantial. The volume averaging corrections are smaller than a few tenths of a percent at shallow depth in a nonmodulated beam and in the plateau of the modulated beam. Nevertheless, the corrections and the variation of the latter might be larger than expected given the flat dose profile of an SOBP. This can be explained by ripples caused by the overlapping Bragg peaks illustrating that the dose at a point is very sensitive to small positioning errors.

13.2.2.2 Stopping Power Ratios

With a graphite calorimeter, one obtains dose to graphite from the measured radiation-induced temperature rise, while the quantity of interest is dose to water. A conversion procedure is thus needed. If the charged particle fluences, differential in energy at equivalent depths z_g and z_{w-eq} in a graphite phantom and a water phantom, respectively, are equal, then dose to graphite in the

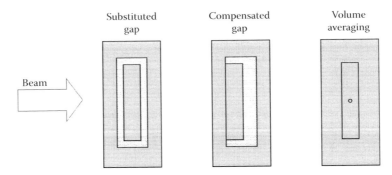

FIGURE 13.2 Side view of the geometries for the Monte Carlo calculation of the substituted gap correction factor, compensated gap correction factor, and volume averaging correction factor. All geometries are axially symmetric around the beam axis; all material is graphite, except the shaded areas, which are either vacuum or graphite.

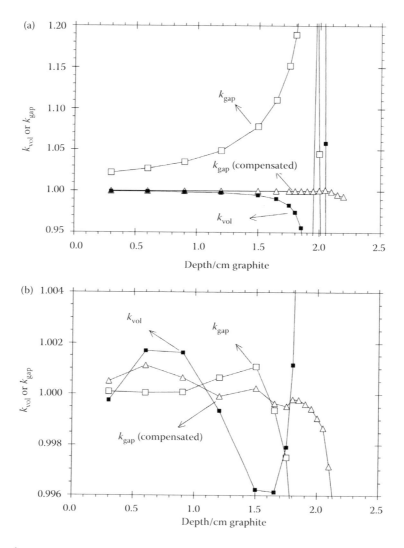

FIGURE 13.3   Gap correction factors and volume averaging correction factors as a function of depth calculated using McPTRAN.CAVITY for a nonmodulated 60 MeV beam (a) and for a full-modulated 60 MeV beam (b).

graphite phantom, $D_g(z_g)$, and dose to water in the water phantom, $D_w(z_{w-eq})$, at the equivalent depth in water, are related by

$$D_w(z_{w-eq}) = D_g(z_g) \cdot s_{w,g}(\Phi_g), \qquad (13.4)$$

where the water-to-graphite mass collision stopping power ratio for the total charged particle fluence distribution, Φ_g, in graphite is

$$s_{w,g}(\Phi_g) = \frac{\sum_i \left[\int_0^{E_{max,i}} \Phi_{E,g,i}(E) \cdot \left(\frac{S_{c,i}(E)}{\rho} \right)_w \cdot dE \right]}{\sum_i \left[\int_0^{E_{max,i}} \Phi_{E,g,i}(E) \cdot \left(\frac{S_{c,i}(E)}{\rho} \right)_g \cdot dE \right]}, \qquad (13.5)$$

and $\Phi_{E,g,i}$ is the fluence distributions differential in energy for charged particle type i in graphite and $S_{c,i}/\rho$ are the mass electronic collision stopping powers for particle type i.

For a given depth in graphite, the equivalent depth in water is given by

$$z_{w-eq} = z_g \cdot \frac{r_{0,w}}{r_{0,g}}, \qquad (13.6)$$

where $r_{0,w}$ and $r_{0,g}$ are the proton ranges in graphite and water, respectively. Here z_{80} is the depth on the distal edge of the Bragg peak where the ionization drops to 80% of the maximum ionization, was used as an estimate of the ranges $r_{0,w}$ and $r_{0,g}$.

Water-to-graphite mass collision stopping power ratios were calculated with Equation 13.5 using charged particle fluence distributions obtained from simulations using the Monte Carlo code Geant4 version 4.9.0 (Agostinelli et al., 2003) for 60 and 200 MeV monoenergetic protons incident on a slab phantom using the low-energy models for electromagnetic interactions, the default ICRU Report 49 stopping power parameterization and the precompound models for nonelastic nuclear

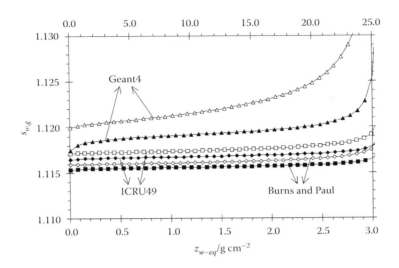

FIGURE 13.4 Water-to-graphite stopping power ratios as a function of depth for 60 MeV protons (hollow symbols, lower horizontal axis) and 200 MeV protons (solid symbols, upper horizontal axis) based on the stopping powers from Geant4 (Agostinelli et al., 2003), those from ICRU Report 49 (1993), and those from Burns (2009) and Paul (2006).

interactions. The stopping powers considered are the ones used in the Geant4 transport simulations, ICRU Report 49 stopping powers and stopping powers calculated using improved values of the mean excitation energies ($I = 82.5 \pm 2.5$ eV for graphite from Burns (2009) and $I = 80 \pm 2$ eV for water from Paul (2006)).

The resulting water-to-graphite stopping power ratios from the full Monte Carlo calculated fluence distributions are shown in Figure 13.4. The ICRU Report 49 data and the data based on the Burns and Paul I values agree within about 0.1% (but the uncertainties for the ICRU data are larger; about 0.9% vs. 0.5% for the Burns and Paul data) while Geant4 results deviate considerably. The Burns and Paul values are almost constant for all depths ($s_{w,g} = 1.116$ with an uncertainty of 0.5%, which can be regarded as the best estimate at present).

13.2.2.3 Fluence Correction Factors

Since the fluence distributions in water and graphite at equivalent depths are not equal, a fluence correction factor k_{fl} needs to be introduced in Equation 13.4:

$$D_w(z_{w-eq}) = D_g(z_g) \cdot s_{w,g}(\Phi_g) \cdot k_{fl}, \qquad (13.7)$$

which, given the relation between dose and stopping power in both materials, can be calculated as

$$k_{fl} = \frac{\sum_i \left[\int_0^{E_{max,i}} \Phi_{E,w,i}(E) \cdot \left(\frac{S_{c,i}(E)}{\rho} \right)_w \cdot dE \right]}{\sum_i \left[\int_0^{E_{max,i}} \Phi_{E,g,i}(E) \cdot \left(\frac{S_{c,i}(E)}{\rho} \right)_w \cdot dE \right]}. \qquad (13.8)$$

In these expressions, $\Phi_{E,w,i}$ and $\Phi_{E,g,i}$ are the fluence distributions differential in energy for charged particle type i in water and graphite, respectively. The fluence correction factor defined in Equation 13.8 can be interpreted as the conversion of dose to water in the graphite phantom, to dose to water at an equivalent depth in the water phantom.

Simulations were performed using Geant4 version 4.9.0 (Agostinelli et al., 2003) using the same geometry and settings as for the stopping power simulations. The calculated fluence correction factors are shown in Figure 13.5. For both energies, there is a small difference between the simulation in which only protons are considered or the one in which all secondary charged particles from nuclear interactions are considered (the difference being dominated by the contribution from alpha particles). The general trends are that the fluence correction factor is below unity at shallow depths and that it increases over the entire range of the beam. For 60-MeV protons, the value for water equivalent depths up to 2 g cm^{-2} is consistent with an experimental value of $k_{fl} = 1.000 \pm 0.003$ determined in a narrow beam setup for the same energy (Palmans et al., 2011). For 200 MeV protons, the fluence correction factor is varying more substantially and is about 2–3% below unity for depths up to about 10 g cm^{-2} beyond which it increases steeply to 6% above unity at a depth of 25 g cm^{-2}.

13.2.3 Faraday Cup Dosimetry

One of the difficulties with Faraday cups is the suppression of signal currents from secondary electrons generated in the cup that manage to escape as well as from secondary electrons and protons generated in the entrance window. While these are all as much as possible suppressed by electromagnetic fields, Monte Carlo simulations can help in understanding and quantifying these corrections as well as feeding into design considerations

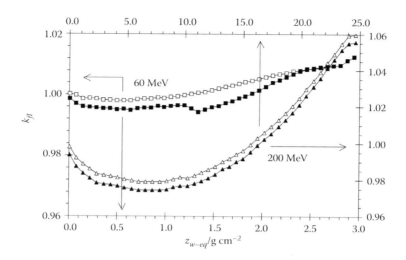

FIGURE 13.5 Monte Carlo calculated fluence correction factors as a function of depth for 60 MeV incident protons based on Equation 13.8 using only the proton fluence (hollow squares) and using all charged particles (solid squares) and for 200 MeV protons using only the proton fluence (hollow triangles) and using all charged particles (solid triangles).

for Faraday cups. Geant4 (Agostinelli et al., 2003) has the capability of combining transport of all particles with accelerations and deflections of charged particles by electromagnetic fields.

Another application of interest is the use of multilayer Faraday cups (MLFC) to quantify the number of nuclear interactions taking place in slabs of nonconducting materials by using the principle of mirror charges. The charge collected in each of a series of thin metallic foils separated by the insulating absorber plates is measured and the method basically determines in which slab protons stop (by nuclear interactions in the entrance plateau and at the end of their range due to electromagnetic interactions in the Bragg peak). This can serve as a test for both nuclear interaction and energy straggling models in the Monte Carlo simulation of the device and the experimental data from such an experiment (Paganetti and Gottschalk, 2003) have been proposed as a benchmark with this purpose (Palmans and Capote-Noy, 2010). Figure 13.6 gives the comparison of the experimental results with Geant4 simulations (Paganetti and Gottschalk, 2003), FLUKA (Ferrari et al., 2005; Battistoni et al., 2007) simulations (Rinaldi et al., 2011), and SHIELD-HIT (Sobolevsky, 2011) simulations (Henkner et al., 2009b). The agreement in the Bragg peak is excellent in all cases but deficiencies in the nuclear models in different codes (or different nuclear interaction models in the same code, Geant4) are visible in the entrance regions. The experimental data points beyond the Bragg peak could have a contribution from measurement noise as well as from knock-on protons generated by neutrons. The substantial disagreement between the measured data and all models cast doubt on the idea that the latter contribution would be significant.

13.3 Ionization Chambers

Reference dosimetry using ionization chambers is recommended to be performed according to IAEA TRS-398 (2000).

Absorbed dose to water, $D_{w,Q}$, in a proton or ion beam with quality Q is derived as

$$D_{w,Q} = M_Q \cdot N_{D,w,Q_0} \cdot k_{Q,Q_0} \qquad (13.9)$$

where M_Q is the ionization chamber reading corrected for influence quantities, N_{D,w,Q_0} is the absorbed dose to water calibration coefficient of the ionization chamber in a calibration beam of quality Q_0, and k_{Q,Q_0} is the beam quality correction factor accounting for the use of the calibration coefficient in a different beam quality Q.

Owing to the limited availability of experimental data, k_{Q,Q_0} values are in practice calculated from

$$k_{Q,Q_0} = \frac{(W_{air}/e)_Q \cdot (s_{w,air})_Q \cdot p_Q}{(W_{air}/e)_{Q_0} \cdot (s_{w,air})_{Q_0} \cdot p_{Q_0}} \qquad (13.10)$$

where, for both beam qualities, W_{air} is the mean energies required to produce an ion pair in dry air, $s_{w,air}$ is the water-to-air mass stopping power ratios for the local charged particle spectrum, and p_Q and p_{Q_0} are the ionization chamber perturbation correction factors in both proton/ion and calibration beam. The beam quality index Q for protons is the residual range R_{res} defined as

$$R_{res} = R_p - z \qquad (13.11)$$

where R_p is the practical range, defined as the depth distal to the Bragg peak at which the dose is reduced to 10% of its maximum value (for ions with a significant fragmentation tail, an estimate for the primary particle peak is made by constructing a tangent to the distal Bragg peak fall-off), and z is the depth of measurement. R_{res} is related to the most probable energy of the highest proton energy peak in the spectrum.

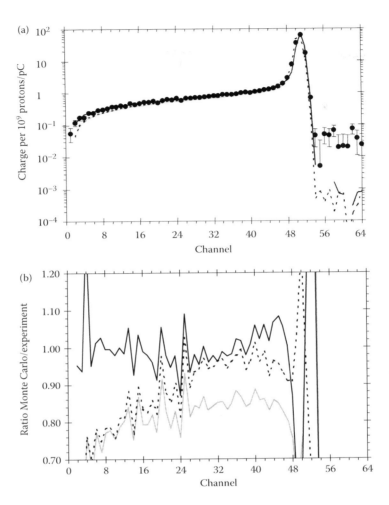

FIGURE 13.6 Comparison of the experimental data from the multilayer Faraday cup (Paganetti and Gottschalk, 2003) (symbols) with the simulations from three different Monte Carlo codes (full black line: Geant4 (Paganetti and Gottschalk, 2003), dotted black line: FLUKA (Rinaldi et al., 2011), full gray line: SHIELD-HIT (Henkner et al., 2009b)). (a) Shows the charge per 10^9 protons collected in each metallic plate, (b) shows the ratios between the Monte Carlo calculated data and the experimental data.

While the product of W_{air} and $s_{w,air}$ is mainly derived from experimental comparisons of calorimeters and ionization chambers, Monte Carlo simulations have mainly aided in calculating accurate values of the water-to-air stopping power ratios and perturbation correction factors, which is the subject of the following subsections.

13.3.1 Stopping Power Ratios

Monte Carlo simulation offers the advantage of in-line calculation of stopping powers and cavity integrals. This was illustrated for the dose conversion between graphite and water before. Medin and Andreo (1997) used this method to calculate water-to-air mass collision stopping power ratios for ionization chamber dosimetry in clinical proton beams. They used the in-house code PETRA, which includes transport of secondary electrons and secondary protons. Figure 13.7 gives the curves as a function

of the difference between the csda range and the measurement depth, which is close to the residual range. The Spencer–Attix stopping power ratios are found to be about 0.5% higher than unrestricted stopping power ratios for the same proton spectra in high-energy protons. Secondary protons have a minimal influence. The variation of the stopping power ratio as a function of energy is not more than 0.5% from $R_{res} = 0.5$ cm to the highest values encountered in clinical proton beams.

SHIELD-HIT (Sobolevsky, 2011) has been used to calculate water-to-air stopping power ratios for ionization chamber dosimetry in clinical carbon ion beams (and for other ions) (Henkner et al., 2009a; Lühr et al., 2011). The results as a function of residual range are shown in Figure 13.8. Both studies for proton and carbon ion beams show that the residual range is a good beam quality specifier corresponding with unique values of the stopping power ratio for depths shallower than the practical range R_p.

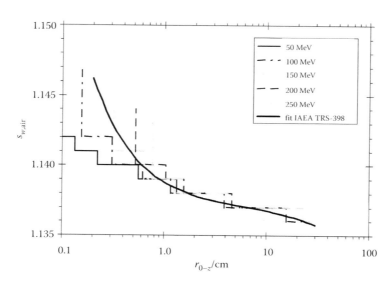

FIGURE 13.7 Water-to-air Spencer–Attix stopping power ratios for monoenergetic protons as a function of the csda range minus the depth as calculated using the Monte Carlo code PETRA (Medin and Andreo, 1997). The smooth black curve represents the equation used in IAEA TRS-398 (2000).

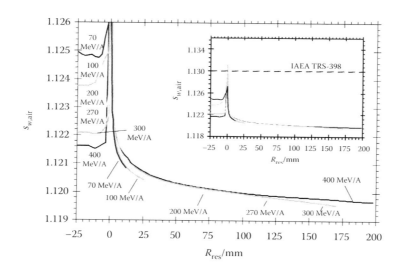

FIGURE 13.8 Water-to-air stopping power ratios for monoenergetic carbon ions as a function of the residual range calculated using the Monte Carlo code SHIELD-HIT (Lühr et al., 2011). The dashed line in the inset (with an extended vertical scale to show the values in the Bragg peak) represents the value used in IAEA TRS-398 (2000). (Adapted from Lühr A et al. Analytical expressions for water-to-air stopping-power ratios relevant for accurate dosimetry in particle therapy. *Phys Med Biol* 2011; 56: 2515–2533. With permission.)

13.3.2 Primary Particle Fluence Perturbation

The change of the proton fluence from a point in water to the cavity volume of an ionization chamber can be modeled as a gradient correction or alternatively as a shift of the effective depth in water from the center of the chamber. For carbon ions, projectile fragments need to be accounted for as well, but it can be assumed that the local changes of these fluence spectra over the dimensions of an ionization chamber are small. A detailed analytical model tracking particles along straight lines through the ion chamber geometry and averaging the

depth corresponding with the kinetic energy they have in the cavity was used to calculate the effective depth of cylindrical ionization chambers in proton and carbon ion beams (Palmans, 2006). Extensive Monte Carlo simulations using McPTRAN.CAVITY confirmed the analytical model. These were performed by simulating dose in the air cavity as a function of depth for all cylindrical ionization chamber types tabulated in IAEA TRS-398 for proton energies from 60 to 200 MeV and compare this with dose to air at the reference points in a homogeneous water phantom. All ionization chambers and all depths were simulated in one run on a distributed computing

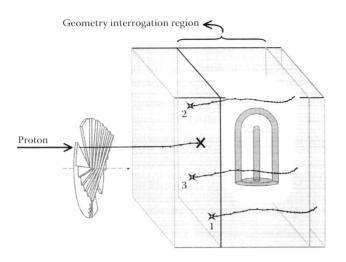

FIGURE 13.9 Geometry used in the Monte Carlo simulation of the effective depth of cylindrical (or thimble) ionization chambers. (Reproduced from Palmans H. Perturbation factors for cylindrical ionization chambers in proton beams. Part I: Corrections for gradients. *Phys Med Biol* 2006; 51(14): 3483–3501. With permission.)

grid system by implementing geometry interrogation, history splitting, and lateral range rejection. The geometry of the calculation is shown in Figure 13.9.

Figure 13.10 shows how the shift of the ionization chamber's effective depth from the chamber center, $(-)\Delta z_{w,tot}$, can be derived from the Monte Carlo simulations by calculating the distance from the center of the ionization chamber to the depth where dose to air in homogeneous water (in a point) equals the average dose to air in the air cavity. This can be done on the proximal slope (plateau) of the depth dose curve provided the curvature is sufficiently small or on the distal edge beyond the Bragg peak region.

Table 13.1 shows the values of $(-)\Delta z_{w,tot}$ obtained from the analytical model and from the Monte Carlo simulations using both methods explained in the two previous paragraphs. The selection of ionization chamber types listed in Table 13.1 have all been reported in the literature as being used as reference instruments. The results are compared with the general recommendation in IAEA TRS-398 to use 0.75 times the radius of the cavity. The last column shows experimental determinations of the effective depth in water showing that in all cases, the agreement is within the experimental uncertainties.

Both the frontal and distal results are in general in good agreement with the analytical results. The deviation from the IAEA recommendation is small for most ionization chambers but can be considerable (up to 1.5 mm) for some chambers commonly used in proton dosimetry (Palmans, 2006.) Among the chambers listed in Table 13.1, the largest deviation is for the FWT IC-18. For Farmer-type ionization chambers as well as for most ionization chambers with a thin wall and a central electrode diameter that is small compared to the cavity diameter, the agreement is in general good (Jakel et al., 2000; Palmans et al., 2000).

The possibility of applying the analytical model to the more challenging problem of converting a depth dose curve obtained with an ionization chamber to the depth dose in homogeneous water is demonstrated in Figure 13.11. A function consisting of an arbitrary set of depth dose values connected by a cubic spline was constructed. This function was convolved with the analytical model and a least square optimization of the set of depth dose values was performed to minimize the difference with the Monte Carlo simulated depth dose function for the ionization chamber. This method is illustrated for the NE2581 in Figure 13.11, which shows the optimized data points and the difference of the cubic spline with the depth dose curve in homogeneous water. This results in differences to the real depth dose curve smaller than 1%

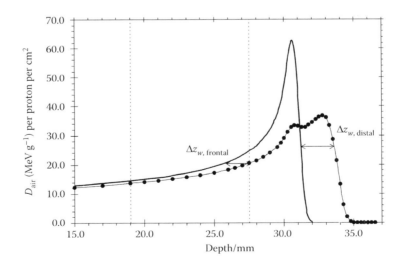

FIGURE 13.10 Depth distribution of dose to air calculated by Monte Carlo in homogeneous water (black curve), the same curve shifted over $\Delta z_{w,tot}$ (gray curve) and depth distribution of average dose to air in the cavity of an NE2581 ionization chamber as a function of depth (symbols connected by thin lines) in a 60 MeV monoenergetic beam. (Adapted from Palmans H. Perturbation factors for cylindrical ionization chambers in proton beams. Part I: Corrections for gradients. *Phys Med Biol* 2006; 51(14): 3483–3501. With permission.)

TABLE 13.1 $(-)\Delta z_{w,tot}$ for a Selection of Ionization Chambers (ICs) Listed in IAEA TRS-398 According to the Analytical Model, the Monte Carlo Simulations (Frontal: MC_f and distal: MC_d), and the General Recommendation of 0.75 Times the Cavity Radius Given in IAEA TRS-398

IC	MC_f	MC_d	Analytical Model	IAEA	Experiment
Capintec PR06C	2.4	2.4	2.4	2.4	2.0 ± 0.4 (Mobit et al. 2000)
Exradin A12	2.2	2.2	2.2	2.3	
FWT IC18	1.3	1.2	1.2	1.7	
NE2571	2.3	2.3	2.3	2.4	
NE2581	1.9	2.0	2.0	2.4	
NE2561	2.6	2.5	2.5	2.8	
PTW30001	2.4	2.4	2.4	2.3	3 ± 1 (Hartmann et al. 1999)
					2.5 ± 0.1 (Kanai et al. 2004)
PTW30004/12	2.1	2.1	2.1	2.3	
PTW30006/10/13	2.4	2.3	2.4	2.3	2.2 ± 0.2 (Jakel et al. 2000)
IBA FC65G	2.2	2.2	2.2	2.3	

Note: The last column gives experimental data from the literature for three Farmer-type chambers (with reference in parentheses).

of the dose maximum in the region frontal to the Bragg peak and 5% in the distal edge, where these differences are less important since they correspond to minor shifts in depth. This indicates that the method will be applicable to measured depth dose curves, which will always exhibit a certain level of fluctuations, as well.

13.3.3 Secondary Electron Perturbation

Several Monte Carlo studies using EGS4 to track secondary electrons in the past have demonstrated that secondary electron perturbations for ionization chambers in proton beams are not negligible and can amount to 1% and more for some wall materials (Palmans and Verhaegen, 1998; Palmans et al., 2001; Verhaegen and Palmans, 2001). In a more recent study, similar calculations were performed given that now better electron transport is implemented in EGSnrc and more computing resources are available enabling the calculation of these effects with lower uncertainties (Palmans, 2011). These are shown here as an example.

The modeled geometry is an air cylinder with a length of 24 mm and a diameter of 6.4 mm surrounded by a wall with a thickness of 0.38 mm, that is, dimensions close to that of a Farmer-type chamber, immersed in water. Both graphite and A150 as wall material were modeled. The broad monoenergetic

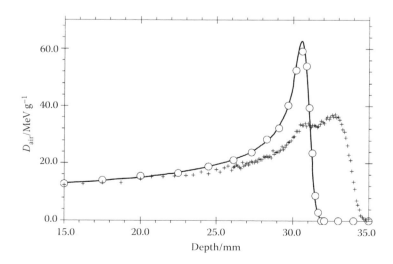

FIGURE 13.11 Reconstruction of a depth dose curve (circles) derived from a Monte Carlo simulated distribution in depth of dose to air in the cavity of a NE2581 ionization chamber (+ symbols) with deliberately added noise (1σ = 2%) in a 60 MeV monoenergetic beam. The full curve is the Monte Carlo calculated depth dose curve in homogeneous water. (Adapted from Palmans H. Perturbation factors for cylindrical ionization chambers in proton beams. Part I: Corrections for gradients. *Phys Med Biol* 2006; 51(14): 3483–3501. With permission.)

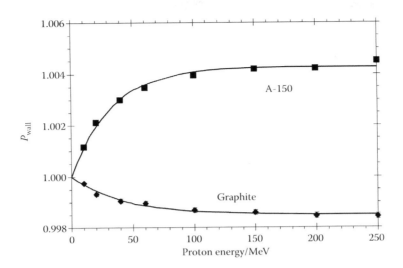

FIGURE 13.12 Wall perturbation factor as a function of proton energy for a Farmer-type chamber with graphite and A150 wall. The full lines are exponentials fitted to the data.

proton beam was incident perpendicular to the chamber axis without any loss of energy over the geometry. Electron recoil spectra in the cavity, wall, and surrounding water were calculated from the Rutherford cross section and the number of interaction centers derived from the electron density in each medium. Isotropic emission of electrons was considered, which is justified given the rotational symmetry of the problem. Electron transport through the geometry was simulated using DOSRZnrc (Rogers et al., 2010) with default transport and boundary crossing settings and cutoff energy for electron transport of 1 keV. Dose to air in the cavity, $D_{air,cav}$, for a broad proton field was evaluated as

$$D_{air,cav} = \Phi_p \cdot \left(\frac{S^\Delta}{\rho}\right)_{air} + \Phi_p \cdot N_A \cdot \left(\frac{Z}{A}\right)_{air}$$

$$\times \left[\int_{\Delta=1keV}^{W_{max}} \sigma_W \cdot dW\right] \cdot E_{e,dep,MC} \quad (13.12)$$

where Φ_p is the (monoenergetic) proton fluence, $(S^\Delta/\rho)_{air}$ the restricted mass stopping power at the energy of the protons, N_A the Avogadro's number, $(Z/A)_{air}$ the average ratio of atomic number and atomic weight of the medium, W_{max} the maximum energy that can be transferred to a secondary electron by the protons, σ_W the Rutherford cross-section differential in energy transfer (the integral thus being the total cross section for emission of an electron), and $E_{e,dep,MC}$ the Monte Carlo calculated dose contributions by electrons per electron set in motion in the cavity.

Wall perturbation correction factors were obtained as the ratio of $D_{air,cav}$ calculated by expression (13.12) for the chamber geometry with the wall set to water and for the chamber geometry with the wall set to the actual wall material. The calculated wall perturbation correction factors as a function of

energy are shown in Figure 13.12. For both wall materials, the correction factors exponentially saturate at high energies, for A150 around 1.0045 and for a graphite wall around 0.985. The ratio for the A150 and graphite-walled chamber at 55 MeV is 1.0045, which is in close agreement with the reported experimental value of 1.005 ± 0.002 in a nonmodulated beam of the same effective energy (obtained by comparing a NE2571 with an in-house built A-150 chamber with the same dimensions) (Palmans et al., 2001).

13.4 Energy Dependence of Relative Dosimeters

Many solid-state dosimeters exhibit an energy-dependent dose response for protons and ions. If this energy dependence is known, it is sufficient to know the local charged particle fluence differential in energy to predict the change in response of the dosimeter in the phantom or to correct its response for this energy dependence.

The relative effectiveness RE_i for charged particle type i can be defined in terms of an isoresponse dose as (Bassler et al., 2008; Herrmann et al., 2011)

$$RE_i = \frac{D_{det,\gamma}}{D_{det,i}} \quad (13.13)$$

where $D_{det,i}$ is the dose given to the detector by charged particle type i, and $D_{det,\gamma}$ is the dose that needs to be delivered in a ^{60}Co calibration beam to induce the same detector reading as the dose $D_{det,i}$. RE_i is energy dependent.

Assuming that only charged particles contribute to the local dose deposition, the average relative effectiveness for the local spectrum can be derived from the charged particle fluence differential in energy $\Phi_{E,i}$ for all particle types i as

$$\overline{RE} = \frac{\sum_i \int_0^{I_{max,i}} RE_i(E) \cdot \varphi_{I,i} \cdot \left(\dfrac{S}{\rho}\right)_{det,i} \cdot dE}{\sum_i \int_0^{I_{max,i}} \varphi_{I,i} \cdot \left(\dfrac{S}{\rho}\right)_{det,i} \cdot dE} \qquad (13.14)$$

where $(S/\rho)_{det,i}$ is the mass stopping power of the detector material for particle type *i*. Monte Carlo is the ideal tool to calculate these fluence distributions. Two examples are given here for radiochromic film and alanine.

13.4.1 Radiochromic Film

The energy dependence of two types of radiochromic film (GafChromic MD-V2-55 and EBT) was measured by Kirby et al. (2010) by comparison of their response with that of a calibrated Markus ionization chamber. Depth dose curves of 15 and 29 MeV proton beams calculated with FLUKA and the predicted underresponse according to Equation 13.14 are shown in Figure 13.13 and compared with the experimental depth response data recorded with MD-V2-55 film. The films were stacked perpendicular to the beam direction and the data points correspond with the depth of the sensitive layer of the films. The energy dependence of the film response was used in a study of laser-induced protons to infer the energy spectrum from the measured depth dose curve (Kirby et al., 2011). A library of depth dose curves and corresponding fluence spectra differential in energy as a function of depth was calculated using FLUKA and the spectrum was deconvolved from the measured depth response curve. Accounting for the reduced relative effectiveness of the film, the proton fluence inferred from this method was found to be 30%

higher than when not taking it into account, as was the practice used in previous papers, demonstrating that it is essential to make those corrections for the energy dependence of the dose response.

13.4.2 Alanine

The relative effectiveness of alanine in protons and ions was obtained from literature data (Palmans, 2003; Bassler et al., 2008). This information was used to predict the underresponse of alanine in the measurement of depth dose distributions in carbon ion beams and compare this with experimental data (Herrmann et al., 2011). The simulations were performed with FLUKA implementing the exact geometry of the stack of alanine pellets with a diameter of 5 mm and nominal thicknesses of 2.5 mm. The predicted response of alanine was derived from the fluence distributions differential in energy according to Equation 13.12. The comparison, exhibiting excellent agreement, is shown in Figure 13.14 for an SOBP established by the use of a ripple filter, which was modeled in detail in the simulation as well (Herrmann et al., 2011). As for radiochromic films, this work shows that it is essential to correct for the relative effectiveness when measuring depth dose curves using alanine.

13.5 Summary and Outlook

Monte Carlo simulations are an important tool in all levels of the dosimetry chain for proton and ion beam radiotherapy.

In primary standard calorimeters that aim to measure dose at a point in a homogeneous medium, Monte Carlo simulations are used to determine perturbations for any deviations from those conditions such as the presence of the high-purity-water-containing glass vessel in water calorimeters and the presence of vacuum gaps in graphite calorimeters as well as the fact that dose

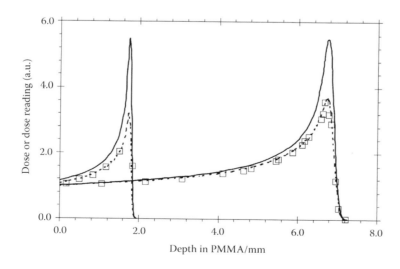

FIGURE 13.13 Depth dose curves for 15 and 29 MeV protons in PMMA simulated with FLUKA (solid lines) and corresponding depth dose reading (or isoeffect dose) curves for GafChromic MD-V2-55 (dashed lines) compared with experimental dose response data (squares). (Reproduced from Kirby D et al. Radiochromic film spectroscopy of laser-accelerated proton beams using the FLUKA code and dosimetry traceable to primary standards. *Laser Part. Beams* 2011; 29:231–239. With permission.)

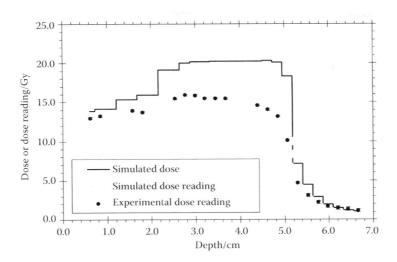

FIGURE 13.14 Depth dose curves (full line) and depth dose reading (or isoeffect dose) curves (dashed line) of alanine in the SOBP of a carbon ion beam (maximum energy 169 MeV/u) compared with experimental data (symbols). (Adapted from Herrmann R et al. *Med. Phys.* 2011; 38: 1859–1866, With permission.)

is measured over a volume in the latter. In graphite calorimetry, the calculation of stopping power ratios and fluence corrections needed for an accurate conversion between dose to graphite and dose to water also rely on Monte Carlo simulations.

For clinical reference dosimetry using ionization chambers, Monte Carlo simulations have made a major contribution to the calculation of water-to-air stopping power ratios and in the calculation of factors to correct for perturbations due to the presence of gradients and due to secondary electrons generated in the wall and the central electrode.

Since most solid-state detectors exhibit an energy-dependent dose response, it is essential to model the influence of the local spectrum when using them for measuring dose distributions. Examples are given of the application of Monte Carlo to this issue for alanine and radiochromic film and similar approaches could be used for other solid-state detectors.

Monte Carlo simulations will continue to play an essential role in all these aspects of dosimetry. The increased use of scanned particle beams will require characterizing calorimeters, ionization chambers, and relative dosimeters for their dose response in these complex dose delivery schemes. The Monte Carlo method is the only approach that can accurately model this complexity. The anticipated increased usage of solid-state detectors for treatment planning verification and auditing purposes in proton and ion beam dosimetry require corrections for the energy dependence of the dose response and thus knowledge of local charged particle spectra for which Monte Carlo is ideally suited. Also, the fluence correction factors potentially required when using plastic phantoms can be assessed by Monte Carlo in the same ways as for the conversion of dose to graphite derived from graphite calorimetry to dose to water.

Nuclear interactions are of importance in various dosimetric problems, such as the fluence corrections discussed above. In most cases, however, their influence is rather modest. An exception to this is PET and prompt gamma imaging (treated in Chapter 4) where the relations between dose distributions and the measured signals are critically dependent on accurate nuclear interaction cross sections, especially for the reaction channels producing the radioactive isotopes. Monte Carlo simulations will play an essential role not only in deriving dose distributions from the measurements but also in the benchmarking of the nuclear interaction models. An IAEA report reviewing the importance of nuclear interactions in dosimetry and providing such benchmark exercises will be forthcoming (Palmans and Capote-Noy, 2010).

References

Agostinelli S et al. GEANT4—A simulation toolkit. *Nucl Instrum Meth A* 2003; 506: 250–303.

Bassler N, Hansen JW, Palmans H, Holzscheiter MH, Kovacevic S and AD-4/ACE Collaboration. The antiproton depth dose curve measured with alanine detectors. *Nucl Instrum Meth B* 2008; 266: 929–936.

Battistoni G, Muraro S, Sala P R, Cerutti F, Ferrari A, Roesler S, Fassò A, Ranft J. The FLUKA code: Description and benchmarking. *Proceedings of Hadronic Shower Simulation Workshop 2006.* AIP Conf. Proc., 2007; 896: 31.

Berger MJ. Proton Monte Carlo transport program PTRAN National Institute for Standards and Technology Report NISTIR 5113. Gaithersburg, MD, USA: NIST; 1993.

Burns DT. A re-evaluation of the I-value for graphite based on an analysis of recent work on W, S(c,a) and cavity perturbation corrections. *Metrologia* 2009; 46: 585–590.

Ferrari A, Sala PR, Fassò A, Ranft J. FLUKA: A multi-particle transport code CERN-2005-10, INFN/TC_05/11, SLAC-R-773, 2005.

Hartmann GH, Jäkel O, Heeg P, Karger CP, Krießbach A. Determination of water absorbed dose in a carbon ion beam using thimble ionization chambers. *Phys Med Biol* 1999; 44: 1193–1206.

Henkner K, Bassler N, Sobolevsky N, Jäkel O. Monte Carlo simulations on the water-to-air stopping power ratio for carbon ion dosimetry. *Med Phys* 2009a; 36(4): 1230–1235.

Henkner K, Sobolevsky N, Jäkel O, Paganetti H. Test of the nuclear interaction model in SHIELD-HIT and comparison to energy distributions from GEANT4. *Phys Med Biol* 2009b; 54: N509–N517.

Herrmann R, Jäkel O, Palmans H, Sharpe P, Bassler N. Dose response of alanine detectors irradiated with carbon ion beams. *Med Phys* 2011; 38: 1859–1866.

http://ptcog.web.psi.ch/.

IAEA. Absorbed dose determination in external beam radiotherapy: An international code of practice for dosimetry based on standards of absorbed dose to water. Technical Report Series no. 398. Vienna, Austria: International Atomic Energy Agency; 2000.

ICRU. Stopping powers and ranges for protons and alpha particles. ICRU Report 49. Bethesda, MD, USA: International Commission on Radiation Units and Measurements; 1993.

ICRU. Nuclear data for neutron and proton radiotherapy and for radiation protection dose. ICRU Report 63. Bethesda, MD, USA: International Commission on Radiation Units and Measurements; 2000.

Jäkel O, Hartmann GH, Heeg P, Schardt D. Effective point of measurement of cylindrical ionization chambers for heavy charged particles. *Phys Med Biol* 2000; 45: 599–607.

Kanai T, Fukumura A, Kusano Y, Shimbo M, Nishio T. Cross-calibration of ionization chambers in proton and carbon beams. *Phys Med Biol* 2004; 49: 771–781.

Karger CP, Jäkel O, Palmans H, Kanai T. Dosimetry for ion beam radiotherapy. *Phys Med Biol* 2010; 5(21): R193–R234.

Kirby D, Green S, Fiorini F, Parker D, Romagnani L, Doria D, Kar S, Lewis C, Borghesi M, Palmans H. Radiochromic film spectroscopy of laser-accelerated proton beams using the FLUKA code and dosimetry traceable to primary standards. *Laser Part. Beams* 2011; 29: 231–239.

Kirby D, Green S, Palmans H, Hugtenburg R, Wojnecki C, Parker D. LET dependence of GafChromic films and an ion chamber in low-energy proton dosimetry. *Phys Med Biol* 2010; 55: 417–433.

Lühr A, Hansen DC, Jäkel O, Sobolevsky N, Bassler N. Analytical expressions for water-to-air stopping-power ratios relevant for accurate dosimetry in particle therapy. *Phys Med Biol* 2011; 56: 2515–2533.

Medin J, Andreo P. Monte Carlo calculated stopping-power ratios, water/air, for clinical proton dosimetry (50–250 MeV). *Phys Med Biol* 1997; 42(1): 89–105.

Medin J, Ross CK, Klassen NV, Palmans H, Grusell E, Grindborg JE. Experimental determination of beam quality factors, kQ, for two types of Farmer chamber in a 10 MV photon and a 175 MeV proton beam. *Phys Med Biol* 2006; 51(6): 1503–1521.

Mobit PN, Sandison GA, Bloch C. Depth ionization curves for an unmodulated proton beam measured with different ionization chambers. *Med Phys* 2000; 27: 2780–2787.

Paganetti H, Gottschalk B. Test of GEANT3 and GEANT4 nuclear models for 160 MeV protons stopping in CH2. *Med Phys* 2003; 30(7): 1926–1931.

Palmans H, Al-Sulaiti L, Andreo P, Thomas RAS, Shipley DR, Martinkovič J, Kacperek A. Conversion of dose-to-graphite to dose-to-water in clinical proton beams. In: *Standards, Applications and Quality Assurance in Medical Radiation Dosimetry—Proceedings of an International Symposium.* Vienna, November 9–12, 2010, Vienna, Austria: IAEA, 2011, pp. 343–355.

Palmans H, Capote Noy R. Summary Report Second Research Coordination Meeting on Heavy Charged-Particle Interaction Data for Radiotherapy. IAEA Report INDC(NDS)-0567 Distr. G+NM, Vienna, Austria: IAEA; 2010.

Palmans H, Kacperek A, Jäkel O. Hadron dosimetry. In: Rogers DWO, Cygler JE, editors. *Clinical Dosimetry Measurements in Radiotherapy.* Madison, WI, USA: Medical Physics Publishing; 2009, pp. 669–722.

Palmans H, Seuntjens J, Verhaegen F, Denis JM, Vynckier S, Thierens H. Water calorimetry and ionization chamber dosimetry in an 85-MeV clinical proton beam. *Med Phys* 1996; 23(5): 643–650.

Palmans H, Thomas R, Simon M, Duane S, Kacperek A, Dusautoy A, Verhaegen F. A small-body portable graphite calorimeter for dosimetry in low-energy clinical proton beams. *Phys Med Biol* 2004; 49(16): 3737–3749.

Palmans H, Verhaegen F. Monte Carlo study of fluence perturbation effects on cavity dose response in clinical proton beams. *Phys Med Biol* 1998; 43: 65–89.

Palmans H, Verhaegen F. On the effective point of measurement of cylindrical ionization chambers for proton beams and other heavy charged particle beams. *Phys Med Biol* 2000; 45: L20–L22.

Palmans H, Verhaegen F, Denis J-M, Vynckier S, Thierens H. Experimental p_{wall} and p_{cel} correction factors for ionization chambers in low-energy clinical proton beams. *Phys Med Biol* 2001; 46: 1187–1204.

Palmans H. Effect of alanine energy response and phantom material on depth dose measurements in ocular proton beams. *Technol. Cancer Res Treat* 2003; 2: 579–586.

Palmans H. McPTRAN.CAVITY and McPTRAN.RZ, Monte Carlo codes for the simulation of proton beams and calculation of proton detector perturbation factors. In: *The Monte Carlo Method: Versatility Unbounded in a Dynamic Computing World.* Illinois, USA: American Nuclear Society; 2005.

Palmans H. Perturbation factors for cylindrical ionization chambers in proton beams. Part I: Corrections for gradients. *Phys Med Biol* 2006; 51(14): 3483–3501.

Palmans H. Secondary electron perturbations in Farmer type ion chambers for clinical proton beams. In: *Standards, Applications and Quality Assurance in Medical Radiation Dosimetry—Proceedings of an International Symposium.* Vienna, November 9–12, 2010, Vienna, Austria: IAEA, 2011; pp. 309–317.

Palmans H. Dosimetry. In: Paganetti H, editor. *Proton Therapy Physics*. London: Taylor & Francis, 2011, pp. 191–219.

Paul H. A comparison of recent stopping power tables for light and medium-heavy ions with experimental data, and applications to radiotherapy dosimetry. *Nucl Instr Meth B* 2006; 247: 166–172.

Pelowitz D.MCNPX User's Manual LANL Report LA-CP-05-369. Los Alamos, NM: Los Alamos National Laboratory; 2005.

Rinaldi I, Ferrari A, Mairani A, Paganetti H, Parodi K, Sala P. An integral test of FLUKA nuclear models with 160 MeV proton beams in multi-layer Faraday cups. *Phys Med Biol* 2011; 56: 4001–4011.

Rogers DWO, Kawrakow I, Seuntjens JP, Walters BRB, Mainegra-Hing E. NRC User Codes for EGSnrc. NRCC Report PIRS-702(revB), National Research Council, Otawa, Canada; 2010.

Ross CK, Klassen NV. Water calorimetry for radiation dosimetry. *Phys Med Biol* 1996; 41(1): 1–29.

Sarfehnia A, Clasie B, Chung E, Lu HM, Flanz J, Cascio E, Engelsman M, Paganetti H, Seuntjens J. Direct absorbed dose to water determination based on water calorimetry in scanning proton beam delivery. *Med Phys* 2010; 37(7): 3541–3550.

Seuntjens J, Van der Plaetsen A, Van Laere K, Thierens H. Study of the relative heat defect and correction factors of a water calorimetric determination of absorbed dose to water in high-energy photon beams. *Proceedings of the Symposium on Measurement Assurance in Dosimetry IAEA-SM-330/6*. Vienna: IAEA; 1993b, pp. 45–59.

Sobolevsky N. SHIELD-HIT Home page. 2011. (http://www.inr.ru/shield/).

Verhaegen F, Palmans H. A systematic Monte Carlo study of secondary electron fluence perturbation in clinical proton beams (70–250 MeV) for cylindrical and spherical ion chambers. *Med Phys* 2001; 28: 2088–2095.

Protons: Clinical Considerations and Applications

Harald Paganetti
Massachusetts General Hospital and Harvard Medical School

14.1 Introduction

14.1.1 Short Introduction to Proton Therapy Beam Delivery

Most proton therapy facilities are currently based on the passive scattering technique. In the future, it is expected that more and more proton therapy patients will be treated using magnetic beam scanning. Both techniques can deliver a spread-out Bragg peak (SOBP) with a homogeneous target dose for each field, while the latter technique also allows intensity-modulated proton beam therapy. The basic principles are outlined in Figure 14.1 (see Paganetti and Bortfeld, 2005 for details).

To deliver a uniform dose to the target for a given treatment field (i.e., beam angle), an SOBP ensures coverage along the thickness of the target volume in the patient. It is created by combining pristine Bragg curves with different beam energies entering the patient. This can be achieved using a rotating absorber consisting of steps of various water-equivalent thicknesses (as in passive scattering) or by discretely changing the beam energy for each energy layer (as in beam scanning). In some scanning systems, the beam energy is adjusted using a range shifter in the treatment head. To cover the target volume laterally, the beam needs to be scanned magnetically in the x and y directions (beam scanning) or the proton beam needs to be broadened by using scatterers and absorbers consisting of various material combinations to achieve a laterally flat dose profile (passive scattering). In passive scattering, further shaping of the field is accomplished by apertures that confine the lateral dimension of the field and range compensators that are sculpted to conform the dose distribution to the distal shape of the target volume. For beam scanning, no patient-specific hardware is needed. However, depending on the width of the delivered pencil beams, an aperture might be used to improve the lateral penumbra of the beam.

14.1.2 Proton Physics

The energy loss of protons can be calculated using the Bethe-Bloch equation. Monte Carlo simulations implement multiple scattering in condensed history class II methods (Kawrakow and Bielajew, 1998). The Moliere theory predicts the scattering angle distribution and the Lewis method (Lewis, 1950) allows calculation of moments of lateral displacement, angular deflection, and correlations of these quantities. The implementation of the multiple scattering theory can differ slightly from Moliere's theory (Andreo et al., 1993; Gottschalk et al., 1993; Urban, 2002). Multiple coulomb scattering is the main cause for a broadening of the beam with depth.

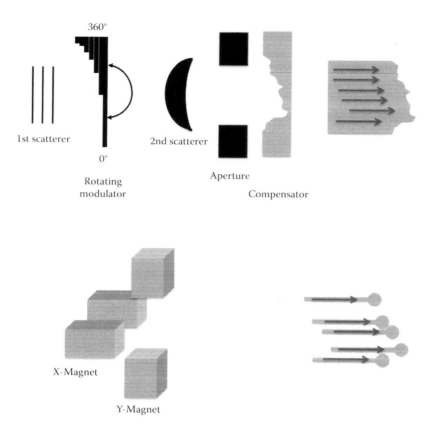

FIGURE 14.1 Illustration of beam delivery in passive scattered proton therapy and beam scanning. Upper, passive scattering: The beam enters from the left and is broadened by a double-scattering system and modulated to an SOBP by a rotating modulator. The field is confined by an aperture. A range compensator modulates the range so that the distal edge of the SOBP follows the distal edge of the target. Lower, beam scanning: Two magnets are used to scan a pencil beam in *x* and *y* directions. The beam energy is ideally modified prior to entering the treatment head (an aperture might still be used to improve the field penumbra).

The majority of dose is deposited via ionization, excitation, multiple coulomb scattering, and nuclear interactions. Depending on the interaction type, the physics is based on models, parameterizations, experimental data, as well as combinations of these. For nuclear interactions, cross sections as a function of proton energy may not be available for all reaction channels and need to be approximated. Although nuclear interactions do not significantly influence the shape of the Bragg peak, they do have a significant impact on the depth dose distribution because they cause a reduction in the proton fluence as a function of depth as well as secondary protons contributing significantly mainly in the entrance region of the Bragg curve (Paganetti, 2002). Note that as a rule of thumb, about 1% of the primary protons undergo a nuclear interaction per centimeter beam range. In pencil beam scanning, secondary particles emitted in nuclear interactions cause a "nuclear halo" around each pencil. The contribution can be significant when adding multiple pencils (Soukup and Alber, 2007; Sawakuchi et al., 2010). Interactions between the proton and a nucleus can be modeled as an intranuclear cascade with the probability of secondary particle emission. Once the energy of the particles in a cascade has reached a lower limit, a preequilibrium model can be applied.

Production cuts for secondary particles can influence energy loss and thus the simulation results (van Goethem et al., 2009). For proton dose calculation, primary and secondary protons account for about 98% of the dose, depending on the beam energy and including the energy lost via secondary electrons created by ionizations (Paganetti, 2002). The maximum range of the delta electrons in water is about 2.5 mm for a 250 MeV proton. The highest energy electrons are preferentially ejected in a forward direction. The energy of most electrons in a proton beam is much less than 300 keV, which corresponds to a range of 1 mm in water. The explicit tracking of secondary electrons can thus be neglected in most cases. For applications other than dose calculation, additional particles might need to be considered, for example, neutrons.

A Monte Carlo code might allow different physics settings from which the user can choose. Some settings might not be tailored for the energy domain of proton therapy and therefore require adjustment (Herault et al., 2005; Stankovskiy et al., 2009; Pia et al., 2010). The dependencies of proton therapy-related simulations on different physics settings can be considerable (Paganetti and Gottschalk, 2003; Kimstrand et al., 2008a; Zacharatou Jarlskog and Paganetti, 2008a). Even with the "correct" settings, values for the mean excitation energy might have uncertainties in the order of 5–15%. Such an uncertainty for beam-shaping materials can lead to up to >1 mm uncertainty in the predicted beam range in water (Andreo, 2009). It is important to consider this uncertainty when simulating energy loss in thick targets (Moyers et al., 2007; Andreo, 2009; van Goethem et al., 2009; Gottschalk, 2010). In particular for nuclear interactions, there are significant uncertainties in physics cross sections or models. For example, according to the international commission on radiation units & measurements (ICRU) (ICRU, 2000), angle-integrated emission spectra for neutron and proton interactions are known only to within 20–30% uncertainty.

14.1.3 Proton Monte Carlo Systems

The Monte Carlo codes widely used in proton therapy are FLUKA (Ferrari et al., 2005; Battistoni et al., 2006), Geant4 (Moyers et al., 2007; Gottschalk, 2010), MCNPX (Waters, 2002; Pelowitz, 2005), VMCpro (Fippel and Soukup, 2004), and Shield-Hit (Dementyev and Sobolevsky, 1999). These codes differ not only in accuracy (presumably this difference is small) but also in the user interface. For example, Geant4 provides only an assembly of object-oriented toolkit libraries with the functionality to simulate different process organized in different functions within a C++ class structure. The codes also differ in the level of control over tracking parameters (e.g., physics settings, step sizes, and material constants).

In addition to the codes mentioned above, there are also programs that serve as interfaces to Monte Carlo codes. For Geant4, these are, for example, PTsim (Aso et al., 2010), GATE (Santin et al., 2003; Jan et al., 2004), GAMOS (Arce et al., 2008), and TOPAS (Perl et al., 2012). TOPAS (TOol for PArticle Simulations), makes Monte Carlo simulation more readily available for research and clinical physicists aiming at modeling passive scattering or scanning beam treatment heads. Furthermore, the user is able to model a patient geometry based on computed tomography (CT). The code provides advanced graphics, and is fully four-dimensional (4D) to handle variations in beam delivery and patient geometry during treatment. All this is achieved with a user-friendly interface that does not require experience in programming.

14.2 Simulating a Radiation Field Incident on a Patient or an Experimental Setup

14.2.1 Characterizing Proton Beams at the Treatment Head Entrance

Proton therapy facilities typically consist of a cyclotron or synchrotron, which is connected to one or several treatment rooms by a complex beamline. The basic beamline elements between the accelerator and the treatment room that might be simulated using Monte Carlo are bending magnets and energy degraders. Monte Carlo beam transport through carbon and beryllium degraders has been performed with the goal of improving beam characteristics (van Goethem et al., 2009). However, beam optics calculations are often done numerically (Kurihara et al., 1983) despite specialized Monte Carlo codes that simulate magnetic beam steering (Brown et al., 1980).

Thus, Monte Carlo simulations typically start either at the exit of the treatment head, if a phase space parameterization or beam model is defined, or at the entrance of the beamline into the treatment room. If the Monte Carlo simulation starts at the treatment room entrance, a reliable parameterization of the beam at this position is needed. The variables are beam energy (E), energy spread (ΔE), beam spot size (σ_x, σ_y), and beam angular distribution ($\sigma_{\theta x}$, $\sigma_{\theta y}$) (Paganetti et al., 2004a; Hsi et al., 2009). There are various methods to measure these parameters. The beam spot size can usually be determined with a segmented transmission ionization chamber located at the treatment head entrance. The size of a proton beam is usually in the order of 2–8 mm in σ_x or σ_y, while the angular spread can be in the order of 2–5 mm-mrad for a cyclotron beam. It can also be parameterized using the emittance of the beam, defined as the product of the size and angular divergence of the beam in a plane perpendicular to the beam direction. The significance of the beam spot size might depend on the design of the treatment head, here in particular the design of the modulator wheel, as a bigger spot size might impact the number of absorber steps covered by the beam at a given time. The energy and energy spread can be obtained with sufficient accuracy by measuring range and shape of Bragg peaks in water (Bednarz et al., 2011) or using an elastic scattering technique (Brooks et al., 1997). Their values might influence the flatness of an SOBP because of the peak-to-plateau ratio of the individual Bragg peaks that form an SOBP. The energy spread of proton beams entering the treatment head from a cyclotron is typically <1% (DE/E) while a synchrotron may extract beams with an energy spread two orders of magnitude lower. Clinically most important is certainly the beam energy as fields are prescribed by the beam range in water.

The parameters describing the beam are typically correlated. Such a correlation, for example, between the particle's position within a beam spot and its angular momentum, needs to be taken into account when modeling proton beam scanning (Dowdell et al., 2012). Consequently, a parameterization of the phase space at treatment head entrance is needed and might be defined based on the knowledge of the magnetic beam steering system, that is, from first principles, or by fitting measured data (Paganetti et al., 2008). Depending on the beam steering, a scanned proton beam can be convergent or divergent. For passive scattered delivery, any correlation will most likely be blurred by the scattering material in the treatment head (Paganetti et al., 2004a). For the same reason, the angular spread at treatment head entrance has little influence on the beam exiting the treatment head in passive scattering.

14.2.2 Monte Carlo Modeling of a Therapy Treatment Head

If we neglect hybrid methods, there are three different types of treatment head concepts in proton therapy. There are treatment heads to deliver beam scanning, passive scattering, or are capable of doing both. Down to which accuracy one needs to model the treatment head in a Monte Carlo system depends on the purpose of the simulation. If the purpose is to calculate phase space distributions for dose calculation in the patient, it is most likely sufficient to have only beam-shaping devices included in the simulation. For beam scanning, these are the scanning magnets plus (if used) a patient-specific aperture. For passive scattering, these include the double- or single-scattering system, the modulator wheel, aperture, and compensator. If dose calculation is not the only purpose of the treatment head model, other devices may have to be included as well, for example, ionization chambers for detector studies or absolute dosimetry, or housing of devices to study scattering or shielding effects. To simulate a realistic dose distribution, simulating treatment head ionization chambers in detail might not be necessary because they cause negligible scattering and energy loss of the beam. Simulating plain or segmented ionization chambers is done for the purpose of designing ionization chambers or studying beam steering, as well as calculating the absolute dose in machine monitor units (Herault et al., 2005, 2007; Paganetti, 2006; Koch et al., 2008). These considerations do not include any devices, for example, detectors, one might want to model for research purposes. Monte Carlo simulations have been performed to design a prompt gamma detector for quality assurance (Kang and Kim, 2009). Another example is the use of Monte Carlo simulations to optimize image reconstruction for proton CT (Li et al., 2006; Schulte et al., 2008). Monte Carlo simulations have also been used to study aperture scattering (van Luijk et al., 2001; Kimstrand et al., 2008b; Titt et al., 2008a) or multileaf collimators to replace patient-specific apertures (Bues et al., 2005).

There are various reports on Monte Carlo treatment head simulations (Paganetti et al., 2004a; Fontenot et al., 2005; Herault et al., 2005, 2007; Newhauser et al., 2005; Polf and Newhauser, 2005; Koch et al., 2008; Titt et al., 2008b; Peterson et al., 2009b; Stankovskiy et al., 2009). Monte Carlo simulations have been used when commissioning planning systems (Koch and Newhauser, 2005; Newhauser et al., 2005, 2007). Furthermore, simulations are very helpful for designing a new facility, for quality assurance, and for supporting everyday operation, for example, to calculate tolerances on the appropriate operational parameters for beam delivery.

Quality assurance in radiation therapy is based on well-defined experimental studies that are repeated frequently. Nevertheless, Monte Carlo simulations can assist clinical quality assurance procedures. By simulating dose distributions and varying beam input parameters, tolerance levels for beam parameters can be defined (Paganetti et al., 2004a). Further, slight uncertainties or misalignments in the treatment head geometry that might occur over time can be simulated.

A treatment planning system defines a field either by specifying the treatment head parameters directly or by specifying the range and modulation width, which are then converted into the treatment head parameters internally by a treatment control system. Other than in photon therapy (where this is true only for collimators and jaws), proton therapy treatment heads are highly patient field specific. For each patient field, there can be a different beam energy (or multiple energies) at treatment head entrance and there are unique settings of devices in the treatment head. For beam scanning, these unique settings might be the setting of the magnetic fields and the scanning pattern, while for passive scattered delivery this refers to settings of the scattering system, choice of the modulator wheel, settings of variable jaws, and patient-specific aperture and compensators. The latter two devices, being mounted at the nozzle exit, might be excluded from the treatment head modeling and incorporated in the patient dose calculation because they are fabricated patient specifically. Typically, at the start of the simulation, a generic treatment head geometry is initialized and then modified using parameters provided via an input file.

The accuracy with which the treatment head elements can be modeled depends on the available information, for example, whether drawings of geometries are provided by the vendor (Paganetti et al., 2004a). In addition, it might depend on our knowledge of exact material compositions and material properties (Bednarz et al. 2011). As in photon Monte Carlo, the modeling of machine-specific components in the treatment head can be done using the manufacturer's blueprints (Paganetti et al., 2004a, 2008; Titt et al., 2008b). Some of the devices can be modeled relatively easy within a Monte Carlo code as the geometry represents a combination of regular geometrical objects. This holds, for example, in the case of ionization chambers or scattering foils. Irregular-shaped objects can be contoured scatterers (Figure 14.2), modulator wheels (Figure 14.3), as well as apertures and compensators. These can be modeled easily if the Monte Carlo code is capable of reading computer-aided design format or if the Monte Carlo allows the definition of objects by importing boundary points in other formats. If these options do not exist, devices have to be programmed by combining regular-shaped objects.

Contoured scatterers consist of two components, one made out of a high-Z and one made out of a low-Z material. They might be modeled by combining cones (Paganetti et al., 2004a). The thickness of the high-Z material decreases radially with distance to the field center while the thickness of the low-Z material increases as a function of radial distance. The scatterer produces a parallel beam and the bimaterial design ensures that the scattering power is independent of the beam energy.

To simulate a modulator wheel, the wheel steps can be modeled as segments out of a circular structure. Each of these segments can then be characterized by thickness, material, minimum and maximum radius, as well as a start and a stop angle. The rotational motion of the modulator can be considered either by adding a large amount of individual Monte Carlo runs based on distinct geometries, or by changing the geometry dynamically,

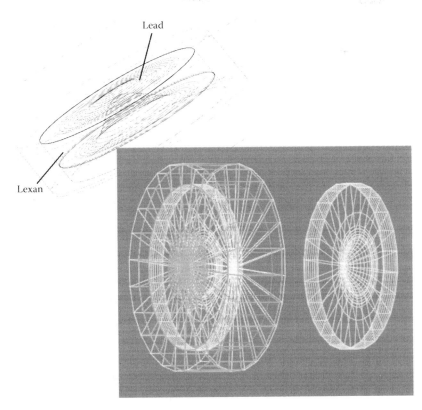

FIGURE 14.2 Monte Carlo simulation (based on the Geant4 code) of one of the contoured scatterers used in a double-scattering system at Massachusetts General Hospital, Boston. Upper left: Schematic drawing. Lower left: Full Monte Carlo representation. Lower right: Subvolume for illustration.

applying a 4D Monte Carlo technique (Paganetti, 2004). Because beam spots often overlap with several wheel steps, it is not sufficient to use one simulation per absorber step. For example, in one simulation of passive scattered delivery, the wheel position was changed in steps of 0.7° (Paganetti et al., 2004a).

Because each patient field is unique in terms of range and modulation width, there is a unique wheel design for each field. However, it is impractical to fabricate a modulator wheel for each field. Consequently, one operates with a small set of modulator wheels (or a set of wheels with a set of tracks where each track resembles a unique step design; see Figure 14.3). Each wheel (or track) is designed to create full modulation to the patient's skin. The desired modulation width is achieved by translating the modulation width into desired rotation angles where the beam current is turned on and off at the cyclotron (beam source) level (Lu and Kooy, 2006). Furthermore, for a

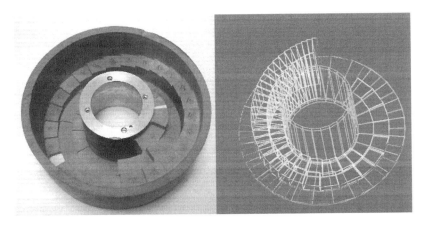

FIGURE 14.3 Left: Picture of one of the modulator wheels consisting of three tracks used at Massachusetts General Hospital, Boston. Right: Simulation of a modulator wheel using the Geant4 Monte Carlo code.

given wheel (track), a flat SOBP is only achieved for a nominal range and modulation width combination. To fine-tune the shape of the SOBP depth dose distribution, the beam current is continuously modulated as a function of rotation angle to ensure that each SOBP satisfies its flatness specification.

Field-specific apertures and compensators are prescribed by the planning system. They are defined in files that parameterize (e.g., by a set of points in space) the geometry for use in a milling machine. How the geometries are imported into the Monte Carlo code depends on the format of the files as well as on the capabilities of the Monte Carlo system. One solution is to use the milling machine files directly. An aperture can be described by a set of points following the inner shape of the aperture opening. If the Monte Carlo code is capable of translating this information into a three-dimensional (3D) object representing the aperture, the information can be used directly. Otherwise, different parameterizations need to be applied, as, for example, a triangulation of the aperture opening (Paganetti et al., 2004a). For the compensator, one can virtually drill holes into a plastic tube to generate the geometry in the Monte Carlo mimicking the actual device. Apertures need to be modeled explicitly because of the secondary radiation they produce (Zacharatou Jarlskog et al., 2008) and the effects of edge scattering (Titt et al., 2008b).

Some Monte Carlo codes are also capable of simulating magnetic beam steering. Treatment head simulations for scanned beams have shown good agreement (Peterson et al., 2009a,b). Depending on the scanning system, accurate description of the field lines as compared to binary fields, and detailed information of the beam emittance are important when simulating dose distributions using Monte Carlo. To simulate pencil beam scanning, the magnetic field settings are prescribed either by the planning system directly or by a treatment control system based on information by the planning system (Paganetti et al., 2005). These settings are typically relative numbers that need to be translated into magnetic strengths in Tesla. The relationship is typically known from results of commissioning measurements. If the Monte Carlo code is capable of modeling magnetic fields, this information can be incorporated by defining the field strengths as a function of position in the treatment head geometry (Paganetti et al., 2004a). This requires a parameterization of the magnetic field lines. One can also use an approximation by defining a constant magnetic field as either on or off in a defined area (i.e., assume perfect dipoles). Attention needs to be paid to the step size while tracking protons through the magnetic field because the field is applied on a step-by-step basis. Large steps lead to considerable uncertainties when simulating the curved path of particles through the field. Figure 14.4 illustrates the simulation of pencil beam scanning where proton pencils have been steered through the treatment head. To simulate an entire scanning pattern, 4D Monte Carlo techniques can be applied (Paganetti, 2004; Paganetti et al., 2005).

As discussed above, for passive scattering as well as beam scanning, the treatment head geometry involves devices with a

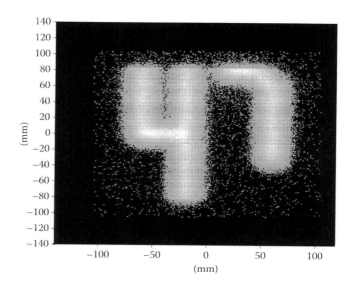

FIGURE 14.4 Proton tracks in a magnetic field as modeled within Geant4: Simulated intensity map in a plane at the treatment head exit using a predefined scanning pattern. (From Paganetti, H., *Physics in Medicine and Biology*, 2004. **49**: N75–N81. With permission.)

time-dependent setting. The rotation of the modulator wheel can be simulated by adding individual runs of pristine Bragg peaks and combining them based on weighting factors obtained from the wheel geometry (Herault et al., 2005; Newhauser et al., 2005). One can also use a slab of a fixed thickness and then change the start position of the beam inside the slab to modulate different thicknesses (Koch et al., 2008). An alternative is to change the geometry dynamically by applying a 4D Monte Carlo technique (Paganetti, 2004; Shin et al., 2012). Furthermore, a time-dependent feature might be the modulation of the beam current in passive scattering systems (Lu and Kooy, 2006). The beam current is continuously modulated as a function of rotation angle. This can be incorporated into the Monte Carlo code by using a finite set of look-up tables correlated to the rotating wheel. To simulate a beam scanning pattern, 4D Monte Carlo can be used to constantly update the magnetic field strength (Paganetti, 2004; Paganetti et al., 2005). This allows studying beam scanning delivery parameters (Paganetti et al., 2005; Peterson et al., 2009a,b; Shin et al., 2012).

14.2.3 Phase Space Distributions

Phase space distributions are commonly used to characterize the beam at or near the treatment head exit. The aim is to minimize calculation time by reusing data in areas that are not varying for each treatment field. In passive scattered proton therapy, phase spaces are less useful than they are in photon therapy. This is because each field typically has a unique setting of the treatment head. The varying parameters are (at least) beam energy, the settings for first scatterer, second scatterer, modulator wheel, utilized wheel solid angle, as well as aperture and compensator.

The same holds for beam scanning as the scanning pattern depends on the treatment field. However, for pencil beam scanning, the definition of a phase space to characterize the radiation field exiting the treatment head is feasible as the variations in beam settings are much lower (beam energy and magnet settings). To scan a particular pattern within the patient, the scanning trajectory for a given energy layer can be discretized. These beam spots, corresponding to magnet settings, can be sampled in a phase space file. For treatment simulation, a predefined number of protons is then simulated for each beam spot.

14.2.4 Beam Models

Beam models allow parameterization of a radiation field to be used instead of a phase space. A model would thus have to incorporate the variety of treatment head settings. Proton therapy treatment head settings are very complex and an accurate beam model might be hard to find. Assuming a homogeneous field impinging on the patient-specific aperture and compensator and making a rough assumption about the angular spread of the field, it is possible to deconvolve an SOBP into its pristine peak contributions (Lu and Kooy, 2006). This method has been suggested for optimizing beam current modulation but could also serve to construct a beam model by optimizing a set of pristine Bragg curves that add up to an SOBP. It is unclear whether such a parameterization can serve as a beam model for patient-related dose calculations because the creation of a given SOBP might not be unique in terms of the underlying pristine Bragg peaks as dose can be parameterized as fluence times linear energy transfer. Thus, there could be more than one solution within certain dose constraints.

Beam models are, however, feasible for pencil beam scanning where a field can be characterized by a fluence map of pencil beams (x, y, beam energy, weight, divergence, and angular spread). Thus, a beam source model can be based on parameterized particle sources for a scanned proton beam using fits of measured data, for example, fluence distributions of pencil beams in air and depth dose distribution measured in water (Kimstrand et al., 2007). The fact that beam models or phase space distributions can easily be constructed for pencil beam scanning has important implications when using Monte Carlo for clinical dose calculation. Owing to the low efficiency of proton therapy treatment heads for passive scattering (typically between 2% and 30%, depending on the field), the majority of calculation time is spent tracking particles through the treatment head. While this makes Monte Carlo for passive scattered delivery less attractive for routine clinical use, the time frame for beam scanning simulations allows the use of Monte Carlo routinely in the clinic (Paganetti et al., 2008). Note that for some scanned beam deliveries, one might want to use an aperture to reduce the beam penumbra. Edge scattering at the aperture can have significant effects on the dose distribution (Titt et al., 2008a) and should thus be included in the beam model.

14.2.5 Uncertainties and Validation

There are various studies comparing experimental depth dose distributions and beam profiles with results from different Monte Carlo codes and for different treatment heads (Paganetti et al., 2004a; Titt et al., 2008b; Clasie et al., 2010). In a well-designed Monte Carlo code, measured dose distributions of an SOBP in water should be reproduced with accuracies within typically ~1 mm in range and ~3 mm in modulation width (Paganetti et al., 2004a). A benchmark example using an inhomogeneous phantom setup is shown in Figure 14.5 with a comparison between measured (ionization chamber) and Monte Carlo predicted dose profile.

Even if an appropriate setting of the physics parameters in the Monte Carlo has been identified, there are still uncertainties because various parameters are not necessarily known down to a sufficient level of accuracy. The outcome of simulations depends, for example, on the known geometry of the treatment nozzle, such as modulator wheel (including the time-dependent settings during particle beam delivery and beam current modulation (Paganetti et al., 2004a)), scatterers, collimators, and patient-specific components (aperture and range compensator).

It has been shown (Bednarz et al., 2011) that the SOBP range is most sensitive to changes in the density of materials in the modulator wheel. Materials commonly used are polyethylene (lexan) and lead. A 10% variation in density could result in range changes on the order of 1 mm. Another material used in scatterers of treatment heads is carbon. Here, the nominal density as specified by the manufacturer can vary substantially from the actual density because carbon is available in various specifications leading to larger uncertainties compared to other materials. This can have an impact on the predicted proton beam range as well.

The uncertainties in the Monte Carlo settings affect not only the beam range but also the modulation width. The flatness of the SOBP might be sensitive to the initial energy spread and spot size at nozzle entrance. These parameters influence the peak-to-plateau ratio of the individual Bragg peaks that form an SOBP. It has been shown that the effect is small for at least certain modulator wheel designs (Bednarz et al., 2011). The impact of the beam spot size depends on the width of each step used on the modulator wheel. Uncertainties in stopping power parameters, especially the mean excitation energy (see above), can also affect dose calculations (Andreo, 2009).

Commissioning of a Monte Carlo treatment head implementation might not be doable based on first principles depending on the complexity of the treatment head or beam settings or the geometry information available. Thus, although good agreement might be found based on first principles (Paganetti et al., 2004a; Bednarz et al., 2011), commissioning of a Monte Carlo-based passive scattering proton beam system might require fine-tuning to ensure accurate dose distributions in the patient. If the simulations differ from experimental data, adjusting parameters can be cumbersome. For example, adjusting nozzle geometries to improve a slight tilt in an SOBP will

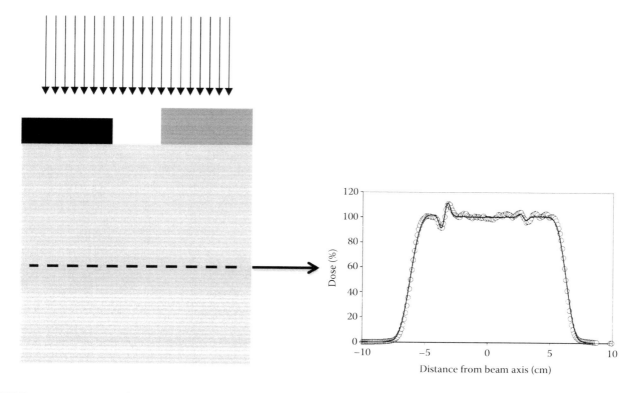

FIGURE 14.5 Comparison of a Monte Carlo predicted dose distribution (open circles) with measured values using an ionization chamber (solid line). The left side depicts the geometry (black: bone equivalent material (thickness: 3 cm); dark gray: lung equivalent material (thickness: 5 cm); light gray: water). The dashed line indicates the plane corresponding to the shown dose distribution. An SOBP (full modulation) was used for the beam profile. (From Paganetti, H. et al., *Physics in Medicine and Biology*, 2008. **53**(17): 4825–4853. With permission.)

correspondingly affect the modulation width. Beam current modulation (beam weight as a function of modulator wheel angle) can be useful to correct the Monte Carlo simulations if the desired SOBP flatness can not be reached based on first principles (Bednarz et al., 2011).

While the comparison of dose distributions is a valuable benchmark for electromagnetic interactions, nuclear interaction models cannot be validated measuring dose alone. A device to study total inelastic cross sections is the multilayer Faraday cup (Gottschalk et al., 1999; Paganetti and Gottschalk, 2003). It measures the longitudinal charge distribution of primary and secondary particles and is capable of separating the nuclear interaction component from the electromagnetic component. Various Monte Carlo physics models have been compared with the measured charge distribution from the Faraday cup (Zacharatou Jarlskog and Paganetti, 2008a; Rinaldi et al., 2011).

A precise modeling of neutron yields is needed when simulating both neutron doses (Zacharatou Jarlskog et al., 2008; Zacharatou Jarlskog and Paganetti, 2008b) or neutron fluence for shielding. These simulations require double differential production cross sections for tissues, beam-shaping devices, and shielding materials. There are often insufficient experimental data to predict inelastic nuclear cross sections. The agreement between experiment and simulation for neutron

doses is typically not as good as with photons, electrons, or protons (Schneider et al., 2002; Polf and Newhauser, 2005; Moyers et al., 2008; Zheng et al., 2008; Chen and Ahmad, 2009; Clasie et al., 2010).

14.3 Simulating Dose to the Patient

14.3.1 Statistical Accuracy of Dose Modeling

Typically, one tries to achieve a statistical uncertainty for dose calculations of less than 2%, at least for the target volume. However, the simulation conditions affecting the uncertainties are slightly different in proton therapy compared to photon therapy. The energy deposited in a patient voxel per proton is higher than in photon Monte Carlo. Thus, typically fewer protons have to be tracked to achieve a targeted accuracy as compared to photon Monte Carlo.

The required number of particle histories depends on the efficiency of the treatment head (determined, for example, by the double-scattering system in passive scattered proton therapy), the required geometrical resolution, and slightly on beam range and modulation width. It normally does not depend on the field size downstream of the patient field-specific aperture because the field size generated in the double-scattering system is typically independent of the aperture opening. For pencil beam scanning, the efficiency of the treatment head is not an issue as it will be almost 100%, while for passive scattering it is typically in the order of 3–30%. For passive scattered delivery, a typical number of proton histories for a given patient field is ~25 million at treatment head entrance if sufficient statistical accuracy is requested for a single field based on the treatment planning grid resolution. The total number of protons entering the patient for simulating dose distributions can be even less than 1 Mio. for certain fields. These numbers are quite low compared to photon therapy. Owing to the proton linear energy transfer (LET), only a few hundred protons are traversing a voxel for a 2% dose variation. Nevertheless, simulations might still be time consuming and several techniques have been presented to improve the computational efficiency of proton beams in patients (Jiang and Paganetti, 2004; Yepes et al., 2009).

14.3.2 CT Conversion

For photon or electron beams, electron density is being used in analytical dose calculation engines because the dominant energy loss process is interaction with electrons. Protons lose energy by ionizations, multiple coulomb scattering, and nonelastic nuclear reactions. Each interaction type has a different relationship with the material characteristics obtained from the CT scan (Matsufuji et al., 1998; Palmans and Verhaegen, 2005). Consequently, for analytical proton dose calculations, instead of electron density or mass density, relative stopping power is being used to define water-equivalent tissue properties and a conversion from CT numbers to relative stopping power is being used in commercial planning systems (Mustafa and Jackson, 1983; Schaffner and Pedroni, 1998). The accuracy of dose calculations is affected significantly by the ability to precisely define tissues based on CT scans (Schaffner and Pedroni, 1998; Jiang et al., 2007).

For Monte Carlo dose calculation, a conversion from CT numbers to material compositions and mass densities is done for each tissue. A conversion scheme to relate CT numbers to human tissues can be deduced by scanning tissue phantom materials or animal tissues. Note that a CT number reflects the attenuation coefficient of human tissues to diagnostic x-rays and may be identical for several combinations of elemental compositions, elemental weights, and mass densities (Schneider et al., 2000). In CT conversion schemes, tissues are grouped into different tissues sharing the same material composition (and mean excitation energy). The number of groups certainly affects the dose calculation accuracy and is typical between 5 and 30 (Jiang and Paganetti, 2004; Jiang et al., 2007; Parodi et al., 2007a). While grouping material compositions is justified, it is typically not sufficient to assign a unique density to each group, that is, the number of densities is typically the same as the number of gray values (CT numbers) (Jiang and Paganetti, 2004). Several tissue classifications for CT-based Monte Carlo dose calculation have been introduced (Schneider et al., 1996, 2000; du Plessis et al., 1998). Most

methods are based on a stoichiometric calibration of CT numbers, which provides a robust division of most soft tissues and skeletal tissues.

Not only can the absolute dose vary but the proton beam range might also depend on the accuracy of the CT conversion (Chvetsov and Paige, 2010; Espana Palomares and Paganetti, 2010). Range uncertainties due to uncertainties in CT conversion are of concern in proton therapy (Unkelbach et al., 2007; Paganetti, 2012). It has been shown that slight discrepancies in mass density assignments play only a minor role in the target region whereas more significant effects are caused by different assignments in elemental compositions (Jiang et al., 2007). The relationship between CT number and material is not unique and various fits can lead to a feasible result (Schneider et al., 2000; Espana Palomares and Paganetti, 2010).

A comparison of two different CT conversion methods (Schneider et al., 2000; Rogers et al., 2002) is shown in Figure 14.6 to illustrate the magnitude of differences one might expect. The largest difference appears at $H = -119$ for carbon and oxygen because of the defined thresholds at the transition of air and adipose tissue. For hydrogen and calcium, the greatest difference is located at $H = +126$ at the transition from soft tissue to bone. To demonstrate the impact these differences might have on proton dose calculation, Figure 14.7 shows the mass stopping power ratio as a function of CT number for the two methods presented in Figure 14.6.

When implementing a conversion scheme for Monte Carlo dose calculation, a normalization to the specific scanner that is being used for imaging in the department is needed to compare the results from a Monte Carlo dose calculation with those from an analytical algorithm in a treatment planning system. The normalization for the Monte Carlo can be found by either doing a separate stoichiometric calibration or, as an approximation, by simulating relative stopping power values based on an existing CT conversion and then compare the results with the planning system conversion curve. Normalization can then be achieved within the Monte Carlo algorithm by slightly adjusting material compositions or, easier, mass densities (Jiang and Paganetti, 2004; Parodi et al., 2007b; Paganetti et al., 2008).

Uncertainties in dose calculation are introduced not only by CT conversion but also because of the imaging resolution, that is, the CT grid size (Chvetsov and Paige, 2010; Espana Palomares and Paganetti, 2010). Treatment planning programs often report dose on a more course grid than the one provided by the patient's CT scan. The Monte Carlo simulation should operate on the actual CT scan resolution, which is typically on the order of 0.5–5 mm. It happens frequently that the CT grid used clinically is nonuniform. For comparison of the results with the planning system (based on the planning grid), the Monte Carlo dose distribution can be interpolated. The speed of the Monte Carlo simulation certainly depends on the grid size. However, assuming that a step size above 1 mm is typically not warranted, a larger grid size does not translate into a huge gain.

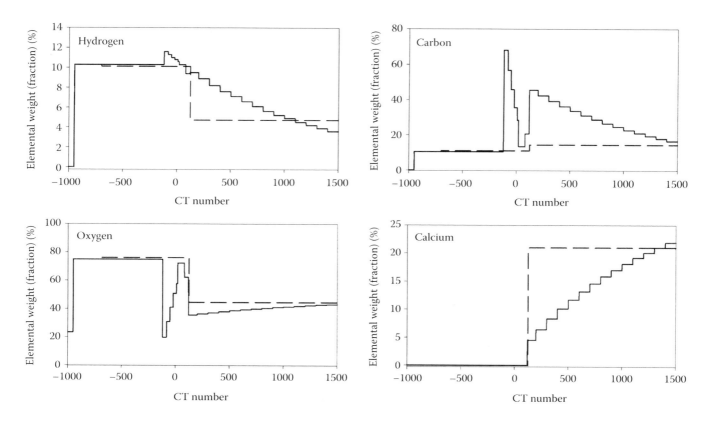

FIGURE 14.6 Elemental weights of four tissue elements (hydrogen, carbon, oxygen, and calcium) as a function of CT number for the CT conversion method of Schneider et al. (Schneider, W., T. Bortfeld, and W. Schlegel, *Physics in Medicine and Biology*, 2000. **45**: 459–478 (solid line)) and by Rogers et al. (Rogers, D.W.O. et al. *BEAMnrc User Manual*. NRCC Report PIRS-0509, 2002 (dashed line)). (From Jiang, H., J. Seco, and H. Paganetti, *Medical Physics*, 2007. 34: 1439–1449. With permission.)

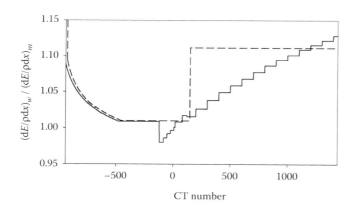

FIGURE 14.7 Water-to-tissue mass stopping power ratio for 100 MeV protons as a function of CT number when applying CT conversion methods by Schneider et al. (solid line) and Rogers et al. (dashed line). (From Schneider, W., T. Bortfeld, and W. Schlegel, *Physics in Medicine and Biology*, 2000. **45**: 459–478 (solid line); Rogers, D.W.O. et al., *BEAMnrc User Manual*. NRCC Report PIRS-0509, 2002 (dashed line). (From Jiang, H., J. Seco, and H. Paganetti, *Medical Physics*, 2007. **34**: 1439–1449. With permission.)

14.3.3 Absolute Dose

The prescribed dose is converted into machine monitor units for patient treatment, that is, absolute doses are reported as cGy per MU. In proton therapy, this is done either by calibration measurements, by using analytical algorithms or by relying on empirical data. A monitor unit typically corresponds to a fixed amount of charge collected in a transmission ionization chamber incorporated in the treatment head whose reading is related to a dose at a reference point in water (Kooy et al., 2003, 2005).

Using Monte Carlo, one can actually simulate the ionization chamber output charge when tracking particles through the treatment nozzle. Here, an exact model of the proton beam delivery nozzle, including devices used for dosimetry, is required (Paganetti, 2006). The disadvantage of this method is that it typically requires a large number of histories to be simulated as the energy deposited in a small reference chambers in air per particle is normally quite small. Nevertheless, Monte Carlo predicted output factors can be part of an extended quality assurance program because the influence of specific devices in the treatment head on the output factor reading can be studied.

Alternatively, absolute dosimetry for Monte Carlo dose calculation can also be done by simply relating the number of protons at nozzle entrance to the dose in an SOBP in water. With an accurate model of the treatment head, this method is equivalent to a direct output simulation because instead of indirectly "measuring" the number of protons at a given plane in the treatment head (in an ionization chamber), one "measures" the number of protons at nozzle entrance.

14.3.4 Dose-to-Water and Dose-to-Tissue

Traditionally, dose in radiation therapy is reported as dose-to-water because analytical dose calculation engines, as the ones incorporated in commercial planning systems, calculate dose by modeling physics relative to water. Furthermore, quality assurance and absolute dose measurements are done in water. There is an ongoing debate about the pros and cons of using the two standards (Liu and Keall, 2002). Monte Carlo dose calculation systems calculate dose-to-tissue because they allow the material properties, as converted from CT numbers, to be modeled using explicit material composition, mass density, and excitation energy. Consequently, to allow a proper comparison between Monte Carlo and pencil beam-generated dose distributions, one has to convert one dose metric to the other.

Based on relative stopping powers and on a nuclear interaction parameterization, a formalism has been described to convert dose-to-medium into dose-to-water for proton beam dose calculations (Paganetti, 2009a). It was found that based on the mean dose to a structure, dose-to-water can be higher by ~10–15% compared to dose-to-tissue in bony anatomy. For soft tissues, the differences were shown to be on the order of 2%. The difference in mean dose roughly scales linear with the average CT number, and thus seems to be clinically insignificant for soft tissues but needs to be taken into account for bony anatomy. It has been shown that in most cases it is sufficiently accurate (within ~1%) to do a conversion to dose-to-water retroactively by simply multiplying the dose with energy-independent relative stopping powers. Furthermore, as the difference in mean dose roughly scales linear with the average CT number, a rough scaling based on the CT numbers might be sufficient, depending on the desired precision (Paganetti, 2009a).

14.3.5 Impact of Nuclear Interaction Products on Patient Dose Distributions

Pencil beam algorithms in treatment planning systems incorporate nuclear interaction contributions intrinsically because they are based on measured or Monte Carlo simulated depth dose curves. It has been shown that nuclear secondaries need to be considered in proton therapy dose calculations as they contribute significantly in particular in the entrance region of the Bragg curve (Carlsson and Carlsson, 1977; Laitano et al., 1996; Medin and Andreo, 1997; Paganetti, 2002; Palmans and Verhaegen, 2005).

Depending on the application, it might not be necessary to track all secondary particles in Monte Carlo dose calculation

and it might be sufficient for some types to deposit the energy locally (to ensure proper energy balance). If secondary doses for out-of-field dose calculations are being studied, one also has to consider neutrons and the protons they cause. For patient-related dose calculations, secondary protons from proton–nuclear interactions solely need to be tracked. Secondary protons deliver up to about 10% of the total dose proximal to the Bragg peak of an unmodulated proton beam. Furthermore, secondary protons cause a dose build-up due to forward emission of secondary particles from nuclear interactions. Because the total nuclear interaction cross section shows a maximum at around 20 MeV and decreases sharply if the energy is decreased, the contribution of dose due to nuclear interactions becomes negligible roughly at the Bragg peak position of a pristine Bragg curve. As a rule of thumb, the average proton energy in the Bragg peak is ~10% of the initial energy. In an SOBP, nuclear interactions can still play a role because dose regions proximal to the Bragg peak also contribute and thus cause a tilt of the dose plateau if their contribution is neglected (Paganetti, 2002). Figure 14.8 demonstrates the magnitude of the dose from secondary protons as well as its 3D distribution in a patient.

Note that the consideration of nuclear interactions is particularly important for pencil beam scanning since each pencil is surrounded by a nuclear halo from secondary protons. Neglecting it will result in significant deviations in absolute dose (Sawakuchi et al., 2008, 2010).

14.3.6 Differences between Proton Monte Carlo and Pencil Beam Dose Calculation

It has been shown that the dosimetric differences between a Monte Carlo and a pencil beam algorithm can be significant (Petti, 1996; Schaffner et al., 1999; Szymanowski and Oelfke, 2002; Soukup et al., 2005; Tourovsky et al., 2005; Soukup and Alber, 2007; Paganetti et al., 2008). In fact, the routine use of Monte Carlo dose calculation in proton therapy could result in a reduction of treatment planning margins, which in turn might reduce treatment side effects (Paganetti, 2012). Monte Carlo simulations have been used to validate analytical algorithms (Sandison et al., 1997; Soukup et al., 2005; Paganetti et al., 2008) or for commissioning of treatment planning systems (Koch and Newhauser, 2005; Newhauser et al., 2007). Overall, some pencil beam algorithms show good agreement with Monte Carlo (Paganetti et al., 2008). Pencil beam algorithms are less sensitive to complex geometries and in-beam density variations, that is, bone–soft tissue, bone–air, or air–soft tissue interfaces (Urie et al., 1984; Petti, 1992, 1996; Pflugfelder et al., 2007). Deviations from the Monte Carlo dose distribution occur specifically in the penumbra if the beam direction is tangential to an interface connecting high and low CT numbers. Not only local difference can occur. Also the overall dose distribution can be affected. If proton beams pass through complex heterogeneous geometries, a phenomenon called range degradation occurs (Urie et al., 1986; Sawakuchi et al., 2008), which is correctly predicted only when using

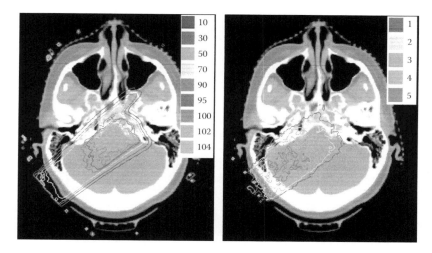

FIGURE 14.8 (**See color insert.**) Impact of nuclear secondaries on the dose distribution in a patient. Left: Dose distribution for one field (out of three for this treatment plan) for a head and neck case (doses in%). Right: Contribution of secondary protons to the dose distribution in% of the target dose.

Monte Carlo. Note also that some pencil beam models do not consider aperture edge scattering (Titt et al., 2008b).

Consequently, one of the areas where pencil beam algorithms specifically show weaknesses is in the presence of lateral heterogeneities (Schaffner et al., 1999; Soukup et al., 2005; Soukup and Alber, 2007). Figure 14.9 demonstrates one of the weaknesses of a pencil beam-generated dose distribution, namely, the inaccurate consideration of interfaces parallel to the beam path. This patient was treated for a spinal cord astrocytoma with three coplanar fields. The figure only shows one of the fields. The Monte Carlo dose calculation was based on a CT with $176 \times 147 \times 126$ slices with voxel dimensions of $0.932 \times 0.932 \times 2.5-3.75$ mm³ (slice thickness varied).

A treatment technique that is being used in passive scattered proton therapy is field patching. Here, a lateral penumbra is matched with a distal fall-off of a second beam to produce an L-shaped dose distribution with just two beams. This places the distal fall-off of one of the beams in the target, making this technique vulnerable to range uncertainties. To minimize the effect of range uncertainties, multiple patch field combinations are being applied. Further, the treatment planners tend to overshoot the field with the distal fall-off ending in the target to avoid a hot spot. Figure 14.10 shows an example for a patch field combination and the impact of differences in the predicted range when using pencil beam or Monte Carlo-based dose calculation. The Monte Carlo dose calculation was based on a CT with $161 \times 177 \times 111$

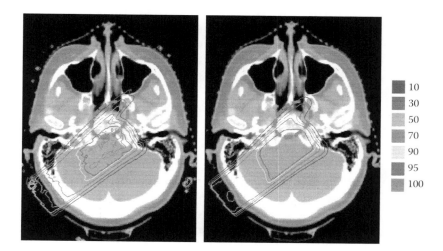

FIGURE 14.9 (**See color insert.**) Axial views of dose distributions calculated using Monte Carlo (left) and a commercial planning system based on a pencil beam algorithm (right). Doses are in%. The red dashed circles show the dosimetric impact of an interface parallel to the beam path. (From Paganetti, H. et al., Clinical implementation of full Monte Carlo dose calculation in proton beam therapy. *Physics in Medicine and Biology*, 2008. **53**(17): 4825–4853. With permission.)

slices with voxel dimensions of $0.65 \times 0.65 \times 1.25$–$3.75$ mm³ (variable slice thickness). The patient had a sphenoid sinus tumor. One can see nicely the differences between pencil beam and Monte Carlo at the end of range, that is, at the patch lines. In this case, the consequence is a hot spot in the target. Also apparent is a discrepancy in the penumbra of the field shown on the right. The reason might be the sharp gradient in the range compensator not modeled exactly in the pencil beam algorithm.

Whether Monte Carlo dose calculation is needed or pencil beam-generated dose distributions are sufficiently accurate is still an area of research. Treatment planners are typically aware of dose calculation uncertainties and take these into account by applying appropriate margins. Furthermore, one avoids pointing the beam toward a critical structure because of uncertainties in the position and gradient of the distal dose fall-off. Because of conservative margins in treatment planning, differences between analytical and Monte Carlo dose calculations are not always clinically significant (Soukup and Alber, 2007; Paganetti et al., 2008). However, more accurate dose calculation might help reducing those margin contributions that are due to dose calculation uncertainties (Paganetti, 2012).

Monte Carlo dose calculations are attractive for studying motion effects in lung dosimetry (Keall et al., 2004; Paganetti et al., 2004b). In proton therapy, 4D Monte Carlo simulations have been used to model respiratory patient motion (Paganetti, 2004; Paganetti et al., 2004b) and to study interplay effects

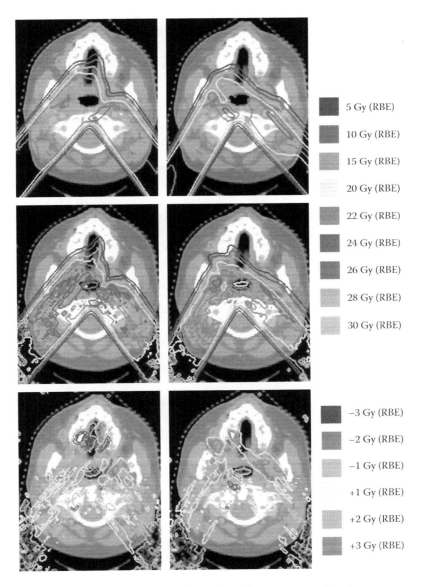

FIGURE 14.10 **(See color insert.)** Dose distributions calculated for a sphenoid sinus tumor patient. The upper row shows two patch field combinations with patch lines across the 50% distal fall-off and 50% lateral penumbra. The middle row shows the respective Monte Carlo dose distributions and the lower row gives the difference maps. (From Paganetti, H. et al., Clinical implementation of full Monte Carlo dose calculation in proton beam therapy. *Physics in Medicine and Biology*, 2008. **53**(17): 4825–4853. With permission.)

between respiratory motion and beam scanning speed (Paganetti et al., 2005).

Obviously, Monte Carlo dose calculations are particularly valuable for studying dosimetric effects in regions with a high gradient in density. Extreme cases are metallic implants in the patient, which are being used to stabilize bony anatomy after surgery or as markers for imaging. Monte Carlo has also been used to study Tantalum markers used to stabilize bony anatomy after surgery or as markers for imaging. These can lead to significant dose perturbations typically not predicted accurately by pencil beam algorithms (Newhauser et al., 2007). Markers implanted in the patient for setup or motion tracking are typically not modeled accurately in pencil beam algorithms due to their high-Z nature.

14.3.7 Clinical Implementation

If Monte Carlo treatment planning is to be done, it needs to be incorporated into a framework for treatment plan optimization (Moravek et al., 2009). Full Monte Carlo treatment planning is time consuming because the dose would have to be simulated multiple times during the optimization process. Alternatively, one might utilize a Monte Carlo dose engine only at a limited number of checkpoints during the optimization process. Monte Carlo can also be used outside of a commercial planning system as a standalone tool to get a second opinion on a dose distribution predicted by an analytical algorithm.

For an in-house-developed Monte Carlo dose calculation engine, treatment head simulation and the capability of the Monte Carlo code to simulate dose on a CT grid are only one step toward clinical implementation. The Monte Carlo simulation has to be based on treatment plan parameters, which have to be imported. How these parameters are extracted depends on the treatment planning program and on the access it provides to those data.

If the radiation field for proton beam scanning is to be simulated, the planning system will most likely provide a matrix of beam spot energy, beam spot weight, and beam spot position, which can be translated into Monte Carlo settings in a straightforward manner. If passive scattered delivery is simulated, there are two ways of how the radiation field is characterized. First, the planning system prescribes the settings of various treatment head devices and the beam at treatment head entrance directly. These settings are beam energy at nozzle entrance, settings of the first and second scatterer (one scatterer if single scattering is used for small fields), and a combination of scatterer steps in a modulator wheel or a ridge filter. Second, the planning system prescribes range and modulation width only, leaving it to the hardware control system of the facility to translate this information into treatment head settings. In this case, it might be warranted to incorporate the control system algorithms into the Monte Carlo code (Paganetti et al., 2004a, 2008). Planning systems also define aperture and compensator for passive scattered delivery. These are typically given as files readable by a milling machine.

Regarding the patient's CT, a Monte Carlo code can either accept a digital imaging and communications in medicine (DICOM) stream directly (Kimura et al., 2004, 2005) or receive the CT information from the planning system (Parodi, K. et al., 2007b; Paganetti et al., 2008). For the latter, the patient geometry is transferred from a departmental patient database to the planning system, which in turn transfers the CT to Monte Carlo using a planning system-specific data format. Within the Monte Carlo, it might be translated into a simple binary format. Parameters required are the number of voxels and the, potentially variable, voxel dimensions in x, y, and z. The latter method is less flexible than using DICOM directly but, depending on the planning system, it might be the only way to allow all simulation-related parameters to be imported from one single source, that is, the planning system.

Additional simulation input parameters needed from the planning system for each field are the gantry angle, the patient couch angle, the air gap between patients and treatment head, the isocenter position, and the prescribed dose or beam weight. Again, to import parameters into the Monte Carlo system, a software link to the treatment planning system needs to be established (Paganetti et al., 2008). Note that the spatial information might be given in different coordinate systems, that is, the treatment head coordinate system, the patient CT coordinate system, or the coordinate system of the planning system.

To facilitate the data flow from the planning system to the Monte Carlo, one might use a user interface where the user selects a specific patient, plan, and treatment field. Once the dose in the patient is calculated, it can be analyzed by a standalone program capable of visualizing CT data and reading/visualizing contours. If the planning system allows, the Monte Carlo results can be imported into the planning system for a side-by-side comparison with the analytical results.

The Francis H Burr Proton Therapy Center at Massachusetts General Hospital uses a standalone script that allows the user to extract treatment-related parameters directly from different planning systems for passive scattered or scanned beam proton therapy and translates the information directly into input parameters for the Monte Carlo system TOPAS (Perl et al., 2012), allowing a convenient interface for clinical personnel to do proton Monte Carlo dose calculation.

14.3.8 Improving Monte Carlo Efficiency

Particle tracking in a voxel geometry can be computationally inefficient. Typically, each particle has to stop when a boundary between two different volumes is crossed. Algorithms have been developed to tackle this problem (Hubert-Tremblay et al., 2006; Sarrut and Guigues, 2008). Such algorithms are based on an image segmentation that compromises the regular voxel geometry. There are several techniques to improve the computational efficiency of Monte Carlo for proton beams in patients (Jiang and Paganetti, 2004; Yepes et al., 2009; Schümann et al., 2012). Some Monte Carlo codes have been specifically designed for fast patient dose calculations using approximations to improve efficiency (Kohno et al., 2002, 2003; Fippel and Soukup, 2004; Tourovsky et al., 2005). One method is to implement a track repeating algorithm in which precalculated proton tracks and their interactions in material are tabulated and used in dose calculations (Li et al., 2005).

To improve calculation efficiency, hybrid methods have been introduced that combine aspects of Monte Carlo simulation and analytical algorithms. There are many ways to improve the accuracy of pencil beam algorithms by adding Monte Carlo components, for example, using spot decomposition. These are often associated with significantly decreased computational efficiency (Soukup et al., 2005; Soukup and Alber, 2007). One might also use Monte Carlo generated kernels as look-up tables in analytical dose calculation algorithms (Soukup and Alber, 2007; Kimstrand et al., 2008b). Simplified Monte Carlo methods using measured depth dose curves in water or other materials as input have also been proposed (Kohno et al., 2002).

A significant speed improvement compared to codes like Geant4 and FLUKA has been reported for VMCpro (Fippel and Soukup, 2004). Here, various approximations were introduced, for example, a simplified multiple scattering algorithm and density scaling functions instead of actual material compositions. Nuclear interactions are treated as a correction to electromagnetic interactions by using parameterizations and distribution sampling and it was shown that these are valid for dose calculations, at least for broad beams.

14.4 Proton Monte Carlo Applications Other Than Treatment Head Simulations and Dose Calculations

14.4.1 Simulating Proton Induced PET Distributions in the Patient

Protons undergo nuclear interactions in the patient, which can lead to the formation of positron emitters. One can use positron emission tomography (PET) imaging for *in vivo* verification of treatment delivery and, in particular, beam range. Monte Carlo simulations are typically used to generate a theoretical PET image based on the prescribed radiation field, which can then be compared to the measured PET distribution for treatment verification (Parodi et al., 2007a,b). Lack of detailed cross-section data is currently one of the limiting factors with respect to accuracy of *in vivo* range verification (Espana et al., 2011). Similarly, the emission of gamma rays from excited nuclear states, the so-called prompt gammas, can be simulated using Monte Carlo (Moteabbed et al., 2011).

14.4.2 Simulating LET Distributions for Radiobiological Considerations

The majority of clinical data in radiation therapy is based on treatments with photon beams, and prescription doses to cancerous tissue as well as dose constraints to organs at risk are based on clinical experience and not necessarily on explicit radiobiological considerations. Proton doses are prescribed relative to photon doses by applying the concept of relative biological effectiveness (RBE) for proton therapy (Paganetti et al., 2002). The RBE of protons depends on the linear energy transfer (among other parameters like dose, biological endpoint, dose rate, etc.).

Distributions of LET (or better dose-averaged and track-averaged LET) can be simulated in the patient using Monte Carlo (Grassberger and Paganetti, 2011). In the target area, the dose is typically delivered homogeneous. This however does not guarantee a homogeneous distribution of LET values (Paganetti and Schmitz, 1996; Paganetti and Goitein, 2000). When considering primary particles only, elevated LET values will appear in the distal end of a proton field. Furthermore, nuclear interaction products may cause elevated LET values even outside of the target volume, that is, in the entrance region of the Bragg curve (Kempe et al., 2007). One might simulate dose-averaged LET distributions in a patient geometry to identify potential hot spots of biological effectiveness (Grassberger et al., 2011).

14.4.3 Simulating Secondary Neutron Doses to the Patient

Monte Carlo simulations have been used for shielding design studies (Binns and Hough, 1997; Fan et al., 2007; Agosteo, 2009). Secondary neutron doses from proton therapy are difficult to measure because neutrons are indirectly ionizing and interact sparsely, causing only low absorbed doses. The MCNPX code (Agosteo et al., 1998; Newhauser et al., 2005; Polf and Newhauser, 2005; Polf et al., 2005; Tayama et al., 2006; Zheng et al., 2007, 2008; Fontenot et al., 2008; Moyers et al., 2008; Taddei et al., 2008, 2009; Perez-Andujar et al., 2009), FLUKA (Agosteo et al., 1998; Schneider et al., 2002), and GEANT4 (Jiang et al., 2005; Zacharatou Jarlskog et al., 2008; Athar and Paganetti, 2009; Athar et al., 2010) were applied to assess secondary doses in proton beams. Neutron production in proton beams was also studied using the Shield-Hit (Gudowska et al., 2002; Gudowska and Sobolevsky, 2005) and PHITS codes (Sato et al., 2009).

There are considerable uncertainties when it comes to simulating secondary particle production because there is insufficient experimental data of inelastic nuclear cross sections in the energy region of interest. Neutron and secondary charged particle emissions from nuclear interactions can be the result of complex interactions. Fast neutrons lose most of their kinetic energy in the initial relatively small number of interactions. In the low/thermal energy region, there are a large number of elastic scatterings in soft tissues, causing the neutron energy distributions in the patient to be dominated by low-energy neutrons (Jiang et al., 2005).

At low doses, the quantity of interest is not the absorbed dose but the equivalent dose. Calculations of the secondary equivalent doses to patients require particle and particle energy-dependent radiation weighting factors to consider the biological effectiveness. This is different from RBE as radiation weighting factors are regulatory quantities at low dose (ICRP, 2003a).

One possible strategy is to calculate the average absorbed dose for the organ under consideration and scale the dose at each interaction point with a radiation weighting factor. Another approach frequently used is to calculate the particle fluences at the surface of a region of interest (organ) and then use energy-dependent fluence-to-equivalent dose conversion coefficients (NCRP, 1973; Boag, 1975; ICRU, 1998; Bozkurt et al., 2000, 2001; Chao et al., 2001a,b; Alghamdi et al., 2005; Polf and Newhauser, 2005; Chen, 2006; Zheng et al., 2007).

Whole-body computational phantoms can play an important role when combined with Monte Carlo dose calculations to simulate scattered or secondary doses to organs outside the area imaged for treatment planning (Paganetti, 2009b). Phantoms have been implemented in many Monte Carlo codes to assess neutron doses in proton therapy (Agosteo et al., 1998; Jiang et al., 2005; Zacharatou Jarlskog et al., 2008; Athar and Paganetti, 2009; Taddei et al., 2009; Athar et al., 2010). Voxel phantoms are largely based on CT images and manually segmented organ contours. For each organ and model, age- and gender-dependent densities, as well as age-dependent material compositions, can be adopted based on ICRU (1992) and organ-specific material composition as a function of age can be based on individuals at the ICRP reference ages (ICRU, 1989; ICRP, 2003b). In addition to standard adult male or female models, models of pregnant patients (Caon et al., 1999; Lee and Bolch, 2003) and the pediatric population have been designed (Zankl et al., 1988; Caon et al., 1999; Nipper et al., 2002; Lee and Bolch, 2003; Staton et al., 2003; Lee et al., 2005, 2006, 2007, 2008). The latest developments in whole-body computational phantoms are hybrid phantoms based on combinations of polygon mesh and nonuniform rational B-spline surfaces (Johnson et al., 2009; Lee et al., 2010). Such phantoms provide the flexibility to model thin tissue layers and allow for free-form phantom deformations for selected body regions and internal organs using patient-specific images and deformable image registration (Tsui et al., 1994; Segars et al., 1999; Segars and Tsui, 2002; Garrity et al., 2003; Lee et al., 2010). Patient body weight can be accommodated through adjustments in adipose tissue distribution (Slyper, 1998; Thirion, 1998; Rueckert et al., 1999). It has been demonstrated that hybrid phantom height matching results in more accurate organ specific neutron dose calculations in proton therapy using Monte Carlo (Moteabbed et al., 2012).

References

Agosteo, S., Radiation protection constraints for use of proton and ion accelerators in medicine. *Radiation Protection Dosimetry*, 2009. **137**(1–2): 167–186.

Agosteo, S. et al., Secondary neutron and photon dose in proton therapy. *Radiotherapy and Oncology*, 1998. **48**: 293–305.

Alghamdi, A.A. et al., Neutron-fluence-to-dose conversion coefficients in an anthropomorphic phantom. *Radiation Protection Dosimetry*, 2005. **115**(1–4): 606–611.

Andreo, P., On the clinical spatial resolution achievable with protons and heavier charged particle radiotherapy beams. *Physics in Medicine and Biology*, 2009. **54**(11): N205–N215.

Andreo, P., J. Medin, and A.F. Bielajew, Constraints of the multiple-scattering theory of Moliere in Monte Carlo simulations of the transport of charged particles. *Medical Physics*, 1993. **20**: 1315–1325.

Arce, P. et al., GAMOS: A GEANT4-based easy and exible framework for nuclear medicine applications. 2008 *IEEE Nuclear Science Symposium and Medical Imaging Conference (2008 NSS/MIC)*, 2008: pp. 3162–3168.

Aso, T. et al., Validation of PTSIM for clinical usage. *2010 IEEE Nuclear Science Symposium and Medical Imaging Conference (2010 NSS/MIC)*, 2010: pp. 158–160.

Athar, B.S. and H. Paganetti, Neutron equivalent doses and associated lifetime cancer incidence risks for head & neck and spinal proton therapy. *Physics in Medicine and Biology*, 2009. **54**(16): 4907–4926.

Athar, B.S. et al., Comparison of out-of-field photon doses in 6-MV IMRT and neutron doses in proton therapy for adult and pediatric patients. *Physics in Medicine and Biology*, 2010. **55**: 2879–2892.

Battistoni, G. et al., The FLUKA code: Description and benchmarking. Proceedings of the Hadronic Shower Simulation Workshop 2006, Fermilab 6–8 September 2006, M. Albrow, R. Raja eds., *AIP Conference Proceeding*, 2007. **896**: 31–49.

Bednarz, B. et al., Uncertainties and correction methods when modeling passive scattering proton therapy treatment heads with Monte Carlo. *Physics in Medicine and Biology*, 2011. **56**(9): 2837–2854.

Binns, P.J. and J.H. Hough, Secondary dose exposures during 200 MeV proton therapy. *Radiation Protection Dosimetry*, 1997. **70**: 441–444.

Boag, J.W., The statistical treatment of cell survival data. *Proceedings of the Sixth L.H. Gray Conference: Cell Survival after Low Doses of Radiation*; Editor T. Alper, 1975: pp. 40–53.

Bozkurt, A., T.C. Chao, and X.G. Xu, Fluence-to-dose conversion coefficients from monoenergetic neutrons below 20 MeV based on the VIP-man anatomical model. *Physics in Medicine and Biology*, 2000. **45**: 3059–3079.

Bozkurt, A., T.C. Chao, and X.G. Xu, Fluence-to-dose conversion coefficients based on the VIP-man anatomical model and MCNPX code for monoenergetic neutrons above 20 MeV. *Health Physics*, 2001. **81**: 184–202.

Brooks, F.D. et al., Energy spectra in the NAC proton therapy beam. *Radiation Protection Dosimetry*, 1997. **70**: 477–480.

Brown, K.L. et al., Transport, a computer program for designing charged particle beam transport systems. CERN 73-16, 1973 & CERN 80-04, 1980.

Bues, M. et al., Therapeutic step and shoot proton beam spot-scanning with a multi-leaf collimator: A Monte Carlo study. *Radiation Protection Dosimetry*, 2005. **115**(1–4): 164–169.

Caon, M., G. Bibbo, and J. Pattison, An EGS4-ready tomographic computational model of a 14-year-old female torso for calculating organ doses from CT examinations. *Physics in Medicine and Biology*, 1999. **44**(9): 2213–2225.

Carlsson, C.A. and G.A. Carlsson, Proton dosimetry with 185 MeV protons. Dose buildup from secondary protons and recoil electrons. *Health Physics*, 1977. **33**: 481–484.

Chao, T.C., A. Bozkurt, and X.G. Xu, Conversion coefficients based on the VIP-Man anatomical model and GS4-VLSI code for external monoenergetic photons from 10 keV to 10 MeV. *Health Physics*, 2001a. **81**: 163–183.

Chao, T.C., A. Bozkurt, and X.G. Xu, Organ dose conversion coefficients for 0.1–10 MeV external electrons calculated for the VIP-Man anatomical model. *Health Physics*, 2001b. **81**: 203–214.

Chen, J., Fluence-to-absorbed dose conversion coefficients for use in radiological protection of embryo and foetus against external exposure to protons from 100 MeV to 100 GeV. *Radiation Protection Dosimetry*, 2006. **118**(4): 378–383.

Chen, Y. and S. Ahmad, Evaluation of inelastic hadronic processes for 250 MeV proton interactions in tissue and iron using GEANT4. *Radiation Protection Dosimetry*, 2009. **136**(1): 11–16.

Chvetsov, A.V. and S.L. Paige, The influence of CT image noise on proton range calculation in radiotherapy planning. *Physics in Medicine and Biology*, 2010. **55**(6): N141–N149.

Clasie, B. et al., Assessment of out-of-field absorbed dose and equivalent dose in proton fields. *Medical Physics*, 2010. **37**(1): 311–321.

Dementyev, A.V. and N.M. Sobolevsky, SHIELD-universal Monte Carlo hadron transport code: Scope and applications. *Radiation Measurements*, 1999. **30**: 553–557.

Dowdell, S.J., B. Clasie, N. Depauw, P. Metcalfe, A. B. Rosenfeld, H. M. Kooy, J. B. Flanz, and H. Paganetti, Monte Carlo study of the potential reduction in out-of-field dose using a patient-specific aperture in pencil beam scanning proton therapy. *Physics in Medicine and Biology*, 2012. **57**: 2829–2842.

du Plessis, F.C.P. et al., The indirect use of CT numbers to establish material properties needed for Monte Carlo calculation of dose distributions in patients. *Medical Physics*, 1998. **25**: 1195–1201.

Espana Palomares, S. and H. Paganetti, The impact of uncertainties in the CT conversion algorithm when predicting proton beam ranges in patients from dose and PET-activity distributions. *Physics in Medicine and Biology*, 2010. **55**: 7557–7572.

Espana, S. et al., The reliability of proton-nuclear interaction cross-section data to predict proton-induced PET images in proton therapy. *Physics in Medicine and Biology*, 2011. **56**(9): 2687–2698.

Fan, J. et al., Shielding design for a laser-accelerated proton therapy system. *Physics in Medicine and Biology*, 2007. **52**(13): 3913–3930.

Ferrari, A. et al., *FLUKA: A Multi-Particle Transport Code.* CERN Yellow Report CERN 2005–10; INFN/TC 05/11, SLAC-R-773 (Geneva: CERN), 2005.

Fippel, M. and M. Soukup, A Monte Carlo dose calculation algorithm for proton therapy. *Medical Physics*, 2004. **31**(8): 2263–2273.

Fontenot, J. et al., Equivalent dose and effective dose from stray radiation during passively scattered proton radiotherapy for prostate cancer. *Physics in Medicine and Biology*, 2008. **53**(6): 1677–1688.

Fontenot, J.D., W.D. Newhauser, and U. Titt, Design tools for proton therapy nozzles based on the double-scattering foil technique. *Radiation Protection Dosimetry*, 2005. **116**(1–4 Pt 2): 211–215.

Garrity, J.M. et al., Development of a dynamic model for the lung lobes and airway tree in the NCAT phantom. *IEEE Transactions in Nuclear Science*, 2003. **50**: 378–383.

Gottschalk, B., On the scattering power of radiotherapy protons. *Medical Physics*, 2010. **37**(1): 352–367.

Gottschalk, B., R. Platais, and H. Paganetti, Nuclear interactions of 160 MeV protons stopping in copper: A test of Monte Carlo nuclear models. *Medical Physics*, 1999. **26**: 2597–2601.

Gottschalk, B. et al., Multiple coulomb scattering of 160 MeV protons. *Nuclear Instruments and Methods in Physics Research*, 1993. **B74**: 467–490.

Grassberger, C. and H. Paganetti, Elevated LET components in clinical proton beams. *Physics in Medicine and Biology*, 2011. **56**: 6677–6691.

Grassberger, C. et al., Variations in linear energy transfer within clinical proton therapy fields and the potential for biological treatment planning. *International Journal of Radiation Oncology, Biology, Physics*, 2011. **80**: 1559–1566.

Gudowska, I., P. Andreo, and N. Sobolevsky, Secondary particle production in tissue-like and shielding materials for light and heavy ions calculated with the Monte-Carlo code SHIELD-HIT. *Journal of Radiation Research (Tokyo)*, 2002. **43 Suppl**: S93–S97.

Gudowska, I. and N. Sobolevsky, Simulation of secondary particle production and absorbed dose to tissue in light ion beams. *Radiation Protection Dosimetry*, 2005. **116**(1–4 Pt 2): 301–306.

Herault, J. et al., Monte Carlo simulation of a proton therapy platform devoted to ocular melanoma. *Medical Physics*, 2005. **32**(4): 910–919.

Herault, J. et al., Spread-out Bragg peak and monitor units calculation with the Monte Carlo code MCNPX. *Medical Physics*, 2007. **34**(2): 680–688.

Hsi, W.C. et al., Energy spectrum control for modulated proton beams. *Medical Physics*, 2009. **36**(6): 2297–2308.

Hubert-Tremblay, V. et al., Octree indexing of DICOM images for voxel number reduction and improvement of Monte Carlo simulation computing efficiency. *Medical Physics*, 2006. **33**(8): 2819–2831.

ICRP, *Basic Anatomical and Physiological Data for Use in Radiological Protection: Reference Values.* International Commission on Radiological Protection (Pergamon Press), 2003a. **89**.

ICRP, *Relative Biological Effectiveness (RBE), Quality Factor (Q), and Radiation Weighting Factor (wR).* International Commission on Radiological Protection (London, UK: Elsevier Ltd), 2003b. **92**.

ICRU, *Tissue Substitutes in Radiation Dosimetry and Measurement.* International Commission on Radiation Units and Measurements, Bethesda, MD, 1989. Report No. 44.

ICRU, *Photon, Electron, Proton and Neutron Interaction Data for Body Tissues.* International Commission on Radiation Units and Measurements, Bethesda, MD, 1992. Report No. 46.

ICRU, *Conversion Coefficients for Use in Radiological Protection against External Radiation.* International Commission on Radiation Units and Measurements, Bethesda, MD, 1998. **57.**

ICRU, *Nuclear Data for Neutron and Proton Radiotherapy and for Radiation Protection.* International Commission on Radiation Units and Measurements, Bethesda, MD, 2000. Report No. 63.

Jan, S. et al., GATE: A simulation toolkit for PET and SPECT. *Physics in Medicine and Biology,* 2004. **49**(19): 4543–4561.

Jiang, H. and H. Paganetti, Adaptation of GEANT4 to Monte Carlo dose calculations based on CT data. *Medical Physics,* 2004. **31:** 2811–2818.

Jiang, H., J. Seco, and H. Paganetti, Effects of Hounsfield number conversions on patient CT based Monte Carlo proton dose calculation. *Medical Physics,* 2007. **34:** 1439–1449.

Jiang, H. et al., Simulation of organ specific patient effective dose due to secondary neutrons in proton radiation treatment. *Physics in Medicine and Biology,* 2005. **50:** 4337–4353.

Johnson, P.B. et al., Hybrid patient-dependent phantoms covering statistical distributions of body morphometry in the U.S. adult and pediatric population. *Proceedings of the IEEE,* 2009. **97**(12): 2060–2075.

Kang, B.-H. and J.-W. Kim, Monte Carlo design study of a gamma detector system to locate distal dose falloff in proton therapy. *IEEE Transactions on Nuclear Science,* 2009. **56:** 46–50.

Kawrakow, I. and A.F. Bielajew, On the condensed history technique for electron transport. *Nuclear Instruments and Methods in Physical Research B,* 1998. **142**(3): 253–280.

Keall, P.J. et al., Monte Carlo as a four-dimensional radiotherapy treatment-planning tool to account for respiratory motion. *Physics in Medicine and Biology,* 2004. **49:** 3639–3648.

Kempe, J., I. Gudowska, and A. Brahme, Depth absorbed dose and LET distributions of therapeutic 1H, 4He, 7Li, and 12C beams. *Medical Physics,* 2007. **34**(1): 183–192.

Kimstrand, P. et al., A beam source model for scanned proton beams. *Physics in Medicine and Biology,* 2007. **52**(11): 3151–3168.

Kimstrand, P. et al., Experimental test of Monte Carlo proton transport at grazing incidence in GEANT4, FLUKA and MCNPX. *Physics in Medicine and Biology,* 2008a. **53**(4): 1115–1129.

Kimstrand, P. et al., Parametrization and application of scatter kernels for modelling scanned proton beam collimator scatter dose. *Physics in Medicine and Biology,* 2008b. **53**(13): 3405–3429.

Kimura, A. et al., DICOM data handling for Geant4-based medical physics application. *IEEE Nuclear Science Symposium Conference Record,* 2004. **4:** 2124–2127.

Kimura, A. et al., DICOM interface and visualization tool for Geant4-based dose calculation. *IEEE Nuclear Science Symposium Conference Record,* 2005. **2:** 981–984.

Koch, N. and W. Newhauser, Virtual commissioning of a treatment planning system for proton therapy of ocular cancers. *Radiation Protection Dosimetry,* 2005. **115**(1–4): 159–163.

Koch, N. et al., Monte Carlo calculations and measurements of absorbed dose per monitor unit for the treatment of uveal melanoma with proton therapy. *Physics in Medicine and Biology,* 2008. **53**(6): 1581–1594.

Kohno, R. et al., Simplified Monte Carlo dose calculation for therapeutic proton beams. *Japanese Journal of Applied Physics,* 2002. **41:** L294–L297.

Kohno, R. et al., Experimental evaluation of validity of simplified Monte Carlo method in proton dose calculations. *Physics in Medicine and Biology,* 2003. **48:** 1277–1288.

Kooy, H. et al., Monitor unit calculations for range-modulated spread-out Bragg peak fields. *Physics in Medicine and Biology,* 2003. **48:** 2797–2808.

Kooy, H.M. et al., The prediction of output factors for spread-out proton Bragg peak fields in clinical practice. *Physics in Medicine and Biology,* 2005. **50:** 5847–5856.

Kurihara, D. et al., A 300-MeV proton beam line with energy degrader for medical science. *Japanese Journal of Applied Physics,* 1983. **22:** 1599–1605.

Laitano, R.F., M. Rosetti, and M. Frisoni, Effects of nuclear interactions on energy and stopping power in proton beam dosimetry. *Nuclear Instruments and Methods A,* 1996. **376:** 466–476.

Lee, C. and W. Bolch, Construction of a tomographic computational model of a 9-mo-old and its Monte Carlo calculation time comparison between the MCNP4C and MCNPX codes. *Health Physics,* 2003. **84:** S259.

Lee, C. et al., The UF series of tomographic computational phantoms of pediatric patients. *Medical Physics,* 2005. **32**(12): 3537–3548.

Lee, C. et al., Whole-body voxel phantoms of paediatric patients— UF Series B. *Physics in Medicine and Biology,* 2006. **51**(18): 4649–4661.

Lee, C. et al., Hybrid computational phantoms of the male and female newborn patient: NURBS-based whole-body models. *Physics in Medicine and Biology,* 2007. **52:** 3309–3333.

Lee, C. et al., Hybrid computational phantoms of the 15-year male and female adolescent: Applications to CT organ dosimetry for patients of variable morphometry. *Medical Physics,* 2008. **35:** 2366–2382.

Lee, C. et al., The UF family of reference hybrid phantoms for computational radiation dosimetry. *Physics in Medicine and Biology,* 2010. **55**(2): 339–363.

Lewis, H.W., Multiple scattering in an infinite medium. *Physical Review,* 1950. **78:** 526–529.

Li, J.S. et al., A particle track-repeating algorithm for proton beam dose calculation. *Physics in Medicine and Biology,* 2005. **50**(5): 1001–1010.

Li, T. et al., Reconstruction for proton computed tomography by tracing proton trajectories: A Monte Carlo study. *Medical Physics,* 2006. **33**(3): 699–706.

Liu, H.H. and P. Keall, D_m rather than D_w should be used in Monte Carlo treatment planning. *Medical Physics*, 2002. **29**: 922–924.

Lu, H.M. and H. Kooy, Optimization of current modulation function for proton spread-out Bragg peak fields. *Medical Physics*, 2006. **33**: 1281–1287.

Matsufuji, N. et al., Relationship between CT number and electron density, scatter angle and nuclear reaction for hadron-therapy treatment planning. *Physics in Medicine and Biology*, 1998. **43**: 3261–3275.

Medin, J. and P. Andreo, Monte Carlo calculated stopping-power ratios, water/air, for clinical proton dosimetry (50–250 MeV). *Physics in Medicine and Biology*, 1997. **42**: 89–105.

Moravek, Z. et al., Uncertainty reduction in intensity modulated proton therapy by inverse Monte Carlo treatment planning. *Physics in Medicine and Biology*, 2009. **54**(15): 4803–4819.

Moteabbed, M., S. Espana, and H. Paganetti, Monte Carlo patient study on the comparison of prompt gamma and PET imaging for range verification in proton therapy. *Physics in Medicine and Biology*, 2011. **56**: 1063–1082.

Moteabbed, M., A. Geyer, R. Drenkhahn, W.E. Bolch, and H. Paganetti, Comparison of wholebody phantom designs to estimate organ equivalent neutron doses for secondary cancer risk assessment in proton therapy. *Physics in Medicine and Biology*, 2012. **57**: 499–515.

Moyers, M.F. et al., Calibration of a proton beam energy monitor. *Medical Physics*, 2007. **34**(6): 1952–1966.

Moyers, M.F. et al., Leakage and scatter radiation from a double scattering based proton beamline. *Medical Physics*, 2008. **35**(1): 128–144.

Mustafa, A.A.M. and D.F. Jackson, The relation between x-ray CT numbers and charged particle stopping powers and its significance for radiotherapy treatment planning. *Physics in Medicine and Biology*, 1983. **28**: 169–176.

NCRP, *Protection against Neutron Radiation*. National Council on Radiation Protection and Measurements Report, 1973. **38**.

Newhauser, W. et al., Monte Carlo simulations for configuring and testing an analytical proton dose-calculation algorithm. *Physics in Medicine and Biology*, 2007. **52**(15): 4569–4584.

Newhauser, W. et al., Monte Carlo simulations of a nozzle for the treatment of ocular tumours with high-energy proton beams. *Physics in Medicine and Biology*, 2005. **50**: 5229–5249.

Newhauser, W. et al., Monte Carlo simulations of the dosimetric impact of radiopaque fiducial markers for proton radiotherapy of the prostate. *Physics in Medicine and Biology*, 2007. **52**(11): 2937–2952.

Nipper, J.C., J.L. Williams, and W.E. Bolch, Creation of two tomographic voxel models of paediatric patients in the first year of life. *Physics in Medicine and Biology*, 2002. **47**(17): 3143–3164.

Paganetti, H., Nuclear interactions in proton therapy: Dose and relative biological effect distributions originating from primary and secondary particles. *Physics in Medicine and Biology*, 2002. **47**: 747–764.

Paganetti, H., Four-dimensional Monte Carlo simulation of time dependent geometries. *Physics in Medicine and Biology*, 2004. **49**: N75–N81.

Paganetti, H., Monte Carlo calculations for absolute dosimetry to determine output factors for proton therapy treatments. *Physics in Medicine and Biology*, 2006. **51**: 2801–2812.

Paganetti, H., Dose to water versus dose to medium in proton beam therapy. *Physics in Medicine and Biology*, 2009a. **54**: 4399–4421.

Paganetti, H., The use of computational patient models to assess the risk of developing radiation-induced cancers from radiation therapy of the primary cancer. *Proceedings of the IEEE*, 2009b. **97**: 1977–1987.

Paganetti, H., Range uncertainties in proton therapy and the role of Monte Carlo simulations. *Physics in Medicine and Biology*, 2012. **57**: R99–R117.

Paganetti, H. and T. Bortfeld, Proton therapy. In: *New Technologies in Radiation Oncology*; Eds. Schlegel, W.; Bortfeld, T.; Grosu, A.L. Berlin, Heidelberg, New York: Springer, ISBN 3-540-00321-5, 2005.

Paganetti, H. and M. Goitein, Radiobiological significance of beam line dependent proton energy distributions in a spread-out Bragg peak. *Medical Physics*, 2000. **27**: 1119–1126.

Paganetti, H. and B. Gottschalk, Test of Geant3 and Geant4 nuclear models for 160 MeV protons stopping in CH_2. *Medical Physics*, 2003. **30**: 1926–1931.

Paganetti, H., H. Jiang, and A. Trofimov, 4D Monte Carlo simulation of proton beam scanning: Modeling of variations in time and space to study the interplay between scanning pattern and time-dependent patient geometry. *Physics in Medicine and Biology*, 2005. **50**: 983–990.

Paganetti, H. and T. Schmitz, The influence of the beam modulation method on dose and RBE in proton radiation therapy. *Physics in Medicine and Biology*, 1996. **41**: 1649–1663.

Paganetti, H. et al., Relative biological effectiveness (RBE) values for proton beam therapy. *International Journal of Radiation Oncology, Biology, Physics*, 2002. **53**: 407–421.

Paganetti, H. et al., Accurate Monte Carlo for nozzle design, commissioning, and quality assurance in proton therapy. *Medical Physics*, 2004a. **31**: 2107–2118.

Paganetti, H. et al., Monte Carlo simulations with time-dependent geometries to investigate organ motion with high temporal resolution. *International Journal of Radiation Oncology, Biology, Physics*, 2004b. **60**: 942–950.

Paganetti, H. et al., Clinical implementation of full Monte Carlo dose calculation in proton beam therapy. *Physics in Medicine and Biology*, 2008. **53**(17): 4825–4853.

Palmans, H. and F. Verhaegen, Assigning nonelastic nuclear interaction cross sections to Hounsfield units for Monte Carlo treatment planning of proton beams. *Physics in Medicine and Biology*, 2005. **50**: 991–1000.

Parodi, K. et al., Clinical CT-based calculations of dose and positron emitter distributions in proton therapy using the FLUKA Monte Carlo code. *Physics in Medicine and Biology*, 2007a. **52**(12): 3369–3387.

Parodi, K. et al., PET/CT imaging for treatment verification after proton therapy: A study with plastic phantoms and metallic implants. *Medical Physics*, 2007b. **34**(2): 419–435.

Parodi, K. et al., Patient study of *in vivo* verification of beam delivery and range, using positron emission tomography and computed tomography imaging after proton therapy. *International Journal of Radiation Oncology, Biology, Physics*, 2007c. **68**(3): 920–934.

Pelowitz, D.B.E., *MCNPX User's Manual, Version 2.5.0.* Los Alamos National Laboratory, 2005. LA-CP-05-0369.

Perez-Andujar, A., W.D. Newhauser, and P.M. Deluca, Neutron production from beam-modifying devices in a modern double scattering proton therapy beam delivery system. *Physics in Medicine and Biology*, 2009. **54**(4): 993–1008.

Perl, J., J. Shin, S. Schuemann, B. A. Faddegon, and H. Paganetti, TOPAS—An innovative proton Monte Carlo platform for research and clinical applications. *Medical Physics*, 2012. **39**: 6818–6837.

Peterson, S. et al., Variations in proton scanned beam dose delivery due to uncertainties in magnetic beam steering. *Medical Physics*, 2009a. **36**(8): 3693–3702.

Peterson, S.W. et al., Experimental validation of a Monte Carlo proton therapy nozzle model incorporating magnetically steered protons. *Physics in Medicine and Biology*, 2009b. **54**(10): 3217–3229.

Petti, P.L., Differential-pencil-beam dose calculations for charged particles. *Medical Physics*, 1992. **19**: 137–149.

Petti, P.L., Evaluation of a pencil-beam dose calculation technique for charged particle radiotherapy. *International Journal of Radiation Oncology, Biology, Physics*, 1996. **35**: 1049–1057.

Pflugfelder, D. et al., Quantifying lateral tissue heterogeneities in hadron therapy. *Medical Physics*, 2007. **34**(4): 1506–1513.

Pia, M.G. et al., Physics-related epistemic uncertainties in proton depth dose simulation. *IEEE Transactions on Nuclear Science*, 2010. **57**(5): 2805–2830.

Polf, J.C. and W.D. Newhauser, Calculations of neutron dose equivalent exposures from range-modulated proton therapy beams. *Physics in Medicine and Biology*, 2005. **50**(16): 3859–3873.

Polf, J.C., W.D. Newhauser, and U. Titt, Patient neutron dose equivalent exposures outside of the proton therapy treatment field. *Radiation Protection Dosimetry*, 2005. **115**(1–4): 154–158.

Rinaldi, I., A. Ferrari, A. Mairani, H. Paganetti, K. Parodi, and P. Sala, An integral test of FLUKA nuclear models with 160 MeV proton beams in multi layer Faraday cups. *Physics in Medicine and Biology*, 2011. **56**: 4001–4012.

Rogers, D.W.O. et al., *BEAMnrc User Manual.* NRCC Report PIRS-0509, 2002.

Rueckert, D. et al., Non-rigid registration using free-form deformations: Application to breast MR images. *IEEE Transactions on Medical Imaging*, 1999. **18**: 712–721.

Sandison, G.A. et al., Extension of a numerical algorithm to proton dose calculations. I. Comparisons with Monte Carlo simulations. *Medical Physics*, 1997. **24**: 841–849.

Santin, G. et al., GATE: A Geant4-based simulation platform for PET and SPECT integrating movement and time management. *IEEE Transactions on Nuclear Science*, 2003. **50**: 1516–1521.

Sarrut, D. and L. Guigues, Region-oriented CT image representation for reducing computing time of Monte Carlo simulations. *Medical Physics*, 2008. **35**(4): 1452–1463.

Sato, T. et al., Fluence-to-dose conversion coefficients for neutrons and protons calculated using the PHITS code and ICRP/ICRU adult reference computational phantoms. *Physics in Medicine and Biology*, 2009. **54**(7): 1997–2014.

Sawakuchi, G.O. et al., Density heterogeneities and the influence of multiple coulomb and nuclear scatterings on the Bragg peak distal edge of proton therapy beams. *Physics in Medicine and Biology*, 2008. **53**(17): 4605–4619.

Sawakuchi, G.O. et al., Monte Carlo investigation of the low-dose envelope from scanned proton pencil beams. *Physics in Medicine and Biology*, 2010. **55**(3): 711–721.

Schaffner, B. and E. Pedroni, The precision of proton range calculations in proton radiotherapy treatment planning: Experimental verification of the relation between CT-HU and proton stopping power. *Physics in Medicine and Biology*, 1998. **43**: 1579–1592.

Schaffner, B., E. Pedroni, and A. Lomax, Dose calculation models for proton treatment planning using a dynamic beam delivery system: An attempt to include density heterogeneity effects in the analytical dose calculation. *Physics in Medicine and Biology*, 1999. **44**: 27–41.

Schneider, U., E. Pedroni, and A. Lomax, The calibration of CT Hounsfield units for radiotherapy treatment planning. *Physics in Medicine and Biology*, 1996. **41**: 111–124.

Schneider, U. et al., Secondary neutron dose during proton therapy using spot scanning. *International Journal of Radiation Oncology, Biology, Physics*, 2002. **53**: 244–251.

Schneider, W., T. Bortfeld, and W. Schlegel, Correlation between CT numbers and tissue parameters needed for Monte Carlo simulations of clinical dose distributions. *Physics in Medicine and Biology*, 2000. **45**: 459–478.

Schulte, R.W. et al., A maximum likelihood proton path formalism for application in proton computed tomography. *Medical Physics*, 2008. **35**(11): 4849–4856.

Schümann, S., H. Paganetti, J. Shin, B.A. Faddegon, and J. Perl, Efficient voxel navigation for proton therapy dose calculation in TOPAS and Geant4. *Physics in Medicine and Biology*, 2012. **57**: 3281–3293.

Segars, W.P. and B.M.W. Tsui, Study of the efficacy of respiratory gating in myocardial SPECT using the new 4-D NCAT phantom. *IEEE Transactions in Nuclear Science*, 2002. **49**: 675–679.

Segars, W.P., D.S. Lalush, and B.M.W. Tsui, A realistic spline-based dynamic heart phantom. *IEEE Transactions in Nuclear Science*, 1999. **46**: 503–506.

Shin, J., J. Perl, J. Schumann, H. Paganetti, and B.A. Faddegon, A modular method to handle multiple time-dependent quantities in Monte Carlo simulations. *Physics in Medicine and Biology*, 2012. **57**: 3295–3308.

Slyper, A.H., Childhood obesity, adipose tissue distribution, and the pediatric practitioner. *Pediatrics*, 1998. **102**(1): e4.

Soukup, M. and M. Alber, Influence of dose engine accuracy on the optimum dose distribution in intensity-modulated proton therapy treatment plans. *Physics in Medicine and Biology*, 2007. **52**: 725–740.

Soukup, M., M. Fippel, and M. Alber, A pencil beam algorithm for intensity modulated proton therapy derived from Monte Carlo simulations. *Physics in Medicine and Biology*, 2005. **50**: 5089–5104.

Stankovskiy, A. et al., Monte Carlo modelling of the treatment line of the Proton Therapy Center in Orsay. *Physics in Medicine and Biology*, 2009. **54**(8): 2377–2394.

Staton, R.J. et al., A comparison of newborn stylized and tomographic models for dose assessment in paediatric radiology. *Physics in Medicine and Biology*, 2003. **48**(7): 805–820.

Szymanowski, H. and U. Oelfke, Two-dimensional pencil beam scaling: An improved proton dose algorithm for heterogeneous media. *Physics in Medicine and Biology*, 2002. **47**: 3313–3330.

Taddei, P.J. et al., Reducing stray radiation dose to patients receiving passively scattered proton radiotherapy for prostate cancer. *Physics in Medicine and Biology*, 2008. **53**(8): 2131–2147.

Taddei, P.J. et al., Stray radiation dose and second cancer risk for a pediatric patient receiving craniospinal irradiation with proton beams. *Physics in Medicine and Biology*, 2009. **54**(8): 2259–2275.

Tayama, R. et al., Measurement of neutron dose distribution for a passive scattering nozzle at the Proton Medical Research Center (PMRC). *Nuclear Instruments and Methods in Physics Research A*, 2006. **564**: 532–536.

Thirion, J.P., Image matching as a diffusion process: An analogy with Maxwell's demons. *Medical Image Analysis*, 1998. **2**: 243–260.

Titt, U. et al., Monte Carlo investigation of collimator scatter of proton-therapy beams produced using the passive scattering method. *Physics in Medicine and Biology*, 2008a. **53**(2): 487–504.

Titt, U. et al., Assessment of the accuracy of an MCNPX-based Monte Carlo simulation model for predicting three-dimensional absorbed dose distributions. *Physics in Medicine and Biology*, 2008b. **53**(16): 4455–4470.

Tourovsky, A. et al., Monte Carlo dose calculations for spot scanned proton therapy. *Physics in Medicine and Biology*, 2005. **50**: 971–981.

Tsui, B.M.W. et al., Quantitative cardiac SPECT reconstruction with reduced image degradation due to patient anatomy. *IEEE Transactions in Nuclear Science*, 1994. **41**: 2838–2844.

Unkelbach, J., T.C. Chan, and T. Bortfeld, Accounting for range uncertainties in the optimization of intensity modulated proton therapy. *Physics in Medicine and Biology*, 2007. **52**(10): 2755–2773.

Urban, L., *Multiple Scattering Model in Geant4.* CERN report, 2002. CERN-OPEN-2002-070.

Urie, M., M. Goitein, and M. Wagner, Compensating for heterogeneities in proton radiation therapy. *Physics in Medicine and Biology*, 1984. **29**(5): 553–566.

Urie, M. et al., Degradation of the Bragg peak due to inhomogeneities. *Physics in Medicine and Biology*, 1986. **31**: 1–15.

van Goethem, M.J. et al., Geant4 simulations of proton beam transport through a carbon or beryllium degrader and following a beam line. *Physics in Medicine and Biology*, 2009. **54**(19): 5831–5846.

van Luijk, P. et al., Collimator scatter and 2D dosimetry in small proton beams. *Physics in Medicine and Biology*, 2001. **46**(3): 653–670.

Waters, L., *MCNPX User's Manual.* Los Alamos National Laboratory, 2002.

Yepes, P. et al., Monte Carlo fast dose calculator for proton radiotherapy: Application to a voxelized geometry representing a patient with prostate cancer. *Physics in Medicine and Biology*, 2009. **54**(1): N21–N28.

Zacharatou Jarlskog, C. and H. Paganetti, Physics settings for using the Geant4 toolkit in proton therapy. *IEEE Transactions in Nuclear Science*, 2008a. **55**: 1018–1025.

Zacharatou Jarlskog, C. and H. Paganetti, Sensitivity of different dose scoring methods on organ specific neutron doses calculations in proton therapy. *Physics in Medicine and Biology*, 2008b. **53**: 4523–4532.

Zacharatou Jarlskog, C. et al., Assessment of organ specific neutron doses in proton therapy using whole-body age-dependent voxel phantoms. *Physics in Medicine and Biology*, 2008. **53**: 693–714.

Zankl, M. et al., The construction of computer tomographic phantoms and their application in radiology and radiation protection. *Radiation and Environmental Biophysics*, 1988. **27**(2): 153–164.

Zheng, Y. et al., Monte Carlo study of neutron dose equivalent during passive scattering proton therapy. *Physics in Medicine and Biology*, 2007. **52**(15): 4481–4496.

Zheng, Y. et al., Monte Carlo simulations of neutron spectral fluence, radiation weighting factor and ambient dose equivalent for a passively scattered proton therapy unit. *Physics in Medicine and Biology*, 2008. **53**(1): 187–201.

15

Application of Monte Carlo Methods to Radionuclide Therapy

Michael Ljungberg
Lund University

15.1 Introduction

When treating patients with systemic malignant diseases, external radiation treatment is sometimes not an option because of the wide spread of the tumors. A possible alternative for local treatment with external radiation is therefore the use of radionuclide therapy. In this modality, a radionuclide with a proper decay scheme is carried to the location of the disease by an appropriate radiopharmaceutical or biomolecule and thereby delivers energy by the emission of charged particles to kill the tumor cells. A classic example of radionuclide therapy is the treatment of cancer in the thyroid where orally administered ^{131}I is taken up naturally by the thyroid as part of its metabolism. If the radionuclide is not active in the metabolism of the tumor/organ, then it can be chemically labeled to a pharmaceutical, antibody, or a peptide. In radionuclide therapy, the major component to the absorbed dose is delivered by radionuclides that emit β-particles or α-particles. However, absorbed dose planning for radionuclide therapy is not a trivial task since the source cannot be turned on–off as is the case in external beam therapy. Instead, the radionuclide decays exponentially with characteristics depending both on the biokinetic properties and on the physical half-life of the radionuclide. On a small-scale level, the radiopharmaceutical is usually heterogeneously distributed, which indicates that the energy deposition may be nonuniform. The biokinetic may also change over time by internal redistribution, which means that activity distribution needs to be

measured at several time points to estimate the total number of decays and the related emitted kinetic energy. The dose rate for radionuclide therapy is also much lower than in external beam therapy, which has radiobiological implications. Since radionuclide treatment is systemic, some circulating activity is unavoidable and may result in uptakes in normal organs and tissues. This poses a problem that may restrict the amount of activity to be possibly administered.

Since the biokinetics varies between patients and also over time, the activity uptake and clearance should therefore be measured for each individual patient to estimate the total number of decays in a particular organ/tissue. The way that this can be performed is either by sequential planar scintillation camera measurements or by single-photon emission computed tomography (SPECT) methods. There are several serious problems associated with measuring activity and activity concentrations with a scintillation camera. First, the spatial resolution is relatively poor and in the order of 1–2 cm. This means that small activity volumes cannot be resolved properly due to the partial volume effect. The spatial resolution also varies with the source-to-detector distance. The statistics in the acquired images is a direct function of the activity administered and time for the acquisition and since the sensitivity of the camera is low, statistically related noise will appear in the image. Also, photon attenuation and contribution from scattered photons are effects that need to be compensated for as will be discussed further on in the chapter. Thus, measuring and characterizing the source distribution *in vivo* from scintillation camera measurements will be subjected to relatively large uncertainties.

15.2 MIRD Formalism

The absorbed dose is defined as the average energy deposit within a mass element. The medical internal radiation dose (MIRD) formalism (Watson et al., 1993; Bolch et al., 2009) has defined the description of internal dosimetry with radionuclides with the following simple equation:

$$ D = \tilde{A} \cdot S \tag{15.1} $$

where $\tilde{A}$ is the cumulative activity (total number of disintegrations during a time interval) and S is a geometry-dependent factor that relates the emitted energy from the source volume to the absorbed energy in the target volume normalized to the mass of the target. The unit of the S value will then be GyMBq^{-1} h^{-1}. The S value can be described in more detail as

$$ S = \frac{\sum_i n_i E_i \phi_i(t \leftarrow s)}{m_t} \tag{15.2} $$

where n_i is the number of particles that have the energy E_i and the summation is made for all possible particles i in the decay scheme. ϕ_i is the fraction of energy released in the source volume s that will be absorbed in the target volume t, and m is the mass of the target volume (usually called absorbed fraction). Since ϕ are generally impossible to measure clinically, the only way to obtain relevant data of ϕ for a particular geometry is to create a computer phantom of the object (patient) and use this phantom in a full Monte Carlo simulation of the radiation transport to score the energy deposited in the target volume for each source organ. By doing this for each source and target combination, the S values can be tabulated.

15.3 Scintillation Camera

As stated in the introduction, the cumulated activity $\tilde{A}$ can be determined using sequential planar scintillation camera measurements or sequential SPECT scans. The scintillation camera consists of a NaI(Tl) crystal so that, when photons are absorbed in the crystal, the energy deposit is converted to visible light through a deexcitation process. This light (proportional to the deposited energy) is guided to several multiplier tubes (PMTs) that each amplifies the weak energy signal, carried by the scintillation light, and generated at the surface of the PMT by a photo cathode. The difference between a counting detector and a camera is that a relevant position of the energy deposition in the crystal can be measured with a good accuracy by calculating the centroid of the measured energy signal obtained from several PMTs. To get an image, however, a collimator is needed. The most commonly used collimator is the parallel-hole collimator, which essentially

is a thick sheet of lead and including thousands of hexagonal holes orthogonal to the normal of the crystal. The purpose of the collimator is to absorb all photons except those impinging in a very narrow acceptance angle. The image created will be in two dimensions (2D) with no depth resolution. The spatial resolution is mainly determined by the dimension of the collimator holes and by the distance between the source and the detector and is of the order of 1–1.5 cm. If parallel-hole collimators are used, the sensitivity (cps/MBq) is close to being constant when moving the source within the field of view (FOV). This makes the calibration of measured counts to units of activity feasible for a source in a scatter- and attenuation-free environment. The energy resolution (i.e., the ability to measure energy) is relatively poor for NaI(Tl) crystals equipping conventional gamma cameras and is in the order of 10% full width at half maximum (FWHM) at 140 keV.

15.4 Activity Quantitation

From Equation 15.1, it is clear that the cumulated activity within the patient needs to be calculated with a high accuracy since this directly affects the absorbed dose. The activity is commonly measured using a scintillation camera that obtains an image of the activity distribution either in 2D mode (planar) or in 3D mode (SPECT). The second of these two modalities should be preferred because the absorbed energy is actually distributed in 3D.

There are several problems associated with measuring activity with 2D imaging methods, so we will focus on 3D tomographic methods. The SPECT technique is based on projections of events coming from photons acquired in multiple views around the patient. From this set of projections, a set of tomographic images can be reconstructed using a mathematical algorithm. The most commonly used reconstruction method today is the ML-EM (maximum likelihood expectation maximization) iterative method and is based on the Poisson statistics inherent in scintillation camera images. The principle is that the imaging process is modeled by a computer algorithm to mimic a real measurement. This means that projections are calculated from an initial guess of the activity distribution that often is a matrix where all voxels are set to unity. The calculated projections are then compared to real measured projections, and in the ML-EM algorithm, it is the ratio between measured and calculated projection that is used. The ratio is calculated for all projection angles and is then back-projected to obtain a reconstructed error image. The initial image is then multiplied with this error image and normalized and the updated image is then used as an input to the same process. This makes the ML-EM reconstruction to be an iterative process. The procedure can be described by the following equation:

$$ f_j^{k+1} = \frac{f_j^k}{\sum_{i=1}^{n} a_{ij}} \cdot \sum_{i=1}^{n} \frac{g_i}{\sum_{j'=1}^{n} a_{ij'} f_{j'}^k} \tag{15.3} $$

Here, a_{ij} is the system matrix that describes the probability for a photon originated from a location j to generate an event in the calculated projection at a location *i* in the measured protection *g*. After performing several iterations, the updated image will converge to a useful image, which means that the numbers in the error image will converge to unity values. The ML-EM method has some attractive mathematical properties but is generally slow in generating an acceptable image since it requires several tens of iterations before convergence. A popular alternative method to the ML-EM method is the OS-EM (ordered-subset expectation-maximization) method (Hudson and Larkin, 1994). This method is very similar to ML-EM but with the important difference in that updates of the activity image are made after a subset of selected projections instead of after all projections have been processed (as is the case for ML-EM).

15.4.1 Photon Attenuation

Photon interactions occur in the patient mainly by photo absorption and Compton and coherent scattering, which means that some photons (otherwise expected to be detected) will be absorbed or scattered in other directions with potentially loss of energy. The scaling from count rate to activity by the use of a single calibration factor (cps/MBq) will then not be valid. The attenuation becomes complex because it also depends on the patient body habitus, the photon energy, and variation in density and tissue composition. In cardiac imaging, nonuniform photon attenuation in regions such as the thorax can introduce artifacts and related false positive defects and therefore correction for this effect is necessary. On modern hybrid SPECT/CT systems, attenuation correction can be made using transmission maps of the patient obtained from a registered CT image. The Hounsfield

numbers from the CT study however need to be converted to attenuation coefficients for the relevant radionuclide. Here, validation with Monte Carlo simulations can be of importance. From a digital phantom, created from a physical phantom, a direct comparison between experimental measurements and simulations can be made. The major advantage of simulations is that the interactions in the phantom can be turned off and the result of such a simulation for SPECT application can be a tomographic image representing a scatter- and attenuation-free situation. These images thus can serve as a reference image when evaluating the accuracy and precision in attenuation compensation.

15.4.2 Scatter Contribution to Image

The problem with scatter contribution in SPECT images is perhaps the area where Monte Carlo simulations have been proven to be most useful. The contribution of scatter has its underlying cause in the poor energy resolution of the NaI(Tl) crystal. Therefore, events from scattered photons with lower energy and change in direction will contaminate the image, leading to a reduction in image contrast. Also, here, the scatter contribution depends on the photon energy, source distribution, and on body composition in addition to camera-specific parameters, such as energy window size and location. The Monte Carlo method allows for a separation of the image into components of events from primary unscattered photons, scattered photons, and even further separation of scatter events into categories of scatter orders. Examples of such simulations are shown in Figure 15.1, where the events are separated as a function of the number of scattering interactions they underwent. These curves are

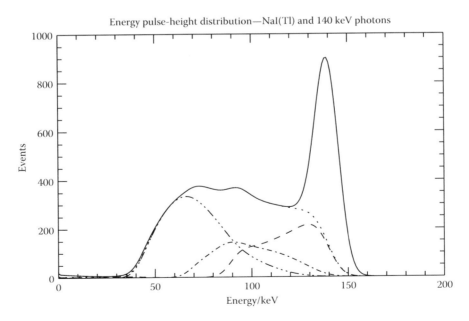

FIGURE 15.1 Energy pulse-height distribution simulated for ^{99}Tc^m and a NaI(Tl) scintillation camera. An energy resolution of 10% (FWHM) at 140 keV has been assumed. The scatter distribution is also shown together with the distribution of first, second, third, or more orders of scattering. Since the spatial distribution of the events in a lower secondary scatter window depends on the location, these types of curves are important to access to optimize the scatter window locations.

impossible to measure experimentally but can provide useful insight of the imaging process.

Many investigators have used such types of Monte Carlo simulation to optimize the locations of additional scatter windows (Holstensson et al., 2007) that are used in energy-window-based scatter compensation methods such as the triple energy window (TEW) method (Ogawa et al., 1991). The TEW method is based on an acquisition in two additional narrow energy windows located in close proximity to the main energy window. The scatter component in the main energy window is then estimated by the average of the images acquired in the two small windows on a pixel-by-pixel basis taking into account the difference in energy window size W. This method can be described as

$$I_{\text{scatter corrected}} = I_{\text{main}} - \left[\frac{I_{\text{upper}}}{W_{\text{upper}}} + \frac{I_{\text{lower}}}{W_{\text{lower}}} \right] \cdot \frac{W_{\text{main}}}{2} \qquad (15.4)$$

where I are images acquired in the upper, the lower, and the main energy window and W are the corresponding energy window sizes in units of keV. The scatter in the main energy window cannot be measured directly but can be estimated by Monte Carlo methods to be used in a direct comparison between the true scatter distribution and the estimated scatter distribution. An example of such a comparison is shown in Figure 15.2 for ^{99}Tc^m. The ranges of the lower and upper scatter windows are 120–126 and 154–160 keV, respectively, and the main energy window range is 126–154 keV. A standard Anger-type scintillation camera with a low-energy high-resolution camera and 10% (FWHM) energy resolution and 140 keV was simulated together with the XCAT anthropomorphic phantom (described below in the chapter).

Scatter compensation can also be included in the forward projector of modern iterative reconstruction methods. This can be either a scatter estimation obtained from a TEW acquisition or a model-based scatter distribution. An example of model-based scatter compensation, including Monte Carlo simulated data is the effective scatter source estimator (ESSE) method that was developed by Frey and colleagues (Frey and Tsui 1996). This method provides spatial-domain scatter estimation where the scatter is modeled from an estimate of the activity distribution. The ESSE calculates an effective scatter source from which the attenuated projection of the scatter source will be the estimate of the scatter component in a SPECT projection. The calculation is based on point source scatter kernels, obtained by Monte Carlo simulations for various depths in a water phantom. The method assumes that the photons are propagating through a uniform media (often water) from point source to the last scattering point before the photon exits the phantom toward the camera. Despite this approximation (that has some limitations when the attenuation is nonuniform), the method has shown to be accurate for other isotopes than ^{99}Tc^m, such as ^{111}In and ^{131}I (Ljungberg et al., 2002, 2003).

Energy-window-based methods are not well suited for bremsstrahlung SPECT of, for example, ^{90}Y (2.21 MeV β-particles) because of the continuous distribution of detected primary events from unscattered bremsstrahlung photons along the energy scale. Yet, the ESSE method together with properly Monte Carlo simulated scatter kernels have shown to be useful for quantitative bremsstrahlung SPECT imaging (Minarik et al., 2008, 2010).

Recent research on scatter compensation methods that have focused on including Monte Carlo simulation directly into the iterative reconstruction process during the iterative process calculate the scatter component within the energy window using the patient's own geometry (taken from a registered CT) and a proper model of the camera. This approach is very promising but has a large drawback in that the method is very computer intensive. For ^{99}Tc^m applications, however, clinically acceptable

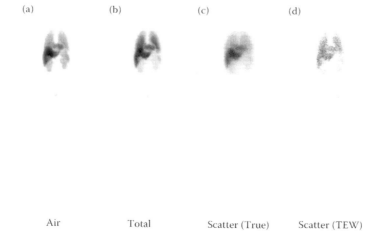

FIGURE 15.2   Simulation of a ^{99}Tc^m whole-body study. Image (a) shows the case with no scatter and attenuation in the phantom. This is not a realistic case but can serve as a reference image when comparing different scatter and attenuation compensation methods. Image (b) shows the total image, including interactions in the phantom. This image corresponds to a real measurement. Image (c) shows the scatter component in image (b) and image (d) shows the scatter as has been estimated by the triple energy method. Note that this image has more statistical noise due to the narrow energy windows.

processing times can be achieved when using variance reduction methods such as "forced detection," which in this case means that the explicit simulation of interactions in the collimator and the camera is replaced by a convolution process using a proper collimator response function (Staelens et al., 2007; deJong and Beekman, 2001). In ^{99}Tc^m imaging, this can be justified since 140 keV photons that interacts in the NaI(Tl) crystal most likely will be absorbed at the first interaction site. For other radionuclides, such as ^{131}I, the septal penetration and the contribution to the image from photons back-scattered behind the crystal is more difficult to model by this approach but investigations have shown it is possible (Liu et al., 2008). If the time required for the calculation is not critical, then a fast Monte Carlo program, such as the SIMIND code (Ljungberg and Strand, 1989), can be used. Dewaraja et al. have shown such an approach for ^{131}I tumor treatment planning using quantitative SPECT (Dewaraja et al., 2006). Recent development in GPU-based computing may also speed Monte Carlo simulation that might lead to a clinical use of such a scatter compensation method (Kim and Ye, 2011).

15.4.3 Estimation of Collimator Response Function

The collimator in SPECT is optimized for a certain photon energy (low-energy, medium–energy, and high-energy collimators), but if there sometimes occur additional high-energy emissions in the decay, such as in ^{131}I and ^{123}I, then septal penetration with resulting characteristic star-like artifacts will occur (Dewaraja et al., 2000; Larsson et al., 2006). This penetration deteriorates the image quality (spatial resolution and image contrast) and makes it difficult to quantify the activity (Figure 15.3).

Compensation for septal penetration can be made by adding the effect in the forward projector in the iterative reconstruction. For ^{99}Tc^m, where penetration is a low-probability effect, the collimator response is accurately described by a geometric Gaussian-shaped component but when photon energies are used were septal penetration becomes important then it necessary to consider explicit Monte Carlo simulations of the actual photon interactions in the collimator to establish the amount and the distribution of events from the penetrating photons but also for photons scattered in the collimator.

15.5 Absorbed Dose Calculations

The absorbed dose can be calculated from a quantitative SPECT image by basically three methods. First, all energy emitted from the decay is located in the voxel. This corresponds to an absorbed fraction, ϕ, in Equation 15.2, equal to 1. Second, the distribution of energy can be modeled by convolving the SPECT images with an appropriate point-dose kernel that radially describes the distribution of energy emitted from a point source. Third, the radiation transport can be modeled in an explicit way within a computer phantom of the patient using coupled photon–electron Monte Carlo programs. Although point dose kernels are usually derived from very accurate and well-validated programs, such as EGS4 (Nelson et al., 1985), EGSnrc, Geant4 (Allison et al., 2006), and MCNP5 (Forster et al., 2004), the procedure assumes that the kernel is invariant in respect to spatial changes in the distribution. Thus, the method is not accurate in boundaries between different densities and/or tissues. A full Monte Carlo calculation can account for this nonhomogeneity and can also use the patient-specific geometry since today a SPECT/CT combination of images is often available for these types of patients. The approach of assuming local energy absorption can be justified when the radial ranges of the particles is less than that of the voxel size of the SPECT image (Ljungberg and Sjogreen-Gleisner, 2011).

15.6 Diagnostic Dosimetry

In diagnostic dosimetry, the aim is to relate the absorbed dose to the risk for late stochastic effects, such as cancer, when exposed to a low dose of ionizing radiation. A population rather than a specific person is therefore of interest. Therefore, more standardized computer phantoms that simulate a "reference man"

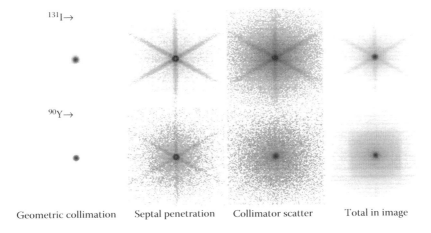

^{131}I→

^{90}Y→

Geometric collimation   Septal penetration   Collimator scatter   Total in image

FIGURE 15.3 Monte Carlo simulation of the collimator response for ^{131}I and ^{90}Y bremsstrahlung imaging. The images are displayed on a logarithmic gray scale.

are used, which will simplify the dosimetry since a set of *S* values can be precalculated and used as representative of the population. *S* values have been calculated by Monte Carlo methods for several configurations of the phantom (male, female, 15y, 10y, 5y, 1y, newborn, and a pregnant female at 3, 6, and 9 month) for 117 relevant radionuclides. The *S* values can be found in specific dosimetry codes, such as the Olinda program (Stabin et al., 2005) or at the RADAR web page (Stabin). The cumulated activity still need to be determined but this is done in work preceding the release of a new radiopharmaceutical.

15.7 Therapeutic Dosimetry

When planning for a radionuclide therapy, the treatment is often preceded by a pretherapy planning phase where a small amount of the radiopharmaceutical is administered to the patient to determine the uptake and expected absorbed dose to tumors and organ at risk per unit administered activity. This estimation requires at least three time point measurements.

Imaging with SPECT is superior because it is a 3D method and the problem with over- and underlapping activity that is common in planar imaging can be avoided and compensation methods are generally more accurate. The time–activity curves for the VOI (volume of interest) are determined by analyzing the SPECT images using user-defined region of interests. If the SPECT images can be registered to each other, then it is possible to obtain suborgan kinetic curves that show activity variations within the organ. If these are integrated over time to calculate the cumulated activity, an absorbed dose image can be obtained by also applying any of the three absorbed dose calculation methods described above. The absorbed dose in the VOI per unit administered activity will then be the base to calculate the activity for therapy expected to give a prescribed absorbed dose.

Often, the therapy radionuclide decays with photon emission not suitable for SPECT imaging. For example, ^{90}Y is a useful β emitter but the decay lacks a gamma photon emission useful for imaging, which makes this radionuclide not easy

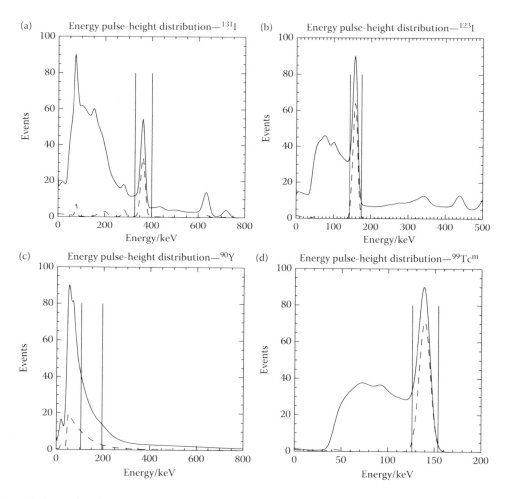

FIGURE 15.4   Monte Carlo simulated energy pulse-height spectrums for ^{131}I (a), ^{123}I (b), ^{90}Y (c), and ^{99}Tc^{m} (d) using the phantom and distribution shown in Figure 15.2. The dashed line represents events from photon unscattered in the phantom and that are geometrically collimated and thus represents the position of the decay.

to use in a pretreatment planning. Therefore, surrogate radionuclides usually are used. In treatment with ^{90}Y-labeled rituximab antibodies for non-Hodgkin's lymphoma, ^{111}In is used as a pretherapy radionuclide. It must here be assumed that the biokinetics is similar between the two compounds. The equivalent absorbed dose for ^{90}Y is then calculated based on the activity measurement of ^{111}In taking into account the differences in decay scheme and half-life. In recent research, focus is on using the bremsstrahlung photons emitted as part of the slowing down of the β-particles. However, the acquired bremsstrahlung spectra by SPECT systems are complicated to analyze since no clear photopeaks appear but rather a continuum of scattered and primary events making the selection of energy window locations for SPECT imaging difficult. However, Monte Carlo simulation of the whole imaging process can be still performed with dedicated software that allow for detailed analysis of the components to build an energy spectrum (Figure 15.4) and a corresponding image. This makes it possible to optimize the size and location of the energy window.

Monte Carlo methods can also be used to create kernels that are used for scatter compensation and collimator-response compensation. Minarik et al. (Minarik et al., 2008; Minarik, 2010) have developed a method for activity quantitation based on Monte Carlo simulated kernels and found good accuracy in a phantom experiment comparing activity values estimated from SPECT images with known values.

15.8 Monte Carlo Codes for Simulating Nuclear Medicine Systems

Several Monte Carlo programs have been developed specially for nuclear medicine imaging simulation, and others are public domain programs mainly developed for complex electron–photon simulation but can be adapted for nuclear medicine imaging simulations.

15.8.1 SIMIND Code

The SIMIND Monte Carlo program has been developed by Michael Ljungberg at the Lund University and is a photon program, dedicated to simulate scintillation camera imaging and SPECT (Ljungberg and Strand, 1989; Ljungberg, 1998). It is written in Fortran-90 and is available for Windows, Mac, and Linux operating systems from the web page (Ljungberg, 2011). It simulates photon interactions and is optimized by the use of several variance reduction methods, such as forced detection, photon splitting, delta scattering, and importance sampling. It simulates both voxel-based phantoms and more simple geometrical shapes. Transmission imaging is also possible to simulate. The program has been used for evaluation of scatter and attenuation compensation method, receiver operation characteristic (ROC) analysis, and to evaluate planar activity quantification methods. This code has been used to generate the examples included in this chapter.

15.8.2 SimSET Code

The software simulation system for emission tomography (SimSET), developed at the University of Washington by Robert Harrison et al. (Harrison and Lewellen, 2012), is designed for both single-photon emitters (SPECT and planar systems) and positron emitters (PET systems). The package includes a photon history generator (PHG), the object editor, a collimator module, and collimator, detector and binning modules. The PHG tracks photons through the tomographic FOV creating a list of photon histories. The object editor helps the user to define the activity and attenuation objects for the PHG expressed as voxel images. The collimator module simulates PET collimators and SPECT collimators. The detector module simulates typical PET and SPECT detectors including energy and (for PET) time-of-flight resolution. The binning module allows for binning in terms of number of scatters, axial position, transaxial distance and angle, and photon energy. The SimSET package does implement techniques to improve efficiency. To gain access to the code, and to be added to a list of users for update notices, interested individuals should send electronic mail to simset@u.washington.edu.

15.8.3 Gate Code

The GATE program (Jan et al., 2004) was originally dedicated to simulation of PET and SPECT systems but from version 6, computed tomography, conventional radiotherapy, and hadron therapy can also be simulated (Jan et al., 2011). This open source simulation software is developed by the OpenGATE collaboration (http://www.opengatecollaboration.org). It is based on the Geant4 Monte Carlo tool kit and uses Geant4's libraries to simulate the particle transport. The user employs an extended version of the Geant4 script language that means that no programming is necessary to set up a simulation, unless the user wants to add or modify the features in the GATE program. The program has a layered architecture with a core layer that defines the main tools and features of GATE with routines written in C++, an application layer with C++ base classes, and at the top, a user layer where the simulations are set up using command-based scripts. An original feature is the possibility of simulating time-dependent phenomena such as source kinetics and movements of geometries, for example, patient motion, respiratory and cardiac motions, changes of activity distribution over time, and scanner rotation. Geant4 does however require static geometries during a simulation. GATE has a very powerful geometry package that allows for simulation of very complex geometries such as modern PET camera and prototypes. It also allows for simulation of analytical or voxel-based geometries and output results to various file formats. GATE is currently the only Monte Carlo simulation tool supporting both imaging and radiotherapy applications. A large number of SPECT and PET scanners have already been modeled in GATE, as well as a number of linear accelerators (see http://www.opengatecollaboration.org for a list of modeled systems). This software is used for assisting detector design, assessing acquisition and quantification protocols, and for tomographic reconstruction, in both clinical and preclinical imaging. GATE also now supports the modeling of CT and optical imaging.

15.9 Voxel-Based Phantoms

A voxel-based phantom is described as images with small voxels representing either activity or density (tissue composition). The main advantage with these types of phantoms is the ability to simulate very complicated geometries from a large number of small voxels. The use of voxel-based phantoms in Monte Carlo program requires some form of ray-tracing method, which makes the computing more demanding as compared to analytical phantoms. Ray-tracing can be made by dividing the particle track into equal spacing and during the step process collect anatomical information about the tissue composition and density. Other methods, such as the delta scattering method (Woodcock et al., 1965), have shown to be effective. The drawback with voxel-based phantoms, derived from, for example, a computer tomography study, is that it a relatively difficult to make internal changes of the phantom, that is, adjusting volumes and location of individual organs. Hybrid phantoms are the latest development that combines the advantages of voxelized and mathematical phantoms. Hybrid phantoms are developed from patient data, from either CT or MRI, which are segmented to an initial voxelized model. These segmented structures are then fit with smooth surfaces, such as nonuniform rational b-splines (NURBS) or different types of meshes based on polygons. Based on patient data, the NURBS surfaces or polygon meshes can accurately model each structure in the body, providing the realism of a voxelized model. They also have the flexibility of a mathematical phantom to model anatomical variations and motion. NURBS surfaces can be easily modified through the control points that define their shape while the polygon meshes can be easily altered through the vertex points that define the polygons. These surfaces can then be redefined in a simple manner by adjusting control points instead of each individual voxel. In the final step, the surfaces are used to define new voxel-based images on arbitrary resolution than then

can be used as input to a Monte Carlo program. An example of such a hybrid phantom is the 4D NURBS-based XCAT phantom developed by Segars (Segars et al., 2010). Figure 15.5 shows rendered images of the XCAT phantom.

The XCAT phantom actually is a software package. The user modifies an input file with a large set of parameters that control the phantom dimension, organ activities, respiratory movements, and so on. Also specified in that file is the resolution and pixel size for the two sets of images (photon attenuation and activity distribution) that the software package produces. These images can then be used in Monte Carlo programs. It should be remembered that the activity is usually uniformly sampled within an organ defined by the voxel phantom. Also, the attenuation is generally treated as uniform. If more realism is needed, then an organ needs to be further segmented into compartments. A comprehensive textbook on the use of voxel phantoms in medical applications has recently been compiled by Xu and colleagues (Xu and Eckerman, 2009).

Figure 15.6 shows an example of a whole-body scintillation camera simulation using the XCAT phantom and the SIMIND Monte Carlo program. The figure illustrates the degradation in image quality caused by collimator penetration, and scatter and attenuation effects in the phantom and camera compartments, but which is possible to evaluate using detailed Monte Carlo simulations.

15.10 Summary

As in many areas in the field of radiation physics, the Monte Carlo method in nuclear medicine has provided a very important insight in how the quality of images depends on the radiation transport and limitation in detector system. The simulation of detector systems in combination with very realistic mathematical phantom is today able to predict experimental measurements to a high degree of accuracy. With increasing development in computing science and hardware development, the method will be very important for "real-time" calculation

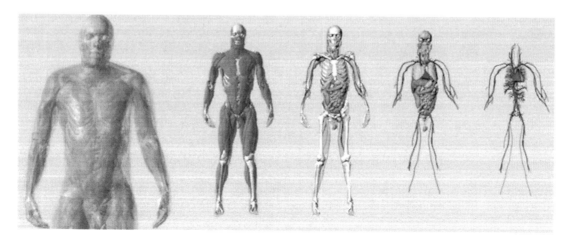

FIGURE 15.5   Rendered images of the XCAT phantom and examples of organs and structures. (Courtesy by Dr. Paul Segars, Duke University, NC, USA.)

FIGURE 15.6 A Monte Carlo simulation of the XCAT phantom. The left image (a) shows a case where no interaction occurs in the phantom. This image thus represents the best quality possible to achieve for the particular camera configuration. Image (b) includes photon attenuation and image (c) includes photon attenuation and scatter contribution from the phantom. Image (d) represents the case, including also septal penetration, scatter in the collimator, and photons back-scattered from the environment behind the crystal. Image (e) represents the case in image (d) but with a realistic noise level added.

of, for example, scatter models in iterative reconstruction algorithms. A field not mentioned here is the possibility of using Monte Carlo simulated data for educational purpose when expected detectability and recognition of possible false-positives due to physical effects, such as attenuation and scatter, could be compiled as a training package.

References

Allison, J., K. Amako, J. Apostolakis, H. Araujo, P. A. Dubois, M. Asai, G. Barrand, R. Capra, S. Chauvie, and R. Chytracek. 2006. Geant4 developments and applications. *IEEE Trans. Nucl. Sci.* 53 (1):270–8.

Bolch, W. E., K. F. Eckerman, G. Sgouros, and S. R. Thomas. 2009. MIRD pamphlet No. 21: A generalized schema for radiopharmaceutical dosimetry—standardization of nomenclature. *J. Nucl. Med.* 50 (3):477–84.

de Jong, H. W. and F. J. Beekman. 2001. Rapid SPECT simulation of downscatter in non-uniform media. *Phys. Med. Biol.* 46 (3):621–35.

Dewaraja, Y. K., M. Ljungberg, and J. A. Fessler. 2006. 3-D Monte Carlo-based scatter compensation in quantitative I-131 SPECT reconstruction. *IEEE Trans. Nucl. Sci.* 53 (1):181–8.

Dewaraja, Y. K., M. Ljungberg, and K. F. Koral. 2000. Characterization of scatter and penetration using Monte Carlo simulation in 131I imaging. *J. Nucl. Med.* 41 (1):123–30.

Forster, R. A., L. J. Cox, R. F. Barrett, T. E. Booth, J. F. Briesmeister, F. B. Brown, J. S. Bull, G. C. Geisler, J. T. Goorley, and R. D. Mosteller. 2004. MCNP (Tm) Version 5. *Nucl. Instrum. Methods Phys. Res. B, Beam Interact. Mater. At.* 213:82–6.

Frey, E. C. and B. M. W. Tsui. 1996. A new method for modeling the spatially-variant, object-dependent scatter response function in SPECT. *IEEE Nucl. Sci. Symp.*, Anaheim, CA, 2:1082–1086.

Harrison, R.L. and T. K. Lewellen. 2012. The SimSET program. In *Monte Carlo Calculation in Nuclear Medicine: Applications in Diagnostic Imaging* 2nd Edition, Boca Raton, FL, USA: CRC Press/Taylor & Francis Group.

Holstensson, M., C. Hindorf, M. Ljungberg, M. Partridge, and G. D. Flux. 2007. Optimization of energy-window settings for scatter correction in quantitative (111)in imaging: Comparison of measurements and Monte Carlo simulations. *Cancer Biother. Radiopharm.* 22 (1):136–42.

Hudson, H. M. and R. S. Larkin. 1994. Accelerated image reconstruction using ordered subsets of projection data. *IEEE Trans. Nucl. Sci.* 13:601–9.

Jan, S., D. Benoit, E. Becheva, T. Carlier, F. Cassol, P. Descourt, T. Frisson et al. 2011. GATE V6: A major enhancement of the GATE simulation platform enabling modelling of CT and radiotherapy. *Phys. Med. Biol.* 56 (4):881–901.

Jan, S., G. Santin, D. Strul, S. Staelens, K. Assie, D. Autret, S. Avner et al. 2004. GATE: A simulation toolkit for PET and SPECT. *Phys. Med. Biol.* 49 (19):4543–61.

Kim, K. S. and J. C. Ye. 2011. Fully 3D iterative scatter-corrected OSEM for HRRT PET using a GPU. *Phys. Med. Biol.* 56 (15):4991–5009.

Larsson, A., M. Ljungberg, S. Jakobsson Mo, K. Riklund, and L. Johansson. 2006. Correction for scatter and septal penetration in (123)I brain SPECT imaging—A Monte Carlo study. *Phys. Med. Biol.* 51 (22):5753–67.

Liu, S., M. A. King, A. B. Brill, M. G. Stabin, and T. H. Farncombe. 2008. Convolution-based forced detection Monte Carlo simulation incorporating septal penetration modeling. *IEEE Trans. Nucl. Sci.* 55 (3):967–74.

Ljungberg, M. 1998. The SIMIND Monte Carlo program. In *Monte Carlo Calculation in Nuclear Medicine: Applications in Diagnostic Imaging*, edited by M. Ljungberg, S. E. Strand, and M. A. King. Bristol and Philadelphia: IOP Publishing.

Ljungberg, M. 2011. *The SIMIND Monte Carlo Program*. Available from http://www.radfys.lu.se/simind/.

Ljungberg, M., E. C. Frey, K. Sjögreen, X. Liu, Y. Dewaraja, and S. E. Strand. 2003. 3D absorbed dose calculations based on SPECT: Evaluation for 111-In/90-Y therapy using Monte Carlo simulations. *Cancer Biother. Radiopharm.* 18 (1):99–108.

Ljungberg, M. and K. Sjogreen-Gleisner. 2011. The accuracy of absorbed dose estimates in tumours determined by Quantitative SPECT: A Monte Carlo study. *Acta Oncol.* 50 (6):981–9.

Ljungberg, M., K. Sjogreen, X. Liu, E. Frey, Y. Dewaraja, and S. E. Strand. 2002. A 3-dimensional absorbed dose calculation method based on quantitative SPECT for radionuclide therapy: Evaluation for 131-I using Monte Carlo simulation. *J. Nucl. Med.* 43 (8):1101–9.

Ljungberg, M. and S. E. Strand. 1989. A Monte Carlo program for the simulation of scintillation camera characteristics. *Comput. Methods Programs Biomed.* 29 (4):257–72.

Minarik, D. 2010. Activity quantification based on scintillation camera imaging—Application to ^{111}In/^{90}Y radioimmunotherapy Department of Medical Radiation Physics, Lund University, Lund.

Minarik, D., K. Sjogreen-Gleisner, O. Linden, K. Wingardh, J. Tennvall, S. E. Strand, and M. Ljungberg. 2010. [90]Y Bremsstrahlung imaging for absorbed-dose assessment in high-dose radioimmunotherapy. *J. Nucl. Med.* 51 (12):1974–8.

Minarik, D., K. Sjogreen Gleisner, and M. Ljungberg. 2008. Evaluation of quantitative ^{90}Y SPECT based on experimental phantom studies. *Phys. Med. Biol.* 53 (20):5689–703.

Nelson, R. F., H. Hirayama, and D. W. O. Rogers. 1985. *The EGS4 Code System*. Stanford, CA: SLAC.

Ogawa, K., Y. Harata, T. Ichihara, A. Kubo, and S. Hashimoto. 1991. A practical method for position-dependent Compton-scatter correction in single photon emission CT. *IEEE Trans. Med. Imag.* 10 (3):408–12.

Segars, W. P., G. Sturgeon, S. Mendonca, J. Grimes, and B. M. W. Tsui. 2010. 4D XCAT phantom for multimodality imaging research. *Med. Phys.* 37 (9):4902–15.

Stabin, M. G. *RADAR—The Radiation Dose Assessment Resource*. Available from http://www.doseinfo-radar.com/.

Stabin, M. G., R. B. Sparks, and E. Crowe. 2005. OLINDA/EXM: The second-generation personal computer software for internal dose assessment in nuclear medicine. *J. Nucl. Med.* 46 (6):1023–7.

Staelens, S., T. de Wit, and F. Beekman. 2007. Fast hybrid SPECT simulation including efficient septal penetration modelling (SP-PSF). *Phys. Med. Biol.* 52 (11):3027–43.

Watson, E. E., M. G. Stabin, and J. A. Siegel. 1993. MIRD formulation. *Med. Phys.* 20:511–4.

Woodcock, E., T. Murphy, P. Hemmings, and S. Longworth. 1965. Techniques used in the GEM code for Monte Carlo neutronics calculations in reactors and other systems of complex geometry. *Proc. Conference on the Application of Computing Methods to Reactor Problems*, ANL-7050, pp. 557–579.

Xu, X. G. and K. F. Eckerman. 2009. *Handbook of Anatomical Models for Radiation Dosimetry*. Boca Raton, FL, USA: CRC Press/Taylor & Francis Group.

Monte Carlo for Kilovoltage and Megavoltage Imaging

George X. Ding
*Vanderbilt University School
of Medicine*

Andrew Fielding
*Queensland University
of Technology*

16.1 Introduction

Since the discovery of x-rays by Rontgen in 1895, x-ray imaging has been widely used in medical applications. The advent of x-ray computed tomography (CT) imaging was one of the most significant advancements in radiography. CT images are obtained by rotating a highly collimated x-ray source around the patient and having detectors record projection x-ray data from a multitude of angles throughout the rotation. Tomographic image reconstruction techniques are then used to create an axial cross-section image of the irradiated volume of the patient. Three-dimensional (3D) images can be constructed by translating the patient through the plane of the x-ray beam in either a stepwise or a continuous motion, resulting in axial or helical scans, respectively, which can be combined in a single 3D image set. Volumetric patient CT images are commonly used in radiotherapy treatment planning. In recent years, a new technique in imaging-guided radiotherapy (IGRT) is kV cone-beam computerized tomography (CBCT) (Uematsu et al., 1996; Jaffray et al., 1999, 2002; Jaffray and Siewerdsen, 2000; Siewerdsen and Jaffray, 2000; Shiu et al., 2003; Yenice et al., 2003; Sharpe et al., 2006) of the patient on the treatment table. CBCT is capable of providing accurate 3D volumetric knowledge about the patient's anatomy for every treatment fraction and is also suitable for adaptive corrections of errors related to interfractional uncertainties of the treatment process. With the availability of in-room kV-CBCT, it becomes practical to monitor the patient treatment positions and tumor shrinkage on a daily or weekly basis as well as to perform potentially adaptive radiotherapy (ART) (Ding et al., 2007b; Pawlowski et al., 2010) and IGRT (Jaffray et al., 1999; Jaffray and Siewerdsen, 2000).

The utilization of x-ray imaging in IGRT has dramatically improved the radiation treatment and improved the lives of cancer patients. X-ray image guidance has emerged as the new paradigm for patient positioning and target localization in radiotherapy. Although widely varied in modality and method, all radiographic guidance techniques have one thing in common, the use of ionizing radiation. Monte Carlo techniques have been extensively applied in both imaging and radiotherapy physics in recent decades. It is well accepted that Monte Carlo methods offer the most powerful tool for modeling radiation transport for radiotherapy applications (Verhaegen and Seuntjens, 2003). There are several advantages of using Monte Carlo techniques in radiation dosimetry and in the development of new medical devices for treatment delivery and for x-ray imaging. In the following, the application of Monte Carlo techniques will be described.

16.2 Monte Carlo and Kilovoltage X-Ray Imaging

Monte Carlo techniques play an important role in the development of new image devices and radiation dose calculations to patients resulting from an imaging procedure using kilovoltage x-rays.

The knowledge of the characteristics of x-ray beam, including the scattered particles, is important for optimizing a scanner design geometry (Siewerdsen and Jaffray, 2000). Although theoretical scatter models involving both experimental measurements and mathematical algorithms have generally been used in many studies (Tofts and Gore, 1980; Glover, 1982; Johns and Yaffe, 1982; Joseph and Spital, 1982; Merritt and Chenery, 1986;

Siewerdsen and Jaffray, 2001; Endo et al., 2006), the Monte Carlo method is becoming more common due to widespread availability of high-speed computers and simulation codes (Klein et al., 1983; Kanamori et al., 1985; Endo et al., 2001; Aviles Lucas et al., 2004; Ay et al., 2004; Colijn et al., 2004; Colijn and Beekman, 2004; Ay and Zaidi, 2005; Malusek et al., 2005). It has been recognized that the Monte Carlo method is the ideal research tool for scatter modeling and evaluation of scatter correction techniques (Kanamori et al., 1985; Endo et al., 2001; Aviles Lucas et al., 2004; Ay et al., 2004; Colijn and Beekman, 2004; Ay and Zaidi, 2005; Malusek et al., 2005).

For patient dose calculations resulting from x-rays in diagnostic energy range, Monte Carlo technique is the most accurate method. Current available model-based dose calculation algorithms used in the dose calculation for radiation treatment megavoltage photon beams are incapable of calculating radiation dose from kilovoltage x-rays, and can lead to dose errors of up to 300% due to the increased photoelectric effect in doses to bone that are up to a factor of 3–4 higher than those in surrounding soft tissue (Ding et al., 2008a; Alaei et al., 2010). Monte Carlo techniques offer the most accurate imaging dose calculations (Ding et al., 2008a,b,c, 2010; Faddegon et al., 2008; Ding and Coffey, 2009; Downes et al., 2009; Spezi et al., 2009; Dixon et al., 2010).

16.2.1 Simulations of Realistic X-Ray Sources

With many improvements in the Monte Carlo codes, such as in EGSnrc (Kawrakow and Rogers, 2002), it is possible to accurately model low-energy x-ray beams in the diagnostic energy range. In addition, with ever-increasing computing speed and decreasing cost of a computer, it is possible to perform the simulation within an acceptable time. Furthermore, the development of a special-purpose Monte Carlo code, such as BEAM (Rogers et al., 1995), to simulate radiotherapy beams has made it practical to simulate not only radiotherapy beam but also beams from x-ray tube. The BEAM code has been extensively used to simulate the electron and photon beams in commercial medical accelerators (Nelson et al., 1985b; Rogers et al., 1995; Kawrakow and Rogers, 2002). The BEAM code (Rogers et al., 1995) has been used extensively to characterize both megavoltage electron and photon beams as well as kilovoltage x-rays (Ding and Rogers, 1995; Verhaegen et al., 1999; Ding, 2002; Sheikh-Bagheri and Rogers, 2002a,b; Ding et al., 2006, 2007a; Jarry et al., 2006). In the Monte Carlo simulation, one can obtain the information that typically includes data pertaining to the particle's position, direction, charge, and interactions that it has undergone. In the BEAM code, this information is stored in the phase-space file using a variable, LATCH. The ability to trace a particle's history allows the fluence of particles that reach the detector panel to be broken down into components: fluence of particles that have interacted only in certain regions, fluence of particles that have not interacted in certain regions, and fluence arising from particles with other user-specified LATCH bit settings. By using LATCH, the Monte Carlo simulation is capable of separating the scattered particles from the primaries even when they have the same incident angle

and energy at the surface of the detector. This is the most effective method to obtain details of interactions within the phantom when x-ray passes through a phantom or a patient. By analyzing the simulated results, we cannot only obtain the x-ray properties before and after an x-ray beam interacts with imaged target but also reveal the characteristics of both primary and scattered photons and their dependence on the size of the imaged target.

The characteristics of x-ray beams depend on scan protocols in which both kVp and filters differ; the x-ray source with detailed x-ray tube geometry for different scan protocols has to be modeled in detail.

Figure 16.1 shows examples of the geometry of a Varian OBI system for the two kV-CBCT scan modes where both bow-tie filters and x-ray collimations differ. The Monte Carlo technique is capable of simulating each realistic x-ray source based on a selected parameter of each image acquisition procedure. The Monte Carlo simulation provides realistic beam details as shown in Figures 16.2 and 16.3 in which energy spectra and the mean energy distributions as a function of the central axis for the five different kV-CBCT beams are presented. For each kV-CBCT beam, two curves are presented: one is analyzed photons within the field defined by the X–Y collimator blades openings (within beam field) and the other is within a small 4×4 cm^2 field near the beam central axis (within 4×4 cm^2). It is seen that the beam spectra near the central axis (where the thickness of the bow-tie filter is thinner) contain more low-energy photons that is consistent with the observation shown in Figure 16.3. The accuracy of each Monte Carlo simulated beam can be validated against measurements as shown in Figure 16.4. It has been shown that the information provided from Monte Carlo simulation plays a significant role in configuring a kV-CBCT beam in a treatment planning system (Alaei et al., 2010).

Low-energy x-rays are often used in imaging guidance techniques. Unfortunately, low-energy x-ray dose calculation methods are not a part of current radiotherapy treatment planning systems and therefore, this additional imaging dose cannot be included in the calculation of total dose to radiotherapy patients.

The Monte Carlo simulations are not only capable to accurately calculate radiation doses to patients resulting from an imaging guidance procedure but also provide beam data for configuration of an x-ray beam in a commercial treatment planning systems (Alaei et al., 2010). The information of additional imaging dose to patient's sensitive organs helps clinicians to make informed decisions on patient selection and the frequency of imaging guidance for each patient. This can lead to improved radiotherapy outcomes and reduced serious complications and poor outcomes of radiation treatment.

The x-ray source parameters obtained from Monte Carlo simulations shown in Figures 16.2 and 16.3 are difficult to obtain through the experimental methods. In addition, the methodology has developed to calculate and predict absolute dose to patients based on CT images from any x-ray imaging procedure (Ding et al., 2008a).

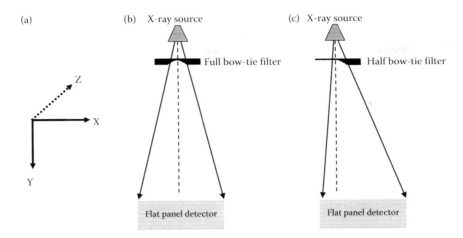

FIGURE 16.1 Schematic drawing to illustrate: (a) Coordinates used in this study, (b) the OBI scan geometry in full fan mode where the detector is centered, and (c) the OBI scan geometry in half fan mode, where the detector is off-set toward the positive X-direction. The isocenter of the CBCT is at 100 cm from the x-ray source and the source-to-detector distance can be selected from among 140, 150, and 170 cm distances. The standard focus to detector distance of 150 cm is employed in our study. The CsI flat panel detector with an active area of 40×30 cm^2 has an anti-scatter grid and the grid lamellas run only along the long axis (X-direction), which is mounted directly above the detector. (Reproduced from Ding G X and Coffey C W 2010. *Phys Med Biol* **55** 5231–48. With permission.)

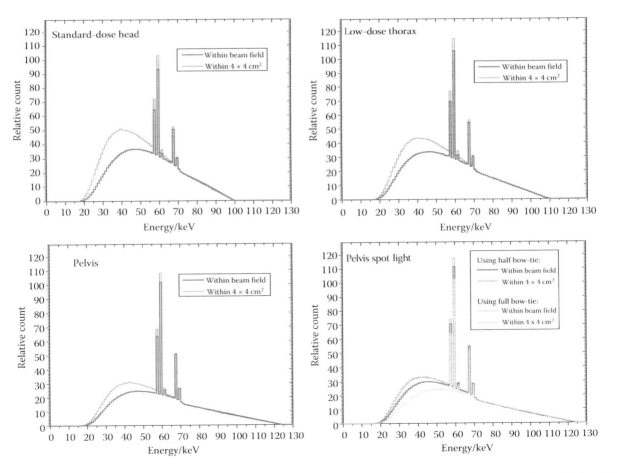

FIGURE 16.2 The energy spectra of five different kV-CBCT beams. (Reproduced from Ding G X and Coffey C W 2010. *Phys Med Biol* **55** 5231–48. With permission.)

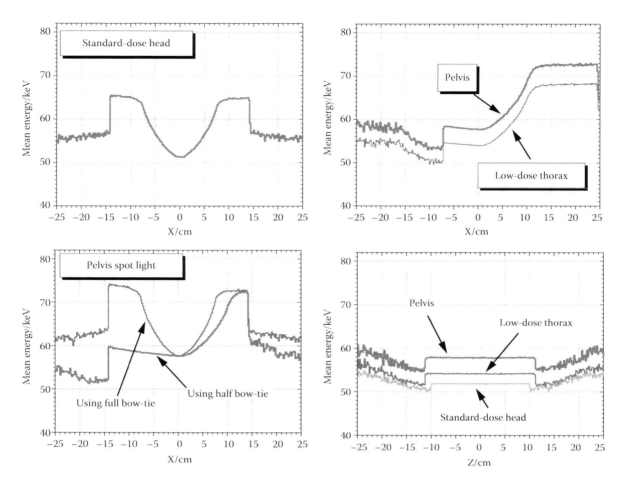

FIGURE 16.3 The mean energy distributions as a function of distance in X- and Z-direction at the SSD = 100 cm for five different kV-CBCT beams. (Reproduced from Ding G X and Coffey C W 2010. *Phys Med Biol* **55** 5231–48. With permission.)

16.2.2 Patient Dose Resulting from Imaging Procedures

The benefits of CBCT imaging are immense because it is possible to see the target and organs at risk before treatment. However, the increased use of x-ray imaging, such as CBCT in radiotherapy, for daily imaging procedures for patient setup, will add significantly to the doses to the normal tissues of the patient (Islam et al., 2006; Gayou et al., 2007; Wen et al., 2007). Accurate dosimetry of radiation dose from image guidance is becoming increasingly important for radiation oncologists to make informed decisions regarding the increased dose to radiosensitive organs. Unlike the general situation with diagnostic imaging, IGRT adds the imaging dose to an already high level of therapeutic radiation. The radiotherapy community is not alone in its search for an appropriate solution for the rapid development of increased utilization of x-ray imaging. The benefits of diagnostic imaging have revolutionized the practice of medicine; however, it has resulted in a significant increase in the population's cumulative exposure to ionizing radiation. The American College of Radiology (ACR) convened an ACR Blue Ribbon Panel on Radiation Dose in Medicine to address this issue. A recent white paper publication

(Amis et al., 2007) has detailed a proposed action plan for diagnostic radiology community. In diagnostic imaging, the primary concern is that the expanding use of imaging modalities using ionizing radiation may eventually result in an increased incidence of cancer in the exposed population. In radiotherapy application of IGRT, the concerns include both the secondary cancer incidence and the increased dose to radiosensitive organ. Hence, the management of imaging dose during radiotherapy will be different from that of routine diagnostic imaging procedures. Recent publication of the AAPM TG-75 report (Murphy et al., 2007) has provided estimated imaging dose data for a variety of imaging guidance techniques and recommended optimization strategies to trade off imaging dose with improvements in treatment delivery. Recent ICRP Publication-102 (ICRP-102, 2007) regarding patient dose in multidetector CT has emphasized that it is important that patient dose is given careful consideration, particularly with repeated or multiple examinations. Experimental methods (Islam et al., 2006; Gayou et al., 2007; Wen et al., 2007) have been used to estimate the radiation dose to patients from CBCT. However, it is difficult to accurately measure dose to different parts of the body, especially when there are large variations in anatomic numbers between bone and soft

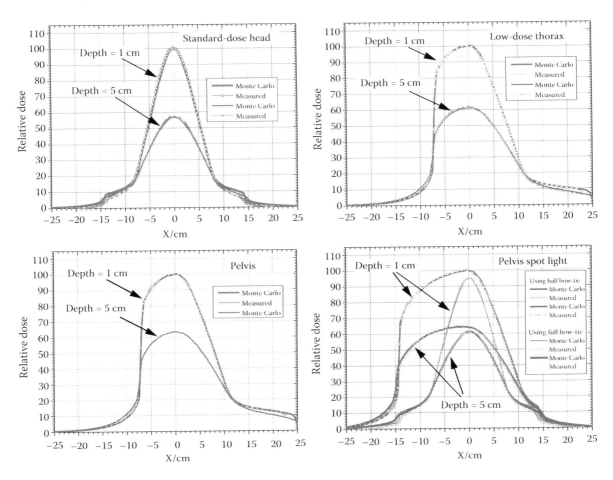

FIGURE 16.4 The comparison of dose profiles at depth = 1 cm and 5 cm in water between the Monte Carlo calculated and measured in water phantom at SSD = 100 cm. The relative dose profiles were normalized to 100% at the central axis and 1 cm depth. (Reproduced from Ding G X and Coffey C W 2010. *Phys Med Biol* **55** 5231–48. With permission.)

tissues. The Monte Carlo method has its advantage in providing detailed information on the overall shape of the dose distributions in a real patient anatomy. In the last section, it is demonstrated that the Monte Carlo technique is able to accurately simulate kV x-ray sources. By using the Monte Carlo simulated realistic x-ray source, the dose distributions to a patient can be calculated using CT-based volumetric images.

Although new model-based calculation algorithms and methodologies are being developed for radiotherapy treatment planning system to model the radiation dose from low-energy x-ray beams (Ding et al., 2008c; Pawlowski and Ding, 2011), the Monte Carlo technique is the most accurate calculation method in the dose calculations for kV beams. Because of the presence of the higher-Z materials in the body, such as bone, the current available model-based dose calculation algorithm in commercial treatment planning systems is adequate for patient dose calculations due to the increased photoelectric effect at low x-ray energies. Figure 16.5 illustrates the photoelectric effect on the calculated dose distributions for a patient treated for head and neck cancer. The isodose lines are normalized in such a way that the maximum dose is 50%. It is seen that the isodose lines above

20% are in the regions of bony anatomy. The dose to the bone is approximately 3–4 times the dose to soft tissues. It is evident that density correction alone is not adequate for dose calculations for x-ray in diagnostic energy range. Before the new accurate model-based dose calculation algorithms are developed, Monte Carlo techniques remain to be the most effective method available.

To accurately calculate the absolute imaging dose for a patient resulting from an image acquisition procedure, the simulated x-ray source has to be calibrated. The calibration of simulated source is necessary for absolute dose calculation to patients. The calibration algorithm allows the user to calculate both relative and absolute radiation doses to patients. In the next section, a method to calibrate simulated beams will be described.

16.2.3 Calibration of X-Ray Beam Output

To accurately calculate absolute dose received by a real patient during a CBCT image acquisition, a technique was developed to calibrate Monte Carlo calculated dose (Ding et al., 2008a). It is accomplished in two steps. First, the absolute dose to a point in a phantom with known geometry is measured by using an ionization chamber based on the following equation (Ma et al., 2001):

$$D_w = MN_k P_{Q,\text{cham}} \left(\frac{\mu_{en}}{\rho} \right)^w_{\text{air}} \qquad (16.1)$$

where D_w is the dose-to-water at the point of measurement in a water phantom, M is the leakage-corrected chamber reading in coulombs (C), N_k is the air kerma calibration factor for the given beam quality, Q, in Gy/C, $P_{Q,\text{cham}}$ is the overall chamber correction factor that accounts for the change in the chamber response due to the displacement of water by the ionization chamber and the presence of the chamber stem, the change in the energy, and so on, and $\left(\mu_{en}/\rho \right)^w_{\text{air}}$ is the water-to-air ratio of the mean mass energy-absorption coefficients. The air kerma calibration factor is traceable to national standards, that is, from an Accredited Dosimetry Calibration Laboratory (ADCL), National Institute for Standards and Technology (NIST), or National Research Council of Canada (NRCC). In this study, air kerma calibration factors for the specified kV-CBCT beam are obtained from an ADCL.

The simulated realistic CBCT beam is stored in a phase-space file (Rogers et al., 1995). The calibration of a Monte Carlo simulated beam involves two steps. First, using the simulated beam, the dose to a point in the phantom from a CBCT scan is calculated to be D_{MCcal}, which is in units of Gy per incident particle (on the x-ray tube). Using an ionization chamber, the dose to the same point in the phantom is experimentally determined based on the Equation 16.1 to be, D_{exp}, in Gy for the same CBCT acquisition. Second, we introduce a Monte Carlo calibration factor, f_{MCcal}, in the following equation:

$$D_{\text{exp}} = f_{\text{MCcal}} D_{\text{MCcal}} \qquad (16.2)$$

The factor, f_{MCcal}, is unique to each Monte Carlo simulated beam stored as a phase-space file. A simulated x-ray beam is called "calibrated" when the beam's calibration factor, f_{MCcal}, has been determined through Equation 16.2. Using the calibrated CBCT beam, the Monte Carlo calculations can be used to determine the absolute dose, D_{MC}, to any point in a different phantom from the same simulated beam using the following equation:

$$D_{\text{MC}} = f_{\text{MCcal}} D_{\text{MCcal}} \qquad (16.3)$$

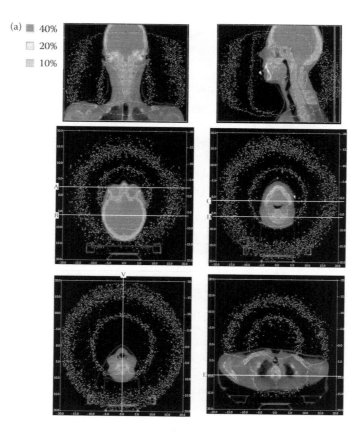

FIGURE 16.5 **(See color insert.)** (a) Monte Carlo calculated dose distributions (isodose lines) on CBCT image slices for a patient treated for head and neck cancer. (b) Dose profiles along lines A, B, C, D, E, and V. Note that line A crosses the eyes and the bony part of the nose, and B crosses the brain. Line C crosses the mandible. Lines D, E, and V cross the spinal cord at different locations. (Reproduced from Ding G X, Duggan D M, and Coffey C W 2008a. *Med Phys* **35** 1135–44. With permission.)

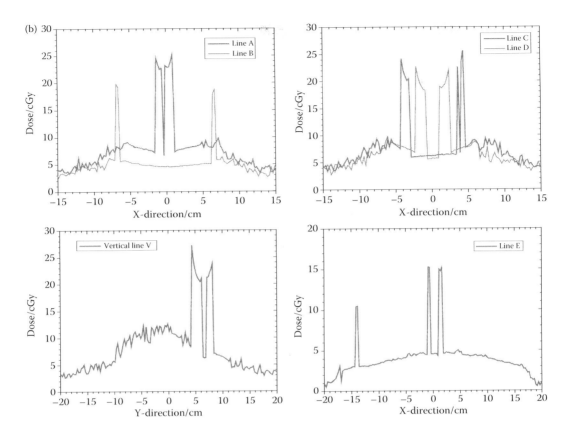

FIGURE 16.5 Continued.

where D_{MC} is the Monte Carlo predicted dose in Gy. The method developed here is not limited to kV-CBCT beam and can be applied to megavoltage photon and electron beams as well. However, the factor f_{MCcal} is specific to the control quantity. In this case, it is controlled by the specific CBCT acquisition parameters, for example, kV beam's mA, bow-tie filters, and so on. When these acquisition parameters remain the same, Equation 16.2 can be used to predict the dose accurately for imaged target that is different from the calibration phantom.

16.2.4 Method to Determine an X-Ray Source Output for Imaging Dose Calculations in Treatment Planning

In addition to kV-CBCT beam parameters (such as dose profile, beam spectrum, etc.), the determination of the beam output is required to commission a radiation beam in a commercial treatment planning system. This is similar to the method used in the last section to calculate the dose to the patient resulting from a kV-CBCT procedure. In a commercial treatment planning system, the beam outputs for a specific kV-CBCT scan determined by measured values in phantom were entered in the system as cGy/min during the beam configuration (Alaei et al., 2010), with 1 min being equivalent to a single kV-CBCT scan. This type of output configuration is treatment planning system specific.

It is worth noting that there is a conceptual difference between the radiation dose calculation to a patient using a therapeutic beam and a kV-CBCT beam in a treatment planning system. The dose calculation from a therapeutic beam is to determine the monitor units, and the shape and the intensity of beams to cover the prescribed radiation dose to the targeted tumor volume, whereas the dose calculation from a kV-CBCT beam is to determine the dose to a patient resulting from a specified kV-CBCT scan. For a given kV-CBCT scan, the dose differences between different patients (or between different scanned sites of the same patient) result from the geometry differences of the scanned patient body; the kV-CBCT beam is unchanged.

The purpose of the measurement and the calculation of the reference dose resulting from a kV-CBCT scan is to determine "the kV-beam output" for a certain reference situation in a treatment planning system so that the calculated dose at the point of measurement equals the measured dose to the reference phantom.

Unlike MV photon beams where dominant interactions are Compton effects, the dose from kV beams is sensitive to phantom atomic number Z due to the increased photoelectric effect. Therefore, the water phantom will be the natural choice for the reference dose measurement; however, plastic phantoms are more practical and convenient in experimental setups. Since Monte Carlo calculations are generally not available for photon beams in commercial treatment planning systems, it is important to find the error in the dose calculations introduced without considering medium dependence. In a commercial treatment planning system that uses model-based dose calculation algorithms (Mackie et al., 1985; Mohan and Chui, 1987; Ahnesjo and Aspradakis, 1999), the dose calculations are based on electron density, which is converted from CT number. Therefore, the accuracy of dose calculations without considering the medium dependency affects the accuracy of beam output determination of the kV-CBCT beams. It has been demonstrated that in Monte Carlo simulations the composition of medium must be taken into account and it is important to select a suitable water substitute material in order to reduce the error in the absorbed dose determination (Ding and Coffey, 2010). This was done by comparing calculated doses at the point of measurement in a plastic phantom between Type I and Type II calculations.

Several investigations have demonstrated that the Monte Carlo simulation could provide details of an x-ray beam data that were essential for commissioning an x-ray beam in a radiotherapy treatment planning system (Alaei et al., 2010). The Monte Carlo generated beam parameters and method of beam output calibration make it possible to commission a kV-CBCT beam in a radiotherapy treatment planning system (Alaei et al., 2010), which facilitates reporting and accounting the additional radiation doses to radiotherapy patients resulting from repeated kV-CBCT procedures (Ding and Coffey, 2008, 2009).

16.2.5 Applications

16.2.5.1 Developing and Validating New Model-Based Dose Calculation Algorithms

The increasingly intensive imaging procedures for IGRT now obligates the clinician to evaluate the total dose to the target and radiosensitive organs from both therapeutic and imaging doses to avoid significant underestimation of dose to radiosensitive organs. This requires that radiotherapy treatment planning systems be capable of calculating dose from kilovoltage photon beams, since low-energy x-rays are often used in imaging guidance procedures. Unfortunately, low-energy x-ray dose calculation methods are not a part of current radiotherapy treatment planning systems and therefore, this additional imaging dose cannot be included in the calculation of total dose to radiotherapy patients. Attempts have been made to incorporate the imaging dose into total patient dose from diagnostic x-rays in treatment planning by Alaei et al. (2000). Owing to the increased photoelectric effect, a kV-CBCT procedure results in doses to bone that are up to a factor of 3–4 higher than those in surrounding soft tissue (Verhaegen et al., 2004; Ding et al., 2008b). It has been shown that the model-based (e.g., convolution/superposition) algorithm is capable of computing dose within phantoms of density equal to or less than that of water and that its limitations introduce inaccuracies for calculations through moderately high-atomic-number materials such as bone for kV x-rays (Alaei et al., 1999, 2000, 2001; Seuntjens and Ma, 1999; Verhaegen et al., 2004). Currently, the Monte Carlo technique is the most accurate dose calculation method for x-rays in the diagnostic energy range in media where

there are large atomic number variations, such as a patient body. However, the lack of availability of the Monte Carlo method in commercial radiation treatment planning (RTP) systems and its long computation time limits its usage for routine patient organ dose calculations. Hence, there is a need to develop a more accurate dose calculation method that is suitable for the implementation in treatment planning systems. Monte Carlo techniques can serve as gold standards in developing and validating new model-based dose calculation algorithms. The progress has been made to develop model-based methods to overcome the deficiencies of existing algorithms in kilovoltage dose calculations, especially for CT-based human volumetric images used in RTP (Ding et al., 2008c; Pawlowski and Ding, 2011).

16.2.5.2 Improving the Design of New X-Ray Tube or Detector to Reduce the Imaging Dose to Patients

Monte Carlo techniques have been shown to be a powerful tool to analyze the beam characteristics and to assist the development of radiation treatment machines (Rogers et al., 1995) as well as x-ray units used in the IGRT (Ding et al., 2007a; Spezi et al., 2009; Ding and Coffey, 2010; Ding and Munro, 2011). It has been shown that the Monte Carlo calculations can be used in modifying and selecting image acquisition parameters to minimize radiation exposures to the patient's sensitive organs (Ding et al., 2010). In addition, Monte Carlo techniques are capable of calculating the dose to organs such bone marrow (Walters et al., 2009) in which the experimental methods are difficult or impossible for low-energy x-rays.

16.3 Monte Carlo and MV Imaging

16.3.1 General Introduction

Monte Carlo techniques have a long history of being used to model the detection of ionizing radiation. Many of the Monte Carlo codes that are applied to medical physics problems were developed at large nuclear and high-energy particle physics laboratories such as CERN, Los Alomos, and the SLAC National Accelerator Laboratory (Menlo Park, CA) (Nelson et al., 1985a; Solberg et al., 2001; Agostinelli et al., 2003). The use of Monte Carlo in this context is useful for two main purposes: First, in the design and development stage, the radiation detector can be cost-effectively optimized for a particular application. Second, Monte Carlo can be used to predict the response of a radiation detector in a particular experiment for direct comparison with experimental data (Apostolakis et al., 2009; Dunn and Shultis, 2009). This capability is also of use in medical physics applications, including diagnostic and radiation therapy imaging (Rogers, 2006). Verification of radiotherapy treatments is extremely important for ensuring the accurate and precise delivery of the prescribed dose to the correct volume and Monte Carlo techniques have had a role to play in this. In the following sections, some of the ways in which Monte Carlo can be used to verify the different aspects of a radiotherapy treatment will be briefly described. The focus will be on the use of

Monte Carlo modeling of megavoltage electronic portal imaging devices (EPIDs) for verifying patient position and dosimetry.

16.3.2 Technology of Electronic Portal Imaging

EPIDs were developed as a digital replacement for radiographic film to assist in minimizing the patient setup error in a radiotherapy treatment. The detectors are attached to the linear accelerator gantry and can be moved to a position behind the patient for imaging the patient when required. The imaging can be performed immediately before the start of a treatment fraction or even during a treatment fraction. The image of the patient in the treatment position can then be compared with a reference image of the patient position. The reference image was traditionally a simulator radiographic film but is now more likely to be a digitally constructed radiograph calculated using the treatment planning CT. Three distinctly different EPID systems have been developed for commercial use over the last 20 years; the scanning liquid ion chamber array, camera-based systems, and more recently amorphous silicon flat panel detectors (Antonuk, 2002; Langmack, 2001; Kirby and Glendinning, 2006).

16.3.2.1 Scanning Liquid Ion Chamber Arrays

The first widely available commercial EPID was the scanning liquid ionization detector (Meertens et al., 1990). The detector consists of two planes of 256 electrodes separated by a gap of 0.8 mm. The gap between the electrode planes is filled with a liquid, 2,2,4-trimethylpentane, that is ionized when irradiated. The electrode spacing is 1.27 mm giving an active area of 32.5×32.5 cm^2. One plane of electrodes is connected to 256 independent electrometers for reading the collected ionization charge and the other plane of 256 electrodes are connected to a 500 V supply that can apply a high voltage to each electrode individually. During irradiation, the voltage is applied to each electrode in succession (on the voltage plane) with an integration time of around 5 ms and the signal recorded for each of the electrodes giving a total readout time of 1.25 s. The scanning readout is synchronized to the linac pulse signal to avoid introducing artifacts due to any changes in the dose rate during acquisition. The system is compact and is geometrically accurate. The main disadvantage of the system is that during acquisition only a single electrode has a high voltage applied at any one time and as a result is relatively inefficient. The slow ion recombination rate of ~0.5 s compensates for this somewhat but a relatively high patient image dose is still required to obtain a useful image. The detector also has relatively poor spatial resolution. Spezi showed that full Monte Carlo simulations of the linear accelerator and detector response were possible and that measured and simulated dose images agreed to within 2%/2 mm (Spezi and Lewis, 2002).

16.3.2.2 Camera-Based Systems

Several different camera-based systems were developed to increase the active detector area, and improve the sensitivity and spatial and contrast resolution (Althof et al., 1996; de Boer et al.,

2000). The camera-based systems incorporate an x-ray to optical light converter; typically, a metal plate and a phosphor screen or solid-state scintillator crystal; a mirror and a camera and lens then capture the optical image. These systems are generally quite large and bulky and have to be physically removed from the linac gantry when not in use. The camera-based systems are also not particularly efficient at collecting the light produced by the phosphor screen and reflected into the camera lens system by the mirror, typically only 0.1–0.01%. This poor light collection efficiency introduces noise into the image. Light collection efficiency can be improved by increasing the thickness of the phosphor screen, thereby increasing the amount of generated light or increasing the solid angle cone of light collected using a large aperture lens and digital camera with a large-area CCD chip (Mosleh-Shirazi et al., 1998). Monte Carlo studies have been performed of the screen–metal plate combination with the aim of optimizing the detector for imaging with a radiotherapy beam produced from a low-Z aluminum target (Flampouri et al., 2002). Yeboah and Pistorius investigated the sensitivity of the detector response to different metal plate–screen combinations and changes in the exit beam spectra from a phantom (Yeboah and Pistorius, 2000). The response of the detector was shown to depend on a combination of phantom–patient thickness, air gap between the detector and phantom–patient, and thickness of the metal plate–screen.

16.3.2.3 Flat Panel Amorphous Silicon Detectors

Owing to the limitations of the liquid ionization chamber and camera-based systems, flat panel amorphous silicon (a-Si:H)-based detectors are now the system of choice on most modern linear accelerators for MV imaging with the first commercial systems becoming available in 2000 (Antonuk et al., 1998). They combine a foldaway compact design with improved sensitivity, and contrast and spatial resolution, and allow high-quality patient images to be acquired with only a few cGy.

The first two components are similar to the camera-based systems. High-energy x-rays interact in a metal plate, usually 1 mm of copper, producing electrons and lower-energy x-rays. The electrons and x-rays then interact in the ~133 mg cm^{-2} phosphor screen, typically gadolinium oxysulfide doped with terbium (Gd_2O_2S:Tb), generating optical photons, some of which escape the exit plane of the screen. Behind the screen, each pixel in the detector array consists of a photodiode on a thin amorphous silicon layer that converts each light photon to an electron–hole pair that is stored in a capacitor element.

Each pixel is connected to a thin film transistor switch that is used to control the readout of the stored charge signal. The signal is read out by applying a bias voltage to the transistor, making it conducting. Data are usually read out row by row with each row comprising a data line. The current commercially available systems consist of an array of either 1024×1024 or 1024×768 pixels corresponding to an area of 41×41 cm^2 or 40×30 cm^2, respectively. The two most common clinical a-Si EPID's are the Elekta iView GT and the Varian Portal Vision as 500/as1000. The a-Si:H detectors have far superior image quality for a low patient imaging dose compared to the liquid scintillator and camera-based

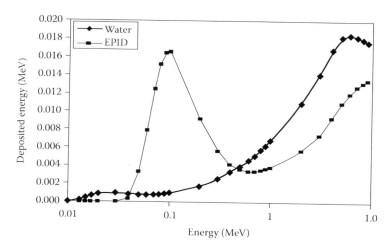

FIGURE 16.6 The energy response of an a-Si flat panel EPID calculated using Monte Carlo compared to the response of a water-equivalent slab of material. (From Parent L 2006. In: *Joint Department of Physics, Institute of Cancer Research* (London: University of London). With permission.)

systems. This is mainly due to the high efficiency of the light collection into the photodiodes. The collected charge signal in each pixel $I_{meas}(x,y)$ is amplified and digitized. The signal is first corrected for dark current $I_{dark}(x,y)$ and then for pixel sensitivity or gain using a flood field $I_{flood}(x,y)$. The corrected signal is

$$I_{corr}(x,y) = \frac{I_{meas}(x,y) - I_{dark}(x,y)}{I_{flood}(x,y) - I_{dark}(x,y)} \quad (16.4)$$

This correction also removes position-dependent variations in the image due to the beam profile from the measured image, and although this will improve anatomical image quality, it is undesirable for accurate dosimetry. The detectors have also been shown to have an overresponse to low-energy x-rays (compared to water) due to an increased photoelectric effect in the high-atomic-number phosphor screen. This is shown in Figure 16.6. This complicates the conversion of the measured signal into dose-to-water for dosimetry applications where Monte Carlo studies have shown that there is significant position-dependent variation in the beam spectrum (Parent et al., 2004). However, the a-Si detector has many properties that make it well suited to dosimetric verification, including a linear dose response (McCurdy et al., 2001; Greer and Popescu, 2003). Two general approaches have been used for using EPIDs for dosimetric verification. The first is known as transit dosimetry where the energy fluence exiting the patient or phantom is determined from a measured image. The energy fluence may then be back-projected through a CT dataset of the patient or phantom and converted to dose before comparing with the planned dose distribution (McNutt et al., 1996a; Partridge et al., 2002; Ansbacher, 2006; Wendling et al., 2009). The second approach involves predicting a portal dose image and comparing it with the measured portal dose image (McCurdy et al., 2001; Siebers et al., 2004; Van Esch et al., 2004; Kairn et al., 2008; Chytyk and McCurdy, 2009). Methods for predicting the portal dose image include a full Monte Carlo method that includes the radiation transport through the

head of the linear accelerator and the subsequent deposition of dose in a model of the detector (Spezi and Lewis, 2002; Siebers et al., 2004; Parent et al., 2006b). These methods suffer from long computation times, making them difficult to implement clinically. Another approach is to use precalculated dose kernels that are convolved with the fluence incident on the detector. The dose kernels can be determined from Monte Carlo simulations (McNutt et al., 1996b; McCurdy and Pistorius, 2000) or alternatively analytical kernels can be used (Greer et al., 2007).

16.3.3 Monte Carlo Modeling and MV Imaging

In this section, the use of the Monte Carlo technique for modeling MV imaging devices and the physical processes involved in MV imaging will be discussed. The focus will be on a-Si:H detectors as these are now the most prevalent in clinical use.

16.3.3.1 Modeling Electronic Portal Imaging Devices

The first stage in modeling an EPID is building the model in the appropriate Monte Carlo code. This requires the specifications of the actual device. These details can be obtained from the manufacturer under a nondisclosure agreement. The model is generally quite simple, consisting of a number of layers or "slabs" of different materials with each layer having a thickness and composition specified by the manufacturer. Figure 16.7 shows the layered structure of a model of the Elekta iView GT a-Si:H detector. A model of the slab geometry Elekta a-Si:H detector can be built using the DOSXYZnrc user code, part of the EGSnrc code. This will be used as an example but the same principles can be used to construct a model for any of the currently available commercial a-Si:H EPIDs. DOSXYZnrc has been developed to calculate dose in a rectilinear voxel-based phantom and a number of authors have applied it to the modeling of a flat panel detector (Spezi and Lewis, 2002; Siebers et al., 2004; Parent et al., 2006b). The active area of the detector is 41×41 cm^2. The actual detector has 1024×1024 pixels but a lower-resolution model can be used to

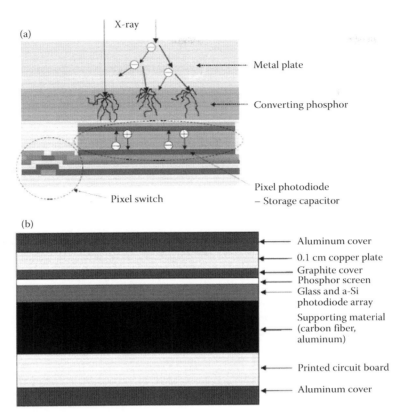

FIGURE 16.7 (a) Schematic diagram of the main components of an indirect detection amorphous silicon detector (Antonuk, 2002). (b) The layered structure of the Elekta iView GT a-Si panel. (From Parent L et al. 2006b. *Med Phys* **33** 4527–40. With permission.)

speed up the calculation. An array of 256×256 pixels can be used in the model, giving a model pixel pitch of 1.6 mm (this corresponds to a projected pixel size of ~1 mm at the isocenter plane). The voxel thickness in the z direction will be different for each

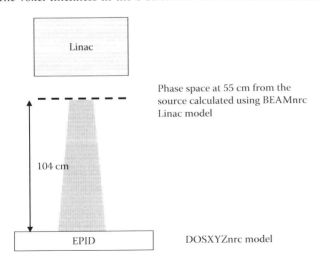

FIGURE 16.8 Diagram showing the geometry of the Monte Carlo models of the Elekta linear accelerator and iView GT EPID. The EPID model is made up of $256 \times 256 \times N_l$ voxels, where N_l is the number of layers of different materials in the detector.

layer and will depend on the thickness of each layer. The number of voxels in the z direction will correspond to the number of layers in the detector. The calculated portal dose image is taken to be the dose scored in the $Gd_2O_2S:Tb$ layer of the EPID model. Munro and Bouius showed that for an a-Si EPID, made only of a 0.1-cm Cu buildup plate, a phosphor screen, and the a-Si components (diodes and TFTs) on a glass substrate with no casing or backscatter material, the signal detected by the light sensor is almost entirely due to photon and electron interactions within the copper and the phosphor layers (Munro and Bouius, 1998). Furthermore, Antonuk et al. showed that the response of the light sensor is proportional to the energy deposition in the phosphor (Antonuk et al., 1998). The geometry of the detector model with respect to the linear accelerator is shown in Figure 16.8.

The Monte Carlo model of the EPID should be commissioned to validate that it is an accurate model of the actual detector. This is done by comparing measured EPID data with data simulated using the EPID model to ensure good agreement. The following is a list of suggested data for commissioning an EPID Monte Carlo model:

- Comparison of measured and simulated in-plane and cross-plane profiles for a flood field (26×26 cm^2).
- Comparison of measured and simulated output factors for a range of different regular field sizes (1×1, 2×2, 5×5, 10×10, 15×15, 20×20, and 26×26 cm^2).

- To ensure the model EPID accurately models the response to radiation scattered from the patient, measured and simulated profiles with different thickness of slab phantom material (e.g., PMMA or solid water) should be compared.
- Finally, comparisons of more complex fields with and without phantom material in the field should be performed, for example, intensity-modulated radiotherapy fields.

Correct normalization of the measured and simulated EPID images is important. A relatively simple approach is to perform a cross calibration:

$$\frac{I_{MC}(x,y)}{I_{MC}^{10\times10}} \cong \frac{I_{meas}(x,y)}{I_{meas}^{10\times10}} \tag{16.5}$$

where $I_{MC}^{10\times10}$ and $I_{meas}^{10\times10}$ represent the mean pixel values in a 1×1 cm^2 central region of simulated and measured EID images for a 10×10 cm^2 field, respectively. The measured and simulated EPID response can then be compared directly. For simulations of Varian accelerators, a correction is also required for backscatter from the collimators into the monitor chambers (Liu et al., 2000).

16.3.3.2 Pixel Sensitivity Variations in the Measured Image

There is an inherent difference between the measured EPID image and the simulated EPID image, which must be considered before an accurate comparison can be performed. The measured images are flood field corrected to remove the pixel sensitivity variations of the detector; this procedure also removes the beam profile due to the flattening filter. There are two ways of correcting for this difference. One is to introduce a pixel sensitivity map into the simulated images and a second is to subtract only the pixel sensitivity from the measured image. Both methods require separation of the pixel sensitivity from the beam profile. Parent et al. (2006a) compared three previously used methods with a new method for doing this; the methods were

- Use of the standard flood field image
- Use of a flood field image obtained with a slab of plastic water in the beam to remove the beam profile
- Use of a series of small overlapping 10×10 cm^2 fields (considered to be uniform) to construct a pixel sensitivity map
- Use of an analytical function fitted to the Monte Carlo beam profile

The work showed that the above methods gave similar results but were all superior to using the standard flood field image. Greer previously showed that multiple overlapping fields could be used to derive the pixel sensitivity map for amorphous silicon detectors (Greer, 2005). The method involves delivering a radiation field (10×10 cm^2) assumed to be uniform to the whole detector using a series of translations of the detector. Each pixel in the detector receives a uniform dose and variations in the pixel values can be used to derive a pixel sensitivity map $I_{pixel}(x,y)$. Dividing subsequent images $I(x,y)$ by this pixel sensitivity map

generates an image that is corrected for pixel sensitivity variations $I_{corr}(x,y)$.

$$I_{corr}(x,y) = \frac{I(x,y)}{I_{pixel}(x,y)} \tag{16.6}$$

The method of using a thickness of solid or plastic water above the detector to flatten the beam was used by Siebers et al. (2004) to calibrate the Varian as500 detector. The thickness of plastic water to remove the beam profile can be determined using Monte Carlo simulation. The resulting image is a map of the detector pixel sensitivity variation that can be used in Equation 16.4.

16.3.3.3 Modeling Off-Axis Response

The methods discussed previously for removing the pixel sensitivity from the measured EPID images allow direct comparison of dose to the detector with simulated EPID images. In radiotherapy, the dosimetric quantity of interest is dose-to-water. The high-atomic-number components in a-Si flat panel imagers cause a response that is nonwater equivalent. They have been shown to overrespond (compared to water) to low-energy photons due to an increase in x-ray photoelectric effect interactions (Kirkby and Sloboda, 2005). Since the clinical radiotherapy beam has a spatially varying energy spectrum, the response of the detector will also vary spatially. This is a particular problem for treatment techniques involving multiple small irregular off-axis fields such as IMRT or VMAT where the energy spectra can vary significantly from the calibration beam; Figure 16.9 shows the example of the variation in the energy spectra for a 10×10 cm^2 field on both the central axis and off-axis. The off-axis field energy spectrum shows an increase in the number of

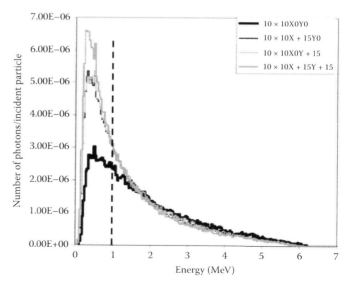

FIGURE 16.9 Calculated energy spectra of 10×10 cm^2 fields on the central axis (black) and off-axis (gray). (From Parent L et al. 2006b. *Med Phys* **33** 4527–40. With permission.)

low-energy photons below 1 MeV. Parent et al. (2007) showed that this can result in variations in response of up to 29% if the same IMRT field is delivered on- and off-axis. The spatially varying response can be accounted for by measurement of the beam profile using ion chambers or film (Van Esch et al., 2004; Greer, 2005; Wendling et al., 2006) and calculation-based methods (Li et al., 2006; Winkler et al., 2007). The problem with using ion chambers or film to measure the beam profile is that they both have a different response to the EPID and therefore can introduce errors.

16.3.3.4 Modeling Patient Scatter

Scattering of radiation from the patient into the EPID has implications for both anatomical imaging and dosimetry. Scatter will degrade the quality of the image, reducing contrast and spatial resolution. Dosimetry applications generally require the primary dose or fluence in the EPID and scatter will increase the measured signal. A number of authors have used Monte Carlo simulations to investigate the effect of scattered radiation from the patient into the EPID. Swindell and Evans developed a physical model for the scatter-to-primary ratio (SPR) in portal images that agreed well with Monte Carlo calculations for a range of field sizes, phantom thicknesses, and phantom-detector distances (Swindell and Evans, 1996). Jaffray et al. (1994) investigated how the scatter and primary photon fluence is influenced by the scatter geometry and energy spectrum of the incident beam. For a particular geometry, they showed that for a quantum noise-limited detector system (an a-Si EPID can be considered quantum noise limited), the differential signal-to-noise ratio was decreased by 10–20% due to the scatter fraction. Spies and Bortfeld carried out a study of the effect of scattered radiation on megavoltage images for different field sizes and phantom thicknesses (Spies and Bortfeld, 2001). Using Monte Carlo simulations, they developed analytical models that modeled the scatter as originating from a point source at a position within the phantom that varied as a function of field size. Spies et al. (2001) also demonstrated the use of scatter kernels calculated using Monte Carlo to correct for scattered radiation in megavoltage cone-beam CT reconstructions. The kernels take account of both the spectrum of the incident beam and the dimensions of the phantom being scanned. The scatter corrections reduced the effect of cupping artifacts in the reconstructed data and reduced deviations in reconstructed electron densities from 30% to 8%.

16.3.3.5 Modeling Detector Backscatter

A number of authors have shown the Varian a-Si EPID to have a nonuniform response due to the support arm below the active layers in the detectors. Modeling the detector as a homogeneous slab of material will therefore introduce errors. Ko et al. (2004) carried out a Monte Carlo study to investigate the effect of placing water, lead, or copper slabs behind the active layers to reduce the nonuniform response due to the support structure. An experimental study was also performed by the same group to verify the findings of the Monte Carlo and confirmed that 5 mm of lead reduced the nonuniform backscatter to less than 0.5%.

There were also improvements in contrast and spatial resolution (Moore and Siebers, 2005). Cufflin et al. (2010) showed how the nonuniform response could be modeled by including a nonuniform structure of water-equivalent material behind the active layers of the EPID model.

16.3.4 Applications

There are two main applications of Monte Carlo for portal dosimetry. The first is pretreatment verification of IMRT treatments and the second is the verification of the delivered dose during a patient treatment, known as *in vivo* or transmission dosimetry.

16.3.4.1 Pretreatment IMRT Verification

IMRT patient treatment plans are typically verified by measurement before treatment can begin (Ezzell et al., 2003; Galvin et al., 2004). This can be a time-consuming process and usually requires delivering the treatment fields to a suitable phantom so that ion chamber or film dosimetry measurements can be compared with the treatment plan dosimetry. The two-dimensional (2D) fluence for each field can also be measured using either 2D ion chamber or diode arrays that can then be compared to the planned fluence using a gamma analysis (Low et al., 2011). The commercially available 2D dosimetry devices have relatively poor spatial resolution that makes accurate verification of intensity-modulated fields difficult. EPID devices can be used as 2D high-resolution dosimeters that are conveniently available on most modern linear accelerators, making them suitable for IMRT pretreatment fluence verification. A full Monte Carlo simulation of the linear accelerator output and EPID can be used to predict the 2D dose in the EPID, obviating the need to account for the non-water-equivalent energy response of the detector. The predicted response of the EPID can then be compared directly with the measured response. Despite the relatively long calculation times, this has been shown to be an accurate method for validating the delivered fluence for an IMRT treatment (Spezi and Lewis, 2002; Siebers et al., 2004; Parent et al., 2006b; Cufflin et al., 2010). Figure 16.10 shows a comparison of two different measurement techniques of 2D fluence for an IMRT treatment field, a 2D diode array (a) and EPID (b), and two different calculation techniques, treatment planning system (c) and Monte Carlo (d). The circled region in the 2D diode array image highlights a difference where the limited resolution of the detector is unable to detect the small high-dose gradient region. The pixel sizes of the EPID and Monte Carlo model EPID are both 1 mm (at the isocenter plane). A gamma analysis of the EPID measurements against the Monte Carlo calculated fluence images for all seven fields in this particular prostate IMRT treatment had greater than 96% of pixels passing with acceptance criteria of 3% and 3 mm.

16.3.4.2 *In Vivo* Patient Dosimetry

The methodology of using Monte Carlo-based EPID dose prediction for pretreatment verification can be extended to

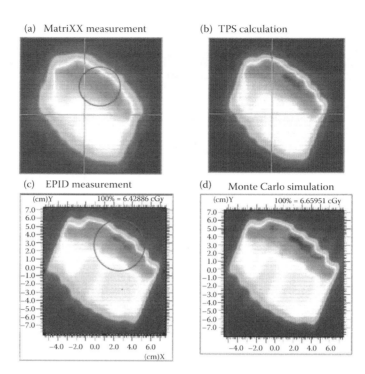

FIGURE 16.10 **(See color insert.)** Comparison of a prostate IMRT field for (a) the matriXX 2D ion chamber array, (b) the treatment planning system (dose to water), (c) EPID measurements, and (d) the Monte Carlo (dose to EPID).

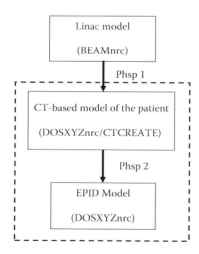

FIGURE 16.11 Schematic diagram of the stages of a Monte Carlo transmission dosimetry simulation.

transmission dosimetry. A model of the patient or phantom can be included in the simulation. Figure 16.11 illustrates schematically how the different components of the treatment can be simulated. First, the linear accelerator model is used to produce phase-space file(s) of the treatment field. Second, the phase-space file(s) are then used as input to model the energy deposition in the patient and phase-space file(s) produced at the exit plane of the patient. Third, this phase space is then used to calculate the energy deposition in a model EPID. The standard distribution of DOSXYZnrc does not generate an output phase-space file,

so a modification of the code is necessary (Spezi et al., 2001). An alternative that avoids writing out the intermediate phase-space file is to combine the CT-based model of the patient and the DOSXYZnrc model of the EPID; the dashed box in Figure 16.11 shows the components to be combined. This also requires some scripting or coding effort to combine the two DOSXYZnrc phantoms, particularly for treatment fields with nonzero gantry angles where the two phantoms with rectilinear voxel geometries have to be rotated with respect to one another (Chin et al., 2003; Kairn et al., 2011).

References

Agostinelli S, Allison J, Amako K, Apostolakis J, Araujo H, Arce P, Asai M et al. 2003. Geant4—A simulation toolkit. *Nucl Instrum Methods Phys Res A, Accel Spectrom Detect Assoc Equip* **506** 250–303.

Ahnesjo A and Aspradakis M M 1999. Dose calculations for external photon beams in radiotherapy. *Phys Med Biol* **44** R99–155.

Alaei P, Ding G X, and Gua H 2010. Inclusion of the dose from kilovoltage cone beam CT in the radiation therapy treatment plans. *Med Phys* **37** 244–8.

Alaei P, Gerbi B J, and Geise R A 1999. Generation and use of photon energy deposition kernels for diagnostic quality x rays. *Med Phys* **26** 1687–97.

Alaei P, Gerbi B J, and Geise R A 2000. Evaluation of a model-based treatment planning system for dose computations in the kilovoltage energy range. *Med Phys* **27** 2821–6.

Alaei P, Gerbi B J, and Geise R A 2001. Lung dose calculations at kilovoltage x-ray energies using a model-based treatment planning system. *Med Phys* **28** 194–8.

Althof V G M, deBoer J C J, Huizenga H, Stroom J C, Visser A G, and Swanenburg B N 1996. Physical characteristics of a commercial electronic portal imaging device. *Med Phys* **23** 1845–55.

Amis E S, Jr., Butler P F, Applegate K E, Birnbaum S B, Brateman L F, Hevezi J M, Mettler F A et al. 2007. American College of Radiology white paper on radiation dose in medicine. *J Am Coll Radiol* **4** 272–84.

Ansbacher W 2006. Three-dimensional portal image-based dose reconstruction in a virtual phantom for rapid evaluation of IMRT plans. *Med Phys* **33** 3369–82.

Antonuk L E 2002. Electronic portal imaging devices: A review and historical perspective of contemporary technologies and research. *Phys Med Biol* **47** R31–65.

Antonuk L E, El-Mohri Y, Huang W D, Jee K W, Siewerdsen J H, Maolinbay M, Scarpine V E, Sandler H, and Yorkston J 1998. Initial performance evaluation of an indirect-detection, active matrix flat-panel imager (AMFPI) prototype for megavoltage imaging. *Int J Radiat Oncol Biol Phys* **42** 437–54.

Apostolakis J, Asai M, Bogdanov A G, Burkhardt H, Cosmo G, Elles S, Folger G et al. 2009. Geometry and physics of the Geant4 toolkit for high and medium energy applications. *Radiat Phys Chem* **78** 859–73.

Aviles Lucas P, Dance D R, Castellano I A, and Vano E 2004. Monte Carlo simulations in CT for the study of the surface air kerma and energy imparted to phantoms of varying size and position. *Phys Med Biol* **49** 1439–54.

Ay M R, Shahriari M, Sarkar S, Adib M, and Zaidi H 2004. Monte Carlo simulation of x-ray spectra in diagnostic radiology and mammography using MCNP4C. *Phys Med Biol* **49** 4897–917.

Ay M R and Zaidi H 2005. Development and validation of MCNP4C-based Monte Carlo simulator for fan- and cone-beam x-ray CT. *Phys Med Biol* **50** 4863–85.

Chin P W, Spezi E and Lewis D G 2003. Monte Carlo simulation of portal dosimetry on a rectilinear voxel geometry: A variable gantry angle solution. *Phys Med Biol* **48** N231–8.

Chytyk K and McCurdy B M C 2009. Comprehensive fluence model for absolute portal dose image prediction. *Med Phys* **36** 1389–98.

Colijn A P and Beekman F J 2004. Accelerated simulation of cone beam X-ray scatter projections. *IEEE Trans Med Imaging* **23** 584–90.

Colijn A P, Zbijewski W, Sasov A, and Beekman F J 2004. Experimental validation of a rapid Monte Carlo based micro-CT simulator. *Phys Med Biol* **49** 4321–33.

Cufflin R S, Spezi E, Millin A E, and Lewis D G 2010. An investigation of the accuracy of Monte Carlo portal dosimetry for verification of IMRT with extended fields. *Phys Med Biol* **55** 4589–600.

de Boer J C J, Heijmen B J M, Pasma K L, and Visser A G 2000. Characterization of a high-elbow, fluoroscopic electronic portal imaging device for portal dosimetry. *Phys Med Biol* **45** 197–216.

Ding G and Munro P 2011. Reduced CBCT imaging dose due to the new x-ray source in TrueBeam™. *Med Phys* **38** 3372–3.

Ding G X 2002. Energy spectra, angular spread, fluence profiles and dose distributions of 6 and 18 MV photon beams: Results of Monte Carlo simulations for a Varian 2100.EX accelerator. *Phys Med Biol* **47** 1025–46.

Ding G X and Coffey C W 2008. Is it time to include imaging guidance doses in the reportable total radiation doses of radiotherapy patients? *Int J Radiat Oncol Biol Phys* **72**, Supplement S145-S6.

Ding G X and Coffey C W 2009. Radiation dose from kilovoltage cone beam computed tomography in an image-guided radiotherapy procedure. *Int J Radiat Oncol Biol Phys* **73** 610–7.

Ding G X and Coffey C W 2010. Beam characteristics and radiation output of a kilovoltage cone-beam CT. *Phys Med Biol* **55** 5231–48.

Ding G X, Duggan D M, and Coffey C W 2006. Commissioning stereotactic radiosurgery beams using both experimental and theoretical methods. *Phys Med Biol* **51** 2549–66.

Ding G X, Duggan D M, and Coffey C W 2007a. Characteristics of kilovoltage x-ray beams used for cone-beam computed tomography in radiation therapy. *Phys Med Biol* **52** 1595–615.

Ding G X, Duggan D M, and Coffey C W 2008a. Accurate patient dosimetry of kilovoltage cone-beam CT in radiation therapy. *Med Phys* **35** 1135–44.

Ding G X, Duggan D M, and Coffey C W 2008b. Accurate patient dosimetry of kilovoltage cone-beam CT in radiation therapy. *Med Phys* **35** 1135–44.

Ding G X, Duggan D M, Coffey C W, Deeley M, Hallahan D E, Cmelak A, and Malcolm A 2007b. A study on adaptive IMRT treatment planning using kV cone-beam CT. *Radiother Oncol* **85** 116–25.

Ding G X, Munro P, Pawlowski J, Malcolm A, and Coffey C W 2010. Reducing radiation exposure to patients from kV-CBCT imaging. *Radiother Oncol* **97** 585–92.

Ding G X, Pawlowski J M, and Coffey C W 2008c. A correction-based dose calculation algorithm for kilovoltage x rays. *Med Phys* **35** 5312–6.

Ding G X and Rogers D W O 1995. Energy spectra, angular spread, and dose distributions of electron beams from various accelerators used in radiotherapy. National Research Council of Canada, Report No. PIRS-0439, Ottawa; see also http://www.irs.inms.nrc.ca/inms/irs/papers/PIRS439/pirs439.html.

Dixon R L, Anderson J A, Bakalyar D M et al. 2010. *AAPM Task Group Report TG-111: The future of CT dosimetry.* College Park, MD: AAPM.

Downes P, Jarvis R, Radu E, Kawrakow I, and Spezi E 2009. Monte Carlo simulation and patient dosimetry for a kilovoltage cone-beam CT unit. *Med Phys* **36** 4156–67.

Dunn W L and Shultis J K 2009. Monte Carlo methods for design and analysis of radiation detectors. *Radiat Phys Chem* **78** 852–8.

Endo M, Mori S, Tsunoo T, and Miyazaki H 2006. Magnitude and effects of x-ray scatter in a 256-slice CT scanner. *Med Phys* **33** 3359–68.

Endo M, Tsunoo T, Nakamori N, and Yoshida K 2001. Effect of scattered radiation on image noise in cone beam CT. *Med Phys* **28** 469–74.

Ezzell G A, Galvin J M, Low D, Palta J R, Rosen I, Sharpe M B, Xia P, Xiao Y, Xing L, and Yu C X 2003. Guidance document on delivery, treatment planning, and clinical implementation of IMRT: Report of the IMRT subcommittee of the AAPM radiation therapy committee. *Med Phys* **30** 2089–115.

Faddegon B A, Wu V, Pouliot J, Gangadharan B, and Bani-Hashemi A 2008. Low dose megavoltage cone beam computed tomography with an unflattened 4 MV beam from a carbon target. *Med Phys* **35** 5777–86.

Flampouri S, Evans P M, Verhaegen F, Nahum A E, Spezi E, and Partridge M 2002. Optimization of accelerator target and detector for portal imaging using Monte Carlo simulation and experiment. *Phys Med Biol* **47** 3331–49.

Galvin J M, Ezzell G, Eisbrauch A, Yu C, Butler B, Xiao Y, Rosen I et al. 2004. Implementing IMRT in clinical practice: A joint document of the American Society for Therapeutic Radiology and Oncology and the American Association of Physicists in Medicine. *Int J Radiat Oncol Biol Phys* **58** 1616–34.

Gayou O, Parda D S, Johnson M, and Miften M 2007. Patient dose and image quality from mega-voltage cone beam computed tomography imaging. *Med Phys* **34** 499–506.

Glover G H 1982. Compton scatter effects in CT reconstructions. *Med Phys* **9** 860–7.

Greer P B 2005. Correction of pixel sensitivity variation and off-axis response for amorphous silicon EPID dosimetry. *Med Phys* **32** 3558–68.

Greer P B and Popescu C C 2003. Dosimetric properties of an amorphous silicon electronic portal imaging device for verification of dynamic intensity modulated radiation therapy. *Med Phys* **30** 1618–27.

Greer P B, Vial P, Oliver L, and Baldock C 2007. Experimental investigation of the response of an amorphous silicon EPID to intensity modulated radiotherapy beams. *Med Phys* **34** 4389–98.

ICRP-102 2007. Managing patient dose in multi-detector computed tomography (MDCT). *Ann ICRP* **37** 1–79.

Islam M K, Purdie T G, Norrlinger B D, Alasti H, Moseley D J, Sharpe M B, Siewerdsen J H, and Jaffray D A 2006. Patient dose from kilovoltage cone beam computed tomography imaging in radiation therapy. *Med Phys* **33** 1573–82.

Jaffray D A, Battista J J, Fenster A, and Munro P 1994. X-ray scatter in megavoltage transmission radiography—Physical characteristics and influence on image quality. *Med Phys* **21** 45–60.

Jaffray D A, Drake D G, Moreau M, Martinez A A, and Wong J W 1999. A radiographic and tomographic imaging system integrated into a medical linear accelerator for localization of bone and soft-tissue targets. *Int J Radiat Oncol Biol Phys* **45** 773–89.

Jaffray D A and Siewerdsen J H 2000. Cone-beam computed tomography with a flat-panel imager: Initial performance characterization. *Med Phys* **27** 1311–23.

Jaffray D A, Siewerdsen J H, Wong J W, and Martinez A A 2002. Flat-panel cone-beam computed tomography for image-guided radiation therapy. *Int J Radiat Oncol Biol Phys* **53** 1337–49.

Jarry G, Graham S A, Moseley D J, Jaffray D J, Siewerdsen J H, and Verhaegen F 2006. Characterization of scattered radiation in kV CBCT images using Monte Carlo simulations. *Med Phys* **33** 4320–9.

Johns P C and Yaffe M 1982. Scattered radiation in fan beam imaging systems. *Med Phys* **9** 231–9.

Joseph P M and Spital R D 1982. The effects of scatter in x-ray computed tomography. *Med Phys* **9** 464–72.

Kairn T, Cassidy D, Sandford P M, and Fielding A L 2008. Radiotherapy treatment verification using radiological thickness measured with an amorphous silicon electronic portal imaging device: Monte Carlo simulation and experiment. *Phys Med Biol* **53** 3903–19.

Kairn T, Warne D, Kenny J, and Dwyer M 2011. Accurately simulating the production of radiotherapy portal images using non-zero beam angles. *Radiat Meas* **46** 1967–70.

Kanamori H, Nakamori N, Inoue K, and Takenaka E 1985. Effects of scattered X-rays on CT images. *Phys Med Biol* **30** 239–49.

Kawrakow I and Rogers D W O 2002. The EGSnrc Code System: Monte Carlo Simulation of Electron and Photon Transport. (Ottawa: Ionizing Radiation Standards, National Research Council of Canada, NRCC Report PIRS-701).

Kirby M C and Glendinning A G 2006. Developments in electronic portal imaging systems. *Br J Radiol* **79** S50–65.

Kirkby C and Sloboda R 2005. Consequences of the spectral response of an a-Si EPID and implications for dosimetric calibration. *Med Phys* **32** 2649–58.

Klein D J, Chan H P, Muntz E P, Doi K, Lee K, Chopelas P, Bernstein H, and Lee J 1983. Experimental and theoretical energy and angular dependencies of scattered radiation in the mammography energy range. *Med Phys* **10** 664–8.

Ko L, Kim J O, and Siebers J V 2004. Investigation of the optimal backscatter for an aSi electronic portal imaging device. *Phys Med Biol* **49** 1723–38.

Langmack K A 2001. Portal imaging. *Br J Radiol* **74** 789–804.

Li W D, Siebers J V, and Moore J A 2006. Using fluence separation to account for energy spectra dependence in computing dosimetric a-Si EPID images for IMRT fields. *Med Phys* **33** 4468–80.

Liu H H, Mackie T R, and McCullough E C 2000. Modeling photon output caused by backscattered radiation into the monitor chamber from collimator jaws using a Monte Carlo technique. *Med Phys* **27** 737–44.

Low D A, Moran J M, Dempsey J F, Dong L, and Oldham M 2011. Dosimetry tools and techniques for IMRT. *Med Phys* **38** 1313–38.

Ma C M, Coffey C W, DeWerd L A, Liu C, Nath R, Seltzer S M, and Seuntjens J P 2001. AAPM protocol for 40–300 kV x-ray beam dosimetry in radiotherapy and radiobiology. *Med Phys* **28** 868–93.

Mackie T R, Scrimger J W, and Battista J J 1985. A convolution method of calculating dose for 15-MV x rays. *Med Phys* **12** 188–96.

Malusek A, Seger M M, Sandborg M, and Alm Carlsson G 2005. Effect of scatter on reconstructed image quality in cone beam computed tomography: Evaluation of a scatter-reduction optimisation function. *Radiat Prot Dosim* **114** 337–40.

McCurdy B M C, Luchka K, and Pistorius S 2001. Dosimetric investigation and portal dose image prediction using an amorphous silicon electronic portal imaging device. *Med Phys* **28** 911–24.

McCurdy B M C and Pistorius S 2000. A two-step algorithm for predicting portal dose images in arbitrary detectors. *Med Phys* **27** 2109–16.

McNutt T R, Mackie T R, Reckwerdt P, and Paliwal B R 1996a. Modeling dose distributions from portal dose images using the convolution/superposition method. *Med Phys* **23** 1381–92.

McNutt T R, Mackie T R, Reckwerdt P, Papanikolaou N, and Paliwal B R 1996b. Calculation of portal dose using the convolution/superposition method. *Med Phys* **23** 527–35.

Meertens H, Vanherk M, Bijhold J, and Bartelink H 1990. 1st clinical-experience with a newly developed electronic portal imaging device. *Int J Radiat Oncol Biol Phys* **18** 1173–81.

Merritt R B and Chenery S G 1986. Quantitative CT measurements: The effect of scatter acceptance and filter characteristics on the EMI 7070. *Phys Med Biol* **31** 55–63.

Mohan R and Chui C S 1987. Use of fast Fourier transforms in calculating dose distributions for irregularly shaped fields for three-dimensional treatment planning. *Med Phys* **14** 70–7.

Moore J A and Siebers J V 2005. Verification of the optimal backscatter for an aSi electronic portal imaging device. *Phys Med Biol* **50** 2341–50.

Mosleh-Shirazi M A, Evans P M, Swindell W, Webb S, and Partridge M 1998. A cone-beam megavoltage CT scanner for treatment verification in conformal radiotherapy. *Radiother Oncol* **48** 319–28.

Munro P and Bouius D C 1998. X-ray quantum limited portal imaging using amorphous silicon flat-panel arrays. *Med Phys* **25** 689–702.

Murphy M, Balter J M, BenComo J, Das I, Jiang S, Ma C M, Olivera G et al. 2007. The management of imaging dose during image-guided radiotherapy: Report of the AAPM Task Group 75. *Med Phys* **34** 4041–63.

Nelson W R, Hirayama H, and Rogers D W O 1985a. The EGS4 code system. (Stanford: Stanford Linear Accelerator Center).

Nelson W R, Hirayama H, and Rogers D W O 1985b. The EGS4 Code System. Report SLAC-265, Stanford Linear Accelerator Center, Stanford, California).

Parent L 2006. The use of Monte Carlo methods to study the effect of x-ray spectral variations on the response of an amorphous silicon electronic portal imaging device. In:

Joint Department of Physics, Institute of Cancer Research (London: University of London).

Parent L, Evans P, Dance D R, Fielding A, and Seco J 2004. Effect of spectral variation with field size on dosimetric response of an amorphous silicon electronic portal imaging device. *Radiother Oncol* **73** S154–5.

Parent L, Fielding A L, Dance D R, Seco J, and Evans P M 2007. Amorphous silicon EPID calibration for dosimetric applications: Comparison of a method based on Monte Carlo prediction of response with existing techniques. *Phys Med Biol* **52** 3351–68.

Parent L, Seco J, Evans P M, Dance D R, and Fielding A 2006a. Evaluation of two methods of predicting MLC leaf positions using EPID measurements. *Med Phys* **33** 3174–82.

Parent L, Seco J, Evans P M, Fielding A, and Dance D R 2006b. Monte Carlo modelling of a-Si EPID response: The effect of spectral variations with field size and position. *Med Phys* **33** 4527–40.

Partridge M, Ebert M, and Hesse B M 2002. IMRT verification by three-dimensional dose reconstruction from portal beam measurements. *Med Phys* **29** 1847–58.

Pawlowski J M and Ding G X 2011. A new approach to account for the medium-dependent effect in model-based dose calculations for kilovoltage x-rays. *Phys Med Biol* **56** 3919–34.

Pawlowski J M, Yang E S, Malcolm A W, Coffey C W, and Ding G X 2010. Reduction of dose delivered to organs at risk in prostate cancer patients via image-guided radiation therapy. *Int J Radiat Oncol Biol Phys* **76** 924–34.

Rogers D W O 2006. Fifty years of Monte Carlo simulations for medical physics. *Phys Med Biol* **51** R287–301.

Rogers D W O, Faddegon B A, Ding G X, Ma C M, We J, and Mackie T R 1995. BEAM: A Monte Carlo code to simulate radiotherapy treatment units. *Med Phys* **22** 503–24.

Seuntjens J and Ma C 1999. Dose conversion factors and depth scaling for tissue dose calculations in kilovoltage x-ray beams (abs). *Med Phys* **26** 1421.

Sharpe M B, Moseley D J, Purdie T G, Islam M, Siewerdsen J H, and Jaffray D A 2006. The stability of mechanical calibration for a kV cone beam computed tomography system integrated with linear accelerator. *Med Phys* **33** 136–44.

Sheikh-Bagheri D and Rogers D W 2002a. Monte Carlo calculation of nine megavoltage photon beam spectra using the BEAM code. *Med Phys* **29** 391–402.

Sheikh-Bagheri D and Rogers D W 2002b. Sensitivity of megavoltage photon beam Monte Carlo simulations to electron beam and other parameters. *Med Phys* **29** 379–90.

Shiu A S, Chang E L, Ye J S, Lii M, Rhines L D, Mendel E, Weinberg J et al. 2003. Near simultaneous computed tomography image-guided stereotactic spinal radiotherapy: An emerging paradigm for achieving true stereotaxy. *Int J Radiat Oncol Biol Phys* **57** 605–13.

Siebers J V, Kim J O, Ko L, Keall P J, and Mohan R 2004. Monte Carlo computation of dosimetric amorphous silicon electronic portal images. *Med Phys* **31** 2135–46.

Siewerdsen J H and Jaffray D A 2000. Optimization of x-ray imaging geometry (with specific application to flat-panel cone-beam computed tomography). *Med Phys* **27** 1903–14.

Siewerdsen J H and Jaffray D A 2001. Cone-beam computed tomography with a flat-panel imager: Magnitude and effects of x-ray scatter. *Med Phys* **28** 220–31.

Solberg T D, DeMarco J J, Chetty I J, Mesa A V, Cagnon C H, Li A N, Mather K K, Medin P M, Arellano A R, and Smathers J B 2001. A review of radiation dosimetry applications using the MCNP Monte Carlo code. *Radiochim Acta* **89** 337–55.

Spezi E, Downes P, Radu E, and Jarvis R 2009. Monte Carlo simulation of an x-ray volume imaging cone beam CT unit. *Med Phys* **36** 127–36.

Spezi E and Lewis D G 2002. Full forward Monte Carlo calculation of portal dose from MLC collimated treatment beams. *Phys Med Biol* **47** 377–90.

Spezi E, Lewis D G, and Smith C W 2001. Monte Carlo simulation and dosimetric verification of radiotherapy beam modifiers. *Phys Med Biol* **46** 3007–29.

Spies L and Bortfeld T 2001. Analytical scatter kernels for portal imaging at 6 MV. *Med Phys* **28** 553–9.

Spies L, Ebert M, Groh B A, Hesse B M, and Bortfeld T 2001. Correction of scatter in megavoltage cone-beam CT. *Phys Med Biol* **46** 821–33.

Swindell W and Evans P M 1996. Scattered radiation in portal images: A Monte Carlo simulation and a simple physical model. *Med Phys* **23** 63–73.

Tofts P S and Gore J C 1980. Some sources of artefact in computed tomography. *Phys Med Biol* **25** 117–27.

Uematsu M, Fukui T, Shioda A, Tokumitsu H, Takai K, Kojima T, Asai Y, and Kusano S 1996. A dual computed tomography linear accelerator unit for stereotactic radiation therapy: A new approach without cranially fixated stereotactic frames. *Int J Radiat Oncol Biol Phys* **35** 587–92.

Van Esch A, Depuydt T, and Huyskens D P 2004. The use of an aSi-based EPID for routine absolute dosimetric pretreatment verification of dynamic IMRT fields. *Radiother Oncol* **71** 223–34.

Verhaegen F, Nahum A E, Van de Putte S, and Namito Y 1999. Monte Carlo modelling of radiotherapy kV x-ray units. *Phys Med Biol* **44** 1767–89.

Verhaegen F, Schulze C, Seuntjens J, Gosselin M, Hristov D, and Svatos M 2004. Heterogeneity corrected convolution dose calculation for kv x-rays versus Monte Carlo simulation (abs). *Med Phys* **31** 1770.

Verhaegen F and Seuntjens J 2003. Monte Carlo modelling of external radiotherapy photon beams. *Phys Med Biol* **48** R107–64.

Walters B R, Ding G X, Kramer R, and Kawrakow I 2009. Skeletal dosimetry in cone beam computed tomography. *Med Phys* **36** 2915–22.

Wen N, Guan H, Hammoud R, Pradhan D, Nurushev T, Li S, and Movsas B 2007. Dose delivered from Varian's CBCT to patients receiving IMRT for prostate cancer. *Phys Med Biol* **52** 2267–76.

Wendling M, Louwe R J W, McDermott L N, Sonke J J, van Herk M, and Mijnheer B J 2006. Accurate two-dimensional IMRT verification using a back-projection EPID dosimetry method. *Med Phys* **33** 259–73.

Wendling M, McDermott L N, Mans A, Sonke J-J, van Herk M, and Mijnheer B J 2009. A simple backprojection algorithm for 3D *in vivo* EPID dosimetry of IMRT treatments. *Med Phys* **36** 3310–21.

Winkler P, Hefner A, and Georg D 2007. Implementation and validation of portal dosimetry with an amorphous silicon EPID in the energy range from 6 to 25 MV. *Phys Med Biol* **52** N355–65.

Yeboah C and Pistorius S 2000. Monte Carlo studies of the exit photon spectra and dose to a metal/phosphor portal imaging screen. *Med Phys* **27** 330–9.

Yenice K M, Lovelock D M, Hunt M A, Lutz W R, Fournier-Bidoz N, Hua C H, Yamada J et al. 2003. CT image-guided intensity-modulated therapy for paraspinal tumors using stereotactic immobilization. *Int J Radiat Oncol Biol Phys* **55** 583–93.

Monte Carlo Calculations for PET-Based Treatment Verification of Ion Beam Therapy

Katia Parodi
Heidelberg Ion Beam Therapy Center

17.1 Introduction

The favorable physical selectivity of ion beams offers superior tumor-dose conformality with better sparing of surrounding critical organs and healthy tissue with respect to external radiotherapy modalities based on conventional electromagnetic radiation. However, these ballistic advantages come at the expense of increased sensitivity to conventional sources of treatment uncertainties during the fractionated course of radiation therapy, such as patient positioning errors or other geometrical/anatomical variations (e.g., weight loss) in the treatment situation with respect to the planning one. Moreover, the validity of the analytical pencil-beam algorithms typically employed for ion therapy treatment planning and, in particular, the accuracy of the semiempirical calibration curve converting the patient computed tomography (CT) data into water-equivalent ion range (Schaffner and Pedroni, 1998) can only be carefully assessed via measurements in tissue-equivalent substitutes but cannot be directly verified *in vivo*. Therefore, to account for all the above-mentioned random or systematic sources of errors, cautious safety margins are still used in clinical routine of ion therapy treatment planning. Moreover, when defining the beam directions, it is a common practice to avoid placement of the distal dose fall-off in front of critical organs. Instead, it is preferred to use the less sharp but more reliable lateral penumbra of the ion beam, often sacrificing the utmost achievable dose conformality for the sake of safety. Therefore, tools for the *in vivo* validation of the ion range and the actually delivered treatment during the fractionated course of radiation therapy would be highly beneficial and might promote full clinical exploitation of the improved balistic selectivity offered by ion beams for highly conformal tumor therapy.

Nowadays, positron emission tomography (PET) is the only technically feasible method fulfilling the requirement for a three-dimensional (3D), noninvasive, *in vivo* monitoring of the delivered ion treatment and, in particular, of the beam range in the patient during or shortly after irradiation (Enghardt et al., 2004a). The unconventional application of this nuclear imaging technique to ion beam therapy is based on the detection of the transient pattern of β+-activation, which is produced in fragmentation reactions between the incoming ions and the irradiated tissue. Imaging can be performed during beam delivery ("in-beam") or minutes afterward ("in-room" or "offline") depending on the availability of the dedicated or commercial PET installation (Shakirin et al., 2011). Owing to the different underlying physical processes, the irradiation-induced β+-activity originating from nuclear interactions can only represent a surrogate signal correlated in a not straightforward way to the electromagnetically deposited dose. Therefore, treatment verification is based on the comparison of the measured PET images with a corresponding expectation based on the planned treatment and the fraction-specific time course of irradiation and imaging (Enghardt et al., 2004a; Parodi et al., 2007c). This comparison may enable detection of unpredictable discrepancies between prescribed and actual irradiation, thus offering an independent verification of the entire therapy chain from treatment planning to delivery as well as opening the possibility for intervention prior to the application of the next therapeutic session (Enghardt et al., 2004b). Therefore, PET-based treatment verification can potentially enable safe reduction of the planning margins and exploitation of the distal ion dose fall-off for more conformal treatments with related dose escalation. Moreover, it might offer an important tool for assuring the correct treatment delivery in the increasingly considered application of ion beams to high-dose hypofractionated therapy (Ertner and Tsujii, 2007; Matsufuji et al., 2007), where less or even no subsequent fractions are available for compensation of errors. Finally, it can improve confidence of the successful implementation of special beam delivery strategies like gating or tracking in the challenging presence of organ motion, as supported by first encouraging phantom experiments (Parodi et al., 2009).

The calculation of the expected pattern of activation plays a central role in the clinical implementation of PET-based treatment verification. Although promising analytical approaches are being proposed and evaluated by several groups for fast computation (Parodi and Bortfeld, 2006; Miyatake et al., 2011; Priegnitz et al., 2011; Remmele et al., 2011), the clinical investigations reported so far have relied on full-blown Monte Carlo (MC) simulations of the interaction between the primary beam and the irradiated tissue. In particular, different approaches have been pursued to address the two main steps of modeling the irradiation-induced positron emitter yield and the subsequent imaging process in dependence of the considered ion species and available PET instrumentation, as addressed in the following sections.

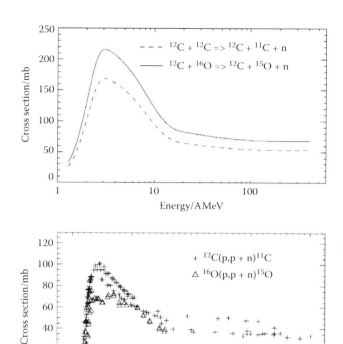

FIGURE 17.1 Partial cross sections for the main production of ^{11}C and ^{15}O nuclei in the interaction of ^{12}C ions (top) and protons (bottom) on ^{12}C and ^{16}O nuclei, respectively, according to calculations from Sihver et al. (1993); Silberberg and Tsao (1973) (top) and experimental data from IAEA Nuclear Data Section (2000) (bottom), respectively. (Adapted from Parodi, K. On the feasibility of dose quantification with in-beam PET data in radiotherapy with 12C and proton beams, PhD thesis, Dresden University of Technology, in Forschungszentrum Rossendorf Wiss-Techn-Ber FZR-415, 2004.)

17.2 Monte Carlo Modeling of Ion-Induced β+-Emitters

Reliable reproduction of the spatial distribution and quantitative yield of irradiation-induced positron emitters requires correct handling of several physical quantities and processes. These comprise the initial characterization of the therapeutic beam and its slowing down in the traversed medium including not only the dominant electromagnetic interactions but also the nuclear processes that may result in positron emitter production, besides causing attenuation of the primary beam and build-up of secondary radiation in depth. While some of these aspects are common to the problem of MC dosimetric calculations in ion beam therapy (cf. Chapter 14), it should be noticed that the partial reaction cross sections yielding positron emitters seldom exceed values of 200 mb (Figure 17.1). Therefore, whereas in terms of nuclear processes the dosimetric calculations are mostly sensitive to a realistic reproduction of the total reaction cross sections and of the main yield of secondary charged hadrons, the PET calculations are extremely sensitive to the specific β+-activation channels of typically low probability. Moreover, depending on the primary ion species, the pattern of irradiation-induced activity either is entirely ascribed to positron-emitting target fragments or exhibits a peaked contribution from β+-emitting projectile fragments, superimposed onto a pedestal of the activated target fragments (Figure 17.2). Especially the latter remarkable difference in the origin of the activity signal historically resulted in a quite different approach for MC calculations of ion-induced β+-emitters, as separately reviewed in the following sections for proton and carbon ion therapy.

17.2.1 Proton Therapy

In the case of primary proton beams, the treatment-induced activity is limited to the target fragments of the irradiated tissue. Indeed, only a full MC simulation including generation and transport of all secondary charged and uncharged particles would be capable of accounting for all possible β+-activity production channels. However, it has been estimated in Parodi et al. (2007a) that the positron emitter yield from secondary particles except protons is negligible (<1.5%) even when including the additional neutron background from passively shaped treatment fields. Moreover, although all the major general-purpose MC codes (e.g., Geant4 (Agostinelli et al., 2003), FLUKA (Ferrari et al., 2005; Battistoni et al., 2007), MCNPX (Briesmeister, 1997), and PHITS (Niita et al., 2006)) are capable of predicting proton-induced activation using their own internal nuclear models, it has been shown in Seravalli et al. (2012) that the resulting positron emitter production may considerably differ in terms of both quantitative yield and spatial distribution.

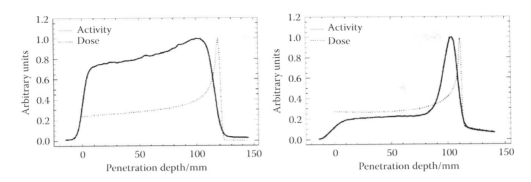

FIGURE 17.2 Measured β+-activity depth profiles (solid line) for proton (left) and carbon ion (right) irradiation of PMMA targets at 140 MeV and 259.5 AMeV initial energy, respectively. The dotted line additionally depicts the calculated dose distributions. (From Parodi, K. On the feasibility of dose quantification with in-beam PET data in radiotherapy with 12C and proton beams, PhD thesis, Dresden University of Technology, in Forschungszentrum Rossendorf Wiss-Techn-Ber FZR-415, 2004.)

A more accurate approach proposed in Parodi et al. (2002, 2007a,b) and already clinically applied in Parodi et al. (2007c) makes use of available experimental or evaluated cross sections $\sigma_{X \to Y}$ combined during runtime with the proton fluence Φ. According to this formalism, the total amount N_Y of positron emitters Y produced in nuclear interaction with a nucleus X in a detecting volume ΔV can be obtained from integration over the primary and secondary proton energy spectrum as

$$N_Y = \int \frac{d\Phi(E)}{dE} \frac{f_X \rho N_0}{A_X} \sigma_{X \to Y}(E) \Delta V \, dE \tag{17.1}$$

where ρ is the medium density, f_X is the fraction by weight of the considered nucleus X of atomic weight A_X, and N_0 is the Avogadro number. Depending on the considered cross-section data, correction for the isotopic abundance of the target nucleus X (e.g., 96.94% ^{40}Ca in natural Ca) may be required (Parodi et al., 2007a). When performing calculations for clinical cases, the less straightforward information on the elemental tissue composition can be deduced from a proper stoichiometric calibration of the CT data, which are typically used for the patient model in the treatment planning system and in the MC input geometry (cf. Chapter 14), for example, relying on the proposal of Schneider et al. (2000).

Examples of external reaction cross sections used within the FLUKA MC code according to Equation 17.1 for application to PET imaging during and after proton irradiation of phantoms (Parodi et al., 2002, 2005, 2007a,b) and patients (Parodi et al., 2007c) at the Helmholtzzentrum für Schwerionenforschung (GSI) in Darmstadt and Massachusetts General Hospital (MGH) in Boston, respectively, are shown in Figure 17.3. In human tissue, according to the main elemental abundance and the magnitude of the reaction cross sections, the most frequently formed isotopes are ^{11}C (half-life $T_{1/2} = 20.39$ min), ^{15}O ($T_{1/2} = 2.03$ min), ^{13}N ($T_{1/2} = 9.97$ min), and ^{38}K ($T_{1/2} = 7.63$ min). An example of their calculated spatial distribution is shown in Figure 17.4 for a patient suffering from a clivus chordoma tumor, when taking into

account only one of the two orthogonal treatment fields delivering a planned dose of about 0.9 Gy each (Parodi et al., 2007a). It is evident that in addition to the cross-section dependence, the pattern of activated target fragments reflects the estimated tissue composition with dominant carbon abundance and thus ^{11}C production in fatty tissue, while prevalent oxygen content with related high ^{15}O yield in soft tissue and calcium content with resulting ^{38}K activation in bone. As statistical fluctuations are reduced in this approach convolving fluence and cross sections at the voxel level with respect to a full MC computation explicitly simulating pointwise positron emitter production (Parodi et al., 2007a), sufficient counting statistics can be achieved within few hours when using several parallel runs on a high-performance computer cluster.

The final contribution of the simulated β+-emitter distributions to the measurable activity signal varies in dependence of the course of irradiation and PET imaging, due to differences up to one order of magnitude in the half-life of the produced isotopes as well as in the timing of in-beam, in-room, or offline data acquisition. Therefore, for accurate PET-based treatment verification, it is strongly recommended to validate the MC prediction of positron emitter yields for the reaction channels that are most relevant to the considered imaging application via dedicated measurements in phantom of known composition (Parodi et al., 2002, 2005; Miyatake et al., 2011). If necessary, the used cross-section data could be tuned to better reproduce the measurements acquired at the own facility. For this purpose, the approach based on Equation 17.1 is indeed more straightforward than relying on the often not accessible internal nuclear models of the codes. The remaining steps for simulation of the irradiation-induced β+-activity starting from the predicted positron emission map will be addressed in Section 17.3.

17.2.2 Carbon Ion Therapy

In the case of primary carbon ion beams, the treatment-induced activity includes contributions from both projectile

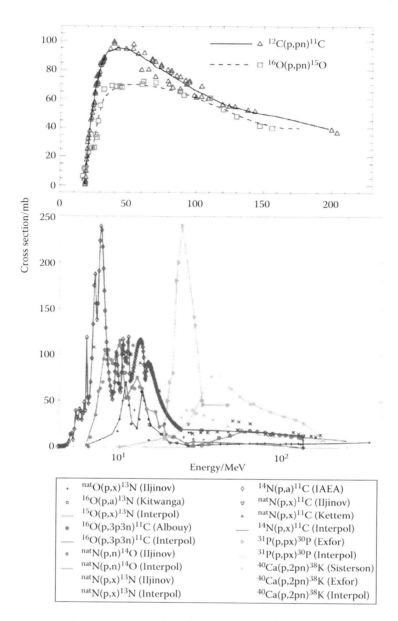

FIGURE 17.3 Example of cross-section values (lines) interpolated from experimental or evaluated data (symbols) and used in the general-purpose MC code FLUKA for calculation of positron emitter production according to Equation 17.1. (Adapted from Parodi, K et al. *Phys Med Biol* 2007a;52:3369–3387; Parodi, K, Enghardt, W, and Haberer, T. *Phys Med Biol* 2002;47:21–36; Parodi, K. On the feasibility of dose quantification with in-beam PET data in radiotherapy with 12C and proton beams, PhD thesis, Dresden University of Technology, in Forschungszentrum Rossendorf Wiss-Techn-Ber FZR-415, 2004. With permission.)

and target fragments. Therefore, the simplified approach proposed in Equation 17.1 for calculation of proton-induced target fragmentation is no longer applicable, and reliable internal nuclear models are required to properly account for the alteration of the primary beam quality and the generation and transport of the secondary charged fragments, which give a nonnegligible contribution to the β+-activity production. Nevertheless, a simplification has been suggested in Pönisch et al. (2004) for clinical routine application to in-beam PET monitoring of scanned carbon ion therapy, to replace the

otherwise too time-consuming full MC simulation approach. The basic underlying idea is that the most relevant information on the *in vivo* beam range mainly arises from the spatial distribution of the positron-emitting projectile fragments (cf. Figure 17.2), which is less sensitive to the specific tissue stoichiometry in comparison to the target fragment production. Therefore, the prediction of the irradiation-induced positron emitter yield can be split into a two-step process, with a one-time patient-independent generation of a positron emitter database in a reference homogeneous medium followed by

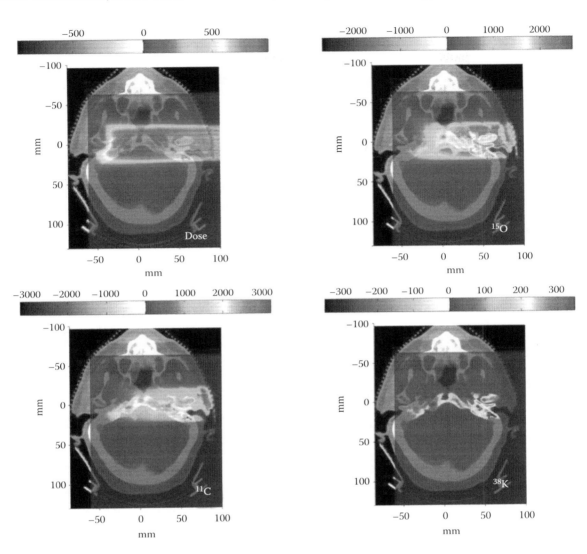

FIGURE 17.4 **(See color insert.)** FLUKA MC calculated dose deposition (upper left panel, color bar in units of mGy) and corresponding positron emitter yields (^{15}O in the upper right panel, ^{11}C and ^{38}K in the lower left and right panels, respectively, color bars in units of produced isotopes for the prescribed dose) for a lateral proton field irradiating a clivus chordoma. The black–white color bar represents the CT number map arbitrary rescaled for display purposes. (Adapted from Parodi, K et al. *Phys Med Biol* 2007a;52:3369–3387. With permission.)

a patient- and fraction-specific adaptation of the generated database to the daily treatment and imaging situation. In this way, the time-consuming detailed simulation of the beam transport and interaction has been performed only once in a homogeneous target (PMMA, $C_5H_8O_2$) containing the most relevant elements of human tissue for pencil-like carbon ion beams at each available therapeutic energy. An example is shown in Figure 17.5. The specific implementation has relied on a dedicated condensed history MC code described in detail in Pönisch et al. (2004) and Hasch (1996). In particular, the more complex handling of nuclear processes has been based on the parametrization of total and partial reaction cross sections proposed in Sihver et al. (1993) and Silberberg and Tsao (1973) and validated against experimental measurements of irradiation-induced activation.

The second step relies on a fast ray-tracing pencil-beam propagation through the patient CT with the same stretching in depth as implemented in the treatment planning system (Kramer et al., 2000). Positron emitter production is obtained via a random MC sampling from the generated database in PMMA repeated as many times as in proportion to the number of delivered ions for each considered energy, adjusting the longitudinal position to the estimated penetration depth in the patient and scaling the lateral position to the actual beam dimension. Moreover, an approximate correction is applied to the target fragment yields to better account for the patient tissue stoichiometry in comparison to PMMA. This correction consists in enhancing the ^{15}O to ^{11}C target fragment ratio of the database to account for the average higher content of oxygen than carbon in real tissue with respect to PMMA, and in

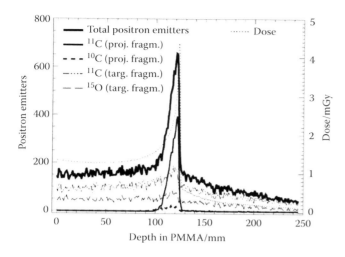

FIGURE 17.5 Example of the positron emitter depth distributions produced in PMMA by 10^5 ^{12}C ions at 270.55 AMeV according to the dedicated simulation code (Hasch, 1996; Pönisch et al., 2004), which uses cross-section calculations from Sihver et al. (1993); Silberberg and Tsao (1973) (cf. Figure 17.1, top). The main contributions from projectile and target fragments are shown separately. The expected dose depth distribution is additionally depicted by the dotted line. (From Parodi, K. On the feasibility of dose quantification with in-beam PET data in radiotherapy with 12C and proton beams, PhD thesis, Dresden University of Technology, in Forschungszentrum Rossendorf Wiss-Techn-Ber FZR-415, 2004.)

linearly scaling the calculated target positron emitter yields according to the tissue mass density as estimated from the CT data (Pönisch et al. 2004). For typically used counting statistics in simulation of patient treatments, this second step calculation relying on the already existent database takes only from seconds up to few minutes of computational time on a single state-of-the-art computer.

Despite the successful application of the simplified two-step approach tailored to in-beam PET monitoring at GSI Darmstadt, more recent investigations have been reported to assess the performances of general-purpose codes such as Geant4 and FLUKA for simultaneous ion transport and generation of positron emitters in arbitrary media (Pshenichnov et al., 2007; Sommerer et al., 2009). In particular, the approach reported in Sommerer et al. (2009) thoroughly validated against phantom experiments is being clinically used for offline imaging at the Heidelberg Ion Beam Therapy Center (Parodi et al., 2011), with a still acceptable computational time of about 5–8 h in 30 parallel runs for positron emitter generation on the patient CT. The advantage of a full MC approach over the dedicated two-step modeling is the more realistic description of the positron emitting target fragments and of the lateral beam spreading in depth, which gain importance in PET-based geometrical localization of the treatment field. A major drawback, besides the increased computational time, is the entire dependence on the internal models of the code. This prevents the possibility to fine tune the MC predictions to data measured at the individual facilities, unless reaction cross sections

can be accessed during runtime or dedicated adjustments can be introduced by the code developers upon request.

17.3 Monte Carlo Modeling of Imaging Process

Two approaches have been pursued so far to account for the imaging process in clinical applications of PET monitoring, mostly dependent on the used detector instrumentation regardless of the method followed for calculation of positron emitter production.

The dual-head configuration of the dedicated in-beam PET installation at GSI Darmstadt is known to introduce limited-angle reconstruction artifacts, which become particularly pronounced when moving away from the center of the imaging field of view. Therefore, a trustful comparison of measured and simulated activity distributions can only be achieved when processing the simulated data with the same software used for the reconstruction of the measured data. This requires the explicit simulation of all the steps of the imaging process, from the radioactive decay of the positron emitters to the moderation of the emitted positron until its annihilation into two opposed gamma rays, which are then propagated through the surrounding medium up to the coincident detection by the PET scanner. All these steps have been implemented in the same dedicated MC code of Pönisch et al. (2004), starting from the positron emitter yield in the patient CT as obtained from the described adaptation of the initial database in PMMA (cf. previous section). The whole process takes into account the recorded fraction-specific time course of irradiation and imaging as follows. The production time of the positron emitters is identified with the time at which the considered pencil-beam is applied in the dose delivery. Starting from this time, random radioactive decays are simulated according to the isotope half-lives, but only those potentially leading to coincident detection during the registered time course of the PET data acquisition are considered for further processing. The emission and moderation of the positron in the medium are performed starting from the random sampling of the positron range in water from an isotope-specific look-up table. A random direction is then selected and the positron is tracked until the stopping point, as determined by properly stretching the range in water to the patient anatomy according to the CT information. Unless the stopping position is in air (within or outside the patient), the emission of the annihilation photons is explicitly simulated taking into account their angular distribution probability. The propagation of the annihilation photons is then performed until they reach the detector or escape the simulation space, accounting for their main physical interactions (Rayleigh and Compton scattering) according to look-up tables and including established variance reduction techniques to increase computational efficiency (Pönisch et al., 2004). This is particularly important taking into account the extremely low geometrical acceptance of the detector, covering only about 9% of the entire solid angle. No explicit interaction with the crystal is simulated, but acceptance of a valid hit on the detector is decided by considering the energy threshold of the positron camera and the energy resolution of the BGO (bismuth germanate) scintillation blocks. The final output is a list mode of detected events in the same format as the measured data, allowing for the usage of the same reconstruction algorithms. An example of experimental validation is shown in Figure 17.6a. With this approach, consistent datasets have been compared throughout the entire pilot project at GSI, enabling clinically useful information to be inferred from the in-beam PET monitoring (Enghardt et al., 2004a,b). According to Pönisch et al. (2004), a simulated PET list mode dataset could be obtained in less than 5 min on a single computer for a typical patient treatment, allowing the results of the PET treatment monitoring to be available within about 30 min after completing the therapeutic irradiation.

More recently, the latter step of the dedicated code (Pönisch et al., 2004) accounting for the response of the GSI in-beam PET detector (in terms of geometrical and energy acceptance) has been coupled to the output of the general-purpose code FLUKA for simulation of the entire PET imaging process in phantom experiments with carbon and oxygen ion beams. In particular, FLUKA was successfully used for the transport of the ions, the creation of the β+-active nuclei, the decay, the electron–positron annihilation, and the transport of the annihilation photons until escape from the irradiated target (Sommerer et al., 2009). For this study, a new variance reduction technique was introduced in the code for directional biasing of the annihilation photons toward the detector of very poor geometrical coverage, complementing useful available tools such as replication of the β+-decay from the produced isotopes to enhance detection in the considered measuring time window. Despite the reported considerable gain in computational efficiency up to a factor of 10 for the direction biasing and 25 for the decay replication, the execution times were deemed to be still too long for clinical routine application. Nevertheless, this work represented an important first step toward the usage of a general-purpose code for full MC simulation of both the irradiation and imaging processes in PET-based verification of heavy ion therapy.

In the case of full-ring PET detector geometries as used at MGH Boston and HIT Heidelberg, the considerably improved coverage of the solid angle significantly reduces imaging artifacts in comparison to limited-angle dual-head configurations. Therefore, an easier approximation of the combined effects of the image formation and reconstruction process consists in a Gaussian smoothing of the MC simulated positron emitter distributions, properly weighted by isotope-dependent factors that account for the number of decays contributing to the activity signal in the considered acquisition time window (Parodi et al., 2007b,c). The parameters of the Gaussian point spread function can be determined from the specifications of the scanner or be empirically obtained from point source measurements. An example of experimental validation is shown in Figure 17.6b. This simple and very fast approach has been already successfully used for several clinical investigations of PET imaging in proton therapy, where it has been applied to CT-based positron emitter

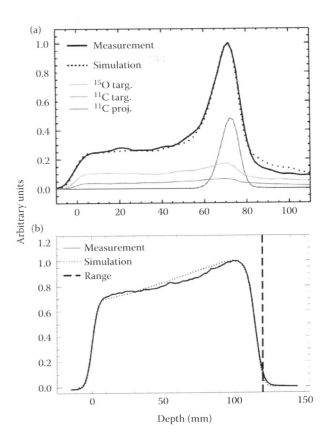

FIGURE 17.6 (a) Depth distributions of β+-activity induced by a 212 AMeV carbon ion beam in a thick PMMA target reconstructed by means of backprojection from measured (thick solid line) and simulated (Pönisch et al., 2004) (dotted line) list mode data acquired in about 5 min long irradiation. The calculated contribution of the most abundant projectile and target fragments is additionally shown by the thin lines of different gray scale. (b) Depth profiles of in-beam PET activity induced by a 10 min irradiation of a thick PMMA target at 140 MeV proton beam energy. The simulation (dotted line) obtained from the Gaussian smoothing of the positron emitter yield calculated with FLUKA on the basis of Equation 17.1 and the cross-section data of Figure 17.3 is compared to the measurement (solid line). The dashed line indicates the expected range. Only relative activity distributions are shown as the in-beam PET camera at GSI could not provide absolute reconstruction. In this example, limited-angle artifacts are not an issue due to the imaging of a compact activity distribution in the central plane of the scanner.

yields generated by full MC codes such as FLUKA (Parodi et al., 2007c) and Geant4 (Zhu et al., 2011).

Indeed, dedicated MC-based tools that were originally developed for characterization of nuclear medicine instrumentation could be coupled to the output of the general-purpose codes for full MC simulation of the imaging process, including the detailed handling of photon interactions in the detector crystals for a complete modeling of the detector response. In particular, the Geant4-based simulation platform GATE (Geant4 Application for Emission Tomography), which was initially developed for research on PET as well as single-photon emission tomography (SPECT), has been recently extended to radiation therapy applications, including in-beam PET modeling for hadron therapy (Jan et al., 2011). However, besides the unavoidable increase of computational time, the feasibility of such a full MC handling requires detailed knowledge of the PET scanner and of its proprietary data formats as well as applicability of the reconstruction software to the simulated data, which is often not available/possible for commercial PET instrumentation. Moreover, it is still controversial whether the realistic addition of statistical noise in the full MC simulation, which is not accounted for by the smoothing approach, is a real benefit for the purpose of PET-based treatment verification due to the resulting degradation of image quality in the simulated image used as reference.

17.4 Remaining Challenges for Clinical Implementation

As reported in the first clinical investigations of in-beam, in-room, and offline PET monitoring of carbon ion and proton therapy (Enghardt et al., 2004a,b; Parodi et al., 2007c; Knopf et al., 2009; Zhu et al., 2011), the main factors for a correct prediction of the irradiation-induced activity are the knowledge of the reaction cross sections and of the tissue composition in the modeling of ion-induced β+-emitters (cf. Section 17.2), as well as the determination of the contribution to the measurable activity in the imaging process (cf. Section 17.3). Indeed, the modeling of ion transport and nuclear processes resulting in β+-activation together with the explicit or approximate description of the image formation process can be extensively validated and possibly fine tuned via irradiation and PET imaging of phantoms of known composition. However, challenges remain in clinical application for the determination of the elemental tissue composition and the modeling of the so-called biological washout, which is responsible for the fact that part of the activity produced in living tissue is washed away by physiological processes such as perfusion.

While the CT-based stoichiometric calibration proposed in Schneider et al. (2000) is typically sufficient for reliable dosimetric calculations (Espana and Paganetti, 2010) (cf. Chapter 14), it may miss details of the relative elemental composition for organs sharing the same radiological properties but exhibiting significant differences in the oxygen and carbon abundance. Especially in the case of proton therapy where the irradiation-induced activity is entirely ascribed to target fragments, a failure in the identification of the elemental composition results in an erroneous modeling of the activity not only in terms of signal strength but also beam range (as deduced from the distal activity fall-off) due to the different energy thresholds of the different activation channels (cf. Figure 17.3). In fact, changes in the carbon-to-oxygen ratio have been shown to result in almost 2 mm differences in the estimated proton range (Parodi et al., 2005), while range variations from 1 to 5 mm were reported in the MC studies of Espana and Paganetti (2010) and Espana et al. (2011) addressing the influence of different CT stoichiometric calibration in

combination with different reaction cross sections for different implementations of PET-based monitoring of proton therapy. In clinical application of offline PET imaging of proton therapy, observable discrepancies between the calculated and the measured activity (predominantly from ^{11}C) were reported and ascribed to an erroneous estimation of carbon abundance in tissue such as bone marrow (Knopf et al., 2009) and white brain matter (Parodi et al., 2007), which show remarkable deviation from the general CT-based calibration of Schneider et al. (2000). A possibility to improve tissue classification using complementary patient-specific information such as magnetic resonance imaging (MRI) has been recently reported (Parodi et al., 2011). For carbon ion therapy, the influence of tissue stoichiometry is less pronounced due to the major contribution from positron-emitting projectile fragments, especially in terms of range verification (Pönisch et al., 2004).

Regardless of the primary ion species, considerable differences between measured and calculated activity may occur due to the physiological processes carrying part of the formed isotopes even far away from the place of production. The relevance of these dynamic processes increases with the time distance between irradiation and imaging, introducing distortions of the physically formed activity that are especially pronounced in offline PET imaging (Figure 17.7). This justifies why washout processes could be in first approximation neglected in the clinical application of in-beam PET imaging of carbon ion treatments at GSI (Enghardt et al., 2004a,b), but had to be considered in the clinical investigations of in-room and offline PET-based verification of proton and carbon ion therapy at MGH (Parodi et al., 2007; Knopf et al., 2009; Zhu et al., 2011) and HIT (Parodi et al., 2011), respectively.

The effects of the biological washout can be included in the modeling of the image formation process, converting the simulated positron emitter yield into the expected β+-activity distribution. As the hot chemistry taking place immediately after the formation of β+-active isotopes in living tissue is very complex and not yet completely understood, compartment models well established in nuclear medicine are not applicable to low-statistics PET imaging of ion therapy. Nevertheless, the washout effects can be accounted for via the introduction of a biological decay in addition to the physical one, as initially suggested in Mizuno et al. (2003) on the basis of animal studies with implanted radioactive ions. For application to patient cases, two different approaches have been proposed in the literature for offline PET verification of proton therapy and in-beam PET monitoring of carbon ion therapy, respectively, both neglecting differences in the biological pathways of the different formed isotopes. The first approach (Parodi et al., 2007c) accounted for tissue-specific biological parameters in the postsimulation weighting of the MC-produced positron emitter yields contributing to the calculated activity in the considered time window (Figure 17.7). The second one (Fiedler et al., 2008) introduced an empirically determined dose-dependent effective half-life to correct the number of decays to be explicitly simulated as potentially giving rise to a detectable signal in the in-beam PET acquisition (Figure 17.8).

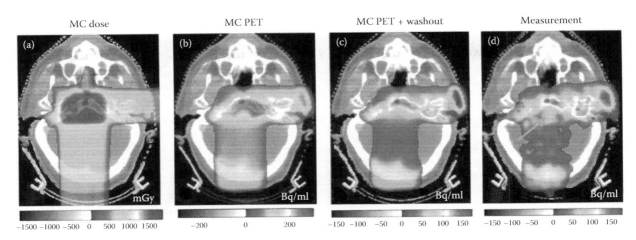

FIGURE 17.7   **(See color insert.)** FLUKA MC calculated dose ((a), color bar units of mGy) on the planning CT for the entire two field irradiation of the same patient of Figure 17.4 treated for a clivus chordoma at MGH. The activity distribution (color bar units of Bq/ml) obtained from the corresponding simulated positron emitter yields (cf. Figure 17.4) with the described weighting and smoothing approach (cf. Section 17.3) is shown when taking into account only the physical (b) or also the biological (c) decay in the calculation of the imaging process (Parodi et al., 2007c). The comparison with the PET/CT measurement (d) clearly indicates the improvement of the modeling in terms of both of spatial distribution and absolute signal intensity when taking biological processes into account. The weighting of the positron emitter yield also accounted for the different delays between completion of irradiation and start of the 30 min PET acquisition, which were 26 and 16 min for the first (posterior–anterior) and second (lateral) portal, respectively. The black–white color bar represents the CT number map arbitrary rescaled for display purposes. The arrow in panel (d) points to the measured activation of the brainstem not reproduced in the MC due to the mentioned shortcomings of the used CT stoichiometric calibration (cf. text). (Adapted from Parodi, K et al. *Int J Rad Onc Biol Phys* 2007c;68:920–934. With permission.)

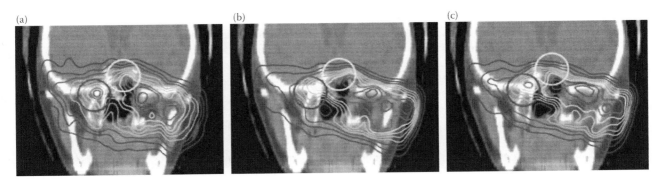

FIGURE 17.8   **(See color insert.)** Distributions of β+-activity superimposed onto the planning patient CT measured in-beam at GSI (a) and compared to the standard clinically used simulation approach of Pönisch et al.(2004) not considering the washout (b) as well as to a further development (Fiedler et al., 2008) introducing an empirically determined dose-dependent biological half-time in the imaging process (c). The qualitative improvement of the normalized distributions when taking biological processes into account is evident, especially in the regions marked by the circles. (Adapted from Fiedler, F et al. *Acta Oncol* 2008;47:1077–1086. With permission.)

Despite the encouraging qualitative results of both approaches illustrated by Figures 17.7 and 17.8, quantitative differences in the simulated activity distributions up to 30% in intensity and 4 mm in range were reported for offline PET imaging of proton therapy in anatomical regions especially subject to perfusion (Parodi et al., 2007; Knopf et al., 2009). Therefore, work is still ongoing at several institutions involved in clinical investigations of PET monitoring to improve the washout modeling, trying to gain patient-specific information from additional imaging modalities. However, these activities go beyond the topic of MC modeling of radiation transport and interaction addressed in this book.

17.5 Conclusion and Outlook

This chapter has reviewed the main implementations and issues of MC calculations for PET-based treatment verification of proton and carbon ion therapy. In particular, it has shown that different computational approaches have been historically pursued, with different approximations introduced in dependence of the considered ion type and PET instrumentation for the sake of fast execution time. Encouraging results could be so far demonstrated in several pilot investigations of in-beam, in-room, and offline PET monitoring of carbon ion and proton therapy at GSI, MGH, and HIT (Enghardt et al., 2004a,b; Parodi et al., 2007c, 2011; Zhu et al., 2011). Moreover, several ongoing research activities are focused toward the establishment of computationally efficient full MC simulation approaches mostly relying on general-purpose codes being thoroughly validated against available experimental measurements (Seravalli et al., 2012; Sommerer et al., 2009; Jan et al., 2011). In this respect, new experimental campaigns of β+-activation in phantoms of known composition would be highly beneficial (Seravalli et al., 2012; Espana et al., 2011). However, despite the promising results and ongoing developments, major challenges remain for clinical application that are mostly related to the knowledge of the tissue elemental composition and physiological properties. The former aspect is especially important for the modeling of β+-activity distributions limited to target fragments as in proton therapy. The latter issue of biological washout currently poses the most serious limitation to the indications which may benefit from PET monitoring especially in data acquisition taken long after irradiation, thus motivating new efforts for dedicated in-beam or in-room instrumentation (Shakirin et al., 2011) as well as the search for alternative range monitoring techniques such as real-time prompt gamma imaging or ion radiography before or in between treatment (cf. Chapter 18).

References

Agostinelli, S, Allison, J, Amako, K. et al. GEANT4—A simulation toolkit. *Nucl Instrum Methods Phys Res* 2003;506:250–303.

Battistoni, G, Muraro, S, Sala, PR, Cerutti, F, Ferrari, A, Roesler, S, Fassò, A, and Ranft, J. The FLUKA code: Description and benchmarking. In M. Albrow, R. Raja, eds. *Proceedings of the Hadronic Shower Simulation Workshop 2006*, Fermilab, 6–8 September 2006, AIP Conference Proceeding 2007;896:31–49.

Briesmeister, JF. MCNP—A general Monte Carlo N-particle transport code. Los Alamos National Laboratory Report No LA-12625-M, 1997.

Enghardt, W, Crespo, P, Fiedler, F, Hinz, R, Parodi, K, Pawelke, J, and Pönisch, F. Charged hadron tumour therapy monitoring by means of PET. *Nucl Instrum Methods A* 2004; 525:284–288.

Enghardt, W, Parodi, K, Crespo, P, Fiedler, F, Pawelke, J, and Pönisch, F. Dose quantification from in-beam positron emission tomography. *Radiother Oncol* 2004;73:S96–S98.

España, S and Paganetti, H. The impact of uncertainties in the CT conversion algorithm when predicting proton beam ranges in patients from dose and PET-activity distributions. *Phys Med Biol* 2010;55:7557–7571.

España, S, Zhu, X, Daartz, J, El Fakhri, G, Bortfeld, T, and Paganetti, H. The reliability of proton-nuclear interaction cross-section data to predict proton-induced PET images in proton therapy. *Phys Med Biol* 2011;56:2687–2698.

Ferrari, A, Fassò, A, Ranft, J, and Sala, PR. FLUKA: A multi-particle transport code. CERN-2005-10, INFN/TC_05/11, SLAC-R-773, 2005.

Fiedler, F, Priegnitz, M, Jülich, R, Pawelke, R, Crespo, P, Parodi, K, Pönisch, F, and Enghardt, W. In-beam PET measurements of biological half-lives of ¹²C irradiation induced β+-activity. *Acta Oncol* 2008;47:1077–1086.

Hasch, BG. Die physikalischen Grundlagen einer Verikation des Bestrahlungsplanes in der Schwerionen-Tumortherapie mit der Positronen-Emissions-Tomographie, PhD thesis, Dresden University of Technology, 1996.

Jäkel, O, Jacob, C, Schardt, D, Karger, CP, and Hartmann, GH. Relation between carbon ion ranges and x-ray CT numbers. *Med Phys* 2001;28:701–703.

Jan, S, Benoit, D, Becheva, E. et al. GATE V6: A major enhancement of the GATE simulation platform enabling modeling of CT and radiotherapy. *Phys Med Biol* 2011;56:881–901.

Knopf, A, Parodi, K, Bortfeld, T, Shih, HA, and Paganetti, H. Systematic analysis of biological and physical limitations of proton beam range verification with offline PET/CT scans. *Phys Med Biol* 2009;54:4477–4495.

Krämer, M, Jäkel, O, Haberer, T, Kraft, G, Schardt, D, and Weber, U. Treatment planning for heavy-ion radiotherapy: Physical beam model and dose optimization. *Phys Med Biol* 2000;45:3299–3317.

Matsufuji, N, Kanai, T, Kanematsu, N, Miyamoto, T, Baba, M, Kamada, T, Kato, H, Yamada, S, Mizoe J, and Tsujii, H. Specification of carbon ion dose at the National Institute of Radiological Sciences (NIRS). *J Rad Res* 2007;48:A81–A86.

Miyatake, A, Nishio, T, and Ogino, T. Development of activity pencil beam algorithm using measured distribution data of positron emitter nuclei generated by proton irradiation of targets containing (12)C, (16)O, and (40)Ca nuclei in preparation of clinical application. *Med Phys* 2011;38:5818–5829.

Mizuno, H, Tomitami, T, Kanazawa, M. et al. Washout measurements of radioisotope implanted by radioactive beams in the rabbit. *Phys Med Biol* 2003;48:2269–2281.

Niita, K, Sato, T, Iwase, H. et al. PHITS—A particle and heavy ion transport code. *Radiat Meas* 2006;41:1080–1090.

Nuclear Reaction Data Center network which is coordinated by the IAEA Nuclear Data Section, 2000. Available at http://www.nndc.bnl.gov/exfor/exfor00.htm

Parodi, K. On the feasibility of dose quantification with in-beam PET data in radiotherapy with 12C and proton beams, PhD thesis, Dresden University of Technology, in Forschungszentrum Rossendorf Wiss-Techn-Ber FZR-415, 2004.

Parodi K and Bortfeld T. A filtering approach based on Gaussian-powerlaw convolutions for local PET verification of proton radiotherapy. *Phys Med Biol* 2006;51:1991–2009.

Parodi, K, Bauer, J, Kurz, C, Mairani, A, Sommerer, F, Unholtz, D, Haberer, T, and Debus, J. Monte Carlo modeling and in-vivo imaging at the Heidelberg Ion Beam Therapy Center. *Conference Record of the IEEE Nuclear Science Symposium and Medical Imaging Conference in Valencia*, Spain, 2011.

Parodi, K, Enghardt, W, and Haberer, T. In-beam PET measurements of β+ radioactivity induced by proton beams. *Phys Med Biol* 2002;47:21–36.

Parodi, K, Ferrari, A, Sommerer, F, and Paganetti, H. Clinical CT-based calculations of dose and positron emitter distributions in proton therapy using the FLUKA Monte Carlo code. *Phys Med Biol* 2007a;52:3369–3387.

Parodi, K, Paganetti, H, Cascio, E, Flanz, J, Bonab, A, Alpert, N, Lohmann, K, and Bortfeld, T. PET/CT imaging for treatment verification after proton therapy—A study with plastic phantoms and metallic implants. *Med Phys* 2007b;34:419–435.

Parodi, K, Paganetti, H, Shih, HA. et al. Patient study on in-vivo verification of beam delivery and range using PET/CT imaging after proton therapy. *Int J Rad Onc Biol Phys* 2007c;68:920–934.

Parodi, K, Pönisch, F, and Enghardt, W. Experimental study on the feasibility of in-beam PET for accurate monitoring of proton therapy. *IEEE Trans Nucl Sci* 2005;52:778–786.

Parodi, K, Saito, N, Chaudhri, N, Richter, R, Durante, M, Enghardt, W, Rietzel, E, and Bert, C. 4D in-beam positron emission tomography for verification of motion-compensated ion beam therapy. *Med Phys* 2009;36:4230–4243.

Pönisch, F, Parodi, K, Hasch, BG, and Enghardt, W. The description of positron emitter production and PET imaging during carbon ion therapy. *Phys Med Biol* 2004;49:5217–5232.

Priegnitz, M, Fiedler, F, Kunath, D, Laube, K, and Enghardt, W. An experiment-based approach for predicting positron emitter distributions produced during therapeutic ion irradiation. *IEEE Trans Nucl Sci* 2011;99:1–11.

Pshenichnov, I, Larionov, A, Mishustin, I, and Greiner, W. PET monitoring of cancer therapy with ³He and ¹²C beams: A study with the GEANT4 toolkit. *Phys Med Biol* 2007;52:7295–7312.

Remmele, S, Hesser, J, Paganetti, H, and Bortfeld T. A deconvolution approach for PET-based dose reconstruction in proton radiotherapy. *Phys Med Biol* 2011;56:7601–7619.

Schaffner, B and Pedroni, E. The precision of proton range calculations in proton radiotherapy treatment planning: Experimental verification of the relation between CT-HU and proton stopping power. *Phys Med Biol* 1998;43: 1579–1592.

Schneider, W, Bortfeld, T, and Schlegel, W. Correlation between CT numbers and tissue parameters needed for Monte Carlo simulations of clinical dose distributions. *Phys Med Biol* 2000;45:459–478.

Schulz-Ertner, D and Tsujii, H. Particle radiation therapy using proton and heavier ion beams. *J Clin Oncol* 2007;25:953–964.

Seravalli, E, Robert, C, Bauer, J. et al. Monte Carlo calculations of positron emitter yields in proton radiotherapy. *Phys Med Biol* 2012;57:1659–1673.

Shakirin, G, Braess, H, Fiedler, F, Kunath, D, Laube, K, Parodi, K, Priegnitz, M, and Enghardt, W. On implementation and workflow for PET monitoring of therapeutic ion irradiation. *Phys Med Biol* 2011;56:1281–1298.

Sihver, L, Tsao, CH, Silberberg, R, Kanai, T, and Barghouty, AF. Total reaction and partial cross section calculations in proton-nucleus (Zt ≤ 26) and nucleus–nucleus reactions (Zp and Zt ≤ 26). *Phys Rev C* 1993;47:1225–1236.

Silberberg, R and Tsao, CH. Partial cross-section in high-energy nuclear reactions, and astrophysical applications: I. Targets with Z ≤28. *Astrophys J Suppl* 1973;25:315–333.

Sommerer, F, Cerutti, F, Parodi, K, Ferrari, A, Enghardt, W, and Aiginger, H. In-beam PET monitoring of mono-energetic ¹⁶O and ¹²C beams: Experiments and FLUKA simulations for homogeneous targets. *Phys Med Biol* 2009;54: 3979–3996.

Zhu, X, España, S, Daartz, J, Liebsch, N, Ouyang, J, Paganetti, H, Bortfeld, TR, and El Fakhri, G. Monitoring proton radiation therapy with in-room PET imaging. *Phys Med Biol* 2011;56:4041–4057.

18

Monte Carlo Studies of Prompt Gamma Emission and of Proton Radiography/Proton-CT

Jerimy C. Polf
Oklahoma State University

Nicholas Depauw
Massachusetts General Hospital

Joao Seco
Harvard Medical School

18.1 State of the Art in Prompt Gamma Research

In addition to the previously discussed methods for using positron emission tomography (PET) imaging techniques to measure induced positron annihilation gamma emission in the patient, researchers have also studied imaging prompt gamma (PG) emission from the patient, as a means of verifying proton and ion beam treatment delivery. During proton or heavy ion therapy, primary particles from the treatment beam interact with elemental nuclei in the tissue, which can leave behind an intact nucleus in an excited state. This excited state quickly ($<10^{-9}$ s) decays by emitting a characteristic PG. Since the PG emission only occurs in regions in which the treatment beam interacts in the tissue, the distribution of PG creation in the patient is correlated to the distribution of dose delivery to the patient. Also, since the excited nuclear states are quantized, each element emits a unique and characteristic PG spectrum. Thus, spectral analysis of PG emitted from tissue may provide useful information about the elemental concentration, composition, and type of the tissues irradiated during treatment.

The possible applications of treatment delivery verification and tissue composition analysis have led to an increasing number of studies, particularly Monte Carlo (MC) studies, of PG emission. In fact, many researchers have made use of the specific

capabilities of many MC codes such as MCNPX (Waters et al., 2005), Geant4 (Agostinelli et al., 2003), and FLUKA (Fasso et al., 2003) to study PG emission during proton beam and heavy ion beam therapy. MC has proven to be a very powerful tool for these studies, as it has allowed the users to study a wide range of topics, including characterization of PG emission from proton-irradiated phantoms and tissues, detector design, image reconstruction, and clinical applications such as proton beam range verification and tissue identification.

18.2 Basic Properties of PG Emission

An important first step in the development of PG imaging techniques is the study and characterization of the basic properties of PG emission. MC is particularly well suited for these studies due to the availability of a wide range of physics models and data cross-section tables. MC also offers a large degree of flexibility for tracking and recording the characteristics of emitted PG gammas.

For this purpose, Polf et al. (2009a) used the MCNPX MC code to perform an initial investigation of PG emission from biological tissues undergoing proton irradiation to investigate its potential use for *in vivo* analysis of irradiated tissues. For this study, small tissue phantoms were defined within the model and irradiated with therapeutic proton beams. The spectra of emitted PG gammas from several different tissue types and several of the major elemental constituents were recorded as a function of position around the phantom. The physics defined in these simulations included energy straggling, multiple Coulomb scattering, and elastic and nonelastic nuclear interactions (including nuclear deexcitation processes). This study showed significant differences in the PG spectra emitted from muscle, lung, adipose, and bone tissues due to the differences in their composition and density. Also, the dependence of the PG emission intensity (production cross section) as a function of proton beam energy was calculated. These MC calculations indicated that a maximum PG emission cross section for all tissue types studied occurred for proton beam energies ranging from 20 to 30 MeV, with the PG production steadily decreasing as the beam energy approached the highest energies (~250 MeV) currently used in clinical proton therapy.

18.3 Range Verification and Tissue Composition Analysis

The main clinical application of PG emission studied to date has been its use in verifying the depth of the proton Bragg peak within the patient during treatment. Many initial studies focused on the correlation between the range of PG production in water and tissue phantoms and the proton beam Bragg peak using MC calculations (Min et al., 2006; Feng et al., 2007, 2010; Kim et al., 2009; Polf et al., 2009a; Le Foulher et al., 2010). In these studies, the researchers have used many flavors of MC, including MCNPX, FLUKA, and Geant4 to track and tally the location within the phantom at which the PG gamma was created and record these data in 1D, 2D, or 3D histogram files. The MC calculated PG emission distributions were compared to the MC calculated distributions of delivered dose. Results of these studies in tissue phantoms have all shown that the range of PG production and the proton beam dose distribution are strongly correlated, with the peak PG emission occurring ~2–4 mm proximal to the Bragg peak.

In an MC study using several clinical cases, Moteabbed et al. (2011) used Geant4 to study the correlations between PG production, annihilation gamma production (used in posttreatment PET imaging), and the dose delivered within the patient during treatment delivery. For these studies, the patient CT data as well as all pertinent information for treatment delivery were imported from a clinical treatment planning system into an MC model of a proton therapy delivery nozzle. MC calculations were performed to simulate the patient treatment delivery and both the prompt and annihilation gamma yields within the patient were tallied. Results from these studies show a strong correlation between PG production and the *in vivo* dose distribution for a wide range of clinical proton therapy cases. As an example, the calculated data for a prostate cancer case is shown in Figure 18.1. For this case, the range and lateral extent of the PG production in the patient correlates well with the distribution of dose delivered by the treatment beam. Along with the PG–dose correlation, the study by Moteabbed et al. showed that the intensity of the PG production was strongly dependent on the density of the tissue (i.e., bone, muscle, and lung) and the presence of large heterogeneities such as bone–air interfaces. Results of this study helped to show the feasibility of using PG emission to verify *in vivo* proton beam range in a clinical setting.

In addition to range verification, Polf et al. (2011) used the MCNPX MC code to study the potential for using PG emission to determine the concentration of the major elemental constituents of irradiated tissues. In this study, an MC model of a clinical proton therapy beam irradiating a large tumor within a patient was used. The CT dataset of a patient with a large lung tumor was loaded into the MC model with the material definition and density of the patient's normal tissue and tumor being defined according to their ICRU definitions (ICRU, 1993) and the tumor defined as ICRU average soft tissue. To simulate compositional changes due to tumor hypoxia, the percent concentration of oxygen within the tumor was then systematically reduced, by adjusting the material definition card in the MCNPX input file accordingly, over a series of MC calculations. For each MC calculation, at a given percent oxygen concentration in the tumor, the characteristic PG emission from oxygen produced in the tumor was tallied. From these calculations, it was shown that the production of PG gammas from oxygen decreased as a logarithmic function of oxygen concentration within the tumor.

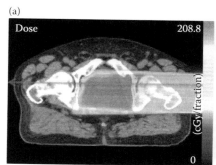
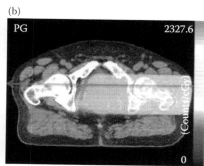

FIGURE 18.1   (a) Proton treatment plan dose on a CT image for a patient treated for prostate cancer. (b) PG emission overlaid on the same CT image. Comparison of the dose and PG distribution shows a strong correlation between dose and PG emission. (Adapted from Moteabbed, M., S. Espana, and H. Paganetti, *Phys. Med. Biol.*, 2011. **59**: 1063–1082.)

18.4 Detector Design Studies

Favorable results from preliminary MC studies of PG production in irradiated tissues have led many researchers to study the design of efficient PG detectors for use in clinical proton and heavy ion therapy applications. Since characteristic PGs emitted from tissue range in energy from 2 to 15 MeV, current gamma ray detectors (mostly designed for lower-energy applications) suffer from low detection efficiency. Since the total number of PGs emitted during radiotherapy is limited due to short patient irradiation times, these low efficiency detectors have not shown the ability to measure enough PG emission data to produce statistically relevant and useful images. In addition, the higher energy of the PG emission makes standard physical collimators used for low-energy medical and nuclear imaging applications ineffective for encoding spatial information about the PG emission within the measured data necessary for reconstructing images. Therefore, a large effort has recently been put forth to design efficient and effective detectors capable of measuring spatial and initial energy information of PG emission from tissue under clinic conditions.

Many recent MC studies have focused on the design and development of detection systems specifically for measuring and imaging PG emission from tissue during proton and ion beam therapy. This includes studies simulating a number of different designs of two-stage (Feng et al., 2008; Kang and Kim, 2009; Frandes et al., 2010; Roellinghoff et al., 2011) and three-stage Compton cameras (Richard et al., 2009; Peterson et al., 2010; Robertson et al., 2011), as well as pinhole and slit cameras (Min et al., 2006; Kim et al., 2009; Bom et al., 2012), and multislit collimated array detectors (Min et al., 2012) using a wide range of materials and detector configurations. The main purposes of these studies were to optimize the PG detection efficiency of the modeled detectors.

For example, Figure 18.2 shows the designs for two different proposed PG detection systems, a three-stage Compton camera and a slit camera, studied by Peterson et al. (2010) and Bom et al. (2012), respectively, using the Geant4 MC toolkit. For these studies, MC calculations were used to determine the feasibility

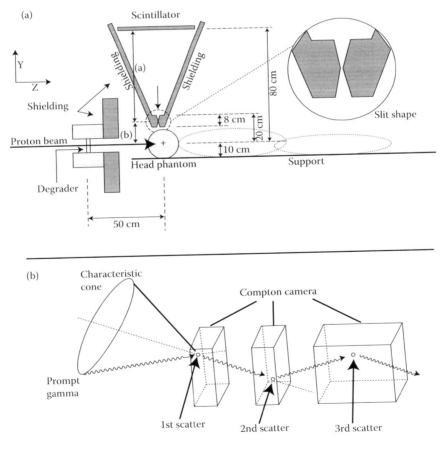

FIGURE 18.2 (a) Cross-section picture of geometry used for an MC design study of a slit camera for PG imaging during proton radiotherapy. (Adapted from Bom, V., F. Joulaeisadeh, and F. Beekman, *Phys. Med. Biol.*, 2012. **57**: 297–308.) (b) Schematic drawing of a three-stage Compton camera design similar to that studied by Peterson et al. (2010) for use in PG imaging during proton radiotherapy. (Adapted from Peterson, S.W., D. Roberts, and J.C. Polf, *Phys. Med. Biol.*, 2010. **55**: 6841–6856.)

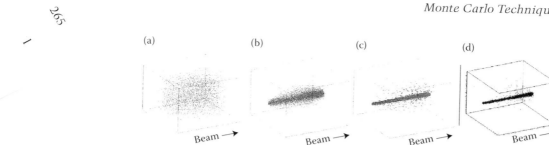

FIGURE 18.3 3D image reconstruction of model PG emission during proton beam irradiation. Detection of PG emission was simulated using Monte Carlo. The calculated data was used as input data to aid in the development of a new iterative image reconstruction code for PG emission. Shown are (a) the initial (0 iterations) image state, (b) the image after 1000, and (c) 10,000 iterations, compared to (d) the actual distribution of PG production in tissue.

of detecting an adequate PG signal under clinical beam delivery conditions to allow for determination of the proton beam range. This was accomplished by using MC simulations to track PGs emitted from the irradiated tissue phantom as they propagate through the modeled detectors and count the number of individual interactions for each PG, as well as record the position and energy deposition of each interaction. For the Compton camera (Figure 18.2b), Peterson et al. performed a series of MC calculations in which they varied the width and thickness of each stage, as well as the spacing between each stage of the camera to determine the specific design that optimized the detected signal from characteristic oxygen (6.13 MeV), carbon (4.44 MeV), and nitrogen (2.31 MeV) PGs. In contrast, Bom et al. used MC to study and optimize the energy deposition thresholds used to trigger a measurement in their detector, which would allow them to discriminate between incident PG emission and secondary neutrons and scattered protons and gammas that interact with the detector during irradiation. From the results of these MC studies, both Peterson et al. and Bom et al. were able to determine optimal design characteristics of their detectors. Using these optimal detector designs, they were then able to calculate the overall detection efficiency of the modeled detectors under clinical proton beam delivery conditions, and thus make initial conclusions as to their feasibility for use in proton beam range verification.

18.5 Image Reconstruction Studies

Additionally, the PG interaction data from the MC calculations have been used by several researchers to study methods of reconstructing images of the PG emission from irradiated phantoms. This includes study of the gamma ray tracking (GRT) technique (Deleplanque et al., 1999; Schmind et al., 1999) and stochastic origin ensemble (SOE) technique (Sitek et al., 2008; Andreyev et al., 2011) used in conjunction with MC calculated data from a Compton camera, and with filtered back projection (FBP) used with MC calculated data from a collimated gamma camera (similar to clinical single-photon emission computed tomography systems). With the MC modeling, the physical characteristics of the detectors can be simulated, including pixel size, Doppler broadening, energy resolution, and polarization effects. The effects these characteristics have on the achievable spatial

resolution and noise levels of PG measurements and the resulting effects on images reconstructed with GRT (Feng et al., 2008; Richard et al., 2009; Frandes et al., 2010) and FBP (Park et al., 2010) have been studied, by varying the their magnitude over a series of MC calculations.

Additionally, simulated PG data from a modeled three-stage Compton camera were used by Mackin et al. (2012) as input data for design/development studies of a new reconstruction method tailored specifically for PG image reconstruction. In this study, the SOE algorithm originally defined by Sitek (2008) for detection of lower-energy (<1 MeV) gammas was adapted to work with the specific characteristics of PG emission and detection (with a Compton camera) during proton radiotherapy. The data files output by the MC model were formatted to mimic the "list-mode" data file output by several experimental Compton cameras. The MC data files contained a list of the position and energy deposition of each PG interaction in each stage of the camera. The MC data was then used in lieu of actual measured data as input data for the SOE reconstruction code. Images of the PG emission reconstructed using the SOE code with input data calculated by the MC model are shown in Figure 18.3. In addition to recording the PG interaction positions in the camera, the MC model also recorded the positions where each PG was emitted (Figure 18.3d). The quality of the images produced by the SOE algorithm from the CC data was quantified by comparing the images to the distribution of the recorded emission positions. In this manner, the effect of each SOE design parameter on the quality of the reconstructed images could be directly tested during each phase of the algorithm development.

18.6 Validation of Monte Carlo-Based PG Calculations

An important consideration for the use of MC in the study of PG emission applications in radiotherapy is the proper tuning of all parameters of the MC model (including the model geometry, material definitions, particle tracking, and physics models) to ensure that the calculated data can accurately reproduce the physical processes occurring during treatment delivery. This validation process is usually performed through the direct comparison of calculated results from the MC model and measured data. Unfortunately, validation studies of MC models used for PG emission studies have been slow to be published. This is most likely due to the scarcity of beamtime available for experimental measurements at proton and carbon ion therapy facilities, leading to a lack of direct measured data of total PG emission rates and PG spectra for use in MC validation studies.

In one such study, Polf et al. (2009b) performed validation studies of an MC model used for calculating the spectrum of PGs emitted from tissue phantoms during proton irradiation. For this study, an MC model of an experimental setup consisting of a shielded high-purity germanium detector setup to measure PG emission from small plastic phantoms irradiated with a proton pencil beam was developed using Geant4. The MC calculations included physics models for gamma interactions ranging from 250 eV to 10 TeV, and both elastic and nonelastic proton–nuclear interactions (including an intranuclear cascade deexcitation model). With this MC model, Polf et al. (2009b) were able to reproduce the measured PG spectra from irradiated soft tissue and bone equivalent plastic phantoms, as shown in Figure 18.4. This included the presence of the measured PG emission lines from the main elemental constituents of the phantoms, namely, oxygen, carbon, and calcium. However, the nuclear scattering models (at the time of this study) were not able to correctly calculate the nuclear Doppler broadening of the PG emission lines present in the measured spectra as seen in the 4.44 MeV emission line from carbon in Figure 18.4.

In another validation study, Le Foulher et al. (2010) compared measured PG count rates and spectral emission from plastic and water phantoms during carbon ion irradiation to calculations from an MC model. In this study, the researchers used Geant4-based MC model that recreated the geometry of the experimental setup (including detectors, shielding, and phantoms) used for the measurements, as well as the parameters of the carbon

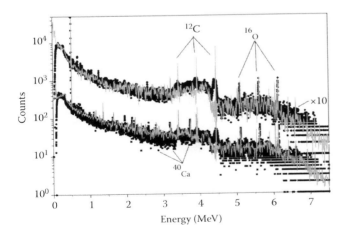

FIGURE 18.4 Comparison of measurement (black symbols) to MC calculations (gray lines) of secondary gamma emission spectra from Lucite (upper curves (×10 for display purposes)) and bone equivalent plastic (lower curves) phantoms irradiated with clinical proton beams. (Adapted from Polf, J.C. et al., *Phys. Med. Biol.*, 2009. **54**: N519–N527. With permission.)

ion beam used for the irradiations. The MC model included all standard electromagnetic processes that dictated the PG propagation and interactions, as well as nuclear physics packages to model interaction processes, including nuclear scattering, fragmentation, and deexcitation of the excited nucleus.

The results of these studies showed that the nuclear physics packages used in their model greatly overestimated the PG production in the phantoms during carbon ion irradiation. However, their investigation and comparison of the fragmentation cross sections within the nuclear physics models showed very good agreement with experimental measurements. Therefore, the discrepancies in the measured and MC calculated PG count rates were attributed to uncertainties in the gamma-ray multiplicity values contained within the nuclear models. These results highlight the importance of continued testing and refinement of the MC physics models to ensure accurate and usable results for MC studies of PG emission during proton and ion beam therapy.

18.7 State of the Art of Proton Radiography

In 1946, Robert Wilson (Wilson 1946) recognized the potential for proton beam therapy due the inverse relationship between the ionization energy loss and the energy of a proton beam passing through tissue. Subsequently, proton radiography was first investigated as a radiologic tool in the early 1970s and 1980s (Steward and Koehler, 1973; Hanson et al., 1982). However, the success of x-ray radiography, which quickly became the most common cancer diagnostic tool in hospitals, rapidly overshadowed proton radiography. Nevertheless, despite a difficult start, proton radiotherapy, as well as heavy ion therapy, is becoming more and more attractive to the radiation oncology community as technology improves. Indeed, protons present a finite range in tissue that can allow for better tumor conformality and therefore better sparing of the surrounding healthy tissue. Although protons' intrinsic imaging capabilities are limited (Schneider and Pedroni, 1994), by using higher energies than the ones for treatment, ion radiography further takes advantage of the aforementioned plateau region and is able to produce patient setup images while delivering significantly less imaging dose than conventional x-rays (Depauw and Seco, 2011).

18.8 Protons versus X-Ray

The subject of proton versus conventional x-rays has been the object of a multitude of studies (Agostinelli et al., 2003; Ryu et al., 2008; Depauw and Seco, 2011; Depauw et al., 2011), especially with MC as it allows a fair comparison between the two modalities without taking into account the current state of the art in detector technology. Current technology is obviously much more advance on the x-ray side as it benefits from more than a hundred years of study, as well as the broad development of photon radiotherapy centers all over the world. When comparing protons with photons for imaging purposes, it is customary to define two main metrics: spatial resolution and density resolution. In

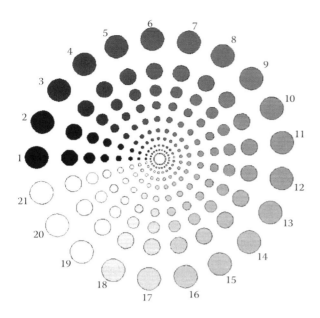

FIGURE 18.5 MC-based water phantom filled with rows of 21 different materials with densities ranging from 0.001 g/cm³ (air—#1) to 1.92 g/cm³ (bony material—#21). This phantom was designed to properly compare protons with photons for imaging purposes.

a recent study (Depauw and Seco, 2011) using the MC toolkit GEANT4 (Schneider and Pedroni, 2012), an imaging phantom was specifically designed to properly compare these aforementioned resolutions. Figure 18.5 presents the $30 \times 30 \times 8$ or 15 cm³ water phantom with 21 rows of cylindrical inserts made of different materials and densities: air (0.001 g/cm³), lung type tissue (0.1, 0.2, 0.3, 0.4, 0.5, 0.6, 0.7, and 0.8 g/cm³), adipose tissue (0.9 g/cm³), soft tissue (1 g/cm³), and bony type tissues inserts (1.1, 1.2, 1.29, 1.39, 1.49, 1.58, 1.67, 1.77, 1.86, and 1.92 g/cm³). For each material, the phantom contains a total of 10 inserts with diameters ranging from 20 to 0.88 mm, decreasing in size. Different simulations considered different phantom thickness (8 or 15 cm), different insert thickness (2, 4, or 8 cm), as well as different incident proton or photon energy (200, 300, 400, and 490 MeV, and 50 keV, 100 keV, 1 MeV, and 2 MeV, respectively). Similar imaging doses were considered across modalities and energies, thus giving an adequate point of reference for an imaging capabilities comparison study. Figure 18.6 presents three radiographs with three modalities for similar delivered doses.

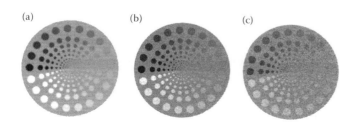

FIGURE 18.6 Reconstructed phantom radiographs using (a) a 200-MeV proton beam, (b) a 50-keV x-ray beam, and (c) a 1-MeV x-ray beam for similar absorbed dose.

18.9 Proton Radiography Image Reconstruction Differences

The processes for generating radiographs from protons and from photon are very different. Proton radiographs are "energy maps." In each pixel, the imaged value is the range of the absorbed energy R_{abs} corresponding to the range in water of the mean energy loss of all the protons going through this particular pixel. This mean absorbed energy E_{abs} has to be computed on a proton-by-proton basis, and different techniques are considered to evaluate the path of each particle such as "the most likely path" (Schneider and Pedroni, 1994). When considering a proton-by-proton system, one would be confronted with the problem of secondary protons generated through nuclear interaction in the patient (~1%/cm). These secondary protons present large scattering angles as well as lower energies and therefore constitute an important source of noise, as shown in Figure 18.7. By evaluating radiographs with energy and angular cuts, it is shown that the effect of secondary protons can be reduced. Another recent study suggests that the air gap between the subject and the radiography system would also play an important role in the final image quality that can be achieved (Schulte, 2005). On the other hand, x-ray radiographs are simple "fluence maps": in each pixel, the imaged value is the number of particles going through this

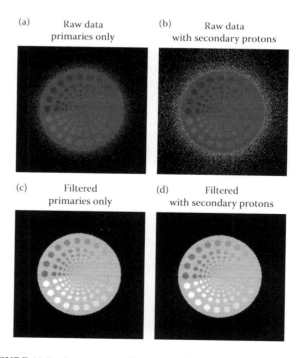

FIGURE 18.7 Reconstructed proton radiographs with, (a) primary protons only, (b) primary and secondary protons, (c) primary protons with angular and energy cuts, and (d) primary and secondary protons with angular and energetic cuts. This shows the importance of correcting for secondary protons to significantly reduce the noise and enhance the output of a proton radiography system, in a way such as the antiscatter grid in an actual x-ray.

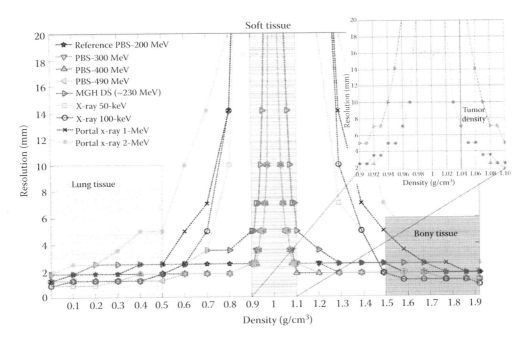

FIGURE 18.8 Spatial resolution for radiographs using different scanned proton pencil beam energies: 200, 300, 400, and 490 MeV, the MGH double scattered (~230 MeV) proton beam, and a 50-keV, a 100-keV, a 1-MeV, and a 2-MeV x-ray beam.

pixel. This study also incorporated a 2° angular cut in order to ameliorate the image quality by appropriately simulating the antiscatter grid of an actual x-ray system, and make the comparison more meaningful.

18.10 Spatial Resolution and Density Resolution

The spatial resolution is defined "broadly" as the minimum size in millimeters needed for a specific region of interest to be visually resolvable. It was therefore retrieved as the smallest visible insert for each substudy.

The authors proceeded by computing the contrast-to-noise ratio (CNR) for each insert in each substudy as

$$CNR = 20 * \log\left(\frac{|f - b|}{\sqrt{\sigma_f^2 + \sigma_b^2}}\right), \tag{18.1}$$

where f is the mean signal value in the region of interest, b is the average background noise, and σ_f and σ_b are their respective standard deviations. The CNR is commonly used as a "figure of merit" in the characterization of imaging systems. The CNR corresponds to the difference in signal-to-noise ratios (SNR) between two adjacent regions of interest and describes the ability to properly distinguish an area of interest relative to background. In the case of radiotherapy, this would define the ability to distinguish cancerous tissue relative to its surrounding normal tissue.

Figure 18.8 depicts the difference in spatial resolution between protons and photons. It includes results from one substudy with a defined phantom thickness and inserts thickness, with the entire set of both proton and x-ray energies.

18.11 Proton Radiography, kV and MV Image Quality for Identical Integral Dose

The image quality of kV and MV x-ray radiographs and proton radiographs was compared for approximately identical integral absorbed doses in a water phantom, using GEANT4 MC (Agostinelli et al., 2003). Five million protons and 50 millions x-rays were simulated through a similar-sized water tank ($30 \times 30 \times 8$ or 15 cm³), and the total energy deposited to the volume was recorded. This simulation permitted the assessment of the number of incident particles per modality required to obtain approximately equal absorbed dose. The authors concluded that 50 millions photons at 100 keV deliver approximately the same absorbed dose to the 15-cm-thick phantom as approximately 30,000 200 MeV protons or 55,000 490 MeV protons.

In Figure 18.9, the contrast-to-noise ratio, CNR, is presented for the proton radiography, kV and MV x-ray images obtained with identical integral dose. Protons provide better CNR results than kV and MV for significantly lower integral doses. For low- and high-density regions ($\rho < 0.6$ g/cm³ or $\rho > 1.5$ g/cm³), CNR values are excellent for protons, and acceptable for x-rays. In the medium density region (0.6 g/cm³ $< \rho <$ 1.5 g/cm³), MV x-rays yield very poor CNR levels.

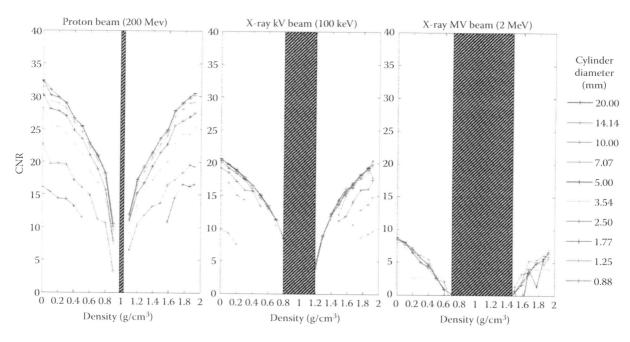

FIGURE 18.9 Contrast-to-noise ratio comparison between radiographs with a scanned pencil 200-MeV proton beam, a pure 100-keV x-ray beam, and pure 2-MeV x-ray beams. Each line corresponds to the same ROI area for each material density (legend is cylinder diameter); the hashed area reflects the density range for which no inserts were visible.

18.12 Proton Radiography: A Step toward Proton-CT

By imaging from different angles, proton radiography opens the door to proton computed tomography (pCT). One difficulty then resides in defining a proper reconstruction algorithm, which was investigated through MC simulations in various studies (Schulte et al., 2005; Li et al., 2006). Various reconstruction techniques

were investigated: (i) straight line path (SLP), (ii) most likely path (MLP), and (iii) cubic spline path (CSP) using an algebraic reconstruction technique (ART). The straight line path is the simplest (though most naïve) solution of the path the proton travels through any media. It assumes a straight line connecting the entrance and exit points of the proton. The MLP approach identifies the likeliest path based on the proton energy, direction of motion, and the entrance and exit positions. The CSP approach used a cubic spline

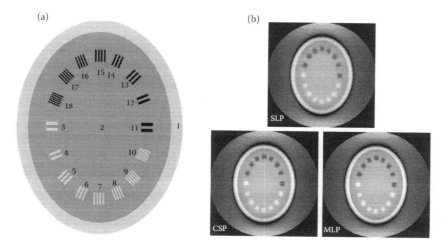

FIGURE 18.10 (a) Phantom used for pCT reconstruction algorithm. It consists of an outer shell with bone density (1), with a water inner filling (2), and strips of either air (black) or bone (white) at different densities: 2 lp/cm (3, 11), 2.5 lp/cm (4, 12), 3 lp/cm (5, 13), 3.5 lp/cm (6, 14), 4 lp/cm (7, 15), 4.5 lp/cm (8, 16), 5 lp/cm (9, 17), and 6 lp/cm (10, 18). (b) Reconstruction of the phantom with the ART algorithm using different proton path estimation approaches.

fit that is applied to the entrance and exit positions to estimate the proton's path length. The MLP and CSP approaches utilize ART reconstruction technique, an iterative solution of the projected relative electron density (or proton stopping power) along a single path. The *N* projections of relative electron density (or proton stopping) are used to generate $N \times N$-dimensional reconstructed vector representing an object slice or slices.

Figure 18.10a shows the elliptical phantom used to test the various reconstruction methods, where the line pairs are composed of either air (black) or bone (white). In Figure 18.10b, the reconstructed elliptical phantom is provided using GEANT4 MC calculation algorithm. The SLP provides the worst estimate of the pCT prediction, where a resolution of 2.5 lp/cm was observed. The spatial resolution of the air density and bone density patterns for the MLP and CPS approaches was 5 and 4.5 lp/cm, respectively. The ART reconstruction with either MLP or CSP generated comparable quality pCT images.

References

Agostinelli, S. et al., Geant4—A simulation toolkit. *Nucl. Instrum. Methods Phys. Res. A*, 2003. **506**: 250–303.

Andreyev, A., A. Sitek, and A. Celler, Fast image reconstruction for Compton camera using stochastic origin ensemble approach. *Med. Phys.*, 2011. **38**(1): 429–435.

Bom, V., F. Joulaeisadeh, and F. Beekman, Real-time prompt gamma monitoring in spot-scanning proton therapy using imaging through a knife-edge-shaped slit. *Phys. Med. Biol.*, 2012. **57**: 297–308.

Deleplanque, M.A. et al., GREAT: Utilizing new concepts in gamma-ray detection. *Nucl. Instrum. Methods Phys. Res. A*, 1999. **430**: 292–310.

Depauw, N. et al., Preliminary study of proton radiography imaging qualities using GEANT4 Monte Carlo simulations. *Nucl. Tech.*, 2011. **175**: 1, 6–10.

Depauw, N. and J. Seco, Sensitivity study of proton radiography and comparison with kV and MV x-ray imaging using GEANT4 Monte Carlo simulations. *Phys. Med. Biol.*, 2011. **56**: 2407–2421.

Fasso, A. et al., *The FLUKA Code: Present Applications and Future Developments*, in *CHEP032003*, La Jolla, CA, USA.

Feng, Y., J.E. Baciak, and A. Haghighhat, Dose verification in proton therapy via imaging gamma ray emission. *Med. Phys.*, 2007. **34**(6): 2574–2576.

Feng, Y., J.E. Baciak, and A. Haghighhat, A design of Compton cameras for imaging gamma emission in proton therapy. *Med. Phys.*, 2008. **35**(6): 2898.

Frandes, M. et al., A tracking Compton-scattering imaging system for hadrom therapy monitoring. *IEEE Trans. Nucl. Sci.*, 2010. **57**(1): 144–150.

Hanson, K. et al. Proton computed tomography of human specimens. *Phys. Med. Biol.*, 1982. **27**: 25.

ICRU, *Clinical Proton Dosimetry Part I: Beam Production, Delivery, and Measurement of Absorbed Dose*, 1993, International Commission on Radiation Units and Measurements: Bethesda, MD.

Kang, B.H. and J.W. Kim, Monte Carlo design study of a gamma detector system to locate distal dose falloff in proton therapy. *IEEE Trans. Nucl. Sci.*, 2009. **56**: 46–50.

Kim, D., H. Yim, and J.-W. Kim, Pinhole camera measurements of prompt gamma rays for detection of beam range in proton therapy. *J. Kor. Phys. Soc.*, 2009. **55**(4): 1673–1676.

Le Foulher, F. et al., Monte Carlo simulations of prompt-gamma emission during carbon ion irradiation. *IEEE Trans. Nucl. Sci.*, 2010. **57**: 2768–2772.

Li, T. et al., Reconstruction for proton computed tomography by tracing proton trajectories: A Monte Carlo study. *Med. Phys.*, 2006. **33**: 3, 699–706.

Mackin, D. et al., Evaluation of a stochastic reconstruction algorithm for use in Compton camera imaging and beam range verification from secondary gamma emission during proton therapy. *Phys. Med. Biol.*, 2012. **57**: 3537–3553.

Min, C.H. et al., Prompt gamma measurements for locating dose falloff region in proton therapy. *App. Phys. Lett.*, 2006. **89**: 183517: 1–3.

Min, C.H. et al., Development of array-type prompt gamma measurement system for *in vivo* range verification in proton therapy. *Med. Phys.*, 2012. **39**(4): 2100–2107.

Moteabbed, M., S. Espana, and H. Paganetti, Monte Carlo patient study on the comparison of prompt gamma and PET imaging for range verification in proton therapy. *Phys. Med. Biol.*, 2011. **59**: 1063–1082.

Park, M.S., W. Lee, and J.M. Kim, Estimation of proton dose ditribution by means of three-dimansional reconstruction of prompt gamma rays. *App. Phys. Lett.*, 2010. **97**: 153705–153707.

Peterson, S.W., D. Roberts, and J.C. Polf, Optimizing a 3-stage Compton camera for measuring prompt gamma rays emitted during proton radiotherapy. *Phys. Med. Biol.*, 2010. **55**: 6841–6856.

Polf, J. et al., Prompt gamma-ray emission from biological tissues during proton irradiation: A preliminary study. *Phys. Med. Biol.*, 2009a. **54**(3): 731–743.

Polf, J.C. et al., Measurement and calculation of characteristic prompt gamma ray spectra emitted during proton irradiation. *Phys. Med. Biol.*, 2009b. **54**: N519–N527.

Polf, J. et al., Measuring prompt gamma ray emission during proton radiotherapy for assessment of treatment delivery and patient response. *AIP Conf. Proc.*, 2011. **1336**: 364–367.

Richard, M.H. et al., Design study of a Compton camera for prompt gamma imaging during ion beam therapy. In *IEEE Nucl. Sci. Sympos. Conf. Record*. 2009. IEEE, Orlando, Florida.

Robertson, D. et al., Material efficiency studies for a Compton camera designed to measure characteristic prompt gamma rays emitted during proton beam radiotherapy. *Phys. Med. Biol.*, 2011. **56**: 3047–3059.

Roellinghoff, F. et al., Design of a Compton camera for 3D prompt gamma imaging during ion beam therapy. *Nucl. Instrum. Methods Phys. Res. A*, 2011. **648**: s20–s23.

Ryu, H. et al., Density and spatial resolutions of proton radiography using a range modulation technique. *Phys. Med. Biol.*, 2008. **53**: 19, 5461–5468.

Schmind, G.J. et al., A gamma-ray tracking algorithm for the Greta spectrometer. *Nucl. Instrum. Methods Phys. Res. A*, 1999. **430**: 69–83.

Schneider, U., and E. Pedroni, Multiple Coulomb scattering and spatial resolution in proton radiography. *Med. Phys.*, 1994. **21**: 11, 1657–1663.

Schneider, U., and E. Pedroni, Technical note: Spatial resolution of proton tomography: Impact of air gap between patient and detector. *Med. Phys.*, 2012. **39**: 798–800.

Schulte, R. et al., Density resolution of proton computed tomography. *Med. Phys.*, 2005. **32**: 4, 1035–1046.

Sitek, A., Representation of photon limited data in emission tomography using origin ensembles. *Phys. Med. Biol.*, 2008. **53**: 3201–3216.

Steward, V., and A. Koehler, Proton beam radiography in tumor detection. *Science*, 1973. **179**: 4076, 913–914.

Waters, L.S., J.S. Hendricks, and G.W. Mckinney, *MCNPX User's Manual*, Version 2.5.0, 2005, Los Alamos National Laboratory: Los Alamos, NM.

Wilson, R.R. Radiological use of fast protons. *Radiology*, 1946. **47**: 487–491.

Monte Carlo for Treatment Device Design

Bruce A. Faddegon
*University of California,
San Francisco*

19.1 Introduction

Monte Carlo simulation has broad application in the field of radiotherapy, beyond the well-known applications in dosimetry and treatment planning. The method allows development and testing of virtual devices using virtual experimentation. This chapter provides a survey of applications of Monte Carlo simulation in the design of treatment devices for conventional linacs used for x-ray and electron therapy. Examples selected from the research, clinical, and consulting experience of the author show the power of the method, and elucidate general principles of treatment device design. The examples are loosely ordered from the exit window of the waveguide, progressing through the different components of the treatment head, through the patient and into the flat panel detector (Figure 19.1).

The EGS4 Monte Carlo system (Nelson et al., 1985) was used for the earlier work, EGSnrc (Kawrakow, 2000), once it was available. The EGS4/EGSnrc user code BEAM (Rogers et al., 1995) was used for treatment head simulation, sometimes with modification, the MCRTP user code (Faddegon et al., 1998) for dose calculation. The results include experimental validation, showing the accuracy of simulation for practical applications. The devices designed have for the most part been fabricated and tested (with the exception of the secondary scattering foils). Many are commercially available.

19.2 Target for Therapy

The accelerated electron beam incident on x-ray targets generates considerable heat. The targets are generally cooled with circulating, chilled water. One design used a gold target cooled by water channeled beneath the target, with water in the beam path. When some of these targets developed hairline cracks, pressurized water would leak out from the cooling channel and drip on components such as the monitor chamber, potentially shorting them out.

Monte Carlo simulation aided in the design of a new target (Figure 19.2). The primary concern was to eliminate the water from the beam path. In addition, the x-ray beam generated in the new target needed to conform closely to the beam from the original target, not just for one beam energy, but for the full range of x-ray beam energies available for treatment on the different accelerator models that used this target. To maintain a comparable dose rate for the same beam current, tungsten was chosen for the target material, being close in atomic number to gold. The tungsten was brazed to copper for heat dissipation. Since x-rays would be generated in both the tungsten and copper, this gave a degree of freedom to match x-ray beam characteristics for the two targets. The question to be answered with Monte Carlo simulation was whether there was a combination of thicknesses of the two target materials that would result in matched beams over the full energy range.

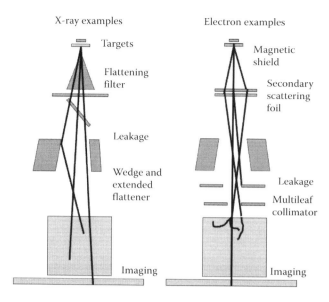

FIGURE 19.1   A generic treatment head configured for x-ray and electron therapy, showing the regions of greatest relevance to the different treatment device design examples (not to scale).

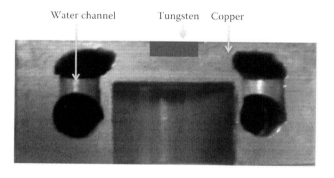

FIGURE 19.2   X-ray target used for radiotherapy, cut in half to show cross section. The water channel for target cooling runs around the outside of the target. The carbon absorber has been removed.

The Monte Carlo method proved of great help in answering this question. The approach taken was to match angular distributions of energy fluence and spectral distributions on the beam axis for both targets, covering the full range of beam energies. Although angular distributions of energy fluence could be measured relatively easily by measuring dose profiles, measuring spectral distributions is rather difficult (see, e.g., Faddegon et al., 1990). Monte Carlo simulation avoided the need to construct a set of targets with different thicknesses of tungsten and copper and then measure the output and full dose distributions covering the clinical range of field size, including those with beam modifiers (physical wedges, in particular). The simulation details and resulting target design were published (Faddegon et al., 2004). It turned out that a specific thickness of tungsten and copper did yield beam details that were a close enough match to the original target to permit replacement while avoiding the tedious process of recommissioning.

19.3 Target for Imaging

Patients are traditionally set up on the treatment couch prior to irradiation by aligning marks placed on the skin with lasers mounted on the wall and ceiling. The volume in the patient targeted with radiation may move in relation to the surface markers. Modern radiotherapy often involves imaging the patient immediately prior to treatment. A relatively straightforward means to do so is to generate the imaging beam with the same accelerator used to generate the treatment beam. In this way, the radiation therapists setting up the patient can observe the position of the internal anatomy of the patient in relation to the beam portal (Figure 19.3). A CT image may also be obtained with this beam.

The energy range of photons in the x-ray beam that is most suitable for imaging is 10–100 keV. Treatment beams have energies in the range 500 keV through 20 MeV. Lower-energy photons are actually produced in the target used to generate the treatment beam, but these are largely absorbed in the target and never reach the patient. The resulting beam has a reduced surface dose, generally preferred in radiotherapy. A low-atomic-number target may be used to retain these photons for the purpose of imaging.

Monte Carlo simulation has proven instrumental in the design of such a target for planar imaging (Ostapiak et al., 1998) and cone-beam CT (Faddegon et al., 2010). The beam energy is lowered, to fully stop the primary electron beam in the target, as those electrons would otherwise leak through and reach the monitor chamber and even the patient, needlessly adding to the skin dose.

Patient images have been successfully taken with this target for patient alignment with dose in the range 0.3–3 cGy (maximum dose to water at 100 cm SSD for a 10 × 10 cm field given the same monitor units used to image the patient), a low dose compared to the daily treatment dose, usually around 200 cGy. An image obtained with a 4 MV x-ray beam generated in a carbon

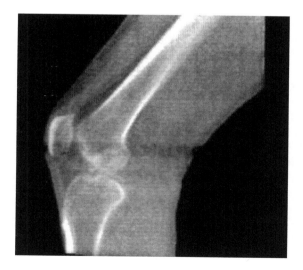

FIGURE 19.3   A sagittal slice of a cone-beam CT of a knee taken with a 4 MV x-ray beam from a carbon target, imaged with a flat panel detector. The dose to the patient was 3 cGy.

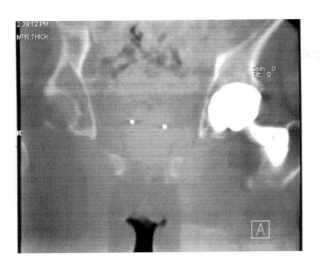

FIGURE 19.4 A coronal slice of a 4 MV carbon target cone-beam CT image taken of a patient immediately prior to radiotherapy.

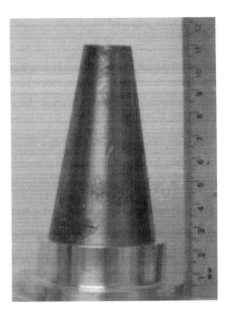

FIGURE 19.5 Shield designed to eliminate the fringe magnetic field from the bending magnet that caused the electron field asymmetry in the inplane direction.

target is shown in Figure 19.4. Two of the three seeds implanted in the prostate to aid in patient positioning are visible on the slice shown. Note that the metallic implant in the left hip results in minimal artifact in this image, a strength of imaging with MV x-ray beams.

19.4 Electron Beam Magnetic Shield

Many modern linacs are equipped with a powerful magnet at the end of the beam line to bend the accelerated electron beam 270°. Thus, the beam emerges from the evacuated beam line at a right angle to the axis of the waveguide. The bending magnet current is very stable and is set to pass accelerated electrons over a narrow range of energies. For the major Siemens linac models, the resulting full widths at half maxima of the energy distribution peaks are in the range of 6–14% of the beam energy (Faddegon et al., 2009). The standard approach for adjusting the beam energy is to monitor the dose rate while adjusting the RF power in the waveguide and maintaining a constant beam pulse frequency. Given the correct bending magnet current, the RF power that gives the peak dose rate also gives the correct beam energy.

At least for one vendor's design, the bending magnet field is known to encroach on the path of the beam between the exit window and monitor chamber. This has no effect on x-rays, since they are uncharged and are not deflected by a magnetic field. However, the fringe magnetic field does deflect electron beams. This field must be included in the simulation to accurately match the known asymmetry in the fluence and dose distributions of electron fields. A study of this effect required modification of the BEAMnrc user code to incorporate asymmetry and spatially varying magnetic fields (Shea et al., 2011).

The knowledge that this fringe magnetic field fully accounted for the asymmetry leads to the possibility that a magnetic field could be shielded to produce symmetric beams. A hollow cone machined from ferromagnetic stainless steel (Figure 19.5) was inserted into the treatment head with the linac operated in electron mode. Dose distributions measured once this magnetic field shield was in place were symmetric, as predicted with the Monte Carlo method.

19.5 X-Ray Standard Flattener and Extended Flattener

The x-ray beam in radiotherapy is produced by highly relativistic electrons moving within 1% of the speed of light. By conservation of momentum, most of the photons generated in the target travel in a tight cone about the beam axis. A cone-shaped flattening filter is often inserted in the line of the x-ray beam to cut the fluence in the beam center down to match the fluence at the beam edge.

Monte Carlo simulation is an established means to design these flattening filters. In one unique application, the technique was used to develop a beam that can be used to image the patient for setup verification and then treat the patient using the exact same beam (Nishimura, 2005). This would permit image-guided radiotherapy using the same beam for imaging and treatment, simplifying the design of the treatment head. Beams generated in low-atomic-number materials are more useful for imaging, having a high proportion of diagnostic energy photons. These photons are low enough in energy to exhibit high contrast to different biological tissues and organs, owing to dramatic differences in the photoelectric effect. These imaging beams could not be fully flattened with a flattener of the same material as the target, since there was insufficient space in the treatment head to accommodate such a flattener. Nevertheless, at the time the study was done, it was felt that partially flattened beams could prove as effective as fully flattened beams in radiotherapy. This

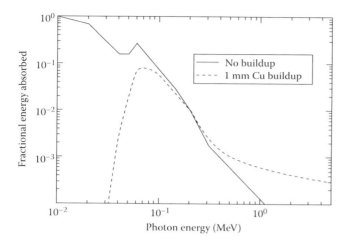

FIGURE 19.6 Calculated response functions for a bare sheet of scintillator and with 1 mm Cu on top of the sheet.

appears to be the case, as unflattened beams are currently under consideration for treatment of even large target volumes, taking advantage of intensity-modulated radiation therapy (IMRT) and modulated arc techniques.

Monte Carlo simulation was used to determine the trade-off of the compromise beams for imaging and therapy by calculating the response of the flat panel detector used for patient imaging (Figure 19.6) and the spatial distribution of photon fluence at the face of the detector as a function of energy. These were used to relate image quality with field flatness for the different beams. Subsequently, one manufacturer incorporated a graphite target into one of the electron foil slots of a multienergy machine (see above), slowing the impetus to further develop a single-beam, imaging-treatment machine.

In a separate application, the existing flattener of a clinical machine was redesigned with the help of Monte Carlo simulation, to fully flatten the x-ray beam out along the diagonals of the largest field (Faddegon et al., 1999). Later, a beveled brass plate was designed as an add-on, placed in the accessory tray slot normally used for wedges, to accomplish the same thing. The plate, designed with Monte Carlo simulation, was used to extend the flat region of the field to cover the largest x-ray field available on the machine: a 40 × 40 cm field at the machine isocenter. The beveled brass plate did produce a nice, flat beam over the whole field, out to 23 cm along the diagonals (Figure 19.7). This plate has been used for total body irradiation (TBI) of patients seated 3 m from the source, as a backup to TBI in a larger room where patients stand up to 4.5 m from the source.

19.6 X-Ray Leakage

Radiation from the linear accelerator target is most intense in the direction along the beam axis, where the region in the patient to be treated is generally positioned for treatment. Still, x-rays emanate in all directions from the target, some outside of the area of

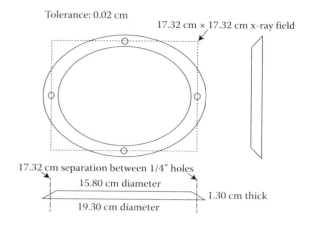

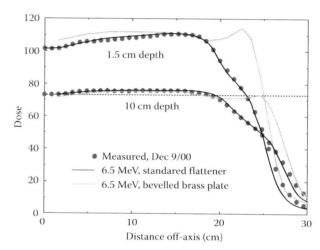

FIGURE 19.7 Brass plate (top), designed to be mounted on the downstream side of a 0.8-cm-thick Al tray at 41.3 cm from the source. Diagonal dose profile for the 6 MV MXE 40 × 40 cm field with and without the brass plate in the beam (bottom).

the treatment beam. Standard practice is for manufacturers to add shielding, usually lead and/or tungsten, to reduce the dose 1 m from the target in all directions to 0.1% of the in-field dose at that same 1 m distance.

The Artiste linear accelerator has a different treatment head than the earlier Oncor model. The distance from the target to the secondary collimators was increased, in part to accommodate full over-travel of the movable jaws that collimate the beam in the Y-direction. Monte Carlo simulation was used to determine the thickness and material of shielding to incorporate in the treatment head. Dose profiles calculated at the depth of maximum dose in the direction of jaw travel with the jaw positioned at full over-travel are shown in Figure 19.8. The results show the magnitude of the leakage from several sources (transmission and scatter from different parts of the treatment head). Note that the study was for a prototype shielding design and does not represent the commercial product.

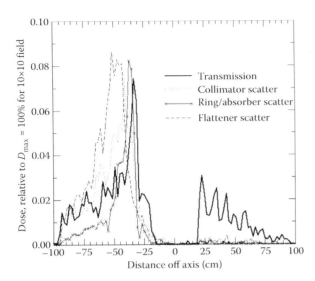

FIGURE 19.8 Calculated leakage profile through isocenter for Artiste 23 MV x-ray beam with full over-travel of jaw and added prototype shielding. Results show contributions from different sources of primary and scattered radiation. The field in this case is completely collimated and would not be used to treat patients.

19.7 Electron Secondary Scattering Foils

Monte Carlo simulation was used to consider what changes in the treatment head were needed to enable precise electron beam collimation with the x-ray multileaf collimator (MLC) (Faddegon et al., 2002). The idea was that by bringing the secondary scattering foil closer to the primary foil, and thus reducing the size of the electron field at the secondary foil, one would obtain a sharper penumbra. The replacement of the air downstream of the secondary foil with the much lower scattering helium atmosphere (at the same pressure as air) was also considered. The secondary foil was positioned at its clinical position, 10.6 cm from the primary foil, and at half that distance, 5.6 cm from the primary foil. Scattering foils were designed for treatment with one of three choices of final collimation: the conventional electron applicator, with the distal surface of the electron applicator at a distance of 95 cm from the nominal source position (close to the primary scattering foil), an add-on electron MLC at 65 cm (eMLC), and the x-ray MLC at 35 cm (xMLC).

Monte Carlo simulation was used to design the secondary scattering foil. The design constraints were

1. Similar penetration depth in water, R_{50} (0.2 MeV energy increase for He)
2. Similar bremsstrahlung tail (it proved sufficient to use the same primary scattering foils)
3. Similar flatness over 30 cm at isocenter without tertiary collimation (40×40 cm field size setting) to produce beams with flatness equivalent to the clinical beams with tertiary collimation, as shown in Figure 19.9 for the 12 MeV beam

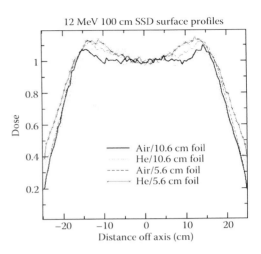

FIGURE 19.9 Surface profiles for the secondary scattering foils designed with Monte Carlo simulation for the 12 MeV electron beam, with air or helium atmosphere, with the foil 5.6 and 10.6 cm from the primary foil.

Dose distributions in water were calculated with the Monte Carlo method for the different atmospheres and foil positions and covering the available energy range (6–21 MeV). The three electron collimators were set to the full range of field sizes. The distance from the collimator to the patient, often referred to as standoff, was the same when comparing the penumbral widths. A standoff of 25 cm was used to allow gantry rotation with the patient in treatment position for x-ray IMRT without retracting the tertiary collimator. The width of the penumbra from the calculated dose distributions was largely independent of leaf position and field size. Representative results are shown in Figure 19.10. The penumbral width from the simulated dose distribution was in reasonable agreement for the two configurations measured, limited to the clinically available foil position of 10.6 cm.

The following conclusions were drawn from these results. Replacement of air with helium sharpened the penumbra for all MLC positions (x-ray MLC, extended electron MLC, and applicator). Halving the distance between the foils was only useful when collimating using the x-ray MLC. With the helium atmosphere, the penumbral width under the x-ray MLC was reduced by 35–40%. Of particular interest, collimation of the electron beam with the x-ray MLC provides a clinically reasonable penumbral width when halving the distance between the scattering foils and using helium downstream of the secondary scattering foil. Further experimental validation would be prudent; however, this would require machining the redesigned secondary foil, placing it 5.6 cm from the primary foil, replacing air with helium in the treatment head, and then measuring profiles for the different beam energies at different standoffs, using the x-ray MLC to collimate the beam.

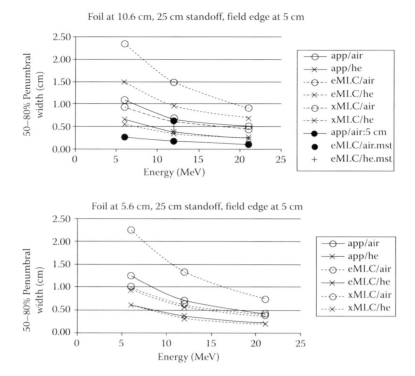

FIGURE 19.10 Calculated penumbral width for the different foil position/atmosphere/electron collimation combinations. The conventional clinical configuration of the foil at 10.6 cm with a 5 cm standoff using an electron applicator is also shown (app/air: 5 cm), along with two points where the penumbral width was measured for the 12 MeV beam (mst).

19.8 X-Ray Wedges

In radiotherapy, a uniform dose distribution is generally sought throughout the clinical target volume. An x-ray field from a single direction produces a uniform dose across the irradiation field with a dose distribution that falls off at a rate of approximately 4% per centimeter. Diametrically opposed fields, known as a parallel-opposed pair, produce a more uniform distribution with depth. A number of situations arise where a wedged-shaped dose distribution is of value, for example, to reduce the dose to structures distal of the tumor by treating at right angles or to adjust for a slope on the patient surface such as in breast treatment (Figure 19.11). The wedge-shaped dose distribution may be produced by placing a wedge made of a dense material such as iron or lead in the beam. The wedge is ideally placed as far away from the patient as possible, to reduce the proportion of

dose to the patient from secondary radiation emanating from the wedge. Monte Carlo simulation is ideal for evaluating wedge design and placement.

The wedges for the Artiste were designed with the help of Monte Carlo simulation. The objective was to match the profile measured at 10 cm depth from wedges on the earlier Oncor model. The wedge on the Artiste was positioned further from the target than on the Oncor model, requiring larger wedges to cover the same field size. The Oncor wedge thickness was used for the preliminary design, using the same wedge thickness, scaled to the increased distance of the wedge from the target. A unique design was used to keep the material in the wedge closer to the target, primarily so that the weight of the wedge could be minimized. Results for the thickest wedges are shown in Figure 19.12. The scatter from the Artiste wedges was different enough from the Oncor wedges that some design iteration was needed to match

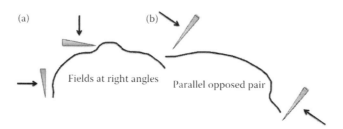

FIGURE 19.11 Examples of treatments using wedged fields: (a) head and neck treatment with beams at right angles, and (b) breast parallel-opposed pair.

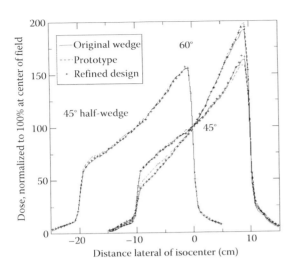

FIGURE 19.12 Prototype wedges based on geometrical divergence alone are compared to the refined design from successive simulations, adjusted to match the wedge profiles for the wedges on the original machine. The calculations are for 0.5 × 2 × 0.5 cm voxels.

the profiles. The final design met manufacturer specification and was adopted for patient treatment.

19.9 Electron Applicator Leakage

Linacs used for radiotherapy often are equipped with scattering foils to provide for treatment with electron beams. Electron beams scatter widely, even in air. Fields are generally collimated close to the patient to sharpen the field edges. Monte Carlo simulation has been used effectively to aid in the design of electron applicators (Janssen et al., 2008). Electrons that hit the applicator can be scattered toward the patient or produce x-rays that can reach the patient. This leakage dose needs to be kept to a low level. The International Electrotechnical Commission document IEC 60621-1 (1998) specifies a limit of 1–2% of the maximum dose in the field, the larger value for beam energies exceeding 10 MeV, with the leakage dose average over the area 4 cm beyond the field edge and out to the area shielded by the primary collimator.

An investigation using Monte Carlo simulation was conducted to design a retrofit to an existing applicator (Sawkey and Faddegon, 2008). The objective was to reduce electron leakage without affecting the field used to treat the patient. The BEAMnrc user code was extended to provide for fluence scoring at the sides of the applicator and to accommodate simulation of the retrofit. A retrofit was designed that indeed reduced leakage (Figure 19.13) without affecting the relative output factors or dose distributions used to treat patients. Thus, this retrofit could be used without recommissioning the six electron beams, a very onerous process.

19.10 Electron Multileaf Collimator

Electrons are charged particles that essentially lose their energy evenly as they penetrate tissue, stopping once they exhaust their energy, with little dose delivered to tissue beyond the electron range. They are widely used in radiotherapy to treat more superficial lesions. They may be used alone or in combination with x-ray therapy, often to boost the dose to the region where there may be residual disease following surgical resection of the tumor, a common approach in treating breast cancer. Electrons, being relatively light compared to the atoms in the tissue, present a challenge for treatment planning as they are scattered widely and as the amount of scattering depends on the material traversed, being very different in air, lung, soft tissue, and bone. Monte Carlo simulation, known to accurately account for electron scatter (Faddegon et al., 2009), provides an accurate means for treatment planning in even the most complex radiotherapy geometries.

The availability of accurate treatment planning with Monte Carlo simulation now extends opportunities for electron therapy. Improved conformation to the tumor with significant reduction in harmful energy deposition to healthy tissue may now be achieved by combining x-ray IMRT with a set of adjacent electron fields of different energies (Ge and Faddegon, 2011). Monte Carlo simulation has been used to aid in the design of the add-on MLC to shape the electron fields for this mixed beam radiotherapy (Hogstrom et al., 2004; Shea et al., 2011).

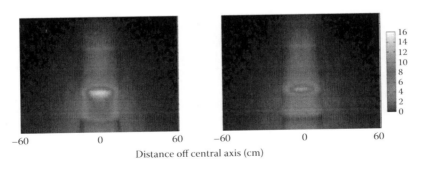

FIGURE 19.13 Dose distribution calculated in air just outside the applicator wall from the isocenter up to 45 cm above the isocenter, without (left) and with (right) additional shielding. The isocenter is situated on the central axis at the bottom of the figure.

19.11 X-Ray Image Detection and Processing

X-rays in both x-ray and electron beams provide useful sources for imaging patients. Imaging is useful for positioning the patient prior to treatment and checking and even adjusting for motion during patient treatment.

Patient images, keeping patient comfort and motion (breathing) in mind, are best taken in seconds rather than minutes. Linac gantry rotation speeds place a lower limit on the time taken for a tomographic CT scan. The time for a single rotation is limited to half a minute or more by regulation (European standard EN 60601-2-1, 1998) and as a consequence of the huge weight they support. Current scanning technologies rely on large fields to cover the region of interest in the patient with fewer rotations of the gantry, using cone-beam CT. The large scatter fractions in kilovoltage cone-beam CT scans reduce image contrast and introduce artifacts, complicating the detection and processing of patient images.

Monte Carlo simulation has proven ideal to characterize the primary beam transmitted through the patient and to characterize the radiation scattered from the patient at the point where that radiation reaches the detector. The technique has provided helpful insight for the design, processing, and interpretation of scatter measurements and for the development of scatter correction algorithms (Maltz et al., 2008a,b).

It is also possible to take images of the patient while being treated with electron beams, using the bremsstrahlung generated in the scattering foils (Aubin et al., 2003; Jarry and Verhaegen, 2005). These x-rays, along with a smaller portion of bremsstrahlung generated in the patient, result in a slow fall-off of dose beyond the electron practical range, as seen in Figure 19.14. The x-rays transmitted through the patient have been used for imaging. The bremsstrahlung dose in the open field, just beyond the practical range, is 0.4% of the maximum dose at 6 MeV, increasing to 1.3% at 12 MeV, and 5% at 21 MeV. Thus, lower-energy electron beams have less bremsstrahlung for imaging. Lower energies also produce a flatter bremsstrahlung intensity profile. Patient images were successfully acquired while undergoing electron therapy on a Primus machine.

19.12 Conclusion

The range of examples shown demonstrates the value of Monte Carlo simulation in treatment device design. A word of caution: Monte Carlo simulation, of great advantage in situations involving complex source and geometry configurations, is necessarily complex and therefore prone to error. Experimental validation should be a general rule. Achieving an accurate match, within the combined experimental uncertainty and statistical precision of the calculation, is gratifying and simplifies the design process, giving confidence in the results for the newly designed device. However, the effort required can dramatically slow the design process. In practice, device modification can be done without the need to obtain accurate matches between measurement and simulation. Knowledge of the match (or mismatch) can be sufficient to redesign the device to a stringent specification, by accounting for this difference as part of the redesign procedure. Experience in simulation is essential to judge whether discrepancies with measured data will unduly affect the geometry and material of the device during the design stage. In any case, once the device is designed, it is prudent to verify the design with measurement. In the end, Monte Carlo simulation can improve the efficiency and cost of the design process by reducing or eliminating the need to machine and make measurements with prototype devices. Indeed, Monte Carlo simulation has become a gold standard for fluence and dose calculation in radiotherapy and is arguably integral to the design process.

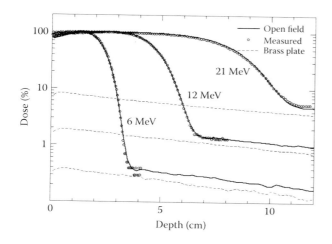

FIGURE 19.14 Measured electron beam depth dose curves (points) compared to Monte Carlo simulation (lines): 6, 12, and 21 MeV Primus electron beams collimated by a 10 × 10 applicator. The solid lines are for the open field and dashed lines for fields fully blocked by a 1.3-cm-thick brass plate.

References

Aubin M., Langen K., Faddegon B., and Pouliot J. 2003. Electron beam verification with an A-Si EPI. *Med. Phys.* 30(6):1475.

European standard EN 60601-2-1 medical electrical equipment Part 2-1: Particular requirements for the safety of electron accelerators in the range of 1 MeV to 50 MeV (IEC 60601-2-1:1998).

Faddegon B. A., Aubin M., Bani-hashemi A., Gangadharan B., Gottschalk A. R., Morin O., Wu V., and Yom S. S. 2010. Comparison of patient megavoltage cone beam CT images acquired with an unflattened beam from a carbon target and a flattened treatment beam. *Med. Phys.* 37:1737–1741.

Faddegon B. A., Balogh J., Mackenzie R., and Scora D. 1998. Clinical considerations of Monte Carlo for electron radiotherapy treatment planning. *Rad. Phys. Chem.* 53:217–227.

Faddegon B., Egley B., and Steinberg T. 2004. Comparison of beam characteristics of a gold x-ray target and a tungsten replacement target. *Med. Phys.* 31:91–97.

Faddegon B. A., Kawrakow I., Kubyshin Y., Perl J., Sempau J., and Urban L. 2009a. Accuracy of EGSnrc, Geant4 and PENELOPE Monte Carlo systems for simulation of electron scatter in external beam radiotherapy. *Phys. Med. Biol.* 54:6151–6163.

Faddegon B. A., O'Brien P. F., and Mason D. L. D. 1999. The flattened area of Siemens linear accelerator x-ray fields. *Med. Phys.* 26:220–228.

Faddegon B. A., Ross C. K., and Rogers D. W. O. 1990. Forward-directed Bremsstrahlung of 10–30 MeV electrons incident on thick targets of Al and Pb. *Med. Phys.* 17:773.

Faddegon B. A., Sawkey D., O'Shea T., McEwen M., and Ross C. 2009b. Treatment head disassembly to improve the accuracy of large electron field simulation. *Med. Phys.* 36:4577–4591.

Faddegon B. A., Svatos M., Karlsson M., Karlsson M., Olofsson L., and Antolak J. A. 2002. Treatment head design for mixed beam therapy. *Med. Phys.* 29:1285.

Ge Y., and Faddegon B. A. 2011. Study of intensity modulated photon-electron radiotherapy using digital phantoms. *Phys. Med. Biol.* 56:6693–6708.

Hogstrom K. R., Boyd R. A., Antolak J. A., Svatos M. M., Faddegon B. A., and Rosenman J. G. 2004. Dosimetry of a prototype retractable eMLC for fixed-beam electron therapy. *Med. Phys.* 31:443.

Janssen R. W. J., Faddegon B. A., and Dries W. J. F. 2008. Prototyping a large field size IORT applicator for a mobile linear accelerator. *Phys. Med. Biol.* 53:2089–2102.

Jarry G., and Verhaegen F. 2005. Electron beam treatment verification using measured and Monte Carlo predicted portal images. *Phys. Med. Biol.*, 50:4977–4994.

Kawrakow I. 2000. Accurate condensed history Monte Carlo simulation of electron transport. I. EGSnrc, the new EGS4 version. *Med. Phys.* 27:485–498.

Maltz J., Gangadharan B., Hristov D. H., Faddegon B. A., Paidi A., Bose S., and Bani-Hashemi A. R. 2008a. Algorithm for X-ray scatter, beam-hardening and beam profile correction in diagnostic (kilovoltage) and treatment (megavoltage) cone beam CT. *IEEE Trans. Med. Imag.* 227:1791–1810.

Maltz J., Gangadharan B., Vidal M., Paidi A., Bose S., Faddegon B., Aubin M. et al. 2008b. Focused beam-stop array for the measurement of scatter in megavoltage portal and cone beam CT imaging. *Med. Phys.* 35:2452–2462.

Nelson W. R., Hirayama H., and Rogers D. W. O. 1985. The EGS4 Code System. SLAC-Report-265, Stanford Linear Accelerator Center, Stanford, California.

Nishimura K. A. 2005. A Monte Carlo study of low-Z target—Flattener combinations for megavoltage imaging. MSc thesis, San Francisco State University.

O'Shea T. P., Foley M. J., and Faddegon B. A. 2011a. Accounting for the fringe magnetic field from the bending magnet in a Monte Carlo accelerator treatment head simulation. *Med. Phys.* 38:3260–3269.

O'Shea T. P., Ge Y., Foley M. J., and Faddegon B. A. 2011b. Characterisation of an extendable multileaf collimator for clinical electron beams. *Phys. Med. Biol.* 56:7621–7638.

Ostapiak O. Z., O'Brien P. F., and Faddegon B. A. 1998. Megavoltage imaging with low Z targets: Implementation and characterization of an investigational system. *Med. Phys.* 25:1910–1918.

Rogers D. W. O., Faddegon B. A., Ding G. X., Ma C.-M., We J., and Mackie T. R. 1995. BEAM: A Monte Carlo code to simulate radiotherapy treatment units. *Med. Phys.* 22:503–524.

Sawkey D., and Faddegon B., 2008. Design of a leakage-reducing electron applicator retrofit. *Med. Phys.* 35:2810.

GPU-Based Fast Monte Carlo Simulation for Radiotherapy Dose Calculation

Xun Jia
University of California, San Diego

Sami Hissoiny
École Polytechnique de Montréal

Steve B. Jiang
University of California, San Diego

20.1 Introduction

Monte Carlo (MC) simulation is considered as the most accurate method for radiotherapy dose calculations due to its capability of faithfully describing the physical processes and flexibly handling complicated geometries. Since the MC simulation is a stochastic method, a large number of particle histories are simulated to achieve a desired statistical accuracy. Despite the vast advancement in computer architecture and the increase of processor clock speed in recent years, the efficiency of the currently available MC dose engines is still not completely satisfactory for routine clinical applications in radiotherapy. One straightforward way to mitigate this issue is to perform the computation in a parallel fashion by taking advantages of advanced parallel computer architectures. By distributing the total computation load to available computing units, it is conceivable that a significant speedup factor can be achieved. Over the years, there has been a considerable amount of research regarding how to implement various MC dose calculation packages on a variety of computing architectures, including multicore central processing unit (CPU), CPU cluster, graphics processing unit (GPU), cloud computing, and so on. The use of those parallel processing techniques for MC dose calculation has offered an attractive approach toward fast MC dose calculation in clinically realistic environments.

The CPU cluster has been the main platform for parallel dose calculations. Under such a platform, one could simply distribute the total number of particle histories to different computing units, which then perform simulations simultaneously and independently of each other. At the end of the calculation, the dose from all the units is accumulated. Although almost all the available MC dose calculation packages were initially developed on a single CPU platform, transferring them to a CPU cluster is straightforward with the help of parallel processing interfaces such as message passing interface (MPI). Currently, parallel versions of a number of available packages, such as MCNP, EGSnrc, PENELOPE, and dose planning model (DPM), have been successfully developed. The expected efficiency gain from such an approach is close to ideal with regard to the number of computing cores used by the simulation (Deng and Xie, 1999; Tyagi et al., 2004; Sutherland et al., 2007).

Lately, a new technology, cloud computing, has also been employed as the parallel processing platform for MC dose calculation. This new platform offers computing resources allocated in an on-demand fashion from a third party, such as Amazon.[*] In terms of parallel computing strategy itself, cloud-based MC dose calculation is not much different from CPU cluster-based calculations, as the virtual platform is essentially a CPU cluster from the end user's point of view. This also makes it straightforward to launch available parallel MC packages developed for CPU clusters on the cloud. As such, many packages such as EGS5 (Wang et al., 2011) have been successfully utilized in a cloud environment for dose calculation. The achieved efficiency improvement is very similar to that on a CPU cluster with a certain amount of overhead due to the initializing of the remote virtual cluster and data transfer (Constantin et al., 2011; Keyes et al., 2011; Pratx and Xing, 2011).

---

[*] http://aws.amazon.com/ec2/.

GPU is a computing architecture originally designed to handle extensive computational tasks in computer graphics, such as rendering images in computer games. Recently, it has been discovered that the highly parallel structure in the GPU makes it very effective for solving scientific computing problems conventionally handled by the CPU. In particular, the GPU has drawn a lot of attention for improving the efficiency of many computationally intensive problems in medical physics and medial image processing with affordable graphic cards such as NVIDIA's GeForce and Tesla series. Among them, GPU-based dose calculation packages with different algorithms, including MC, have lately become available (Hissoiny et al., 2009, 2011a; Men et al., 2009; Jacques et al., 2010; Jia et al., 2010, 2011; Gu et al., 2011).

Among these available parallel computing architectures, the GPU offers a number of advantages for the purpose of MC dose calculation in radiotherapy. Because of the driving forces from many computer graphics applications, such as computer gaming, GPUs are extremely low cost nowadays, much more than CPU clusters. Indeed, it is usually a couple of orders of magnitude lower in cost compared to a CPU cluster with similar computing power. Moreover, the GPU is much easier to install and maintain than a cluster, as it runs on a local desktop computer. Compared to cloud computing, GPU-based MC dose calculations do not require an initialization stage, which can take up to a few minutes for a virtual cluster in the cloud. This extra initialization time limits the further speedup of MC in cloud and hence its application in many time critical tasks such as the dose calculation for online adaptive radiotherapy. In addition, the overhead due to data transfer in GPU computing is usually negligible due to the high bandwidth between the CPU and GPU memory.

However, GPU also has its unique disadvantages. Compared to the CPU, it is a relatively new platform. The previously developed MC dose calculation packages cannot be launched on it without modifications. It is therefore necessary to develop GPU-specific packages, including rewriting codes in a programming language usable on the GPU, and possibly a complete redesign of the parallel simulation scheme. Moreover, the hardware architecture of the GPU makes it extremely suitable for data-parallel problems, but not so for task-parallel problems such as MC dose calculation. Rethinking the parallelization model of the MC simulation is essential to achieve high speedup. These issues have posed a significant amount of challenges for the developers and are active research topics nowadays.

The rest of this chapter will be organized as follows. Section 20.2 will introduce some basic principles of the GPU architecture and GPU programming. Section 20.3 will discuss several issues encountered when applying GPU for the MC dose calculation problem, especially the incompatibilities between the GPU data-parallel architecture and the MC task-parallel paradigm. In Section 20.4, a number of currently available GPU-based MC dose calculation packages, as well as their applications in radiotherapy, will be introduced. Finally, Section 20.5 will conclude this chapter with some future works.

20.2 Graphics Processing Unit: Hardware and Programming

20.2.1 Historical Development of GPU Programming

GPU is a specialized hardware designed to accelerate the processing of graphics information and to write frame buffer that will ultimately be displayed on the screen. The GPU can be present on a discrete graphics card, embedded directly on the motherboard or, more recently, directly on the die of the CPU.

The first GPUs had a fixed pipeline with no control left for the programmer. The graphics primitive information would be sent to the GPU, processed by the fixed pipeline, and sent to the screen. The ability for programmers to have access to the parts of the hardware resources of the GPU directly, rather than through a fixed pipeline, was introduced with the NV20 from NVIDIA (Santa Clara, California, USA) and the R200 from ATI Technologies (now AMD, Sunnyvale, California, USA) in 2001. The programmers could then write pieces of code that would be executed on the vertex and fragment processors, two of the steps present in the graphics pipeline. A distinction, in hardware, was still present between vertex processors and fragment processors (called shaders in computer graphics) and the programmer had the possibility of programming these two units separately. Scientific computing was possible, but difficult, with this generation of the graphics card. Programming was done within a computer graphics context. The programming of the GPU itself was done through languages such as Cg, GLSL, or HSLS, which had the benefit of being at a higher level than the GPU's specific assembly language.

The NV80 and the R600, in 2006, were the first GPUs to have a "unified shader architecture." This new feature meant that the distinction between vertex and fragment shaders was gone, replaced by a single type of processor, more general, capable of handling both tasks. Shortly after the introduction of the unified shader architecture, NVIDIA unveiled the CUDA language and AMD (formerly ATI), its stream computing SDK. These two new application programming interfaces (API) gave access to the GPU directly through the C language, without the need of a computer graphics context such as OpenGL or DirectX. In 2008, the specification of the OpenCL API version 1.0 was unveiled.

20.2.2 Hardware Architecture and Threading Model

In what follows, the OpenCL vocabulary is used and the corresponding CUDA vocabulary is put between parentheses; if no corresponding word is given, the same word is valid for both OpenCL and CUDA.

The GPU follows a single-instruction, multiple-data (SIMD) design. As such, it has many *processing elements (CUDA cores)*, split across a number of *computing units (streaming multiprocessor)*. This architecture allows scalability in the number of computing units per device while preserving the same hardware architecture. High-performance GPUs have more computing units than low-performance parts. The GPU, or the GPU and its graphics card, will henceforth be referred to as the *device*. The system on which the device is installed or, equivalently, the portion of the computer running the C code, will be referred to as the *host*.

In an SIMD architecture, all the cores of a computing unit must execute the same instruction. A task to be run on the GPU requires data parallelism. For example, doing a term-by-term vector addition of two vectors is a data-parallel task since the same operation (the addition) is applied over a range of data (the terms of the vector). This parallelism is expressed by defining a number of *threads* required to accomplish the task. Following our example, this number of threads could be equal to the size of one of the vectors. These threads are then grouped in *work groups* (*thread blocks*), which are in turn grouped to form a *grid*. Finally, one *kernel*, which is a function that runs on the device, is assigned to be executed by one grid. This is illustrated in Figure 20.1. One work group is assigned to one computing unit and multiple work groups can be assigned to one computing unit. If more work groups are required than the computing units can accommodate, the work groups are serialized and the scheduler on the GPU launches them when another work group finishes.

Several memory levels are present on the GPU and on the graphics card in general. Each computing unit has a set of registers. The number of registers is variable over brands and generations. Also present on the computing unit is a pool of local memory (L1 memory), variable in size across brands and generations. This local memory is shared across the threads of one work group and can be used, for example, to store elements needed by several threads in the work group, reducing the need to load from slower memory levels. On newer devices, there is a device-wide data cache (L2 memory) that serves as a hardware-controlled data cache. Finally, outside the GPU but on the graphics card is a pool of global memory. This memory serves as the link between the system memory (system RAM) and the GPU. It is the only memory level writable from the host. Three different types of memory allocation can be done in global memory: linear memory, arrays, or constant memory. Linear memory is assigned by using the equivalent of `malloc` on the GPU and can be read or written directly by the processing elements. Arrays are allocated and written from the host, after which they are bound to the texture. Textures offer the benefit of a multidimensional spatial locality cache (1D, 2D, or 3D) as well as hardware linear interpolation between elements of the texture. Textures are read-only from the processing elements. The memory levels have been listed in ascending order of latency and descending order of size.

A major issue with MC radiation transport on the GPU is divergence. This will be dealt with in more details in Section 20.3 but the reason for this issue is presented here. As mentioned, the GPU uses an SIMD approach. This means that the same instruction needs to be executed by concurrent threads. This limitation is not GPU-wide; instead, threads of a given work group are split, on the hardware level, in *warps*. The warp is therefore the smallest granularity and all threads of a warp must execute the same instruction. However, when writing software, it is almost inevitable that branching will occur and that different threads of a warp will have to execute different instructions. The hardware and compiler handle this in two ways: predication or serialization.

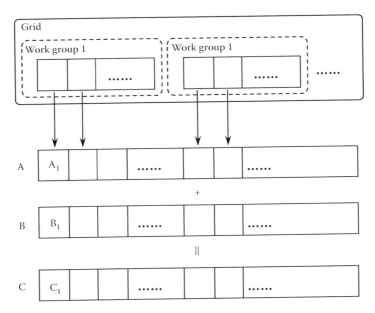

FIGURE 20.1 Illustration of componentwise addition of two vectors A and B. A kernel is executed on the GPU grid, and each thread in it computes the addition of one element.

In predication, all instructions are scheduled for execution. For example, in the following pseudocode:

```
1    if(a<5)
2        a+1;
3    else
4        a+2;
5    end
```

lines 2 and 4 would be scheduled for execution for all threads of the warp, even if for some the condition is true and for others it is false. A per-thread predicate value of true or false is set according to the value of the condition for instructions 2 and 4. Only the instruction with a "true" predicate will write its result. In serialization, the warp is split into subwarps, according to their branching value. Threads with a "true" condition will be placed in one warp while threads with a "false" value are put in another thread. Each warp will be launched, one after the other, to execute the branch and reunite after the branch. In both cases, some computational power is lost: in predication, extra instructions are executed, and in serialization, some processing elements will be idle to handle the subwarps.

20.2.3 Programming the GPU

20.2.3.1 Available APIs

Two major APIs are available for general-purpose GPU programming: OpenCL and CUDA. OpenCL is a vendor-neutral API managed by the Khronos group. Vendors are then responsible for implementing the OpenCL API within their drivers. CUDA has been developed by NVIDIA, one of the two main vendors of GPUs. NVIDIA alone defines the API, which is only usable on NVIDIA hardware. The same high-level concepts of work groups, processing elements, grids, and memory levels apply to both approaches with only a difference in the vocabulary used to define the elements. A third API is provided by Microsoft, DirectCompute, through its DirectX platform. It is, however, tied to DirectX and Windows programming.

Both these APIs provide an extension to the C/C++ language and are not themselves a language. This allows existing C programs to be modified so that a critical, and parallel, section of the code can be reprogrammed to run on the GPU, while leaving the rest of the program intact.

GPU programming is also available to other programming languages through, mostly, user-maintained libraries. For example, the CUDA.NET and OpenCL.NET provide GPU computing for the .NET platform from Microsoft. PyCUDA allows Python programs access to CUDA. A compiler by the Portland Group allows FORTRAN programs to define kernels written in FORTRAN to run on NVIDIA GPUs.

20.2.3.2 A Bird's-Eye View of GPU Programming

The broadness of general-purpose programming on graphics processing units (GPGPU) programs is obviously something

```
__global__ void vecAdd(const float* a, canst float* b, float* c, int n)
{
    int i = blockDim.x * blockldx.x + threadldx.x;
    if (i < n) c[i] = a[i] + b[i];
}

int main(int argc, char** argv)
{
    float *h_a, *h_b, *h_c, *d_a, *d_b, *d_c;

    int N = 1000;
    size_t size = N * sizeof(float);

    h_a = (float*)malloc(size); // Allocate host memory
    h_b = (float*)malloc(size);        h_c = (float*)malloc(size);

    cudaMalloc((void**)&d_a, size); // Allocate device memory
    cudaMalloc((void**)&d_b, size);        cudaMalloc((void**)&d_c, size);

// generate input data in h_a and h_b (not shown)

    cudaMemcpy(d_a, h_a, size, cudaMemcpyHostToDevice); // Copy data from host to device
    cudaMemcpy(d_b, h_b, size, cudaMemcpyHostToDevice);

    int blockSize = 256; int numBiocks = N/blockSize + 1; // Define execution configuration
    vecAdd<<< numBlocks, blockSize >>>(d_a, d_b, d_c, N); // launch kernel

    cudaMemcpy(h_c, d_c, size, cudaMemcpyDeviceToHost); // Copy data from device to host

// at this point, h_c contains the result computed on the GPU, the host C program could then use the data in h_c
}
```

FIGURE 20.2 A sample GPU program that adds two vectors into a third vector.

that cannot be covered in a subsection. However, there are general operations, or steps, that are common to most GPGPU programs. The routine steps are the following:

- Allocate memory on the graphics card and move data from the host to the device
- Define the execution configuration (size of work groups and number of work groups)
- Launch the kernel
- Copy the results from the graphics card to the host

If we apply these steps to our example case of adding two vectors into a third vector, we get the program shown in Figure 20.2.

20.3 Monte Carlo Simulation on GPU for Radiotherapy

Owing to the stochastic nature of the MC method, simulating an enormously large number of particles is necessary for achieving a desired level of precision for dose calculation. As a consequence, the computation time is prolonged, making this dose calculation method impractical in many clinical contexts, such as online adaptive radiation therapy. Even in some applications where time is not a critical factor, for example, treatment plan verification, having a fast MC dose calculation engine is still highly desirable for a smooth workflow.

Parallel computation is a straightforward, and in many contexts, effective approach to speed up MC dose calculations. MC simulations are known as "embarrassingly parallel," because they are readily adaptable for parallel computing without the need to rethink the algorithm or the data structures. Specifically for the dose calculation problem, since each computing unit can work on a portion of the total particle histories without interfering with each other, a roughly linear scalability of computational efficiency with respect to the number of computing units is commonly expected. This is indeed the case for the CPU cluster-based MC simulations. For instance, a dose calculation package, DPM, has been ported onto a CPU cluster and almost a linear speedup has been observed with the number of processors, when up to 32 Intel CPU nodes are used (Tyagi et al., 2004).

Nonetheless, this linear scalability seen on the CPU cluster is hardly achievable on GPU architectures due to an inherent conflict between the special architecture of the GPU and the random nature of MC simulation. In general, the means of performing parallel computation are categorized into *task parallelization* and *data parallelization*. MC simulation, a typical task parallelization problem, is preferable for a CPU cluster using, for example, the MPI. For the MPI implementation on a CPU cluster, all the particle histories simulated in an MC dose calculation can be equally divided and distributed to all the processors, which can then be executed simultaneously without interfering with each other. Only at the end of the computation will the dose distribution be collected from all the processors. This method of parallelization is apparently capable of speeding up the simulation easily on a large number of CPUs. On the other hand, a GPU is suitable for the data parallelization approach. As mentioned previously, a GPU multiprocessor employs an architecture called SIMD (NVIDIA, 2009), under which the multiprocessor executes a program in groups of 32 parallel threads termed *warps*. If the paths for threads within a warp diverge due to, for example, some *if-else* statements, the warp serially executes one thread at a time while putting all other threads in an idle state. Thus, high computation efficiency is only achieved when 32 threads in a warp proceed together along the same execution path. In the aforementioned example of adding up two vectors, we can simply have each GPU thread add two corresponding entries in the two vectors. In this typical data parallelization process, all GPU threads are essentially performing the same operation on a different data set, where no thread divergence occurs and high efficiency can be expected. In contrast, in an MC calculation, the computational work paths on different threads are statistically independent. It is quite hard, if not possible, to explicitly control the operations at each thread within the stochastic nature of the MC simulation. This vast divergence between GPU threads, if not treated carefully, will destroy the parallel nature of GPU computation to a large extent, leading to low computational efficiency. A GPU-based MC simulation package, gDPM v1.0, was developed (Jia et al., 2010), where the GPU threads are simply treated, as if they were independent computational units in a CPU cluster. Despite the large number of GPU threads used in the simulation, only 5~6 times speedup was observed, indicating that special care has to be taken when implementing an MC engine on the GPU.

To mitigate this incompatibility between the GPU data-parallel architecture and the MC task-parallel paradigm, one has to seek for new simulation strategies to achieve a high efficiency. While this issue has not yet been completely resolved and is still an active research topic, there are already some effective approaches to partially relieve the problem.

There are two types of GPU thread divergence that one may encounter in an MC simulation for dose calculation. First, owing to the different particle transport physics for different types of particles, for example, between electrons and photons, one has to perform simulation in different manners. The simultaneous simulations of different types of particles will then result in divergence between threads. Second, even between particles of the same type, thread divergence still occurs because of the randomness of the particle transport process. For example, the particle transport trajectory, number of transport steps, and interaction type vary significantly from one thread to another. While we have no explicit control on the divergence of the second type, by programming the simulation carefully, it is possible to mitigate the first one. Recently, Hissoiny et al. (2011a) have developed an MC dose calculation package, GPUMCD, where the idea of separating the simulations of electrons and photons was first proposed. By smartly placing the particles to be simulated into two arrays holding electrons and photons separately and having the GPU simulate particles in only one array at a time, a considerable amount of speedup has been reported. In another recently developed package, gDPM v2.0 (Jia et al., 2011), such a strategy was also employed and was again found effective.

20.4 Current GPU-Based Monte Carlo Codes

GPU-based MC is a new area of research in radiotherapy and as such not many codes currently exist. They can be divided into two categories: photon codes and coupled electron–photon MC codes.

20.4.1 Photon Monte Carlo Codes

Photon-only MC codes were the first to appear in the literature due to the relative ease of implementation that such codes involve. They have first appeared in the domain of light transport in biomedical optics.

Alerstam et al. (2008b) have the earliest publication relating to MC light transport on GPU. Their work consisted in porting an existing CPU photon migration code to the GPU (Alerstam et al., 2008a). Briefly, photons are launched toward a medium and tracked until the photon escapes the medium or its total time of flight exceeds a user-defined maximum value. In their short letter, they report accelerations of up to 1080× when comparing to a CPU implementation of the same code.

This project has later evolved in the CUDAMCML project, which is a GPU implementation of the Monte Carlo modeling light transport (MCML) photon migration code (Wang et al., 1995). A comparison between the CPU and GPU code yielded up to a 272× acceleration. At that point, CUDAMCML could only handle layered geometries with a maximum of 100 layers, stored in constant memory. The manual of the platform details the first attempt at reducing divergence in GPU MC applications by rethinking the overall problem. Indeed, the implementation used on the GPU deviates from the original MCML implementation's handling of layer crossing, which would not have been efficient on the GPU.

This project again evolved in GPU–MCML (Alerstam et al., 2010). In this version of the work, the authors incorporate the previous developments in light transport (Badal and Badano, 2009; Fang and Boas, 2009; Lo et al., 2009; Ren et al., 2010) and add new developments to reduce divergence and other bottlenecks. One such development is the storing of the portion of the dose grid near the photon source in shared memory. If a photon is absorbed in this region, its contribution is summed in shared memory. With GPU–MCML, they observe an acceleration of up to 869× compared to a CPU version. The shared memory optimization alone increased the acceleration factor from 260× to 620×. Their paper also presents the first multi-GPU MC code implementation. Close to the ideal linear gains with regard to the number of GPUs used are reported.

Finally, GPU–MCML was modified to accept complex three-dimensional (3D) geometries (Lo et al., 2009). The geometry array is however still located in constant memory, which reduces the size of the geometry modeled due to the relatively small size of the constant memory. The authors show an acceleration of 270× compared to a conversion of the GPU program to a serial CPU program.

An independent work by Fang and Boas (2009) was also targeted at light transport in a 3D environment. Their solution departs from storing the geometry information in constant memory and instead places it in global memory, incurring the performance penalty associated with the lower bandwidth and higher latency. The authors also present a study on the error rate of using atomic versus nonatomic writes to global memory to update the scoring grid. They show that at a distance further than three voxels away from the source, the error induced by using nonatomic operations is around 1%. The authors show a speedup of 100× through 250× for their version not using atomic writes.

Badal and Badano have ported the photon transport part of PENELOPE (Salvat et al., 2009) to the GPU (Badal and Badano, 2009), yielding MCGPU code. The implementation is targeted to radiographic image creation. They make use of the Woodcock tracking algorithm (Woodcock et al., 1965). Their test simulation consists of launching 50 keV photons at a 3D phantom and collection results in a two-dimensional (2D) detector. They report a speedup of 27× compared to a modified version of the PENELOPE code using the Woodcock algorithm. The code is made freely available by the authors.

Recently, Jia et al. (2012b) have developed photon transport codes gCTD, which targets at the estimation of the imaging radiation dose to a patient from computed tomography (CT) or cone beam computed tomography (CBCT) scans. As opposed to MC GPU that translates PENELOPE directly into CUDA, the development of gCTD is optimized for GPU programming. A number of sampling algorithms are designed to yield a high efficiency. Moreover, gCTD supports the simulation of CT/CBCT scanners to a high level of realism, including the modeling of source spectrum, fluence map, and so on. Since no corresponding CPU codes are available, the speedup factors are measured against EGSnrc, a commonly used CPU MC simulation package, where electron transport in EGSnrc is turned off for a fair comparison. It was observed that gCTD is about 76 times faster than EGSnrc in a realistic patient phantom case. As for the absolute computation time, the radiation dose to a patient in a CBCT scan can be computed in ~17 s with less than 1% relative uncertainty.

Another package gDRR is also developed by Jia et al. (2012a) for accurate and efficient computations of x-ray projection images in CBCT under clinically realistic conditions. Simulation of x-ray projection images plays an important role in CBCT-related research projects, such as the design of reconstruction algorithms or scanners. gDRR accurately computes the three key components in a CBCT projection image, namely, primary, scatter, and noise signals. In particular, a GPU-based MC simulation is utilized to obtain the scatter signal and to facilitate the calculations of the noise signal. gDRR uses the same MC photon transport functions as in gCTD, with some necessary modifications. It also supports very flexible scanner geometry as well as some details of the scanner, such as a finite x-ray focal spot size, detector response, and so on. For a typical CBCT projection with a polyenergetic spectrum, the MC simulations take 28.1~95.3 s, depending on the voxel size. The MC-simulated scatter signal using gDRR is in agreement with EGSnrc results with a relative difference of 3.8%.

Finally, Hissoiny et al. have developed bGPUMCD, a dedicated GPU package aimed at brachytherapy dose calculations. It is useable for both low-dose-rate (Hissoiny et al., 2011b) and high-dose-rate brachytherapy dose calculations. For LDR calculations, a mixed simulation with both analytical geometrical description of the seeds and voxelized patients is used. TG-43 parameters have been reproduced by GPUMCD and compared to values found in the literature and in the TG-43 report. GPUMCD was found to be able to reproduce radial functions within 1.25% and anisotropy functions within 4% for low-energy single-seed simulations. Phase spaces can also be used to remove the need to track particles through their seed. For HDR dose calculations, bGPUMCD uses a phase space to represent the seed in a purely voxelized geometry. Single dwell position dose kernels as well as full dose calculations using preoptimized plans and real patient geometries have been performed. bGPUMCD is able to compute single dwell position dose kernels within 0.5 s and full dose calculations within 2 s for a 2% statistical uncertainty. Dosimetric parameters, such as D90 and V100, are within 2% when compared to a reference GEANT4 dose calculation.

20.4.2 Coupled Electron–Photon Monte Carlo Codes

Three GPU-coupled electron–photon MC codes have been developed and targeted at radiotherapy applications: GPUMCD (Hissoiny et al., 2011a), gDPM (Jia et al., 2010, 2011), and GMC (Jahnke et al., 2012).

20.4.2.1 GPUMCD

GPUMCD is a new implementation of the MC simulation based on existing physics. The code, the overall algorithm, the inputs/outputs, the memory management, and others have all been tailored specifically for the GPU while the sampling routines come from existing general-purpose MC packages such as EGSnrc (Kawrakow, 2000), EGS5 (Hirayama et al., 2010), and PENELOPE (Salvat et al., 2009).

The photon simulation is done in an analog manner. GPUMCD models Compton scattering, the photoelectric effect, and pair production. Compton scattering uses a free atom approximation and the Klein–Nishina cross section. The sampling routine is derived from the method by Everett et al. (Everett and Cashwell, 1971). The photoelectric effect is modeled simplistically by assuming that every electron is ejected from the K-shell and by ignoring atomic relaxation and fluorescence. The angular sampling of the electron is selected according to the Sauter distribution (Sauter, 1931) following the sampling formula presented in Section 2.2 of the PENELOPE manual (Salvat et al., 2009). Pair production events are modeled by ignoring the production of positrons. Instead, two electrons, with randomly selected

energies within the incoming photon energy, are generated. The angle of both electrons is sampled using the algorithm presented in Equation 2.1.18 of the EGSnrc manual.

The electron simulation is performed using a class II condensed history approach (Berger, 1963). The so-called hard interactions are simulated explicitly while soft, subthreshold, interactions are modeled in a condensed manner. Inelastic collisions are modeled using the Møller cross section, and the sampling routine presented in Section 2.4.3.i of the EGSnrc manual (Kawrakow et al., 2011) is used to sample the energies and orientations of the electrons. The photon created during a bremsstrahlung interaction has its energy sampled following the sampling routine presented in Section 2.4.2.ii of the EGSnrc manual (Kawrakow et al., 2011). The EGS4 approximation to the photon angle is used in which the photon angle is assigned as $\theta = m_0 c^2 / E$, where E is the incoming electron kinetic energy.

Multiple scattering is modeled by using the methods of Kawrakow and Bielajew (1998) to select a scattering angle and PENELOPEs to perform the electron step. The random hinge method is used to transport the particle: a given electron step is randomly split in two parts; the electron advances to the hinge point, is deflected, and advances for the rest of the step length. At the hinge point, the angle selected using the method of Kawrakow and Bielajew is used to deflect the electron.

GPUMCD can handle voxelized or parametric geometries as well as a mix of the two (Hissoiny et al., 2011b). Parametric geometries include secondary quadratic surfaces as well as boxes. GPUMCD also supports a large number of particle sources (monoenergetic, polyenergetic, phase space based, parallel, divergent, sweeping beamlets, multiple beams, arbitrary orientations, etc.) and the C++ approach to particle source definition makes it easy to include a new source type.

The GPUMCD implementation of the MC algorithm is different compared to the other existing MC packages since it had to be redesigned to fit the GPU architecture. The major difference is in the way particle histories are sequenced. In most platforms, a primary particle is launched and its secondary particles are put on top of the particle stack, regardless of their type (electron, photon, or positron). An electron track simulation could then be immediately followed by a photon simulation, and so on. On the GPU architecture, many particles are simulated at once, where the number of particles in parallel depends on the number of processing elements of the GPU. Also, since the GPU is an SIMD machine and all threads of a warp must execute the same instruction flow, simulating electrons and photons at the same time would lead to a high degree of divergence. Similarly, if all secondary particles of a primary particle are simulated on the spot, a large degree of divergence in the number of secondary particles per primary particle could be found. These two factors would lead to a large degree of performance loss on the GPU. For these reasons, GPUMCD implements a dual stack of particles: one for electrons and one for photons. Every secondary particle created during a particle track simulation is placed on its corresponding stack for a later simulation. This way, each thread only simulates one particle. Also, particle stacks are treated individually, which

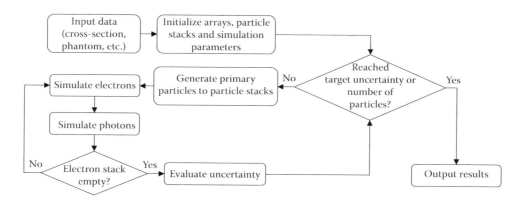

FIGURE 20.3 The workflow of GPUMCD.

ensures that at a given time, only electrons or only photons are being simulated. The overall workflow of GPUMCD is shown in Figure 20.3. The size of these stacks is graphics card dependent and varies with the amount of global memory available on the graphics card. For example, a 1.5 Gb pool of global memory can accommodate 8×2^{20} particles per stack while leaving enough memory for the other elements required by the simulation. Such elements include cross-section data, stored as one-dimensional (1D) textures, phantoms of material identifiers and densities per voxel, stored as 3D textures, pseudorandom number generator states, stored in registers and global memory, and a large number of simulation parameters, stored in constant memory.

GPUMCD has been benchmarked in a number of situations. In voxelized slab geometries with simplistic monoenergetic sources (Hissoiny et al., 2011a), it has been found to be within a 2%–2 mm gamma criteria of EGSnrc in 98% of all voxels or more for most cases. On a per particle basis, it has been benchmarked to be around 1000 times faster than EGSnrc for photons simulation and 1500 times faster for electrons simulations. Compared to the DPM platform (Sempau et al., 2000), its speed is around 400 times faster for photons simulations and 250 times faster for electrons simulations.

In voxelized patient geometries (Hissoiny et al., 2011c), it is able to compute a dose distribution within a 2% statistical uncertainty in less than 8 s per beam or 11 s per plan. GPUMCD has also been adapted to be the dose calculation tool of the magnetic resonance imaging (MRI)-linac by incorporating the effects of the magnetic field on charged particles (Hissoiny et al., 2011c). Within this dose calculation scenario, it is able to compute the dose distribution in less than 15 s per plan within a 1.5T magnetic field.

20.4.2.2 gDPM

gDPM, another GPU-based MC package for coupled photon–electron transport, was developed by Jia et al. (2010, 2011). Such a package tries to port a publically available CPU-based fast MC package, DPM, to the GPU platform. In this process, DPM particle transport physics is maintained and hence the simulation accuracy, whereas the code is optimized for the GPU architecture to improve computational efficiency.

The original sequential DPM MC code was previously developed for fast dose calculations in radiotherapy treatment planning (Sempau et al., 2000). It aims to simulate coupled photon–electron transport with a set of approximations valid for the energy range considered in radiotherapy. Specifically, the photon transport is handled by using Woodcock tracking method, which greatly increases the simulation efficiency of the boundary tracking process (Woodcock et al., 1965). As for the electron transport, DPM implements a condensed history technique. Step-by-step simulation is used for inelastic collisions and bremsstrahlung emission involving energy losses above certain cutoffs. It also employs new electron transport mechanics and multiple scattering distribution functions to allow long transport steps. Continuous slowing down approximation is employed for energy losses below some preset energy thresholds. Positron transport is treated in the same way as for electrons and two photons are created at the end of the positron path to account for the annihilation process. The accuracy of DPM has been demonstrated to be within ±2% of measurements for both clinical photon and electron beams (Chetty et al., 2002, 2003).

The first version of gDPM was developed in 2009 (Jia et al., 2010). In this version, all GPU threads are treated as if they were independent computational units. All the source particles are distributed to the GPU threads and each thread tracks the entire history of a source particle as well as all the secondary particles it generates. It is found that only 5.0~6.6 times speedup can be achieved due to the intrinsic conflicts between the randomness of the MC simulation and the GPU SIMT architecture mentioned previously.

Lately, the gDPM v2.0 was released with a higher computational efficiency (Jia et al., 2011). gDPM executes the simulations in a batch fashion, where the total number of particle histories are divided evenly into a number of batches. The simulations are conducted for each batch sequentially and the dose depositions are recorded. Statistical analysis is performed based on the results in all batches to obtain the average dose to each voxel and the corresponding uncertainties. The overall workflow is shown in Figure 20.4. Within each batch, the simulation details are presented by the workflow in the right panel of Figure 20.4. Since photons and electrons undergo different physics and hence attain different

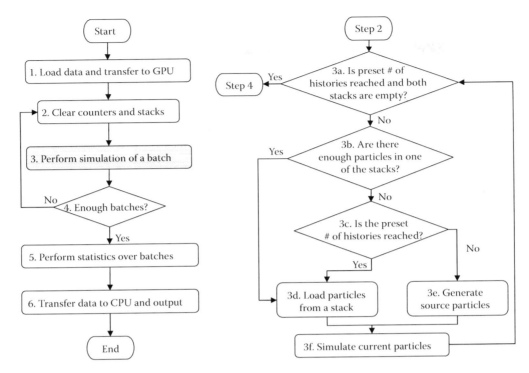

FIGURE 20.4 The flowchart of gDPM. Detailed steps of the batch simulation part (step 3 in the left panel) are shown on the right panel. (Adapted from Jia et al. 2011. *Phys Med Biol* **56** 7017–1031.)

execution paths, gDPM v2.0 uses a simulation scheme where photon transport and electron transport are separated to partially relieve the thread divergence issue, as first proposed by Hissiony et al. Specifically, a particle array of length N is allocated to store all the particles currently being simulated. The particles in this array can be either photons or electrons, but all of the same type at any time during the simulation. The size of this particle array should be large enough, so that the GPU can fully exploit its parallelization ability, while not too large to fit in the GPU memory. Moreover, since the GPU executes simulation in warps, that is, a group of 32 threads run simultaneously on a multiprocessor, it is beneficial to choose N to be a multiple of warp size to avoid wasting resources, for example, $N = 65,536$ in a typical setup. Meanwhile, two more arrays are allocated as stacks to store any secondary particles generated during the simulation, one for photons and one for electrons. Their lengths are large enough to hold all secondary particles. During simulation, a total number of N GPU threads are launched simultaneously to transport N particles of the same type, while putting all secondary particles into the corresponding stacks, which effectively separates the simulations of the photon and the electron transport. Besides this simulation strategy, a high-performance random number generator supported by NVIDIA (NVIDIA, 2010) and hardware supported linear interpolation are also utilized to further speed up calculations.

The development of gDPM v2.0 also emphasizes on its clinical practicality by integrating various key components necessary for dose calculation in radiotherapy. An interface of the gDPM package is built to load clinical IMRT or VMAT treatment

plans in the format of DICOM RT. This includes the functions of loading voxelized patient CT data, organ structure information, fluence map of a treatment plan, and so on. Computations related to the geometry of a treatment plan are enabled to take into account the rotations of linac gantry, multileaf collimator (MLC), collimator, and couch.

As for source modeling, two simple aspects have been considered in gDPM in the current released version. First, it is supported to generate source particles according to a designed photon fluence map in IMRT or VMAT plans. Such a function is achieved by using Metropolis sampling algorithm (Hastings, 1970), which is favored in GPU due to its infrequent memory access. Second, photon source particles are generated according to a given energy spectrum. Specifically, the total number of photons is distributed to a few energy intervals according to the spectrum and simulation is performed for each interval sequentially. This strategy ensures that, at any moment of the simulation, all GPU threads are dealing with particles of similar energies, removing the possibility of losing efficiency due to the variance of simulation time between GPU threads handling photons of different initial energies.

Yet, accurate source modeling of a linear accelerator is not a trivial problem. gDPM v2.0 leaves the interface of source particle generation part open and users can supplement their own functions to generate source particles according to their own source model or simply by using a phase space file.

As for the validation of gDPM, simulations in various phantom cases indicate that results from CPU DPM and gDPM are in good agreement for both the electron and the photon sources. In particular, statistical *t*-tests are performed and the dose differences between the CPU and the GPU results are found not statistically significant in over 96% of the high-dose region and over 97% of the entire phantom region. Speedup factors of 69.1~87.2 have been observed against a 2.27 GHz Intel Xeon CPU processor. An IMRT or a VMAT plan dose calculation using gDPM can be achieved in 36.1~39.6 s with a single GPU for the average standard deviation less than 1% of the prescription dose. Moreover, multi-GPU implementation of gDPM has also been developed. Another speedup factor of 3.98~3.99 compared to a single GPU has been observed using a four-GPU system.

20.4.2.3 GMC

Recently, another MC simulation package, GMC, has been developed (Jahnke et al., 2012). It is a GPU implementation of the electromagnetic part of the Geant4 MC code (Agostinelli et al., 2003). The main difference between this code and the aforementioned two packages is the GPU implementation of lepton (electron and positron in this code) transport. A full history of a lepton consists of a large number of small steps separated by voxel boundaries or discrete interaction sites. As opposed to having a GPU kernel simulate, a full history of a lepton as in GPUMCD and gDPM, a GPU kernel in GMC loads leptons from a stack in global memory, transports each of them by only one step, and writes them back to the stack. Such a kernel is repeatedly invoked to move the leptons forward till they are absorbed or are outside the phantom region. The advantages of this simulation scheme are the reduction of thread divergence due to different history lengths among particles. In particular, if a GPU thread handles an entire particle history, those threads with short-lived particles will wait for others, resulting in reduced efficiency.

Another unique feature of GMC is random number generation. GMC utilizes a modified Mersenne twister algorithm for pseudorandom number generation due to its very long period. The random numbers are generated by a dedicated kernel and are not on the fly. As such, a number of random numbers are pregenerated and stored in a pool residing in the global memory. Other particle transport kernels read these numbers whenever necessary. Once those pregenerated random numbers are exhausted, a GPU kernel is invoked to refresh the random number pool. It was reported that a 3~4 times of speedup in terms of total computation time has been achieved by this method compared to the on-the-fly generation of random numbers. This can be attributed to the fact that removing random number generations in particle transport simulations reduces the degree of GPU thread divergence.

GMC has been validated in a number of contexts ranging from a simple homogeneous water phantom with an open beam to a real patient IMRT case with a high degree of fluence modulation. A phase space source model is used in those testing cases, which is generated by simulating an Elekta Synergy linac head using Geant4 on the CPU. In all the phantom cases, the agreement between the GMC results and the Geant4 results is satisfactory with over

97.5% region passing the 2%/2 mm gamma index test. The computational speed is assessed by executing GMC on an NVIDIA GeForce GTX580 card. On an average, GMC processes 657.60 histories per millisecond. Compared with the CPU execution of Geant4 on a 2.13 GHz Intel Core2 processor, a speedup factor of 4860 is achieved. This large speedup factor can be partly ascribed to the slow Geant4 simulations on the CPU. Furthermore, a linear speedup with respect to the number of GPUs is also observed.

20.5 Summary

In this chapter, we have discussed the applications of GPU for fast MC dose calculations in cancer radiotherapy. Some basic principles of the GPU architecture and GPU programming have been introduced, as well as those difficulties when applying GPU for MC simulation. It has been seen that the GPU architecture is not ideal for conducting MC simulations in radiotherapy. Developing GPU-friendly packages for MC simulations therefore requires a redesign of the simulation algorithms. A number of GPU-based MC packages developed for a variety of purposes in radiotherapy have become available. The achieved high performance with affordable GPU cards has undoubtedly demonstrated the unique advantages of the GPU in radiotherapy.

At the end of this chapter, we would like to present a few directions to be explored in the near future for GPU-based MC simulations in radiotherapy. Generally speaking, there are two aspects one needs to keep in mind to improve the practicality of a GPU-based MC simulation package in clinical environments: efficiency and accuracy. Those future directions essentially target the improvements of those two aspects.

First, it is highly desirable to further boost the simulation efficiency. Despite the accomplishments achieved so far, MC simulation can only preliminarily satisfy the time requirements in radiotherapy for the purposes of dose calculation and treatment verifications. In many other applications where repeated dose calculations are needed, such as computing dose deposition coefficients for IMRT planning, the current GPU-based MC simulation is still far away from being clinically applicable. There are a number of ways for potentially increasing the MC simulation efficiency besides those trivial approaches of using more and better GPUs. In particular, it is, and will continue to be, an active research topic to design GPU-friendly simulation schemes to further alleviate the branching issue and to use the GPU's memory more efficiently. Also included in this direction are utilizing variation reduction techniques, and designing new ones suitable for GPU if needed, to increase the simulation convergence rate and hence the overall efficiency. Those variation reduction algorithms might be of particular use on GPU, as they usually tend to force the simulation be performed in a more predictable fashion and therefore may encounter less branching problems among different GPU threads.

Another direction is performing simulation with a higher accuracy in the underlying physics to obtain better simulation results. Previously, constrained by the limited computing power, a number of components in MC simulations were simplified or even ignored. Armed with the powerful GPUs, it becomes possible for us to retrieve those components, if needed, to yield simulation processes that are closer to the realistic cases and hence potentially more accurate results. For example, the simulation accuracy is largely dependent on how realistically the source particles are generated. In many current MC packages, it is popular to empirically model the linear accelerator head, that is, energy spectrum, fluence distribution, angular distribution, and so on, and generate source particles accordingly. Detailed simulations of MLC transmission using particles from a phase space file obtained from a separate MC simulation are another possible but complicated approach. With the computing power of GPU, we can potentially perform the particle transport starting from the phase space file above the MLC, while considering many effects such as MLC transmission, interleaf leakage, MLC round end leakage, and so on. It is expected that those components will greatly enhance the simulation accuracy and hence facilitate the clinical applications. Yet, it is a legitimate question regarding to what extent of realism one needs to perform MC simulations. It is highly likely that the improvements of accuracy from many additional detailed simulations are incremental, especially considering there exist other error sources on the simulation results, such as statistical uncertainties from the MC method itself. Nonetheless, the GPU at least offers us a means to consider those components that are necessary but was simplified or ignored previously. It is an interesting research topic to identify those crucial components responsible for the accuracy and implement them in the simulation.

References

Agostinelli S, Allison J, Amako K et al. 2003. GEANT4—A simulation toolkit. *Nucl Instrum Methods Phys Res A, Accel Spectrum Detect Assoc Equip* **506** 250–303.

Alerstam E, Andersson-Engels S, and Svensson T 2008a. White Monte Carlo for time-resolved photon migration. *J Biomed Opt* **13** 041304.

Alerstam E, Lo W C, Han T D et al. 2010. Next-generation acceleration and code optimization for light transport in turbid media using GPUs. *Biomed Opt Express* **1** 658–75.

Alerstam E, Svensson T, and Andersson-Engels S 2008b. Parallel computing with graphics processing units for high-speed Monte Carlo simulation of photon migration. *J Biomed Opt* **13** 060504.

Badal A and Badano A 2009. Accelerating Monte Carlo simulations of photon transport in a voxelized geometry using a massively parallel graphics processing unit. *Med Phys* **36** 4878–80.

Berger M J 1963. Monte Carlo calculation of the penetration and diffusion of fast charged particles. *Methods Comput Phys* **1** 135–215.

Chetty I J, Charland P M, Tyagi N et al. 2003. Photon beam relative dose validation of the DPM Monte Carlo code in lung-equivalent media. *Med Phys* **30** 563–73.

Chetty I J, Moran J M, McShan D L et al. 2002. Benchmarking of the dose planning method (DPM) Monte Carlo code using electron beams from a racetrack microtron. *Med Phys* **29** 1035–41.

Constantin M, Sawkey D, Mansfield S, and Svatos M 2011. The compute cloud, a massive computing resource for patient?Independent Monte Carlo dose calculations and other medical physics applications. *Med Phys* **38** 3392.

Deng L and Xie Z S 1999. Parallelization of MCNP Monte Carlo neutron and photon transport code in parallel virtual machine and message passing interface. *J Nucl Sci Technol* **36** 626–9.

Everett C J and Cashwell E D 1971. *A New Method of Sampling the Klein–Nishina Probability Distribution for All Incident Photon Energies above 1 keV*. (Los Alamos: Los Alamos Scientific Laboratories).

Fang Q and Boas D A 2009. Monte Carlo simulation of photon migration in 3D turbid media accelerated by graphics processing units. *Opt Express* **17** 20178–90.

Gu X J, Jelen U, Li J S, Jia X, and Jiang S B 2011. A GPU-based finite-size pencil beam algorithm with 3D-density correction for radiotherapy dose calculation. *Phys Med Biol* **56** 3337–50.

Hastings W K 1970. Monte-Carlo sampling methods using Markov chains and their applications. *Biometrika* **57** 97–109.

Hirayama H, Namito Y, Bielajew A F, Wilderman S J, and Nelson W R 2010. The EGS5 code system. Stanford Linear Accelerator Center Report SLAC-R-730.

Hissoiny S, Ozell B, Bouchard H, and Despres P 2011a. GPUMCD: A new GPU-oriented Monte Carlo dose calculation platform. *Med Phys* **38** 754–64.

Hissoiny S, Ozell B, and Després P 2009. Fast convolution-superposition dose calculation on graphics hardware. *Med Phys* **36** 1998–2005.

Hissoiny S, Ozell B, Despres P, and Carrier J F 2011b. Validation of GPUMCD for low-energy brachytherapy seed dosimetry. *Med Phys* **38** 4101–7.

Hissoiny S, Raaijmakers A J, Ozell B, Despres P, and Raaymakers B W 2011c. Fast dose calculation in magnetic fields with GPUMCD. *Phys Med Biol* **56** 5119–29.

Jacques R, Taylor R, Wong J, and McNutt T 2010. Towards real-time radiation therapy: GPU accelerated superposition/convolution. *Comput Methods Programs Biomed* **98** 285–92.

Jahnke L, Fleckenstein J, Wenz F, and Hesser J 2012. GMC: A GPU implementation of a Monte Carlo dose calculation based on Geant4. *Phys Med Biol* **57** 1217–29.

Jia X, Gu X, Graves Y J, Folkerts M, and Jiang S B 2011. GPU-based fast Monte Carlo simulation for radiotherapy dose calculation. *Phys Med Biol* **56** 7017–7031.

Jia X, Gu X, Sempau J et al. 2010. Development of a GPU-based Monte Carlo dose calculation code for coupled electron-photon transport. *Phys Med Biol* **55** 3077.

Jia X, Yan H, Cervino L, Folkerts M, and Jiang S B 2012a. A GPU tool for efficient, accurate, and realistic simulation of cone beam CT projections. To be appeared in *Med. Phys.*

Jia X, Yan H, Gu X, and Jiang S B 2012b. Fast Monte Carlo simulation for patient-specific CT/CBCT imaging dose calculation. *Phys Med Biol* **57** 577–90.

Kawrakow I 2000. Accurate condensed history Monte Carlo simulation of electron transport. I. EGSnrc, the new EGS4 version. *Med Phys* **27** 485–98.

Kawrakow I and Bielajew A F 1998. On the representation of electron multiple elastic-scattering distributions for Monte Carlo calculations. *Nucl Instrum Methods Phys Res B, Beam Interact Mater At* **134** 325–36.

Kawrakow I, Mainegra-Hing E, Rogers D W O, Tessier F, and Walters B R B 2011. The EGSnrc code system: Monte Carlo simulation of electron and photon transport, NRCC. http://irs.inms.nrc.ca/software/egsnrc/documentation/pirs701/.

Keyes R, Arnold D, Reynaud A, and Luan S 2011. McCloud: Toward 10 million Monte Carlo primaries in 5 min for clinical use. *Med Phys* **38** 3648.

Lo W C, Redmond K, Luu J et al. 2009. Hardware acceleration of a Monte Carlo simulation for photodynamic therapy [corrected] treatment planning. *J Biomed Opt* **14** 014019.

Men C, Gu X, Choi D et al. 2009. GPU-based ultra fast IMRT plan optimization. *Phys Med Biol* **54** 6565–73.

NVIDIA 2009. NVIDIA CUDA computing unified device architecture, Programming guide. http://developer.download.nvidia.com/compute/DevZone/docs/html/C/doc/CUDA_C_Programming_Guide.pdf.

NVIDIA 2010. CUDA CURAND Library. https://developer.nvidia.com/curand.

Pratx G and Xing L 2011. Monte Carlo simulation in a cloud computing environment with MapReduce. *Med Phys* **38** 3869.

Ren N, Liang J, Qu X et al. 2010. GPU-based Monte Carlo simulation for light propagation in complex heterogeneous tissues. *Opt Express* **18** 6811–23.

Salvat F, Fernández-Varea J M, and Sempau J 2009. *PENELOPE-2008: A Code System for Monte Carlo Simulation of Electron and Photon Transport* (Issy-les-Moulineaux, France: OECD-NEA.)

Sauter F 1931. Über den atomaren photoeffekt in der K-scale nach der relativistischen wellenmechanik diracs. *Annalen der Physik* **403** 454–88.

Sempau J, Wilderman S J, and Bielajew A F 2000. DPM, a fast, accurate Monte Carlo code optimized for photon and electron radiotherapy treatment planning dose calculations. *Phys Med Biol* **45** 2263–91.

Sutherland K, Miyajima S, and Date H 2007. A simple parallelization of GEANT4 on a PC cluster with static scheduling for dose calculations. *J Phys: Conference Series* **74** 012020.

Tyagi N, Bose A, and Chetty I J 2004. Implementation of the DPM Monte Carlo code on a parallel architecture for treatment planning applications. *Med Phys* **31** 2721–5.

Wang H, Ma Y, Pratx G, and Xing L 2011. Toward real-time Monte Carlo simulation using a commercial cloud computing infrastructure. *Phys Med Biol* **56** N175–81.

Wang L, Jacques S L, and Zheng L 1995. MCML—Monte Carlo modeling of light transport in multi-layered tissues. *Comput Methods Programs Biomed* **47** 131–46.

Woodcock E, Murphy T, Hemmings P, and Longworth S 1965. Techniques used in the GEM code for Monte Carlo neutronics calculations in reactors and other systems of complex geometry. In: *Applications of Computing Methods to Reactor Problems*, Argonne National Laboratories Report, pp. ANL-7050.

Monte Carlo for Shielding of Radiotherapy Facilities

Peter J. Biggs
Massachusetts General Hospital

Stephen F. Kry
*University of Texas MD Anderson
Cancer Center*

21.1 Introduction

Shielding is an important and challenging aspect of radiotherapy management. The purpose of shielding radiotherapy machines is to reduce the dose equivalent to a point outside the treatment room to an acceptably low level. Failure to design shielding adequately can result in overexposures to personnel or expensive barrier retrofitting.

The methods for calculating treatment vault shielding for conventional radiotherapy accelerators are documented in the National Council on Radiation Protection and Measurements reports (NCRP Report Nos. 51, 79, and 151). The approach relies on transmission, leakage, scatter, and albedo factors based on standard room layouts and designs. The transmission factors are based solely on ordinary concrete (NCRP Report No. 151) with a density of 2.35 g/cm^3 (147 lb/ft^3), and have been generated for radiotherapy beams of only a few nominal energies. Between these and other approximations detailed below, they can only be considered as approximations of the true shielding requirements of a given medical accelerator. Different compositions of concrete or other shielding materials are often used, and different accelerators have notably different spectra, even at the same nominal energy (Sheikh-Bagheri and Rogers, 2002). Despite these reservations, the calculations have been shown to be quite conservative in practice (IAEA, Low 2003). However, there are many situations in radiotherapy shielding that cannot be calculated with NCRP formalisms, or where it is of interest to determine shielding requirements

more precisely. In addition, there are now many nonstandard accelerators (e.g., Tomotherapy®, Cyberknife®, and Gammaknife®) as well as nonstandard beams (e.g., flattening filter free (FFF)). For example, Kry et al. (2009a) have shown, via Monte Carlo, substantial differences in 10th-value layers (TVLs) between flattened and FFF beams. There are also nonstandard geometry rooms for which the standard NRCP formalisms do not easily apply, for example, a primary beam striking the outer maze wall. Additionally, while the NCRP offers guidance on proton and light-ion particle therapy vault shielding (NCRP Report Nos. 51 and 144), the novelty of such equipment makes shielding studies of considerable current interest. For these types of scenarios, the formalisms of the NCRP reports may be insufficient and Monte Carlo methods are a viable option to examine the shielding requirements of a given situation.

There are two approaches to solving radiation shielding problems using the Monte Carlo method. First, Monte Carlo may be directly applied to the solution of a shielding problem by modeling the specific accelerator, vault, and potentially even the patient or phantom of interest. This approach offers the most accurate solution for a given problem by accounting for the specifics of the radiation beam and environment. However, attaining sufficient details of all aspects of the radiotherapy system and its environment is extremely challenging. Sufficient details should be readily achievable for the primary and scattered radiation; however, for simulations of head leakage, this can be very demanding. As head leakage is often the dominant factor for secondary barriers, particularly for intensity modulated radiation therapy (IMRT) treatments, and is important for direct door shielding, its accurate calculation is essential. Monte Carlo models that might be sufficiently accurate to calculate head leakage levels are complex and therefore uncommon.

The second approach using Monte Carlo is to solve for general shielding parameters and from these derive an analytic solution that allows for the shielding requirements of particular cases to be solved. This is the approach presented in NCRP Report No. 151, where the Monte Carlo method has been used to generate shielding parameters, such as patient scatter fractions, for a general situation. This second approach has the advantages of requiring a less complex model and being less computationally intense. However, this is not a general solution and only solves part of the overall problem.

This chapter reviews the rationale and special considerations for performing shielding-relevant calculations using the Monte Carlo technique. Section 21.2 describes the Monte Carlo techniques as applied to shielding, Section 21.3 details photon shielding calculations, Section 21.4 deals with neutron calculations, Section 21.5 addresses applications of Monte Carlo for nonstandard room geometries, Section 21.6 addresses the application of the Monte Carlo approach for nonstandard beams and accelerators, including protons and light-ion accelerators, and Section 21.7 provides a summary of this chapter and future considerations for Monte Carlo research in therapy shielding. Finally, while considerable Monte Carlo work has been carried out for radiation shielding in the field of nuclear reactors (e.g., Jaeger et al., 1975), this chapter concerns itself only with the work that is specific to radiotherapy installations.

21.2 Monte Carlo Codes

21.2.1 Specific Programs

A wide range of different Monte Carlo codes have been used for shielding calculations in radiotherapy, including EGS/EGSnrc (Kawrakov et al., 2010), MCNP/MCNPX (Briesmeister 1993), GEANT4 (Agostinelli et al., 2003), ITS (Halbleib and Mehlhorn 1998), and others. Typically, any of these codes is a viable option. Naturally, different codes have different strengths and weaknesses and some considerations are presented below.

Computation time is an important consideration. For primary barrier attenuation calculations, calculation of the third or fourth TVLs may require the simulation of considerably more particles than would be typically needed for patient dose calculations. Calculations involving secondary barriers may be even more time consuming as they will typically involve relatively low-frequency scattering events or head leakage and only a tiny minority of generated and tracked particles may even impinge on the barrier. Codes that handle radiation transport most expediently are clearly advantageous. Moreover, variance reduction techniques will typically be an important part of such simulations.

The flexibility of the geometry specification is also a potentially important consideration. For many shielding applications

(Sections 21.3.1 through 21.3.3), the barrier transmission can be calculated accurately using a detailed simulation of the radiation transported through the beam line accelerator components. Therefore, Monte Carlo systems that readily model beam line components hold an advantage (e.g., BEAMnrc). However, if head leakage or neutrons are to be simulated, modeling of non beam line components of the accelerator head (primary collimator, structural and shielding components, etc.) may also be required. In this case, a Monte Carlo code with more flexible geometry modeling (e.g., MCNP/MCNPX) is likely to be much more useful. Similarly, if arbitrary vault designs are being considered, flexible geometry specifications are again important. The need for a flexible geometry definition clearly depends on the specifics of the shielding problem being evaluated.

Finally, the ability of the code to track a variety of particles may be an important consideration. Codes including MCNP/MCNPX or GEANT can track a wide range of particles (although not all materials may have the available cross sections, e.g., photoneutron production in steel may require additional cross-section data in MCNPX), whereas EGSnrc only tracks photons, positrons, and electrons (thereby neglecting neutrons). In general, only some cross sections will be tabulated in any cross-section library, while other cross sections may be approximated with models. Users of any code should be aware of the limitations in the physics employed in their Monte Carlo system.

21.2.2 Interpretation of Results and Uncertainties

In any Monte Carlo simulation, calculated tally values have an associated uncertainty. This uncertainty will be a combination of both random errors (which are readily estimated by most MC codes and typically quoted) and systematic errors (that reflect uncertainties in the modeling geometry, source specification, cross-section data, etc.). In shielding problems, the random/statistical uncertainty may be minimized with large numbers of histories and the use of variance reduction techniques. In that case, systematic errors may dominate and an estimate of the systematic uncertainty should be incorporated into the uncertainty of the final result.

21.3 Photon Shielding

Shielding of photons must include primary photons, as well as those scattering off a patient or wall onto a secondary barrier (patient scatter and wall scatter, respectively), and leakage from the accelerator head. Both the intensity of the photons and their energy (TVL) affect the shielding burden of each of these sources of radiation. In the following sections, these two approaches to shielding calculations (full Monte Carlo simulation of the linear accelerator and vault or piecemeal determination of individual shielding parameters) will be discussed and compared.

21.3.1 Barriers

For radiotherapy shielding, the most common shielding material is concrete. Concrete is a mixture of cement, water, and aggregate. The quantity and type of aggregate can vary yielding many different types of concrete. Even concrete of the same name can show substantial variation: "ordinary" concrete can vary in density between 2.09 and 2.50 g/cm³ (NCRP Report No. 151). Additionally, high-density aggregates (including iron) can be used to make heavy concretes that offer increased photon attenuation. The atomic composition of several types of concrete (by mass fraction) is shown in Table 21.1. Similar tables are available elsewhere (see, e.g., Kase et al., 2003 or Jaeger 1975). While these tables allow for simple calculations, the reality of a particular concrete wall is invariably more complicated as even nonuniformity in the distribution of aggregate can notably affect the shielding capabilities of concrete (Kase et al., 2003). While this poses challenges for the application of Monte Carlo to concrete shielding, this is also a challenge for analytical shielding approaches. Monte Carlo approaches have a distinct advantage in the determination of TVLs or half-value layers (HVLs) of high-density concretes. The attenuation (and therefore the amount of shielding material) of such concretes cannot simply be scaled by the density of the medium. Because of the high-Z components in high-density concrete, the true TVL of high-density concrete is less than that predicted by density scaling the TVL of low-density concrete. This was noted by Facure et al.

(2007) who found the TVL to be overestimated by up to 20% when the low-density concrete TVL was scaled by the relative density.

For certain applications, such as retrofitting an older room other materials are employed in shielding (or could be relevant to Monte Carlo shielding problems). The composition (by mass fraction) and density of some such materials are presented in Table 21.2.

21.3.2 Primary Radiation and TVLs

21.3.2.1 Direct Calculation

Directly calculating HVLs and TVLs with Monte Carlo has many attractive features. In addition to providing better accuracy than analytic approaches, HVL/TVLs can also be examined under broad-beam versus narrow-beam geometries. This means that the HVL/TVLs can be determined for a variety of field sizes, which may be of particular interest for machines where the field size is limited, for example, Cyberknife.

A Monte Carlo model of the beam line of the accelerator is sufficient for the task of calculating transmission of the primary beam in a barrier. Such a Monte Carlo model can be readily validated against attenuation (such as using percent-depth dose measurements in water). In place of developing a Monte Carlo model of the accelerator head (e.g., beginning with electrons incident on a bremsstrahlung target), it is also possible to

TABLE 21.1   Elemental Composition (By Mass Fraction) of Various Types of Concrete

Type	Portland	Ordinary	Magnetite	Barytes	Magnetite and Steel	Limonite and Steel	Serpentine
Density (g/cm³)	2.30	2.35	3.53	3.35	4.64	4.54	2.1
Element				Mass Fraction			
H	0.010	0.006	0.003	0.004	0.002	0.007	0.016
C	0.001						0.001
O	0.529	0.498	0.331	0.312	0.138	0.156	0.512
Na	0.016	0.017					0.004
Mg	0.002	0.003	0.009	0.001	0.004	0.002	0.135
Al	0.034	0.046	0.023	0.004	0.010	0.006	0.019
Si	0.337	0.315	0.026	0.010	0.016	0.015	0.209
S		0.001	0.001	0.108			
K	0.013	0.019		0.048		0.001	0.004
Ca	0.044	0.083	0.071	0.050	0.056	0.058	0.068
Ti			0.054		0.016		
V			0.003		0.001	0.001	
Cr			0.002				0.001
Mn			0.002				
Fe	0.014	0.012	0.474		0.758	0.755	0.031
Ba				0.463			

Source: Data are from NIST for Portland concrete (http://physics.nist.gov/cgi-bin/Star/compos.pl) and Chilton A.B., Shultis J.K., and Faw R.E. *Principles of Radiation Shielding.* Prentice-Hall, Inc., Englewood, NJ, 1984 otherwise.

TABLE 21.2   Elemental Composition (By Mass Fraction) of Various Materials

Material	Air (Dry, Sea Level)	Polystyrene	Polyvinyl Chloride	Borosilicate Glass (Pyrex)	Lead Glass	Lead
Density (g/cm³)	0.001205	1.060	1.406	2.23	6.22	11.35
Element				Mass Fraction		
H		0.0774	0.0484			
B				0.0401		
C	0.0001	0.9226	0.3844			
N	0.7553					
O	0.2318			0.5396	0.1565	
Na				0.0282		
Al				0.0116		
Si				0.3772	0.0808	
Cl			0.5673			
Ar	0.0128					
K				0.0033		
Ti					0.0081	
As					0.0027	
Pb					0.7519	1.000

Source: Data are from NIST (http://www.nist.gov/pml/data/xraycoef/index.cfm).

begin simply with the final photon field (skipping the transport through the head). Historically, this approach was used to calculate TVL values based on monoenergetic photon beams. This is clearly only an approximate solution. Contemporary studies use polyenergetic spectra. For example, the Mohan spectra (1983) have been used by several groups to calculate the TVL of barriers (Avila-Rodriguez et al., 2005; Facure et al., 2007). This process allows the user to avoid developing even a beam line model of the accelerator, simplifying the modeling as well as the radiation transport. However, it is a less robust modeling approach as it does not account for the specifics of individual treatment heads or changes with treatment parameters. Of note is the change in TVL with field size (Jaradat and Biggs 2007), particularly for ^{60}Co and low-energy x-ray beams. Such evaluation would be particularly useful when designing shielding for a vault with a dedicated stereotactic machine with a limited aperture size.

When the primary beam is obliquely incident on a barrier, the required barrier thickness is less than the calculated value (based on normal incidence) due to the obliquity of the beam. For angles of incidence less than 45°, the effective thickness of the barrier (i.e., the physical thickness times the obliquity factor) is given by $t/\cos(\theta)$, where t is the normal barrier thickness and θ is the angle of incidence of the primary beam. For angles of incidence greater than 45°, this adjusted obliquity factor thickness may not be adequate because more photons are likely to scatter normal to the barrier and, hence, may not travel the oblique path through the concrete. NCRP methodology specifies rules of the thumb to correct this effect by adding one or two HVLs to t, depending on the beam energy and shielding material. This was found by the experimental work of Kirn et al. (1954) and verified by the Monte Carlo work of Biggs (1996).

21.3.2.2 Derivation of General Parameters

For primary beam shielding considerations, determination of the TVL_1 (the first TVL) and TVL_e (the equilibrium TVL) are useful calculations because they allow for arbitrary barriers to be evaluated. In this case, there is little difference between the direct shielding calculation approach (Section 21.3.2.1) and the general parameter approach to determine the shielding metrics of TVL_1 and TVL_e. Of note, the divergence of the beam must be removed from the calculated TVLs so that they can be applied to a barrier at an arbitrary distance from the source. Perhaps, the earliest publication on primary beam spectra and TVLs was that of Nelson and LaRiviere (1984). Their calculation, based on EGS, was generic, that is, a simple target, and did not include specific beam line elements.

21.3.2.3 Inverse Square Law

McGinley (2001) made measurements of the exposure at various distances beyond a primary barrier for several linear accelerator vaults, expecting to see the dose fall off as the inverse square of the distance between the target and the point of interest. He was surprised to find that the effective source position was somewhere between the target and the barrier. He attributed this effect to photon scattering in the primary barrier. Biggs (2002) investigated this effect using the MCNP program and found that indeed the exposure fell off more rapidly than the inverse square, indicating that the effective photon source was not the target. The importance of this may be minimal because the experimentally derived TVLs as well as those using the Monte Carlo method have this factor inherently built into them, based, as they are, on single-point measurements. However, if multiple points beyond a barrier are being examined and compared, some discrepancy from an inverse square relationship would be expected.

21.3.3 Patient-Scattered Radiation Fractions and TVLs

Patient scatter is important only for the secondary barrier adjacent to the primary barrier and it should be noted that this radiation is incident obliquely on the secondary barrier. NCRP Report No. 151 recommends against using an obliquity factor for scattered radiation based on the diffuse nature of the scattering source.

21.3.3.1 Direct Calculation

With a validated primary photon beam, the radiation scattering from a patient/phantom and striking and penetrating a secondary barrier can be calculated using the Monte Carlo procedure (Biggs and Styczynski 2008). The scattering body should be appropriately considered as its size, shape, and composition can affect the amount of scattered radiation. Larger scattering bodies may scatter more radiation, but they will also attenuate more of the scattered radiation that is produced within them. Similarly, the size of the radiation field should be considered as larger field sizes will generate more scatter and the scatter will originate at different locations relative to the edge of the phantom (altering the amount of attenuation of the scattered radiation).

21.3.3.2 Derivation of General Parameters

Patient scatter fractions (NCRP Report No. 151 based on Taylor et al. (1999, 2000)) are the ratio of the scattered kerma to the dose at d_{max} in a water sphere. For an arbitrary radiation beam, this approach can be applied to determine scattered radiation fractions and TVLs (e.g., Kry et al., 2009a). Such values can then be applied to general shielding problems. While this approach is well suited for solving shielding problems for unique radiation beams, it leaves substantial uncertainties for conditions such as irregularly shaped treatment vaults where a simple scatter fraction may not be sufficient information.

As with the direct calculation approach, care should be given to the size of the phantom and the radiation field (scatter is typically reported per unit area of the field, although a maximum field size would also be a reasonable approach).

Owing to an extended source size, the NCRP Report No. 151 specifically prohibits the use of an obliquity factor for scattered radiation. However, the use of an obliquity factor for scattered radiation is used in one of the examples. Because of this, Biggs and Styczynski (2008) used the MCNP code to determine the TVLs for 30° scattered radiation obliquely incident on the secondary barrier. Their findings showed that in contradistinction to NCRP recommendations, an obliquity factor could be justified.

21.3.4 Wall-Scattered Radiation Fractions and TVLs

Wall scatter is generally the least important component in determining secondary barrier or door thickness requirements.

21.3.4.1 Direct Calculation

As with patient scatter, once a primary beam is validated, radiation, as it scatters off the treatment vault walls, can be modeled relatively easily in a Monte Carlo calculation. Calculation of radiation scattered around a vault is likely to be largely limited by the geometrical definition of the vault and the number of particle histories.

It is unreasonable and unnecessary to model every detail of a clinical treatment vault, such as cupboards, sinks, and other ancillary equipment. However, large, dense structures should be modeled, or at least investigated to determine their importance in scattering radiation. Additionally, it is important to note the difference between the locations of the surface of the wall and the surface of the underlying concrete. Particularly in the ceiling, the concrete may be a substantial distance away from the false ceiling. Construction blueprints are generally a good source of information of the shielding-critical geometry.

Because wall-scattered radiation may have undergone multiple scatters, and usually has traveled large distances from the source, the particle flux is very low. Variance reduction will almost certainly be necessary to account for contributions to the shielding burden of radiation scattered throughout the vault. This may be a challenging step, and as with all variance reduction, care must be used to not suppress potentially relevant contributions to scatter throughout the treatment vault.

12.3.4.2 Derivation of General Parameters

General wall-scatter parameters can be readily calculated by impinging the primary beam on a wall (usually at both 90° (normal incidence) and 45°) and scoring the quantity and quality of the radiation scattered as a function of the angle.

Raso (1963) first investigated the scattering of gamma rays up to 10 MeV using the Monte Carlo approach for various angles of incidence and slab thicknesses. He demonstrated the sharp fall-off in scattered dose with increasing angle of scatter with respect to normal incidence. He also investigated inplane and out-of-plane scatter. Lo (1992) determined the scattering albedos for 4, 10, and 18 MV x-ray beams normally incident on ordinary concrete, iron, and lead using the ITS code (ITS 2.1). His test results from ^{60}Co and ^{137}Cs agreed very well with measured values. Two interesting results developed from this paper. The first is that the thickness for achieving saturation in the number of reflected photons is about 47 g/cm^2 indicating deep penetration before backscattering out of the medium. The second is that the average energy of the backscattered photons from concrete is around 0.15 MeV, whereas for iron it is ~0.25 MeV and for lead ~0.55 MeV, independent of incident energy over a wide range of megavoltage energies.

There has been little Monte Carlo work regarding second scatters, as would be important for maze calculations. Only direct maze calculations using Monte Carlo have been performed by Biggs (1991) and Al-Affan (2000). In the case of Biggs, the calculation was performed to validate the NCRP calculations for primary beams hitting the maze wall, whereas Al-Affan calculated for the more general maze situation. In both cases, they determined the x-ray fluence spectra striking the maze door (see Section 21.3.6).

21.3.5 Leakage Radiation and TVLs

Head leakage can be the most important component in defining secondary barrier thickness, particularly for treatment vaults that use IMRT. It is also particularly important for critical secondary barriers such as direct-shielded doors; however, it is also typically the most challenging component of the photon field to calculate due to the complex and irregular nature of the head shielding, and the corresponding spatial variations in leakage fluence and spectra.

21.3.5.1 Direct Calculation

Although there have been Monte Carlo models of accelerators designed with extensive detail of the entire treatment head (Kase et al., 1998, Howell et al., 2005, Kry et al., 2007, Bednarz and Xu, 2009), no model has yet been built that has been validated for simulation of the leakage radiation from a linear accelerator head. Given the importance of head leakage in defining the secondary barrier shielding requirements, careful and well-validated Monte Carlo modeling is necessary. For such a simulation, in-field validation of an accelerator model would clearly not be sufficient to verify accurate head leakage calculations. Most commonly, an analytic solution is sought for this part of the shielding question.

21.3.5.2 Derivation of General Parameters

Because it is hard to simulate head leakage levels accurately, general parameters for analytic models tend to be crudely defined. Because head leakage levels are mandated for accelerator manufacturers to be less than 0.1% of the primary beam, this is typically taken as the relative dose rate of head leakage (although it may be scaled by the amount of modulation used in IMRT treatments). Nelson and LaRiviere (1984) performed simple calculations to determine the penetration of leakage radiation in concrete, using a uniform spherical lead shell as a shielding model. Kase et al. (2003) and Li et al. (2006) performed TVL measurements for leakage radiation and found good agreement with Nelson and Lariviere's (1984) calculations. Consequently, although it has not been employed to date, a simple Monte Carlo model (e.g., a simple lead shell) may therefore be sufficient for calculating the shielding burden of leakage radiation. Of course, the acceptability of this approach depends on the desired level of accuracy.

21.3.6 Mazes

There is relatively little Monte Carlo work performed on mazes in low-energy rooms where neutrons are not a problem, suggesting that analytical calculations suffice, at least for standard room geometries. Al-Affan (2000) used the MCNP package to calculate the dose and energy spectrum of photons at the door of a maze for a 6 MV bremsstrahlung beam. He showed that the photon energy spectrum at the door reached values up to 300 keV, but the most probable energy was <100 keV. However, although he presented the dose at the door in comparison with the dose at the isocenter (given in units of Gy/photon), he did not compare the results with the standard NCRP calculation. Biggs (1991) calculated the dose at the door of a maze for 4 and 10 MV x-ray beams (ignoring neutrons) using the ITS Monte Carlo code. Two types of rooms were considered: (i) a conventional room where the gantry rotation axis is parallel to the inner maze wall and (ii) a room where the primary beam is directly incident on the outer maze wall. It is difficult to calculate the dose in the latter configuration under the NCRP formalism because the coefficients are not well defined in this circumstance. He found that the Monte Carlo results were in agreement with the NCRP calculation at 10 MV for the conventional room, but a factor of 2 lower at 4 MV. For the nonstandard room, the Monte Carlo predicted a lower value by a factor of 3 for 4 MV and 1.5 for 10 MV. He also showed that the most probable photon energy at the door was 70–80 keV, irrespective of configuration, although the nonstandard room had a larger high-energy component.

21.3.7 Direct Door Shielding

There has been little to no interest in using the Monte Carlo technique to aid the calculation of direct door shielding for low-energy machines. There are two reasons for this. The first is that the most important source of radiation for direct-shielded doors is leakage radiation and a successful model that describes the leakage fluence and energy as a function of polar and azimuthal angles has not yet been developed. Second, because neutrons are not involved in the calculation, the "door edge" effect.* (Figure 2.10, NCRP Report No. 151) is much less problematic as only lead is used for shielding and door overlap can be generous.

21.3.8 HVAC Ducts

Depending on the size of the rooms and the required air flow rate, the heating, ventilation and air conditioning (HVAC) ducts can be quite large and are therefore a source of radiation leakage to a point just outside the barrier. Monte Carlo investigation of HVAC ducts is therefore of interest, particularly for rooms without mazes. However, no work on this subject has yet been conducted. In an unpublished work, Biggs and Styczynski found that for a typical HVAC duct located in the nonprimary plane, the 10th-value attenuation factor is approximately 60–70 cm, depending on the location and primary beam energy. Further study of this issue is warranted.

---

* This effect refers to the overlap region between the shielding barrier and the door. Because the door is very thick for high-energy direct-shielded doors, rays passing obliquely through the barrier/door junction will see a reduced shielding thickness, and therefore result in an elevated dose at the edge of the door. While this can be solved by making the door considerably wider, the weight and cost escalate prohibitively and would require a much heavier duty door controller.

21.4 Neutrons from High-Energy Photon Therapy

Neutrons are relevant for the shielding of accelerators operated at 10 MV or above. Because neutrons undergo multiple reflections from the walls, the vault must be included in the calculation. Calculations may be more sensitive to other ancillary equipment, which has the potential to degrade the neutron energy.

21.4.1 Neutron Production

Photon production of neutrons has a minimum threshold energy that varies with atomic number. For example, ^{208}Pb has a threshold of 7.4 MeV, ^{56}Fe has a threshold of 11.2 MeV, and ^{184}W has a threshold of 7.4 MeV. Thus, given the standard photon energies offered by the vendors, beam energies must be at least 10 MV for photoneutrons to be produced in lead and 15 MV for steel. However, even at high energies, the main concern is the door of the vault or HVAC ducts in a mazeless room. Other barriers (primary and secondary barriers) that adequately shield photons will also generally shield the neutrons.

Neutrons are produced primarily through photon interactions with high-Z materials in the accelerator head (Howell et al., 2005). Once generated, these neutrons largely exit into the treatment vault and scatter throughout the room as high-Z materials are not efficient at reducing their energy or fluence. Neutrons scatter and will cross the treatment vault >2 times, on average, before finally being captured (NCRP Report No. 79). Neutrons lose energy primarily through elastic scatter interactions with low-Z materials, particularly hydrogen. Once thermalized, neutrons may be captured. Neutron capture cross sections vary greatly between elements, but are particularly high for boron and cadmium. After the capture of the thermal neutron, a photon is emitted by the excited nucleus. These gamma rays are typically of high energy; for example, after capturing a neutron, boron emits a photon of $E = 2.4$ MeV. Neutron shielding must therefore address all these stages of neutron absorption. First, thermalization of the neutrons (by, e.g., polyethylene), followed by absorption of the neutrons (by, e.g., a boron layer), and followed by attenuation of the emitted capture gamma rays (by, e.g., lead). This layered approach to neutron shielding is standard in the doors of radiotherapy treatment vaults.

Neutron source strength values vary between accelerator manufacturers and nominal energies. Followill et al. (2003) provide the broadest overview of source strengths, while Howell et al. (2009) provide detailed spectral measurements for common types and energies of accelerators currently available. Electron therapy has less neutron production than photon therapy because (e,n) cross sections are lower than those for (γ,n), and because electron therapy target currents are less than those for photon therapy (Lin et al., 2001). Electron-only accelerators, such as mobile linear accelerators used for intraoperative radiation therapy, are an exception to this, but there has been no Monte Carlo work to date on neutron production from these machines.

21.4.2 Neutron Spectra

Many different codes and modeling approaches have been used to study neutrons produced by medical accelerators. Agosteo et al. (1993) calculated the spatial distribution of the neutron flux inside an accelerator vault and confirmed the results using the MCNP code. However, their results did not lead to a practical application of neutron fluence calculation. Mao et al. (1997) and Kase et al. (1998) simulated a detailed head model of a clinical accelerator and calculated the production of neutrons and their transport through the head shielding into a typical therapy vault using EGS for photon production and MORSE for neutron production. They determined the neutron spectra outside the primary beam and identified the sources of neutrons in the accelerator head. Zanini et al. (2004) used MCNP4B-GN, which includes a routine to model photoneutron production, to investigate in-field neutron spectra for various collimation configurations. They only modeled the main components of the head, although they included a multileaf collimator (MLC). For an 18 MV linac, the most probable neutron energy was ~500 keV with a sharp drop in the spectrum above 1 MeV. Garnica-Garza (2005) also modeled only the main head components of a 15 MV linac using MCNP4C to generate the in-field spectra. The spectrum of room-degraded neutrons, with an average energy of 0.35 MeV was far different from the primary spectrum, with an average energy of 1.6 MeV. Howell et al. (2005) compared measured and calculated neutron spectra for an 18 MV linac, completely modeling the linac head. In this case, there was good agreement for a closed MLC field, but, for nonzero fields, the agreement was relatively poor. Barquero et al. (2005) modeled the accelerator head in MCNPX as a ball of tungsten with an isotropic neutron source. Facure et al. (2006) used a similar approach to study the scattered and thermal neutron fluence inside a room with the MCNP code. There are three publications from an Iranian group (Zabihzadeh et al., 2009, Ghiasi and Mesbahi 2010, Mesbahi et al., 2010). In the first paper, they used the same approach as Facure et al. (2006) and showed spectra of scattered neutrons for 10, 15, and 18 MV x-ray beams. Their spectra showed the typical fast and thermal neutron peaks. In the other two papers, the main components of the linac head were simulated. The first of these two papers examined the in-room spectra for 18 MV for three field sizes, while the second was primarily concerned about neutrons in the maze, but showed a neutron spectrum at 1 m from the source in the absence of room shielding.

Several authors have looked at the impact of field size on neutron flux or dose equivalent. Interestingly, seemingly contradictory results have been found. Mao et al. (1997) found that the neutron yield decreased with increasing field size, consistent with the decreased fluence calculated by Facure et al. (2006). Garnica-Garza (2005) found an increase in neutron yield with increasing field size, consistent with Kry et al. (2009b) who calculated an increase in dose equivalent with increasing field size. These disparate results likely reflect differences in neutron yield (neutrons produced in the head) versus flux in the treatment vault/patient plane, and differences in the extent of modeling details employed in different studies.

Caution must be used in interpreting some Monte Carlo results. Many authors have used simple models of accelerators for simulating neutrons, including only beam line structures. However, because of the extensive production and scattering processes throughout the accelerator head, studies have found that this approach produces results that may be in sizeable error as compared to results determined with complete models of the accelerator head (Kase et al., 1998).

In summary, there has been considerable effort to determine the neutron spectra from high-energy linacs over the past 20 years, but there is not the same degree of agreement with measurements as there is with primary photon beams. This is, in no small part, due to the difficulties in measuring neutrons, and also due to limitations in many of the simulations. It is also difficult to compare various results since each evaluates neutron spectra under different conditions. The prime need for having the neutron spectra is to assess direct door shielding, HVAC ducts for mazeless rooms, and neutron dose at the end of a maze. There is clearly room for further research.

21.4.3 Barriers

Concrete, by virtue of its high hydrogen content, is a particularly good neutron shielding material. In addition to the variations in aggregate distribution as described in Section 21.3.1, Monte Carlo simulations of concrete may pose an additional challenge for neutrons because, as concrete ages, the moisture level decreases as it dries out, which can change the hydrogen content. Concrete can lose 1–2% of its weight through lost water due to evaporation, affecting its ability to attenuate fast neutrons (Chilton 1984). Unfortunately, it is very difficult to account for this phenomenon and it is routinely neglected in Monte Carlo simulations, thereby adding to the overall uncertainty in calculations. Kase et al. (2003) has most thoroughly evaluated neutron TVLs for various concretes. They found reasonable agreement with measurements.

When space becomes an issue, either in new room construction or in room remodeling, high-atomic-number materials such as lead and steel are frequently used as shielding material to conserve space. However, for photon beams with energies >10 MV, neutrons will be produced where the primary beam strikes the high-*Z* barrier material. Some of these neutrons will subsequently produce capture gamma rays having an energy greater than 2 MeV. McGinley and Butker (1994) investigated this effect experimentally and derived an empirical formula for the photoproduction of neutrons from lead at 15 and 18 MV and for steel at 18 MV. They also showed that the contribution of capture gamma rays can be conservatively estimated to be 1.7 times the transmitted photon dose. However, these data were restricted to existing vaults. Thus, there is no data for steel and concrete laminated barriers at 15 MV and no data at all at 10 MV (though note that since the photoneutron

threshold for steel is 11.2 MeV (see Section 21.4.1), there is no neutron production on steel for 10 MV beams). Facure et al. (2008) sought to remedy this by performing Monte Carlo calculations using MCNP5 for a bremsstrahlung beam with a peak energy of 11 MeV. They were able to show that steel generates no photoneutrons since the reaction thresholds are too high. They were also able to show that lead generates no substantial capture gamma rays, but there is sizeable neutron production in lead. Because McGinley's 1.7 factor is very conservative (he took the worst case and there was considerable spread in the data), a Monte Carlo approach should be able to determine the dependence of this factor on other parameters such as thickness and relative location of the lead.

21.4.4 Mazes

The first empirical formula to determine the neutron dose at the end of a maze was due to Kersey (1979). This formula, or a variant of it, has been recommended by the NCRP (NCRP Report No. 79, 1984). Several authors have used the Monte Carlo method to solve this problem. The first difficulty is to generate a neutron spectrum that represents the neutron fluence coming from the head. Agosteo et al. (1995) simplified the concept by employing a point copper target at the center of a hollow tungsten sphere and adopting the method of Tosi et al. (1991) for the photoneutron spectrum. Carinou et al. (1999) used the same approach for generating the neutron spectrum and used this to investigate the neutron dose at various points along the maze and compared the results with empirical models. The results showed that the Monte Carlo results predicted lower values than the Kersey approach. Falcao et al. (2007) used the MCNP Monte Carlo code to examine the neutron fluence along the maze of multiple rooms housing accelerators with energies of 15, 18, and 25 MV. They also used Tosi's technique to generate the neutron spectrum using a 10 cm tungsten sphere. However, unlike Carinou et al., they found that the Kersey method appeared to substantially underestimate the dose at the door. However, it was later shown (Wang et al., 2011) that this was due to a discrepancy in the ICRP conversion factors. Wu and McGinley (2003) modified the Kersey formula based on measurements along the maze of several typical linear accelerator rooms.

High-energy accelerators also have photon components to the maze dose. McGinley et al. (2000) evaluated the relative contributions of photon sources to the dose at the end of a maze, considering capture gamma rays, leakage, and scattered radiation. They found that capture gamma rays were generally the dominant source of photon dose, and that empirical equations could overestimate the leakage and scattered radiation by as much as a factor of 2.9.

21.4.5 Direct Door Shield (High Energy)

There has been little work done in this area, despite the pressing need for a Monte Carlo approach to this problem. The reason for this need is that this is a difficult problem to solve using hand calculations and usually requires considerable time and effort to trace both photon and neutrons through the ends of the door, the so-called door edge effect (see Figure 2.10, NCRP Report No. 151). For Monte Carlo calculations, there is no adequate model to describe the spectrum of leakage photons from the head of the machine—unlike scattered radiation—as this depends very heavily on the shielding design of the head and is both vendor and model dependent. In addition, the spectrum of neutrons has to be taken into account and, while the neutron fluence spectrum has been derived, there is no assurance that the same spectrum will apply at angles of 45° to 135° to the beam axis. This is an area, which if solved by Monte Carlo, would bear considerable fruit since it would not only speed up calculations but it would also permit fast comparison between different door designs.

21.4.6 HVAC Ducts

As with low-energy machines, this is an important problem for mazeless rooms. For high-energy rooms, the problem is exacerbated by the need to shield against neutrons. Depending on the location of the ducts, Monte Carlo input needs both photon (leakage and scatter) and neutron spectra at the location of the duct for different gantry angles. Even for rooms with mazes, NCRP Report No. 151 provides no guidance in terms of analytical calculations, instead relying on the empirical evidence of McGinley (NCRP Report No. 151). This empirical evidence states that for an 18 MV linear accelerator, a dose equivalent reduction of four for neutrons and two for photons will be produced by a 1.2-m-long duct wrap consisting of 2.5 cm Borated polyethylene (BPE) and 1-cm lead in a 3.6-m-long maze. It is clear that this data cannot be extrapolated easily to a general situation and therefore a Monte Carlo-based approach would allow some generalizations to be made. The merit of Monte Carlo simulations for such situations is well known; the NCRP formalism for handling neutrons from high-energy linear accelerators was historically based on a large number of Monte Carlo simulations performed by McCall.

21.5 Application of Monte Carlo for Nonstandard Room Geometries

The NCRP-recommended calculation technique for door shielding for rooms with mazes (NCRP Report Nos. 49, 51, and 151) assumes a model where the primary beam does not impinge upon the outer maze wall. In fact, the assumption is made that the primary beam does not point in the direction of the maze wall (see, e.g., Figure 2.8, NCRP Report No. 151). However, for many rooms, this is not the case. While this may be a simpler calculation because only one scatter calculation is needed, the dose albedo for photons incident on a concrete surface and scattering at near 90° is not well known and is sharply dependent on the scattering angle (Chilton 1984). Thus, a large uncertainty in the value for the wall scatter component (α) may exist for these nonstandard rooms. Biggs applied the Monte Carlo approach to this problem and compared it with NCRP calculations (Section 21.3.5).

Facure et al. (2010) looked at unusual beam geometries in a standard room, such as orienting the beam toward the maze (this would be a possible orientation for a Cyberknife machine, for example). They were able to show that, although the Monte Carlo calculations agreed with NCRP calculations for standard geometries (the latter being the more conservative), "where the axis of gantry rotation is redirected at, for example, 45° with respect to the walls of the room, the photon doses at the entrance can reach values up to seven times higher than those obtained under the standard conditions, depending on the energy of the primary beam."

Other situations where nonstandard geometries may arise include cases where the gantry rotation axis is parallel to the maze wall (standard geometry), but the maze opening is very large (>2.0–2.5 m). In such a case, one must integrate the NCRP equations over this broad geometry—something that may be handled more easily using Monte Carlo.

Finally, there are rooms that have a curved maze entrance or the vault itself has an unusual configuration or the gantry rotation plane is not parallel to any wall. Again, NCRP calculations can be performed, but with difficulty, and a Monte Carlo approach would have an advantage.

21.6 Nonstandard Beams and Accelerators

The C-arm linac is the most common device used to deliver radiotherapy. Correspondingly, these machines have been most thoroughly studied. Nevertheless, there are novel variants of the x-ray beams from such devices, as well as several other modes of delivering radiation therapy. These scenarios are less common clinically but are often of greater interest because there is less documentation on how to evaluate the shielding requirements. The applications of Monte Carlo to solving these specific problems are detailed below.

FFF beams are available on many new models of C-arm accelerators. These beams have substantially different spectra and profiles (Vassiliev et al., 2006), and therefore have different shielding characteristics. Kry et al. (2009a) used BEAMnrc to reproduce analytic parameters for shielding C-arm FFF linacs to mimic the data in the NCRP 151 report (including primary beam TVL, patient scatter fractions, etc.). They also evaluated neutron production in these beams using MCNPX and found an ~80% reduction in dose equivalent per course of 18 MV IMRT with a Varian accelerator (Kry et al., 2008). They solved for the shielding of a low-energy and mixed-energy accelerator in a typical radiotherapy vault and found reduced thickness of primary and secondary barriers by 10–20% compared to traditional beams. These simulations were confirmed by measurements by Vassiliev et al. (2007).

Radiosurgery uses small treatment fields, which are of particular interest because they do not meet the broad-beam assumption of the NCRP-151 guidelines. This issue was explored by Rodgers (2007) for Cyberknife beams incident on concrete, lead, and steel, who found that the TVL for these beams changes substantially with field size, particularly for field diameters smaller than 8 cm where the TVL can be more than 20% less than the broad-beam TVL.

Brachytherapy has moved toward high-dose rate procedures in dedicated bunkers. Analytic models exist for calculating concrete wall thicknesses (ICRP Report No. 33, NCRP Report No. 49); however, these exist only for specific shielding materials and sources. Additionally, as with most analytic models, maze and vault door shielding are more uncertain. Lymperopoulou et al. (2006) and Papagiannis et al. (2008) used GEANT4 to expand the available data by calculating attenuation through four TVLs for a wide variety of both shielding materials and brachytherapy sources. Perez-Calatayud et al. (2004) used GEANT4 to evaluate the kerma and x-ray spectra in an high dose rate (HDR) maze and at the vault door. They also compared their results to analytic solutions using albedo coefficients, showing generally good agreement, but also some cases where the analytic solution underestimated the dose by 30%.

Neutron and hadron therapy facilities have also received attention for bunker shielding. Chen et al. (2008) used two coupled Monte Carlo codes to solve for dose and neutron and photon spectra in the room, maze, and outside the bunker of a boron neutron capture therapy (BNCT) facility. Porta et al. (2005) did similar analysis of shielding requirements for a hadron therapy bunker.

Proton therapy facilities are becoming increasingly popular. While NCRP Report No. 144 and Report No. 51 address shielding issues for such facilities, the novelty and variety of these units make additional evaluations essential. This is particularly true because the neutron fluence and spectral data are much more challenging to determine and therefore more scarce for proton therapy facilities than for high-energy photon beams. Nevertheless, this spectrum is very important in determining correct shielding values (Avery et al., 2008). Kim et al. (2005) therefore conducted a Monte Carlo shielding evaluation of their facility under construction to confirm their analytic solution. Of interest, while the analytic solution typically overestimated the dose, this was not always the case. Several other Monte Carlo studies have investigated the neutron spectra associated with proton therapy in detail, and treatment parameters that affect the spectra, particularly as they pertain to vault shielding. Zheng et al. (2007, 2009) showed that the neutron spectra consisted of two main peaks, one at ~1 MeV (isotropic neutrons produced by evaporation processes) and a high-energy peak at the maximum proton energy (forward-peaked neutrons produced by intranuclear cascade reactions), and related neutron doses to treatment parameters (2007). Perez-Andujar et al. (2009) showed that the bulk of these neutrons were produced in the range-modulator wheel in passive scattering systems. Agosteo (2009) simulated neutron spectra secondary to proton therapy, but also investigated the vault maze and room activation. As proton therapy continues to evolve, shielding

considerations will require more refinement and additional study. For example, Fan et al. (2007) evaluated shielding requirements for a laser-accelerated proton therapy system based on a unique proton production process. Proton shielding will certainly be an area of active research for many years.

21.7 Summary and Future Directions

Numerous Monte Carlo codes have been successfully used to conduct shielding evaluations of radiotherapy treatment vaults. Such studies have answered questions about direct shielding needs and capabilities, they have tested analytic models, and they have been used to develop new analytic models. However, there is still a great need for further Monte Carlo research in this area. More refined models of leakage and neutron spectra are important for solving problems in mazeless rooms, such as direct door and HVAC duct shielding, and would eliminate long, tedious calculations and ray tracing by generating simplified rules. In addition, modeling the beams from newer, nonstandard accelerators, such as Cyberknife, tomotherapy, or FFF linear accelerators would be of substantial benefit.

References

Al-Affan I.A.M. Estimation of the dose at the maze entrance for x-rays from radiotherapy linear accelerators. *Med. Phys.* 27:231–238;2000.

Agosteo S. Radiation protection constraints for use of proton and ion accelerators in medicine. *Radiat. Prot. Dosim.* 137:167–186;2009.

Agosteo S., Foglio A.P, Maggioni B. Neutron fluxes in radiotherapy rooms. *Med. Phys.* 20:407–414;1993.

Agosteo S., Foglio Para A., Maggioni B., Sangiust V., Terrani S., and Borasi G. Radiation transport in a radiotherapy room. *Health Phys.* 68:27–34;1995.

Agostinelli S., et al. GEANT4—A simulation toolkit. *Nucl. Instrum. Methods Phys. Res. B. A* 506:250–303;2003.

Avery S., Ainsley C., Maughan R., and McDonogh J. Analytical shielding calculations for a proton therapy facility. *Radiat. Prot. Dosim.* 131:167–179;2008.

Avila-Rodriguez M.A., DeLuca P.M. Jr, and Bohm T. Simulation of medical electron linac bremsstrahlung in typical shielding materials. *Rad. Prot. Dosim.* 116:547–552;2005.

Barquero R., Edwards T.M., Iniguez M.P., and Vega-Carrillo H.R. Monte Carlo simulation estimates of neutron doses to critical organs of a patient undergoing 18 MV x-ray LINAC-based radiotherapy. *Med. Phys.* 32:3579–3588;2005.

Bednarz B. and Xu G.X., Monte Carlo modeling of a 6 and 18MV Varian C linac medical accelerator for in-field and out-of-field dose calculations: Development and validation. *Phys. Med. Biol.* 54:N43–N57;2009.

Biggs P.J. Calculation of the shielding door thickness for radiation therapy facilities using the ITS Monte Carlo program. *Health Phys.* 61:465–472;1991.

Biggs P.J. Obliquity factors for ^{60}Co and 4, 10 and 18 MV x rays for concrete, steel and lead and angles of incidence between 0° and 70°. *Health Phys.* 70:527–536;1996.

Biggs P.J. Does the dose outside a primary barrier follow the inverse square law? *Med. Phys.* 29:1236;2002.

Biggs P.J. and Styczynski J.R. Do angles of obliquity apply to 30° scattered radiation from megavoltage beams? *Health Phys.* 95:425–432;2008.

Briesmeister J.F. MCNP—A general Monte Carlo N-particle transport code, Version 4A. *Los Alamos Laboratory Report No. LA-12625*, 1993.

Carinou E., Kamenopoulou V., and Stametaletos I.E. Evaluation of neutron dose in the maze of medical electron accelerators. *Med. Phys.* 26:2520–2525;1999.

Chen A.Y., Liu Y.W.H., and Sheu R.J. Radiation shielding evaluation of the BNCT treatment room at THOR: A TORT-coupled MCNP Monte Carlo simulation study. *Appl. Radiat. Isot..* 66:28–38;2008.

Chilton A.B., Shultis J.K., and Faw R.E. *Principles of Radiation Shielding.* Prentice-Hall, Inc., Englewood, NJ, 1984.

Facure A. and da Silva A.X. The use of high-density concretes in radiotherapy treatment room design. *Appl. Rad. Isot.* 65:1023–1028;2007.

Facure A., da Silva A.X., and Falcao R.C. Monte Carlo simulation of scattered and thermal photoneutron fluences inside a radiotherapy room. *Rad. Prot. Dosim.* 123:56–61;2006.

Facure A., da Silva A.X., da Rosa L.A.R., Cardoso S.C., and Rezende G.F.S. On the production of neutrons in laminated barriers for 10 MV medical accelerator rooms. *Med. Phys.* 35:3285–3292;2008.

Facure A., Cardoso S.C., da Rosa L.A.R, and da Silva A.X. Photon dose at the entrance of ^{60}Co and low energy medical accelerator rooms under unusual irradiation conditions. *Radiat. Prot. Dosim.* 138:251–256; 2010.

Falcao R.C., Facure A., Silva A.X., and Brazil R.J. Neutron dose calculation at the maze entrance of medical linear accelerator rooms. *Rad. Prot. Dosim.* 123:283–287;2007.

Fan J., Luo W., Fourkal E., Lin T., Li J., Veltchev I., and Ma C.M. Shielding design for a laser-accelerated proton therapy system *Phys. Med. Biol.* 52:3913–3930;2007.

Followill D.S., Stovall M.S., Kry S.F., and Ibbott G.S. Neutron source strength measurements for Varian, Siemens, Elekta, and general electric linear accelerators. *J. Appl. Clin. Med. Phys.* 4(3):189–194;2003.

Garnica-Garza H.M. Characteristics of the photoneutron contamination present in a high-energy radiotherapy treatment room. *Phys. Med. Biol.* 50:531–539;2005.

Ghiasi H. and Mesbahi A. Monte Carlo characterization of photoneutrons in the radiation therapy with high energy photons: A comparison between simplified and full Monte Carlo models. *Iran. J. Radiat. Res.* 8:187–193;2010.

Halbleib J.A. and Mehlhorn T.A. ITS, the integrated TIGER series of coupled electron/photon Monte Carlo transport codes; CCC-467. Oak Ridge, TN; Radiation Shielding Information Center, Oak Ridge National Laboratory; 1988.

Howell R.M., Ferenci M.S., Hertel N.E., and Fullerton G.D. Investigation of secondary neutron dose for 18 MV dynamic MLC IMRT delivery. *Med. Phys.* 32:786–793;2005.

Howell R.M., Kry S.F., Burgett E., Hertel N.E., and Followill D.S. Secondary neutron spectra from modern Varian, Siemens, and Elekta linacs with multileaf collimators. *Med. Phys.* 36:4027–4038;2009. Erratum: *Med. Phys.* 38:6789;2011.

IAEA. Radiation Protection in Radiotherapy. Part 7: Design of Facilities and Shielding, Lecture 2: Shielding. https://rpop.iaea.org/RPOP/RPoP/Content/InformationFor/HealthProfessionals/2_Radiotherapy/index.htm.

ICRP (International Commission on Radiological Protection). Protection against ionizing radiation from external sources used in medicine, *ICRP Publication 33 (Ann. ICRP 9(1))* (New York: ICRP), 1982.

Jaeger R.G., Blizard E.P., Chilton A.B., Grotenhuis M., Honig A., Jaeger T.H.A., and Eisenlohr H.H. *Engineering Compendium on Radiation Shielding—Volume II: Shielding Materials.* Berlin/Heidelberg: Springer–Verlag, 1975.

Jaradat A.K. and Biggs P.J. Tenth value layers for 60Co gamma rays and for 4, 6, 10, 10, 15 and 18 MV x rays in concrete for beams of cone angles between 0° and 14° calculated by Monte Carlo simulation. *Health Phys.* 92:456–463;2007.

Kase K.R., Mao X.S., Nelson W.R., Liu J.C., Kleck J.H., and Elsalim M. Neutron fluence and energy spectra around the Varian 2100C/2300C medical accelerator. *Health Phys.* 74:38–47;1998.

Kase K.R., Nelson W.R., Fasso A., Liu J.C., Mao X, Jenkins T.M., and Kleck J.H. Measurements of accelerator-produced neutron and photon transmission through concrete. *Health Phys.* 84:180–187;2003.

Kawrakov I., Mainegra-Hing E., Rogers D.W.O., Tessier, F., and Walters B.R.B. The EGSnrc code system: Monte Carlo simulation of electron and photon transport, National Research Council, Canada. Report No. PIRS-0701, 2010.

Kersey R.W. Estimation of neutron and gamma ray doses in the entrance mazes of SL75–20 linear accelerator treatment rooms. *Medica. Mundi* 24:151–155;1979.

Kim J.W., Kwon J.W., and Lee J. Design of radiation shielding for the proton therapy facility at the national cancer center in Korea *Radiat. Prot. Dosim.* 115:271–275;2005.

Kirn F.S., Kennedy R.J., and Wyckoff H.O. Attenuation of gamma rays at oblique angles. *Radiology* 63:94–104;1954.

Kry S.F., Titt U., Followill D.S., Ponisch F., and Vassiliev O.N., White R.A., Stovall M., and Salehpour M. A Monte Carlo model for out-of-field dose calculation from high-energy photon therapy *Med. Phys.* 34:3489–3499;2007.

Kry S.F., Howell R.M., Polf J., Mohan R., and Vassiliev O.N. Treatment vault shielding for a flattening filter-free medical linear accelerator. *Phys. Med. Biol.* 54:1265–1273;2009a.

Kry S.F., Howell R.M., Salehpour M., and Followill D.S. Neutron spectra and dose equivalents calculated in tissue for high-energy radiation therapy *Med. Phys.* 36:1244–1250;2009b.

Kry S.F., Howell R.M., Titt U., Salehpour M., Mohan R., and Vassiliev O.N. Energy spectra, sources, and shielding considerations for neutrons generated by a flattening filter-free C linac. *Med. Phys.* 35:1906–1911;2008.

Li Z.L., Mutic S., and Low D. Measurement of linac 90° head leakage radiation TVL values. *Med. Phys.* 33:3541–3545;2006.

Lin J.P., Chu T.C., Lin S.Y., and Liu M.T. The measurement of photoneutrons in the vicinity of a Siemens Primus linear accelerator. *Appl. Radiat. Isot.* 55:315–321;2001.

Lo Y.C. Albedos for 4-, 10- and 18 MV bremsstrahlung x-ray beams on concrete, iron and lead—Normally incident. *Med. Phys.* 19:659–666;1992.

Low D. Shielding for IMRT. AAPM 2003 Summer School on Intensity Modulated Radiation Therapy, June 22–26, Colorado College, Colorado.

Lymperopoulou G., Papagiannis P., Sakelliou L., Georgiou E., Hourdakis C.J., and Baltas D. Comparison of radiation shielding requirements for HDR brachytherapy using ^{169}Yb and ^{192}Ir sources. *Med. Phys.* 33:2541–2547;2006.

Mao X.S., Kase K.R., Liu J.C., Nelson W.R., Kleck J.H., and Johnsen S. Neutron sources in the Varian C linac 2100C/2300C medical accelerator calculated by the EGS4 code. *Health Phys.* 72:524–529;1997.

McGinley P.H. Dose rate outside primary barriers. *Health Phys. Oper. Radiat. Saf.* S7–8, February, 2001.

McGinley P.M. and Butker E.K. Laminated primary ceiling barriers for medical accelerator rooms. *Phys. Med. Biol.* 39:1331–1336;1994.

McGinley P.H., Dhaba'an A., and Reft C. Evaluation of the contribution of capture gamma rays, x-ray leakage, and scatter to the photon dose at the maze door for a high energy medical electron accelerator using a Monte Carlo particle transport code. *Med. Phys.* 27:225–230;2000.

Mesbahi A., Ghiasi H., and Mahdavi S.R. Photoneutron and capture gamma dose equivalent for different room and maze layouts in radiation therapy. *Rad. Prot. Dosim.* 140:242–249;2010.

Mohan R., Chui C., and Lidofsky L. Energy and angular distributions of photons from medical linear accelerators. *Med. Phys.* 12:592–597;1985.

Nelson W.R. and LaRiviere P.D. Primary and leakage radiation calculations at 6, 10 and 25 MeV. *Health Phys.* 47:811–818;1984.

NCRP 31 (National Council on Radiation Protection and Measurements). Shielding for high energy electron accelerator installations. *NCRP Report No. 31* (Bethesda, MD: NCRP), 1964.

NCRP 49 (National Council on Radiological Protection and Measurements). Structural shielding design and evaluation for medical use of x rays and gamma rays of up to 10 MeV. *NCRP Report No. 49* (Bethesda, MD: NCRP), 1976.

NCRP 51 (National Council on Radiation Protection and Measurements).Radiation protection design guidelines for 0.1–100 MeV particle accelerator facilities. *NCRP Report No. 51* (Bethesda, MD: NCRP), 1976.

NCRP 79 (National Council on Radiation Protection and Measurements). Neutron contamination from medical electron accelerators. *NCRP Report No. 79* (Bethesda, MD: NCRP), 1984.

NCRP 144 (National Council on Radiation Protection and Measurements). Radiation protection for particle accelerator facilities. *NCRP Report No. 144* (Bethesda, MD: NCRP), 2003.

NCRP 151 (National Council on Radiation Protection and Measurements). Structural shielding design and evaluation for megavoltage x- and gamma-ray radiotherapy facilities. *NCRP Report No. 151* (Bethesda, MD: NCRP), 2005.

Nelson W.R. and LaRiviere P.D. Primary and leakage radiation calculations at 6, 10 and 25 MeV. *Health Phys.* 47:811–818;1984.

Papagiannis P., Baltas D., Pérez-Calatayud J., Granero D., Gimeno J., Ballester F., and Venselaar J.L.M. Radiation transmission data for radionuclides and materials relevant to brachytherapy facility shielding *Med. Phys.* 35:4898–4906;2008.

Perez-Andujar A., Newhauser W.D., and DeLuca P.M.Jr. Neutron production from beam-modifying devices in a modern double scattering proton therapy beam delivery system. *Phys. Med. Biol.* 54:993–1008;2009.

Perez-Calatayud J., Granero D., Ballester F., Casal E., Crispin V., Puchades V., Leon A., and Verdu G. Monte Carlo evaluation of kerma in an HDR brachytherapy bunker *Phys. Med. Biol.* 49:N389–N396;2004.

Porta A., Agosteo S., and Campi F. Monte Carlo simulations for the design of the treatment rooms and synchrotron access mazes in the CNAO hadron therapy facility. *Radiat. Prot. Dosim.* 113:266–274;2005.

Raso D.J. Monte Carlo calculations on the reflection and transmission of scattered gamma ray. *Health Phys.* 17:411–418;1963.

Rodgers J.E. Analysis of tenth-value layers for common shielding materials for a robotically mounted stereotactic radiosurgery machine. *Health Phys.* 92:379–386;2007.

Sheikh-Bagheri D. and Rogers D.W.O. Monte Carlo calculation of nine megavoltage photon beam spectra using the BEAM code. *Med. Phys.* 29:391–402;2002.

Taylor P.L., Rodgers J.E., and Shobe J. Scatter fractions from linear accelerators with x-ray energies from 6 to 24 MV. *Med. Phys.* 26:1442–1446;1999.

Taylor P.L., Rodgers J.E., and Shobe J. Erratum: Scatter fractions from linear accelerators with x-ray energies from 6 to 24 MV. *Med. Phys.* 27:2000;2000.

Tosi G., Torresin A., Agosteo S., Foglio Para A., Sanguist V., Zeni L., and Silari M. Neutron measurements around medical electron accelerators by active and passive detection techniques *Med. Phys.* 18:54–60;1991.

Vassiliev O.N., Titt U., Kry S.F., Ponisch F., Gillin M.T., and Mohan R. Monte Carlo study of photon fields from a flattening filter-free clinical accelerator. *Med. Phys.* 33:820–827;2006.

Vassiliev O.N., Tott U., Kry S.F., Mohan R., and Gillin M.T. Radiation safety survey on a flattening filter-free medical accelerator. *Rad. Prot. Dosim.* 124:187–190;2007.

Wang X., Esquivel C., Nes E., Shi C., Papanikolaou N., and Charlton M. The neutron dose equivalent evaluation and shielding at the maze entrance of a Varian C linac 23EX treatment room. *Med. Phys.* 38:1141–1149;2011.

Wu R.K. and McGinley P.H. Neutron and capture gamma along the mazes of linear accelerator vaults. *J. Appl. Clin. Med. Phys.* 4:162–171; 2003.

Zabihzadeh M., Reza Ay M. Allahverdi M., Mesbahi A., Mahdavi S.R., and Shahriari M. Monte Carlo estimation of photoneutron contamination from high-energy x-ray medical accelerators in treatment room and maze: A simplified model. *Rad. Prot. Dosim.* 135:21–32;2009.

Zanini A. et al. Monte Carlo simulation of the photoneutron field in linac radiotherapy treatments with different collimations systems. *Phys. Med. Biol.* 49:571–582;2004.

Zheng Y., Newhauser W., Fontenot J., Taddei P., and Mohan R. Monte Carlo study of neutron dose equivalent during passive scattering proton therapy. *Phys. Med. Biol.* 52:4481–4496;2007.

Zheng Y., Newhauser W., Klein E., and Low D. Monte Carlo simulation of the neutron spectral fluence and dose equivalent for use in shielding a proton therapy vault. *Phys. Med. Biol.* 54:6943–6957;2009.

Index